Conference Proceedings of the Society for Experimental Mechanics Series

Series Editor

Tom Proulx
Society for Experimental Mechanics, Inc.
Bethel, CT, USA

More information about this series at http://www.springer.com/series/8922

Michael Mains
Editor

Topics in Modal Analysis, Volume 10

Proceedings of the 33rd IMAC, A Conference and Exposition on Structural Dynamics, 2015

Springer

Editor
Michael Mains
Brüel & Kjær North America
Cincinnati, OH, USA

ISSN 2191-5644 ISSN 2191-5652 (electronic)
Conference Proceedings of the Society for Experimental Mechanics Series
ISBN 978-3-319-15250-9 ISBN 978-3-319-15251-6 (eBook)
DOI 10.1007/978-3-319-15251-6

Library of Congress Control Number: 2015935654

Springer Cham Heidelberg New York Dordrecht London

Printed on acid-free paper

Springer International Publishing AG Switzerland is part of Springer Science+Business Media (www.springer.com)

Preface

Topics in Modal Analysis represents one of ten volumes of technical papers presented at the 33rd IMAC, A Conference and Exposition on Structural Dynamics, 2015, organized by the Society for Experimental Mechanics, and held in Orlando, Florida, February 2–5, 2015. The full proceedings also include volumes on Nonlinear Dynamics; Dynamics of Civil Structures; Model Validation and Uncertainty Quantification; Dynamics of Coupled Structures; Sensors and Instrumentation; Special Topics in Structural Dynamics; Structural Health Monitoring & Damage Detection; Experimental Techniques, Rotating Machinery & Acoustics; and Shock & Vibration, Aircraft/Aerospace, Energy Harvesting.

Each collection presents early findings from experimental and computational investigations on an important area within Structural Dynamics. *Topics in Modal Analysis I* represents papers on enabling technologies for Modal Analysis measurements and applications of Modal Analysis in specific application areas. Topics in this volume include:

Experimental Techniques
Processing Modal Data
Rotating Machinery

The organizers would like to thank the authors, presenters, session organizers, and session chairs for their participation in this track.

Cincinnati, Ohio, USA
Bruel & Kjaer North America

Michael Mains

Contents

Chapter 1
An Efficient Treatment of Parameter Identification in the Context of Multibody System Dynamics Using the Adjoint Method

Karim Sherif, Karin Nachbagauer, Stefan Oberpeilsteiner, and Wolfgang Steiner

Abstract In multibody system dynamics, a wide range of parameters can occur where some of them may not be known a priori. Therefore, this work presents an efficient adjoint method for parameter identification that can be utilized in multibody simulation software. Compared to standard system sensitivity based approaches the adjoint method has the major advantage of being independent on the number of parameters to identify. Especially when dealing with large and probably flexible multibody systems this characteristic is crucial. Formulating parameter identification as an automatable procedure, of course, leads to a complicated structure of the involved matrices and equations. However, during a forward simulation of the system many of the matrices needed for solving the so called "adjoint system equations" are already evaluated. Adopting the functionality of the forward solver for the adjoint system solver therefore results in little additional effort. In order to illustrate the performance of the adjoint method two examples are presented. A planar example shows the possibility of identifying non-linear control parameters and a three-dimensional example is presented for identifying time-invariant inertia parameters.

Keywords Adjoint method • Parameter identification • Optimal control • Inverse dynamics • Multibody systems

1.1 Introduction

The adjoint method is probably the most efficient way to solve a variety of optimization problems in engineering sciences. Much attention to this approach has been paid recently in the context of continuous systems for sensitivity analysis (see, e.g., [1–4]). The class of dynamic programming methods for the computation of gradients in an optimization problem includes the adjoint method, which has a long history in optimal control theory [5]. In the explanations of Giles and Pierce [6], the adjoint method is seen as a special case of linear duality, in which the dual problem has to solve only a single linear system. Obviously, huge benefits can be achieved by solving the dual formulation. The adjoint method utilizes this powerful aspect of duality to dramatically improve the efficiency of the computation.

Previous work on the adjoint method in multibody dynamics can be found in the work of Bottasso et al. [7], where the solution of inverse dynamics and trajectory optimization problems for multibody systems is reflected by an indirect approach combining optimal control theory with control and adjoint equations and transversality conditions. The design of the indirect method for solving optimal control problems for multibody systems presented by Bertolazzi et al. [8] seems to be familiar with the idea of the adjoint method. The work by Schaffer [9] presents a numerical algorithm, the piecewise adjoint method, which formulates the coordinate partitioning underlying ordinary differential equations as a boundary value problem, which is solved by multiple shooting methods.

The group around Petzold, Cao, Li and Serban [10–12] describe forward and adjoint methods for sensitivity analysis for differential-algebraic equations and partial differential equations, and state that the results of sensitivity analysis have wide-ranging applications in science and engineering, including model development, optimization, parameter estimation, model simplification, data assimilation, optimal control, uncertainty analysis and experimental design [10]. In the work of Eberhard [4], the adjoint method is used for sensitivity analysis in multibody systems interpreted as a continuous, hybrid form of automatic differentiation.

Despite of the great potential of the adjoint method, in multibody dynamics the adjoint method is rarely applied, since the structure of the equations of motion is usually extremely complicated, in particular if flexible bodies are included, and the effort to obtain the set of adjoint equations seems tremendously high. Hence, dealing with the adjoint method is obviously unattractive for most developers of multibody simulation software, as far as they are familiar with it at all. The main goal

K. Sherif (✉) • K. Nachbagauer • S. Oberpeilsteiner • W. Steiner
University of Applied Sciences Upper Austria, Wels Campus, Stelzhamerstr. 23, 4600 Wels, Austria
e-mail: karim.sherif@fh-wels.at; karin.nachbagauer@fh-wels.at; stefan.oberpeilsteiner@fh-wels.at; wolfgang.steiner@fh-wels.at

M. Mains (ed.), *Topics in Modal Analysis, Volume 10*, Conference Proceedings of the Society for Experimental Mechanics Series, DOI 10.1007/978-3-319-15251-6_1

of the present paper, which is a shortened form of [13], is to show how the adjoint method can be embedded efficiently to a multibody system described by a system of differential-algebraic equations of index 3 for optimal control problems or parameter identification applications. The present paper shows the potential of the adjoint method for solving classical optimization problems in multibody dynamics and presents applications for optimal control problems and a parameter identification.

1.2 The Adjoint Method in Multibody Dynamics

In this section we discuss the application of the adjoint gradient computation to multibody dynamics. The adjoint equations for the equations of motion of a multibody system will be derived and a flowchart for embedding the adjoint method in multibody dynamics is illustrated.

1.2.1 Adjoint Equations

In the augmented formulation the dynamics of a multibody system is described by a set of differential-algebraic equations in the following form:

$$\dot{\mathbf{q}} = \mathbf{v} \tag{1.1}$$

$$\mathbf{M}\dot{\mathbf{v}} = \mathbf{f}(\mathbf{q}, \mathbf{v}, \mathbf{u}) - \mathbf{C}_{\mathbf{q}}^{\mathsf{T}}\boldsymbol{\lambda} \tag{1.2}$$

$$\mathbf{C}(\mathbf{q}) = \mathbf{0} \tag{1.3}$$

where $\mathbf{q}$ and $\mathbf{v}$ denote the generalized coordinates and velocities. Moreover, $\mathbf{u}$ may either describe a vector of time-invariant parameters or a vector of time-dependent control signals actuating the system. For simplicity we assume that $\mathbf{u}$ appears only in the vector of generalized forces and gyroscopic forces $\mathbf{f}$, which is e.g. the case if $\mathbf{u}$ is a stiffness or damping parameter or an actuating force. The matrix $\mathbf{M}$ represents the symmetric mass matrix of the system. The algebraic constraint equation $\mathbf{C}(\mathbf{q}) = \mathbf{0}$ influences the equations of motion via the Jacobian $\mathbf{C}_{\mathbf{q}}$ and the vector of Lagrange multipliers $\boldsymbol{\lambda}$.

The key idea of the adjoint method (see [4, 10–12] for example) is to determine the parameter/control $\mathbf{u}$ such that a functional of the form

$$J(\mathbf{u}) = \int_0^T h(\mathbf{q}, \mathbf{v}, \mathbf{u})\, dt + S(\mathbf{q}, \mathbf{v})\Big|_{t=T} \tag{1.4}$$

is minimized, in which the function $h(\mathbf{q}, \mathbf{v}, \mathbf{u})$ and the end point term $S(\mathbf{q}, \mathbf{v})|_T$ have to be defined appropriately. The end point term $S(\mathbf{q}, \mathbf{v})|_T$ is often also denoted as *scrap function*. Numerous methods are available to compute the argument for which a function or a functional attains a minimum. We just refer to the method of the steepest descent, the conjugate gradient method, the Gauss-Newton method or quasi Newton methods like the BFGS algorithm when estimating the Hessian or the DFP algorithm when estimating the inverse of the Hessian. Some authors embed these methods in a homotopy continuation to obtain a global minimum [14]. In any cases the gradient of $J(\mathbf{u})$ must be determined. For this purpose several strategies can be pursued again. On the one hand, if $\mathbf{u}$ is a vector of N_u parameters, the sensitivity equations for $\mathbf{x}_{\mathbf{u}} = \partial\mathbf{x}/\partial\mathbf{u}$ are usually considered (see [15, 16] or [14] for example). The computational effort for this approach is equal to solving N_u linear sets of equations with the same dimension as Eqs. (1.1)–(1.3). On the other hand, if $\mathbf{u}$ represents a vector of time-dependent control signals, these signals are often discretized in order to transform the problem into a finite dimensional one. The adjoint method is a powerful alternative to compute the gradient of $J(\mathbf{u})$ in both cases.

First, we notice that $J(\mathbf{u})$ does not change if we incorporate Eqs. (1.1)–(1.3) to the integrand

$$J(\mathbf{u}) = \int_0^T \left[h(\mathbf{q}, \mathbf{v}, \mathbf{u}, t) + \mathbf{p}^{\mathsf{T}}(\dot{\mathbf{q}} - \mathbf{v}) + \mathbf{w}^{\mathsf{T}}(\mathbf{M}\dot{\mathbf{v}} - \mathbf{f} + \mathbf{C}_{\mathbf{q}}^{\mathsf{T}}\boldsymbol{\lambda}) + \boldsymbol{\mu}^{\mathsf{T}}\mathbf{C}\right] dt + S(\mathbf{q}, \mathbf{v})\Big|_{t=T} \tag{1.5}$$

no matter how the adjoint variables $\mathbf{p}(t)$, $\mathbf{w}(t)$ and $\boldsymbol{\mu}(t)$ are chosen.

Let us now consider a forward solution $\mathbf{q}(t)$, $\mathbf{v}(t)$ and $\boldsymbol{\lambda}(t)$ of the system Eqs. (1.1)–(1.3) for a set of parameters or control variables $\mathbf{u}$. A variation of $\mathbf{u}$ will result in variations of the variables $\mathbf{q}$, $\mathbf{v}$ and $\boldsymbol{\lambda}$, and, moreover, in a variation of the functional J. Thus, the variation of the functional J can be written in the form

$$\begin{aligned}\delta J = &\int_0^T \Big\{ h_{\mathbf{q}}\delta\mathbf{q} + h_{\mathbf{v}}\delta\mathbf{v} + h_{\mathbf{u}}\delta\mathbf{u} + \mathbf{p}^{\mathsf{T}}(\delta\dot{\mathbf{q}} - \delta\mathbf{v}) + \\ &+ \mathbf{w}^{\mathsf{T}}\Big[(\mathbf{M}\dot{\mathbf{v}})_{\mathbf{q}}\delta\mathbf{q} + \mathbf{M}\delta\dot{\mathbf{v}} - \mathbf{f}_{\mathbf{q}}\delta\mathbf{q} - \mathbf{f}_{\mathbf{v}}\delta\mathbf{v} - \mathbf{f}_{\mathbf{u}}\delta\mathbf{u} + \left(\mathbf{C}_{\mathbf{q}}^{\mathsf{T}}\boldsymbol{\lambda}\right)_{\mathbf{q}}\delta\mathbf{q} + \mathbf{C}_{\mathbf{q}}^{\mathsf{T}}\delta\boldsymbol{\lambda}\Big] + \boldsymbol{\mu}^{\mathsf{T}}\mathbf{C}_{\mathbf{q}}\delta\mathbf{q}\Big\}\, dt \\ &+ S_{\mathbf{q}}\delta\mathbf{q}\Big|_{t=T} + S_{\mathbf{v}}\delta\mathbf{v}\Big|_{t=T}\end{aligned} \tag{1.6}$$

where $h_{\mathbf{q}}$, $h_{\mathbf{v}}$ and $h_{\mathbf{u}}$ denote the partial derivatives of the function h with respect to the components of $\mathbf{q}$, $\mathbf{v}$ and $\mathbf{u}$.

In order to get rid of the variations of $\dot{\mathbf{q}}$ and $\dot{\mathbf{v}}$ we integrate by parts

$$\begin{aligned}\int_0^T \mathbf{p}^{\mathsf{T}}\delta\dot{\mathbf{q}}\, dt &= -\int_0^T \dot{\mathbf{p}}^{\mathsf{T}}\delta\mathbf{q}\, dt + \mathbf{p}^{\mathsf{T}}\delta\mathbf{q}\Big|_{t=T} \\ \int_0^T \mathbf{w}^{\mathsf{T}}\mathbf{M}\delta\dot{\mathbf{v}}\, dt &= -\int_0^T \frac{d}{dt}\left(\mathbf{w}^{\mathsf{T}}\mathbf{M}\right)\delta\mathbf{v}\, dt + \mathbf{w}^{\mathsf{T}}\mathbf{M}\delta\mathbf{v}\Big|_{t=T}\end{aligned} \tag{1.7}$$

where the fact has been used that $\delta\mathbf{q}(0) = 0$ and $\delta\mathbf{v}(0) = 0$ since the initial conditions for the variables $\mathbf{q}$ and $\mathbf{v}$ are given. Using Eq. (1.7) the variation of the functional J given by Eq. (1.6) can be rewritten as

$$\begin{aligned}\delta J = &\int_0^T \Big\{\Big[h_{\mathbf{q}} - \dot{\mathbf{p}}^{\mathsf{T}} + \mathbf{w}^{\mathsf{T}}\left((\mathbf{M}\dot{\mathbf{v}})_{\mathbf{q}} - \mathbf{f}_{\mathbf{q}} + \left(\mathbf{C}_{\mathbf{q}}^{\mathsf{T}}\boldsymbol{\lambda}\right)_{\mathbf{q}}\right) + \boldsymbol{\mu}^{\mathsf{T}}\mathbf{C}_{\mathbf{q}}\Big]\delta\mathbf{q} \\ &+ \Big[h_{\mathbf{v}} - \mathbf{p}^{\mathsf{T}} - \mathbf{w}^{\mathsf{T}}\mathbf{f}_{\mathbf{v}} - \frac{d}{dt}\left(\mathbf{w}^{\mathsf{T}}\mathbf{M}\right)\Big]\delta\mathbf{v} + \left[\mathbf{w}^{\mathsf{T}}\mathbf{C}_{\mathbf{q}}^{\mathsf{T}}\right]\delta\boldsymbol{\lambda} + \left[h_{\mathbf{u}} - \mathbf{w}^{\mathsf{T}}\mathbf{f}_{\mathbf{u}}\right]\delta\mathbf{u}\Big\}\, dt \\ &+ \left[S_{\mathbf{q}} + \mathbf{p}^{\mathsf{T}}\right]\delta\mathbf{q}\Big|_{t=T} + \left[S_{\mathbf{v}} + \mathbf{w}^{\mathsf{T}}\mathbf{M}\right]\delta\mathbf{v}\Big|_{t=T}\end{aligned} \tag{1.8}$$

The key question now is how the variation of J is related to the variation of $\mathbf{u}$. In order to obtain this relationship we choose our adjoint variables $\mathbf{p}(t)$, $\mathbf{w}(t)$ and $\boldsymbol{\mu}(t)$ such that the terms involving $\delta\mathbf{q}$, $\delta\mathbf{v}$ and $\delta\boldsymbol{\lambda}$ vanish from the integrand. Thus, we end up with the following system of adjoint equations:

$$\frac{d\mathbf{p}}{dt} = h_{\mathbf{q}}^{\mathsf{T}} + \mathbf{A}\mathbf{w} + \mathbf{C}_{\mathbf{q}}^{\mathsf{T}}\boldsymbol{\mu} \tag{1.9}$$

$$\frac{d}{dt}(\mathbf{M}\mathbf{w}) = h_{\mathbf{v}}^{\mathsf{T}} - \mathbf{p} - \mathbf{B}\mathbf{w} \tag{1.10}$$

$$\mathbf{0} = \mathbf{C}_{\mathbf{q}}\mathbf{w} \tag{1.11}$$

$$\mathbf{0} = S_{\mathbf{q}}^{\mathsf{T}} + \mathbf{p}(T) \tag{1.12}$$

$$\mathbf{0} = S_{\mathbf{v}}^{\mathsf{T}} + \mathbf{M}(\mathbf{q}(T))\mathbf{w}(T) \tag{1.13}$$

with the following abbreviations

$$\mathbf{A} = (\mathbf{M}\dot{\mathbf{v}})_{\mathbf{q}}^{\mathsf{T}} - \mathbf{f}_{\mathbf{q}}^{\mathsf{T}} + (\mathbf{C}_{\mathbf{q}}^{\mathsf{T}}\boldsymbol{\lambda})_{\mathbf{q}}^{\mathsf{T}} \tag{1.14}$$

$$\mathbf{B} = \mathbf{f}_{\mathbf{v}}^{\mathsf{T}} \tag{1.15}$$

Using this set of adjoint equations, Eq. (1.8) reduces to

$$\delta J = \int_0^T \left[h_\mathbf{u} - \mathbf{w}^\mathsf{T}\mathbf{f}_\mathbf{u}\right] \delta\mathbf{u}\, dt \tag{1.16}$$

which directly relates the independent variation $\delta\mathbf{u}$ to the variation of the objective function. In order to find the minimum of J, we walk along its negative gradient for a finite distance and update the gradient form time to time until we reach a point where the gradient gets zero. If $\mathbf{u}$ is a control signal, it is easy to show that the largest infinitesimal increase of δJ is obtained, if

$$\delta\mathbf{u}(t) = -\kappa \left[h_\mathbf{u} - \mathbf{w}^\mathsf{T}\mathbf{f}_\mathbf{u}\right]^\mathsf{T} \quad \text{(control optimization)} \tag{1.17}$$

where κ is an infinitesimal positive factor. Hence Eq. (1.17) is the gradient formula for the cost functional we are looking for.

Moreover, if $\mathbf{u}$ is a vector of time-invariant parameters, Eq. (1.16) can be written as

$$\delta J = \left(\int_0^T \left[h_\mathbf{u} - \mathbf{w}^\mathsf{T}\mathbf{f}_\mathbf{u}\right] dt\right) \delta\mathbf{u} = \nabla J^\mathsf{T} \delta\mathbf{u} \tag{1.18}$$

where the gradient of the functional J is defined as $\nabla J = \int_0^T \left[h_\mathbf{u} - \mathbf{w}^\mathsf{T}\mathbf{f}_\mathbf{u}\right]^\mathsf{T} dt$. Thus, the variation of $\mathbf{u}$ can be written for the case of time-invariant parameters as

$$\delta\mathbf{u} = -\kappa\nabla J = -\kappa \left(\int_0^T \left[h_\mathbf{u} - \mathbf{w}^\mathsf{T}\mathbf{f}_\mathbf{u}\right]^\mathsf{T} dt\right) \quad \text{(parameter optimization)} \tag{1.19}$$

If κ is sufficiently small, the updated control/parameter $\mathbf{u} + \delta\mathbf{u}$ will always reduce J.

The adjoint equations (1.9)–(1.13) have to be solved backwards in the physical time. Therefore it is advantageous to introduce a new time coordinate τ by the following transformation

$$\tau = T - t, \quad \tau \in [0, T], \quad \frac{d}{dt} = \frac{d}{d\tau}\frac{d\tau}{dt} = -\frac{d}{d\tau}$$

For the numerical solution of the adjoint equations we propose a backward differentiation scheme which approximates the derivative of a function $F(\tau)$ at a time instant τ_n by using the function values at $\tau_n, \tau_{n-1}, \ldots, \tau_{n-k}$. The backward differentiation formula (BDF) reads

$$\left.\frac{dF}{d\tau}\right|_{\tau_n} \approx \frac{1}{\gamma}\sum_{i=0}^{k} \alpha_i F(\tau_{n-i}) \tag{1.20}$$

The coefficients α_i result from differentiating an interpolation polynomial through $F(\tau_n), \ldots, F(\tau_{n-k})$ and are chosen as the standard coefficients presented, e.g., in [17, p. 349].

A closer look at the adjoint equations (1.9)–(1.13) shows that the boundary condition (1.13) of the variable $\mathbf{w}$ is in general incompatible with the adjoint constraint equation (1.11). Only when $S_\mathbf{v} = 0$, i. e. when the scrap function does not depend on $\mathbf{v}$, all equations are satisfied by setting $\mathbf{p}(T) = -S_\mathbf{q}^\mathsf{T}$ and $\mathbf{w}(T) = \mathbf{0}$. For the case that $S_\mathbf{v} \neq 0$ a consistent boundary condition approach has to be applied, see [13] for more details on this issue.

1.2.2 Flowchart of the Adjoint Method Embedded in Multibody Systems

At this point it should be mentioned that only two systems of DAEs must be integrated for computing the direction of the gradient, that is Eqs. (1.1)–(1.3) and (1.9)–(1.13). Subsequently, we give a flowchart of those steps, which are needed for the computation of the gradient of the objective function J with the adjoint method. Let $\mathbf{u}$ denote a given vector of controls or parameters of a multibody system. The flowchart describes how one can successively decrease the cost function J until a user-defined limit is achieved by varying $\mathbf{u}$.

1. Solve the equations of motion Eqs. (1.1)–(1.3) forward in time $t \in [0, T]$ yielding $\mathbf{q}(t)$, $\mathbf{v}(t)$ and $\boldsymbol{\lambda}(t)$. This may be done e.g. by choosing the Hilbert-Hughes-Taylor (HHT) integration scheme, as proposed in [18] and its application for a differential algebraic system given in an index three formulation in [19].
2. Compute the objective function J by inserting $\mathbf{q}(t)$, $\mathbf{v}(t)$ and $\mathbf{u}(t)$ into Eq. (1.4). Note, that the integration must be done numerically. If J is smaller than a user defined minimum then stop.
3. Along the forward simulation of the equations of motion compute the mass matrix $\mathbf{M}$, the constraint Jacobian $\mathbf{C_q}$ and the Jacobian matrices $\mathbf{A}$, $\mathbf{B}$ from Eqs. (1.14) and (1.15).
4. Determine the consistent boundary conditions at $\tau = 0$ for the adjoint variables $\mathbf{w}$, $\mathbf{p}$ as suggested in [13].
5. Solve the adjoint equations (1.9)–(1.13) for $\mathbf{p}(\tau)$, $\mathbf{w}(\tau)$ and $\boldsymbol{\mu}(\tau)$, where $\tau = T - t$.
6. Compute the adjoint variables as functions of the original time by setting $\mathbf{p}(t) = \mathbf{p}(\tau = T - t)$ and $\mathbf{w}(t) = \mathbf{w}(\tau = T - t)$. Moreover, determine $h_{\mathbf{u}}$ and $\mathbf{f_u}$ along the forward simulation.
7. Compute $\mathbf{u} = \mathbf{u} + \delta\mathbf{u}$. Use Eq. (1.17) for the computation of $\delta\mathbf{u}$ if $\mathbf{u}$ is a control signal. Otherwise, if $\mathbf{u}$ is a set of time-invariant parameters, use Eq. (1.19) for the update. A sufficiently small number $\kappa > 0$ has to be chosen for that purpose. Go to step 1.

For the sake of overview, it should be mentioned at this point that the Jacobian matrices $(\mathbf{M}\dot{\mathbf{v}})_\mathbf{q}$, $\mathbf{f_q}$, $\mathbf{f_v}$ and $(\mathbf{C}_\mathbf{q}^\mathsf{T}\boldsymbol{\lambda})_\mathbf{q}$ which are needed in the adjoint equations may be required already for the simulation of the multibody system if an implicit integration scheme such as, e.g., the HHT-algorithm [18, 19] is applied.

1.3 Numerical Examples

Two numerical examples will be presented to illustrate the application of the adjoint method in typical multibody systems. As a first example, a planar overhead crane is considered as an example of an underactuated mechanical system which follows a given trajectory and the optimization process identifies the control force and the control torque in a specific time domain. The second example incorporates a single rigid body which is parametrized by the four redundant Euler parameters. A point of the body is excited in order to follow a specific trajectory and the optimization process identifies all entries of the inertia tensor. A time history of the identified forces and moments as well as convergence analyses for the cost functionals and the inertia parameters are presented.

1.3.1 Planar Overhead Crane

The overhead crane presented in [20, 21] is a classical example of an underactuated mechanical system consisting of the cart A, the hoisting drum B, the cable C and a mass D mounted at the end of the cable, see Fig. 1.1a. The generalized coordinates of this two-dimensional problem are chosen as the position of the cart in x-direction x_c, the length of the cable l and the position of the mass in x- and y-direction, x_m and y_m, respectively, see Fig. 1.1b. The goal of this example is to compute the drive signals F and M, depicted in Fig. 1.1b, such that the point mass D follows a given trajectory defined by a linear path from a given starting point $(x_{m0}, y_{m0}) = (0/4)$ to a given end point $(x_{mf}, y_{mf}) = (5/1)$ within 3 s, see Fig. 1.2. The control force F and control torque M are identified within the time period $t \in [0, 3]$. For the computation we set the mass of the cart and the hoisting drum $A + B$ to 10 kg, the mass of D is set to 100 kg, the moment of inertia of the hoisting drum is defined as 0.1 kg m^2 and the radius of the hoisting drum is given by 0.1 m. As starting point of our drive signals we choose the static solution, i.e. $M_0 = 100 \cdot 9.81 \cdot 0.1 = 98.1$ N m and $F_0 = 0$ N. The results in Fig. 1.3 are computed with a constant step size and show the initial settings for the first iteration and the identified control force F and control torque M after 300 iterations. The optimization process reduces the costs to a factor 10^{-7} within 300 iterations, see Fig. 1.4.

1.3.2 Single Rigid Body Parametrized with Euler Parameters

The second example fall in the category of parameter identification, i.e. $\mathbf{u}$ is time-invariant. The goal of the present example is to identify the components of the inertia tensor ($I_{11}, I_{22}, I_{33}, I_{12}, I_{13}$ and I_{23}) of a single three-dimensional rigid body. The position of the rigid body is described by the coordinates of the center of mass $\mathbf{x} = (x, y, z)$ and for the description of

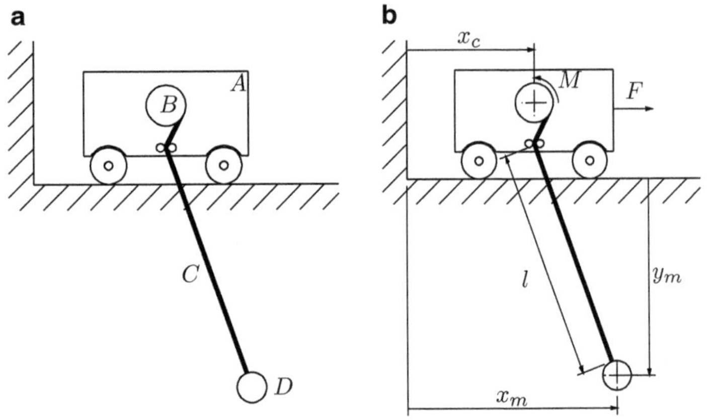

Fig. 1.1 Geometry description of the planar, rigidly modeled overhead crane (**a**) Planar overhead crane (**b**) Parameters of planar overhead crane

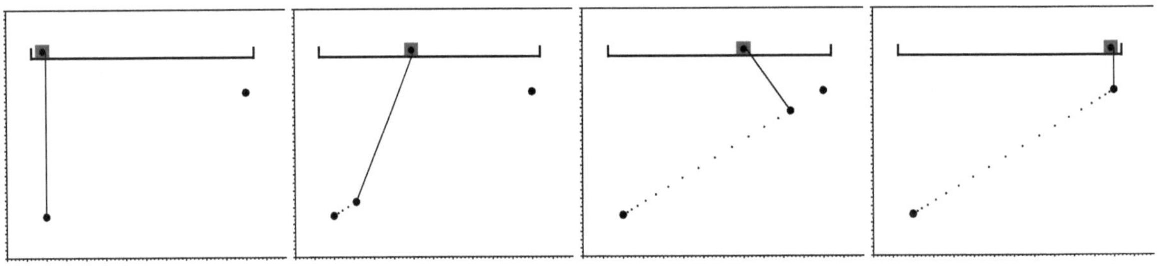

Fig. 1.2 The point mass has to follow a linear trajectory from a specified starting point to a fixed end point

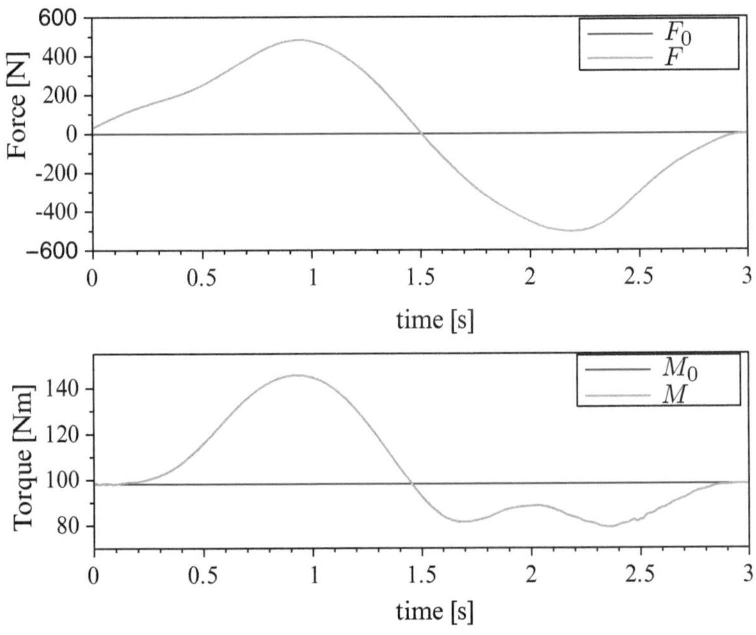

Fig. 1.3 Time history of identified force F and torque M after 300 iterations and initial settings for the first iteration for Sect. 1.3.1. The initial input for the force is set to $F_0 = 0\,\text{N}$ for the first iteration. The initial input for the torque is defined as the static torque $M_0 = 98.1\,\text{N m}$

the orientation of the body Euler parameters $\boldsymbol{\theta} = (\theta_0, \theta_1, \theta_2, \theta_3)$ are used. Figure 1.5 shows the arbitrarily shaped rigid body for which the inertia parameters describing the inertia tensor are not known in advance. The point S of the body follows a prescribed motion realized by a constraint and the velocity of the point P is measured (in global coordinates) in order to identify the entries of the inertia tensor. The geometry used for generating the inertia data is a cuboid with one diagonal congruent with the z-axis. The excitation for the identification process is applied in the point S situated at the origin at time $t = 0$. Velocity measurements are taken in the point P situated in the opposite direction to S along the z-axis. In order to start the optimization, *initial* values for the parameters to identify have to be defined. Table 1.1 shows the correct, the initial values for the first iteration and the final identified values for the moments of inertia after 126 iterations, for which the values are assumed to be converged, see Fig. 1.6. The convergence analysis of the cost functional shows that the optimization process

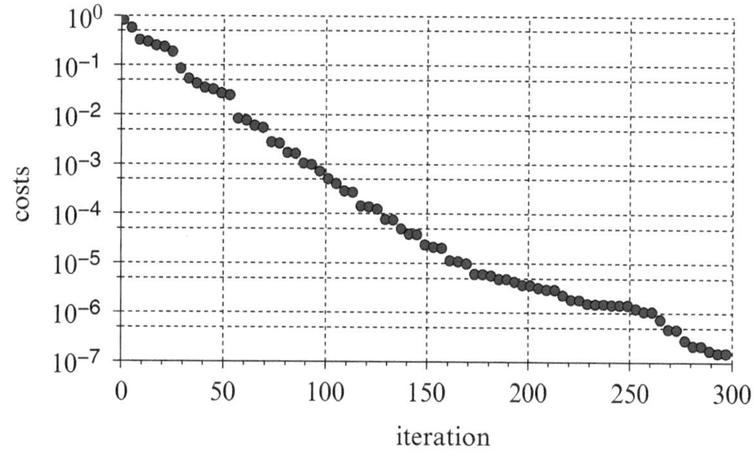

Fig. 1.4 The convergence analysis of the cost functional shows that the optimization process reduces the costs to a factor of 10^{-7} within 300 iterations

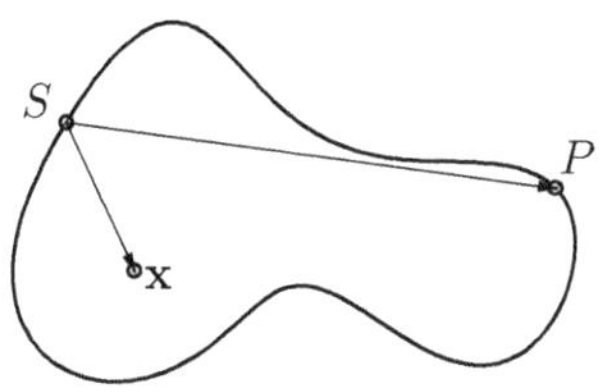

Fig. 1.5 A single rigid body is studied for which the moments of inertia parameters describing the inertia tensor are not known. Point S follows a specified motion and the velocity of point P is measured in order to identify the entries of the inertia tensor

Table 1.1 Moments of inertia: the correct values, the initial values for the first iteration and the final identified values after 126 iterations, for which the values are assumed to be converged

Parameter	Correct value	Initial value	Final identified value	Unit
I_{11}	132.285	1.0	133.441	kg m^2
I_{22}	162.054	1.0	162.073	kg m^2
I_{33}	92.809	1.0	96.815	kg m^2
I_{12}	22.382	0.0	22.198	kg m^2
I_{13}	−42.342	0.0	−44.593	kg m^2
I_{23}	2.689	0.0	3.011	kg m^2

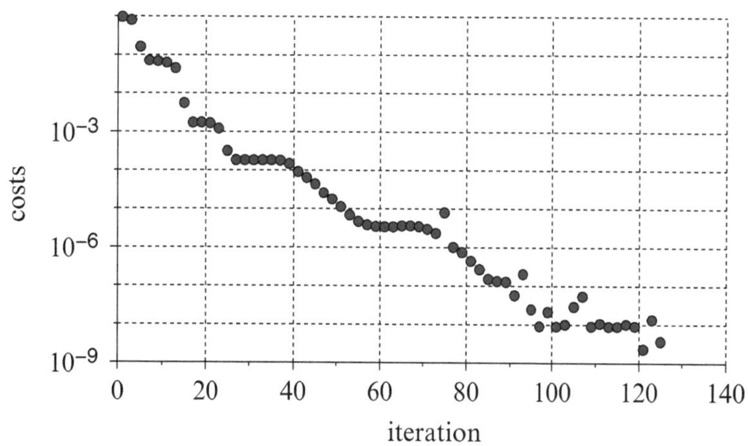

Fig. 1.6 The convergence analysis of the cost functional shows that the optimization process reduces the costs already tremendously within the first 100 iterations. It has to be mentioned that only the costs of every second iteration are depicted here

reduces the costs already within the first 100 iterations to a factor 10^{-8}, see Fig. 1.6. It has to be mentioned that in case of identifying only the diagonal entries of the inertia tensor, I_{11}, I_{22} and I_{33}, the results show higher accordance to the correct values as in case of identifying the whole inertia tensor also incorporating the deviation moments of inertia.

1.4 Conclusions

The adjoint method has been embedded to multibody system dynamics. The presented method can be used for inverse dynamic problems as well as for parameter identification in multibody systems. The computation of the direction of the gradient of the cost functional is based on solving two DAEs. On the one hand, the equations of motion of the multibody system have to be integrated forward in time, and, on the other hand, the adjoint equations have to be integrated backwards. Depending on the chosen forward time integration scheme, almost all necessary matrices for the backward time integration can be reused from the forward time integration and, therefore, a time- and memory-efficient simulation tool for inverse dynamics in the field of multibody dynamics can be constructed. Therefore, the adjoint method shows an efficient way to incorporate inverse dynamics to flexible multibody system applications arising from modern engineering problems.

Acknowledgements This research has been funded by the European Regional Development Fund and the government of the Upper Austria via a Regio 13 project.

References

1. Haug E, Ehle P (1982) Second-order design sensitivity analysis of mechanical system dynamics. Int J Numer Methods Eng 18(11):1699–1717
2. Haug E, Wehage R, Mani N (1984) Design sensitivity analysis of large-scaled constrained dynamic mechanical systems. Trans ASME 106: 156–162
3. Bestle D, Eberhard P (1992) Analyzing and optimizing multibody systems. Mech Struct Mach 20:67–92
4. Eberhard P (1996) Adjoint variable method for sensitivity analysis of multibody systems interpreted as a continuous, hybrid form of automatic differentiation. In: Proceedings of the 2nd international workshop on computational differentiation, Santa Fe. SIAM, Philadelphia, pp 319–328
5. Lions J (1971) Optimal control of systems governed by partial differential equations. Springer, New York
6. Giles M, Pierce N (2000) An introduction to the adjoint approach to design. Flow Turbul Combust 65:393–415
7. Bottasso C, Croce A, Ghezzi L, Faure P (2004) On the solution of inverse dynamics and trajectory optimization problems for multibody systems. Multibody Sys Dyn 11:1–22
8. Bertolazzi E, Biral F, Lio MD (2005) Symbolic-numeric indirect method for solving optimal control problems for large multibody systems. Multibody Sys Dyn 13:233–252
9. Schaffer A (2005) On the adjoint formulation of design sensitivity analysis of multibody dynamics. Dissertation, University of Iowa. http://ir.uiowa.edu/etd/93
10. Petzold L, Li S, Cao Y, Serban R (2006) Sensitivity analysis for differential-algebraic equations and partial differential equations. Comput Chem Eng 30:1553–1559
11. Cao Y, Li S, Petzold L (2002) Adjoint sensitivity analysis for differential-algebraic equations: algorithms and software. J Comput Appl Math 149:171–191
12. Cao Y, Li S, Petzold L, Serban R (2003) Adjoint sensitivity analysis for differential-algebraic equations: the adjoint DAE system and its numerical solution. SIAM J Sci Comput 24(3):1076–1089
13. Nachbagauer K, Oberpeilsteiner S, Sherif K, Steiner W (2014) The use of the adjoint method for solving typical optimization problems in multibody dynamics. J Comput Nonlinear Dyn. doi: 10.1115/1.4028417
14. Vyasarayani CP, Uchida T, McPhee J (2011) Parameter identification in multibody systems using lie series solutions and symbolic computation. J Comput Nonlinear Dyn 6(4):041011. doi: 10.1115/1.4003686
15. Ding JY, Pan ZK, Chen LQ (2012) Parameter identification of multibody systems based on second order sensitivity analysis. Int J Non-Linear Mech 47:1105–1110
16. Özyurt DB, Barton PI (2005) Cheap second order directional derivatives of stiff ODE embedded functionals. SIAM J Sci Comput 26(2): 1725–1743
17. Süli E, Mayers D (2003) An introduction to numerical analysis. Cambridge University Press, Cambridge
18. Hilbert H, Hughes T, Taylor R (1977) Improved numerical dissipation for time integration algorithms in structural dynamics. Earthq Eng Struct Dyn 5:283–292
19. Negrut D, Rampalli R, Ottarsson G, Sajdak A (2005) On the use of the HHT method in the context of index 3 differential algebraic equations of multibody dynamics. In: Goicolea JM, Cuadrado J, Garcia Orden JC (eds) Proceedings of the ECCOMAS conference on advances in computational multibody dynamics, Madrid
20. Blajer W, Kołodziejczyk K (2004) A geometric approach to solving problems of control constraints: theory and a DAE framework. Multibody Sys Dyn 11(4):343–364
21. Betsch P, Uhlar S, Quasem M (2009) Numerical integration of mechanical systems with mixed holonomic and control constraints. In: Arczewski K, Frączek J, Wojtyra M (eds) Proceedings of the ECCOMAS thematic conference on multibody dynamics, Warsaw University of Technology, 29th June–2nd July

Chapter 2
Calibration and Validation of a Car Subframe Finite Element Model Using Frequency Responses

T.J.S. Abrahamsson, F. Bartholdsson, M. Hallqvist, K.H.A. Olsson, M. Olsson, and Å. Sällström

Abstract A finite element model of a car front subframe has been calibrated against test data. Stepped-sine testing has been used to give frequency response function estimates on an ensemble of seemingly identical subframes. Therefore, the deviation between test data and simulation results can be compared in a meaningful way by the outcome of model calibration and cross-validation. Emphasis has been put on the preparation of the test pieces for high fidelity testing and on bettering the chances of getting a calibration outcome that provides insight into the physical processes that govern the subframe dynamics. The front subframe model has more than 200,000 degrees-of-freedom and 17 model calibration parameters. The efficiency of the calibration procedure under these conditions is reported. To achieve efficiency, a calibration with a smooth deviation metric is used together with a damping equalization method that eliminates the need for matching of experimental and analytical eigenmodes. The method is combined with surrogate model frequency response evaluation based on model reduction for increased speed. The Matlab based open-domain software tool FEMcali that employs the Levenberg-Marquardt minimizer with randomized starts has been used for calibration and an unregularized Gauss-Newton minimizer has been used in the cross-validation.

Keywords FEM model calibration • FEM model validation • Large model calibration • Surrogate model • Damping equalization

2.1 Introduction

For car industry, as well as other advanced industry, it is important to shorten the development time and reduce cost for testing. This has increased the need for attribute predictions that are based on computational models with high credibility. Good modeling guidelines are essential to achieve trust in model predictions that are made prior to the availability of physical test objects. Such modeling guidelines should ideally be created based on the outcome of model correlation and validation studies made on test data from already existing components, subsystems or full systems. Since industry-size models tend to be very large, efficient methods for calibration, correlation and validation are needed in this process of increasing the model credibility. This paper concerns the calibration and validation of a subframe model for a Volvo V40 car. This part of the car was selected since it is represented by a reasonable large computational model and was believed to behave linearly under mild loading conditions. A Nastran model with 207,912 degrees-of-freedom is the target for calibration. The subframe and its physical location in the car is indicated in Fig. 2.1.

Many FEM calibration practices are based on modal analysis and modal comparisons. The accuracy of the FEM is determined by comparing modal data from test and analysis. The eigenfrequencies are compared directly, while the corresponding mode shapes are compared with various metrics. The metrics most often employed are based on the orthogonality and cross-orthogonality of the modes with respect to the mass distribution quantified by a FEM mass matrix, see [1]. A significant drawback of these metrics is the fact that there is not an explicit connection between the metric values and a corresponding deviation in a predicted response. Another drawback is that measurement noise, limited spatial resolution in the vibration sensing, and shortcomings that affect the test data quality and processed data often magnify the problems associated with closely spaced modes that produce a spurious coupling sensitivity between the test and FEM mode shapes. This sensitivity makes a modal orthogonality metric problematic for matching test and FEM mode shapes

T.J.S. Abrahamsson (✉)
Department of Applied Mechanics, Chalmers University of Technology, Hörsalsvägen 7A, S-41296 Göteborg, Sweden
e-mail: thomas.abrahamsson@chalmers.se

F. Bartholdsson • M. Hallqvist • K.H.A. Olsson • M. Olsson • Å. Sällström
Volvo Car Corporation, S-40531 Göteborg, Sweden

M. Mains (ed.), *Topics in Modal Analysis, Volume 10*, Conference Proceedings of the Society for Experimental Mechanics Series, DOI 10.1007/978-3-319-15251-6_2

Fig. 2.1 Front subframe seen as placed in the car between the front wheels. The subframe is a vital part in the front wheel suspension and connects the suspension link arms to the car body and engine which makes it an important part regarding transmitted vibrations from road and engine

in the process of FEM calibration for systems with high modal density. Also, as the vibrational wavelengths approach the scale of structural geometry variations, the mode shapes become very sensitive to modelling errors and uncertainties [2–4]. Alternatively, modal approaches could be avoided by using frequency responses for FEM validation. Many researchers, e.g. Grafe [5], have studied model calibration using frequency domain data. The methods developed can loosely be categorized based on their criterion deviation metric as equation error methods or output error methods, or on their parameterization strategy, as direct matrix entity parameterization or physical property parameterization. In frequency domain equation error methods, such as described in [6–12], sensitivities of the FEM impedance matrix with respect to the design variables must be computed. One observed drawback of equation error approaches is that the comparably small number of responses measured in comparison to the large number of dofs in the FE model can produce non-smooth reduced impedance functions that have sharp peaks at the resonances that occur when the sensor dofs are fixed, see [6]. This can lead to a complicated calibration process in which a numerical calibration minimizer can easily be stuck in a local minimum. Output error methods on the other hand, such as described in References [13–18], attempt to minimize the difference between the measured data and the analytical prediction of that data. Balmès [13] minimized the least-squares or log-least-squares deviation in the frequency response functions. Herendeen et al. [14] presented a general multidisciplinary optimization framework for minimizing the deviation in any type of structural response. Dascotte and Strobbe [15] minimized the error in a frequency response function correlation metric that is based on the modal assurance criterion adapted to the frequency domain. The drawback of this approach is that the deviation metric, like many of the frequency response based deviation metrics in use [18], are not physically well motivated. There is no direct connection between a metric deviation and the deviation in the FEM prediction of responses due to loading. The advantage of the output approach is that no model dof reduction is necessary. However, these techniques require the minimization of an objective function that is non-linear in the parameters that can lead to minimizer convergence problems and a large computational effort.

Early developed FEM calibration strategies directly targeted the matrix entities of the FE mass and stiffness matrices by adjusting these to achieve best fit to test data. Recent papers, of more practical use to an engineer in the product development process, deal with model parameters that have clear physical interpretations, such as Youngs' moduli, material densities and thicknesses of shell-like parts. In a paper by Abrahamsson and Kammer [19] a discrete-frequency response based deviation metric was presented. It was shown that, together with a method for damping equalization that makes the calibration problem regularized, the metric gives good convergence properties for the calibration optimizer. The smoothness of the calibration metric was demonstrated to increase with increased system damping. In the damping equalization procedure the system damping can be arbitrarily set to any fictitious level for both the experimental data and the FEM. By assigning the same amount of modal damping to all system modes, modal correlation analysis for mapping of experimentally found damping to FEM damping can be avoided. By considering the FE coefficient matrices to be linear in the parameters around a linearization point, the reduced order model can be easily formed without expensive evaluation of the FEM equation as the model parameter settings vary as the iteration process proceeds to achieve the calibration minimum.

In this paper, the calibration and validation problem of a large scale FE model is treated by the method set out in [19]. The measurement procedure, involving tests of five seemingly identical subframes, is described. The outcome of the calibration, validation and cross-validation is reported and observations made during the process are discussed.

2.2 Theory

The theory presented in [19] is reiterated in condensed form here for reader convenience. We set out the theory from the FE representation of the linear time-invariant mechanical system as the set of ordinary differential equations

$$\boldsymbol{M}\ddot{\boldsymbol{q}} + \boldsymbol{V}\dot{\boldsymbol{q}} + \boldsymbol{K}\boldsymbol{q} = \boldsymbol{Q}(t) \tag{2.1}$$

with $M, V, K \in \Re^{n \times n}$ being the mass, viscous damping and stiffness matrices respectively. The displacement vector is here $\boldsymbol{q}$, and the dot notation is used for time differentiation. The vectorially associated loadings to $\boldsymbol{q}$ is $\boldsymbol{Q}$, for which just but a few entities are normally non-zero. These non-zero entities constitute the stimuli vector $u \in \Re^{n_u}$ from which the full load vector is given by the Boolean transformation matrix $\boldsymbol{P}_{\mathrm{u}}$ as $Q = P_u u$. It is assumed that only some mass and stiffness properties are subjected to calibration and therefore the FE problem is parameterized, with physical property parameters collected in data vector $P \in \Re^{n_p}$, such that the mass and stiffness matrices depend on these, i.e. $K = K(P)$ and $M = M(P)$. The physical parameter setting $\boldsymbol{P}$, with entities that can be of any order, is related to a normalized parameter vector $\boldsymbol{p}$ and some fixed non-zero nominal parameter setting $\boldsymbol{P}_0$ such that $P = P_0 \cdot (1 + p)$. The normalization of the parameters is made to better suit numerical minimizing schemes that often works better with scaled unknowns.

Using the dummy equation $I\ddot{q} - I\ddot{q} = 0$, the first-order differential equation counterpart to Eq. (2.1) can be formed as the state-space equations

$$\dot{x} = \boldsymbol{A}x + \boldsymbol{B}\boldsymbol{u} \quad y = \boldsymbol{C}x + \boldsymbol{D}\boldsymbol{u} \tag{2.2}$$

with the state vector $x^{\mathrm{T}} = \left\{ q^{\mathrm{T}}\ \dot{q}^{\mathrm{T}} \right\}$. It can easily be verified that the coefficient matrices $\boldsymbol{A}$ and $\boldsymbol{B}$ of Eq. (2.2) relate to the mass, damping and stiffness matrices such that

$$\boldsymbol{A} = \begin{bmatrix} 0 & \boldsymbol{I} \\ -\boldsymbol{M}^{-1}\boldsymbol{K} & -\boldsymbol{M}^{-1}\boldsymbol{V} \end{bmatrix} \quad \boldsymbol{B} = \begin{bmatrix} 0 \\ \boldsymbol{M}^{-1}\boldsymbol{P}_u \end{bmatrix} \tag{2.3}$$

and the matrices $\boldsymbol{C}$ and $\boldsymbol{D}$ are the appropriate matrices forming the system output $y \in \Re^{n_y}$ from linear combination of the states in $x \in \Re^{2n}$ and the stimuli in $\boldsymbol{u}$. Using the state-space representation (2.2), the system transfer function, relating the input to the output, can be expressed as

$$\boldsymbol{H}(\boldsymbol{p}) = \boldsymbol{C}(i\omega \boldsymbol{I} - \boldsymbol{A})^{-1} B + D \tag{2.4}$$

The transfer function $\boldsymbol{H}$ established from analytical FE data will here be denoted $\boldsymbol{H}^{\mathrm{A}}$. The experimentally determined system transfer function (denoted $\boldsymbol{H}^{\mathrm{X}}$) is often obtained from vibration testing and the use of signal processing and state-space system identification. The model calibration problem is thus about finding that parameter setting $p = p*$ (called the oracle parameter setting) that minimizes the deviation between $H^A(p)$ and H^X, i.e. $p* = argmin \ \left(f\left(H^A(p) - H^X\right)\right)$ with f being a calibration metric.

2.2.1 A Frequency Response Calibration Metric

A calibration scheme that uses a gradient based minimizer, needs to work with a smooth deviation metric for high likelihood of success. That is to obtain convergence in the search for parameter optimum from multiple start settings of the parameters. A well calibrated model should give high accuracy in simulation of test output quantities, and ideally predictions with high credibility of other output quantities not tested. In a frequency domain context, this often translates to that model which accurately captures the structural resonances and possibly also its anti-resonances. A metric that does not discriminate against deviations at frequencies where the structural response is small is the quadratic functional

$$\delta = \boldsymbol{\varepsilon}^H \boldsymbol{\varepsilon} / N \tag{2.5}$$

which is called the square deviation (*SD*). In this the deviation vector $\boldsymbol{\varepsilon}$ is

$$\boldsymbol{\varepsilon}(\boldsymbol{p}) = \log_{10}\left(\text{vect}\left(\boldsymbol{H}^{A}(\boldsymbol{p})\right)./\text{vect}\left(\boldsymbol{H}^{X}\right)\right) \tag{2.6}$$

where the ./ operator denotes the element-by-element division and vect(.) is the vectorizing operation that makes all frequency response function elements of the $n_y \times n_u$ transfer function, at a dense set of n_f discrete frequencies used for evaluation, into an $n_y n_u n_f \times 1$ column vector. *N* is the number of elements of that vector.

Since finite element model calibration tends to be computationally demanding, calibration criteria that lead to computational efficiency are strongly of the essence. One such efficiency concern targets the sampling strategy for the discrete frequencies that are selected for frequency response function evaluation and the selection is based on the half-band width of the eigenmodes. The half-bandwidth $\Delta\omega_i$ of a damped structural resonance at frequency ω_i is given by $\Delta\omega_i = 2\zeta_i\omega_i$ with ζ_i being the relative modal damping of the i:th mode. One observes that the half-bandwidth increases linearly with increasing resonance frequency. It has been found to be a good frequency sampling strategy to utilize frequency steps that increase linearly with frequency to give a balanced contribution to the deviation metric from various system modes. Such sampling keeps the number of samples over one half-bandwidth constant over the range. That sampling strategy seems reasonable, provided that relative damping of all modes in the range is equal, which rarely happens for experimentally found eigenmodes. However, the damping can be equalized by a procedure that is described next.

2.2.2 Damping Equalization

A central issue for FRF based model calibration is that of model damping. Since in general, damping has been found to be very difficult to model using first principles, it is most often assigned a simple representation for modeling convenience. For modal damping modelling, the model's damping is usually set using the outcome of experimental modal analysis of test data from the structure under investigation. The modal dampings found in experiments are normally used for FE simulation without further attempts to understand their physical background. The nature of the damping mechanisms is normally such that the modal damping varies from mode to mode. That makes a mapping of experimentally obtained modal damping into modal damping of FE modes cumbersome. The difficulty arises since the mapping of modal damping relies on mode shape pairing, meaning that the same amount of modal damping should be assigned to modes that are in some sense similar by their deformation pattern. Such pairing is normally difficult, especially for systems with a high modal density. To overcome the problem of mode pairing, a method of damping equalization has been suggested [19]. The damping equalization is achieved by imposing the same modal damping on all experimentally found system modes by perturbation of a mathematical model of the experimental data found from system identification using raw frequency response function data $\boldsymbol{H}^X_{\text{raw}}$. The experimentally found system transfer function $\boldsymbol{H}^X_{\text{raw}}$ can then be represented by an identified state-space system $\dot{\boldsymbol{x}} = \boldsymbol{A}\boldsymbol{x} + \boldsymbol{B}\boldsymbol{u}$, $\boldsymbol{y} = \boldsymbol{C}\boldsymbol{x} + \boldsymbol{D}\boldsymbol{u}$ such that

$$\boldsymbol{H}^X_{\text{raw}} \approx \boldsymbol{C}(i\omega\boldsymbol{I} - \boldsymbol{A})^{-1}B + D \tag{2.7}$$

The experimental state-space system can be brought to diagonal form by a similarity transformation. Using the mode matrix $\boldsymbol{X}$, pertinent to the eigenvalue problem $\boldsymbol{A}\boldsymbol{X} = \boldsymbol{X}\boldsymbol{\Lambda}$, for transformation we have for the diagonal realization that

$$\boldsymbol{x} = \boldsymbol{X}\boldsymbol{z} \qquad \dot{\boldsymbol{z}} = \overline{\boldsymbol{A}}\boldsymbol{z} + \boldsymbol{X}^{-1}\boldsymbol{B}\boldsymbol{u} \qquad \boldsymbol{y} = \boldsymbol{C}\boldsymbol{X}\boldsymbol{z} + \boldsymbol{D}\boldsymbol{u} \tag{2.8}$$

with

$$\overline{\boldsymbol{A}} = \boldsymbol{X}^{-1}\boldsymbol{A}\boldsymbol{X} = \text{diag}(\lambda_i) \tag{2.9}$$

for which λ_i are the complex-valued system poles as given by the experimental data. For small damping, the relative modal damping ζ_i, obtained from these poles are

$$\zeta_i = -\Re(\lambda_i)/|\Im(\lambda_i)| \tag{2.10}$$

In the process of damping equalization, the real parts of the poles are perturbed such that the damping is made equal for all modes. The modal dampings are then set to a single fixed value ζ_0, i.e. $\zeta_i = \zeta_0 \;\; \forall i$. The effect of such damping equalization is that the oscillatory imaginary part of the poles are preserved and the real damping part is modified such that the perturbed system poles $\tilde{\lambda}_i$ are now

$$\tilde{\lambda}_i = \Im(\lambda_i)(-\zeta_0 + \mathrm{i}), \quad \forall \Im(\lambda_i) > 0, \; \tilde{\lambda}_i = \Im(\lambda_i)(\zeta_0 + \mathrm{i}), \quad \forall \Im(\lambda_i) < 0 \tag{2.11}$$

and the modified state-space realization is then

$$\dot{z} = \tilde{\boldsymbol{A}} z + \boldsymbol{X}^{-1}\boldsymbol{B}\boldsymbol{u} \qquad y = \boldsymbol{C}\boldsymbol{X}z + \boldsymbol{D}\boldsymbol{u} \tag{2.12}$$

with $\tilde{\boldsymbol{A}} = \mathrm{diag}\left(\tilde{\lambda}_i\right)$. This in turn gives us a modified transfer function for the experimental model, such that the transfer function used for calibration with damping equalization is

$$\boldsymbol{H}^X \approx \boldsymbol{C}\boldsymbol{X}\left(i\omega\boldsymbol{I} - \tilde{A}\right)^{-1} X^{-1}B + DE \tag{2.13}$$

At this stage it should be obvious that the application of a system identification procedure on the raw test data $\boldsymbol{H}^X_{\mathrm{raw}}$ has led us to a mathematical model which we can evaluate for any frequency. We are also able to make fictitious modifications of the system under test. A particularly useful such modification is that we can adjust the system damping level, leaving stiffness and inertia properties intact, such that all system modal damping are set equal. The model calibration of the FE model can then be made towards this fictitious experimental model for calibration of parameters that relate to mass and stiffness only.

For the FE based system representation, the modal damping allows for a simple representation. For a system with given mass and stiffness matrices $\boldsymbol{M}$ and $\boldsymbol{K}$ we have the viscous damping matrix $\boldsymbol{V}$ to be [20]

$$V = \boldsymbol{M}\boldsymbol{X}\mathrm{diag}(1/m_i)\,\mathrm{diag}(2\zeta_0 m_i\omega_i)\,\mathrm{diag}(1/m_i)\,\boldsymbol{X}^{\mathrm{T}}\boldsymbol{M} \tag{2.14}$$

with eigenfrequencies ω_i, modal masses m_i, and the modal matrix $\boldsymbol{X}$ given by the undamped system's eigenvalue problem

$$\boldsymbol{K}\boldsymbol{X} = \boldsymbol{M}\boldsymbol{X}\mathrm{diag}\left(\omega_i^2\right) \qquad \mathrm{diag}(m_i) = \boldsymbol{X}^{\mathrm{T}}\boldsymbol{M}\boldsymbol{X} \tag{2.15}$$

In a calibration procedure we are then able to search for the mass and stiffness related parameters $\boldsymbol{p}$ of the FE model $\{K(p), M(p)\}$ that render the transfer function $\boldsymbol{H}^{\mathrm{A}}$ given by Eq. (2.4) and that let the criterion function of Eq. (2.5) to be minimal. The discrete frequencies used to evaluate Eq. (2.5) do not have to match the discrete frequencies used in testing. After a successful calibration, the FE model is in better agreement with test data. The eigenmodes from test and analysis are therefore normally well suited for mode-pair matching by correlation analysis. After such matching, the modal damping found in the experiment can easily be mapped to the finite element model's viscous damping matrix using Eq. (2.14).

2.2.3 Surrogate Modeling

To be practical, an FRF based model calibration requires rapid calculations of the frequency responses of the model as the parameter settings change in the iterative search for minimum deviation to test data. For big models this is undoable without model reduction. To obtain efficiency, a modal reduction scheme is here applied to create a surrogate model used by the calibration procedure. The eigenmodes of the corresponding undamped system at the nominal parameter configuration, belonging to all eigenvalues in a frequency range that significantly overlaps the frequency range of interest are then used for reduction. Let that range be $\omega = \left[\omega_{\mathrm{low}}, \omega_{\mathrm{high}}\right]$. To save computational effort, that reduction basis is kept constant within one full calibration cycle and thus not modified as the parameter settings vary during minimization procedure iterations. Let the eigenvalue problem formulated at the nominal parameter setting $\boldsymbol{p}_0$ be

$$\boldsymbol{K}(\boldsymbol{p}_0)\boldsymbol{T} = \boldsymbol{M}(\boldsymbol{p}_0)\boldsymbol{T}\boldsymbol{\Omega} \qquad \boldsymbol{\Omega} = \mathrm{diag}\left(\omega_i^2\right)\,\forall\boldsymbol{\omega} = \left[\omega_{\mathrm{low}}, \omega_{\mathrm{high}}\right] \tag{2.16}$$

Then the mass and stiffness matrices at any parameter setting $\boldsymbol{p}$ of the reduced model are

$$\overline{\boldsymbol{M}}(\boldsymbol{p}) = \boldsymbol{T}^{\mathrm{T}}\boldsymbol{M}(\boldsymbol{p})\boldsymbol{T} \; \overline{\boldsymbol{K}}(\boldsymbol{p}) = \boldsymbol{T}^{\mathrm{T}}\boldsymbol{K}(\boldsymbol{p})\boldsymbol{T} \tag{2.17}$$

and, in particular at the nominal configuration

$$\overline{\boldsymbol{M}}_0 = \boldsymbol{T}^{\mathrm{T}}\boldsymbol{M}\left(\boldsymbol{p}_0\right)\boldsymbol{T} \; \overline{\boldsymbol{K}}_0 = \boldsymbol{T}^{\mathrm{T}}\boldsymbol{K}\left(\boldsymbol{p}_0\right)\boldsymbol{T} \tag{2.18}$$

We introduce the gradients of the reduced mass and stiffness matrices

$$\overline{\boldsymbol{M}}_{,j} = \boldsymbol{T}^{\mathrm{T}}\left(d\boldsymbol{M}/dp_j\big|_{p=\boldsymbol{p}_0}\right)\boldsymbol{T} \text{ and } \overline{\boldsymbol{K}}_{,j} = \boldsymbol{T}^{\mathrm{T}}\left(d\boldsymbol{K}/dp_j\big|_{p=\boldsymbol{p}_0}\right)\boldsymbol{T} \tag{2.19}$$

with respect to the j:th calibration parameter and note in particular that these may be computed without new eigensolutions since the reduction basis is kept invariant. A reduced order surrogate model that is linear in the parameters is taken as the first order expansion of the Taylor series of $\tilde{\boldsymbol{M}}$ and $\tilde{\boldsymbol{K}}$ about $\boldsymbol{p}_0$ as

$$\tilde{\boldsymbol{M}}(p) = \overline{\boldsymbol{M}}_0 + \sum_{j=1}^{n_p}\left(\boldsymbol{p}_j - \boldsymbol{p}_{j,0}\right)\overline{\boldsymbol{M}}_{,j} \; \tilde{\boldsymbol{K}}(p) = \overline{\boldsymbol{K}}_0 + \sum_{j=1}^{n_p}\left(\boldsymbol{p}_j - \boldsymbol{p}_{j,0}\right)\overline{\boldsymbol{K}}_{,j} \tag{2.20}$$

with $\boldsymbol{p}_{\mathrm{j},0}$ being the j:th parameter at the nominal setting. It can be observed that the surrogate model approximation error is likely to increase with parameter variation from $\boldsymbol{p}_0$ for two reasons. The first reason is that the reduction basis $\boldsymbol{T}$ is kept fixed and the second is that the $\{M(p), K(p)\}$ model itself is not necessarily linear in the parameters. One also notes that, once the reduced order model and its gradients are established from the full size FE mode, no further evaluation of the FE model is required. That leads to fast computations. With the state transformation $\boldsymbol{x} = \boldsymbol{T}\boldsymbol{\xi}$, the reduced order surrogate model fits well into the state-space setting of the system transfer function as given by Eq. (2.4). The corresponding state-space model matrices are

$$\boldsymbol{A} = \begin{bmatrix} 0 & \boldsymbol{I} \\ -\tilde{\boldsymbol{M}}^{-1}\tilde{\boldsymbol{K}} & -\tilde{\boldsymbol{M}}^{-1}\tilde{\boldsymbol{V}} \end{bmatrix} \; \boldsymbol{B} = \begin{bmatrix} 0 \\ \tilde{\boldsymbol{M}}^{-1}\boldsymbol{T}^{\mathrm{T}}\boldsymbol{P}_u \end{bmatrix} \tag{2.21}$$

with the reduced viscous damping matrix $\tilde{\mathbf{v}}$ formed in analogy with Eq. (2.14). With such reduced order surrogate model the frequency response calculations for various parameter settings needed during the calibration minimization procedure can be made very efficiently.

2.2.4 Validation and Cross-Validation

For validation and cross-validation purposes we calculate the model deviation to test data that has not been used for calibration. We establish the validation deviation vector $\boldsymbol{\gamma}$ as

$$\boldsymbol{\gamma}\left(\boldsymbol{p}\right) = \log_{10}\left(\mathrm{vect}\left(\boldsymbol{H}^{\mathrm{A}}\left(\boldsymbol{p}\right)\right)./\mathrm{vect}\left(\boldsymbol{H}^{X}_{\mathbf{raw}}\right)\right) \tag{2.22}$$

in analogy with the calibration metric $\boldsymbol{\varepsilon}$ of Eq. (2.6), but here the transfer function relate to other frequencies or transfer function paths than were used for calibration. Also, the test data are taken as the raw test data and not the data computed from an identified state-space model. We also introduce the square deviation as a validation metric as

$$SD = \boldsymbol{\gamma}^{H}\boldsymbol{\gamma}/N \tag{2.23}$$

in which N is the number of data, i.e. the number of elements of the vector $\boldsymbol{\gamma}$. For a validated model we demand that the square deviation becomes smaller after calibration than before, and also that it becomes smaller than a specific target $\widehat{SD}$ that is established in line with the intended use of the model, i.e.

$$SD\left(\boldsymbol{p}^*\right) < SD\left(\boldsymbol{p}_0\right) \text{ and } SD\left(\boldsymbol{p}^*\right) < \widehat{SD} \tag{2.24}$$

For cross-validation we do statistics on the square deviation to establish the mean squared deviation (often called the residual squared error, *RSE*) and statistics on the calibration parameters after several calibrations in which various parts of the data has been used for repeated calibrations and the rest of the data used for establishing the square deviation. Since the repeated calibration result in varying parameter estimates, the statistics of the calibration outcomes is clearly of strong interest.

2.3 Calibration and Validation of the Subframe Model

The subframe FE model to be calibrated and validated is described first. A discussion on the testing, including considerations made after a pre-test, is made thereafter. The section concludes with results of the calibration, validation and cross-validation.

2.3.1 *Finite Element Model and Its Parameterization*

The subframe model is a Nastran FE model that consists of 39,978 nodes and 41,509 elements. Of these 35,254 are plate elements, 900 are solid elements, 1,782 are bar elements and 3,573 are rigid elements. The model was updated prior to calibration from a preliminary design CAD status to the best-of-knowledge-status of the subframe parts as they leave the production line. The density of the FE model parts were scaled to reflect the true weight of one subframe weighted on a high precision scale to be 16.985 kg. The nominal FE model indicates 18 elastic modes in free-free conditions below 800 Hz. The model and the first seven elastic modes are shown in Fig. 2.2. Seventeen parameters were selected for calibration. These were

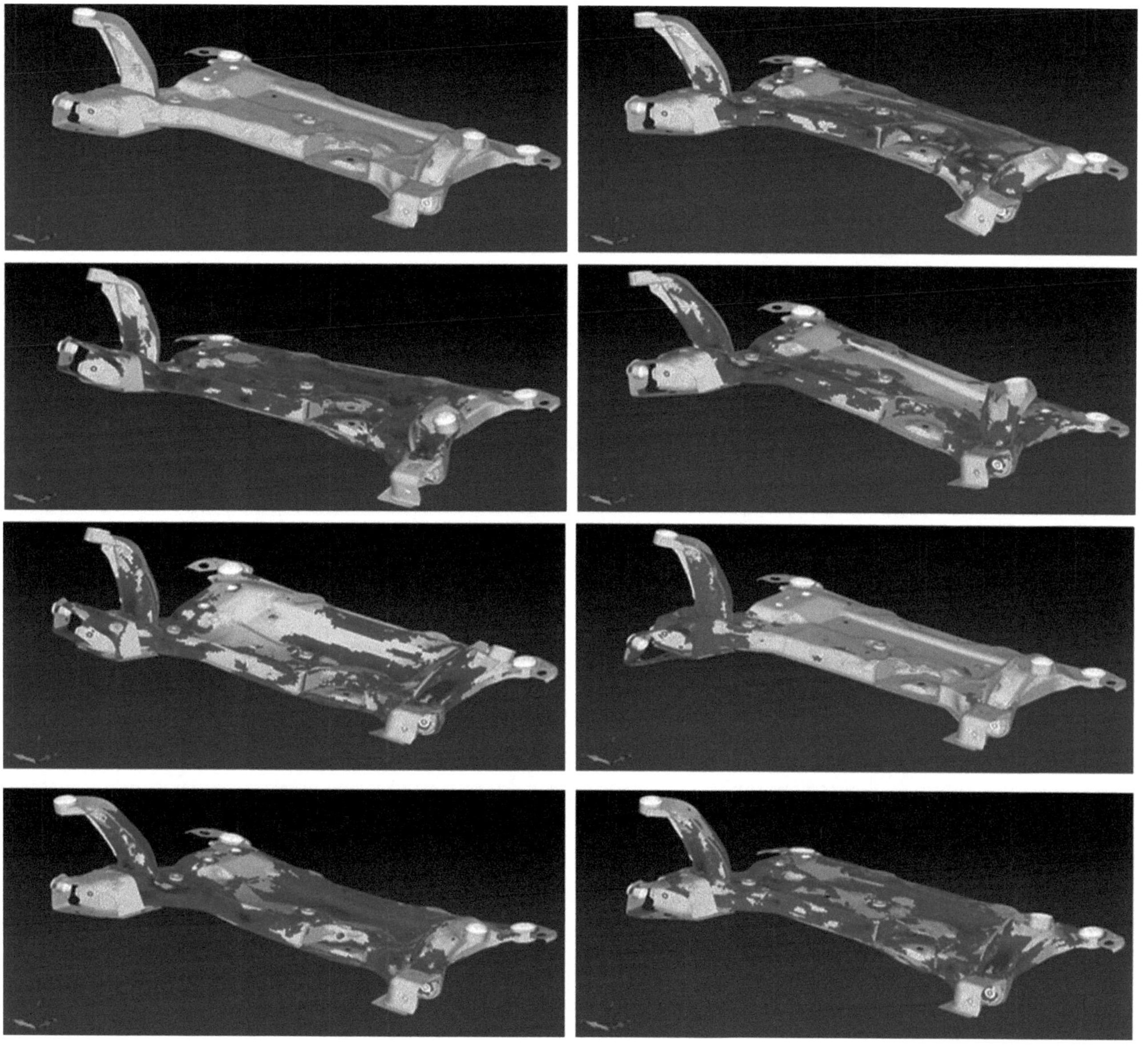

Fig. 2.2 Undeformed FE model (*upper left*) and seven flexible body modes in frequency range of calibration at 129, 174, 222, 266, 307, 348, 358 Hz. Eigenfrequencies are before calibration. The contour plots indicate regions of high strain (*blue/cyan/yellow*) and low strain (*magenta*)

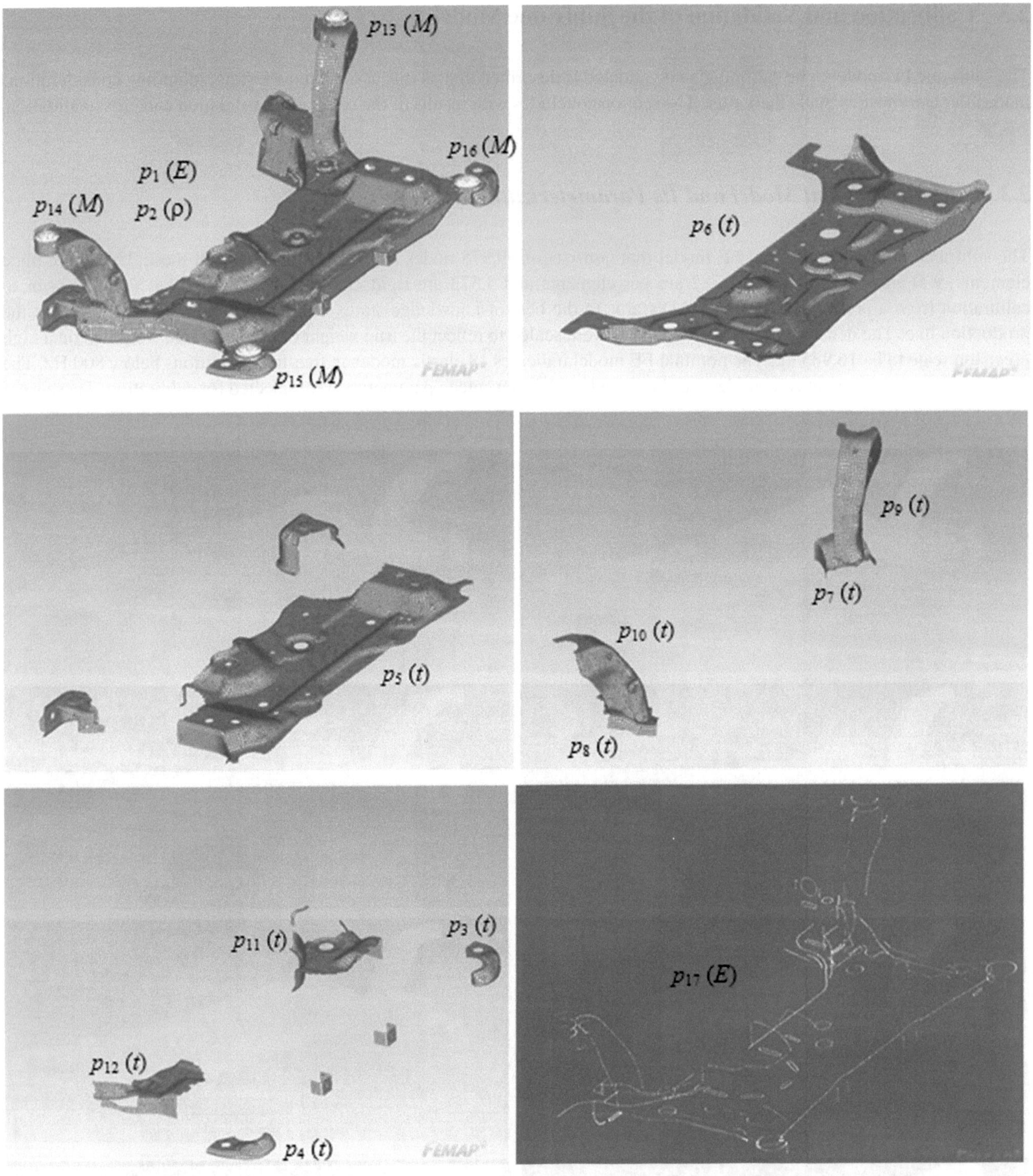

Fig. 2.3 Subframe domains affected by the parameters. *Top left*: parameters 1 and 2 (p_1 and p_2) are Youngs' modulus and density of overall subframe bulk material (steel). Four bushing masses are represented by parameters p_{13} through p_{16}. *Top right*: p_6 is thickness of bottom plate. *Mid left*: p_5 is thickness of top plate. *Mid right*: p_9 and p_{10} are thicknesses of LHS and RHS arms, and p_7 and p_8 are thicknesses of their base plates. *Bottom left*: parameters p_3, p_4, p_{11} and p_{12} represent thicknesses of various smaller parts of subframe. *Bottom right*: p_{17} is parameter representing elasticity (Youngs' modulus) of the welds

related to material stiffness and density, plate thickness, bushing masses and weld elasticity. The parameters are specified in Fig. 2.3. Many parameters, in particular the thickness parameters, can be seen as modeling placeholders for other physical properties such as geometry and thickness variations that results from the sheet metal forming of the various subframe parts.

2.3.2 Test Procedure and System Identification

Prior to testing, the accelerometer layout was decided based on the Effective Independence pretest planning scheme [21] targeting the first 20 elastic modes. A set of 20 single-axis accelerometers was used and two shaker locations used in two repeated SIMO tests between which the position of the shaker was moved. Data from one such SIMO test were used in this work. The shaker configurations were such that the actuation forces were collocated and co-oriented with accelerometers, see Fig. 2.4. A stepped-sine procedure was used with 1 N load magnitude and frequency steps in the range of a few tenths of a Hertz. Low-weight IEPE sensors, 0.5 g PCB Piezotronic type 352C22, were used to measure the accelerations and a Brüel&Kjær force sensor (type 8203 with IEPE converter 2647B) was used to measure the excitation force. The sensor layout is shown in Fig. 2.4. A significant effort was put on isolating the subframe test-pieces from the surrounding with a flexible supporting system. Two doubled very thin fishing lines (0.16 mm diameter and 1.8 m long), ending in a locking ring (see Fig. 2.4) at existing bolt holes on the subframe at one end and at flexible bungee cords at the fixture end resulted in a supporting system that gave six quasi-rigid body modes all below 4 Hz.

Five subframe units were tested in total. The test results separated the subframes into two groups with very similar results within each group, with two subframes in on group and three subframes in the other. A close inspection revealed that while two subframes had a clearly visible air gap between two particular sheet metal parts, the same air gap was closed and the sheet metal parts in tight contact for the other three subframes. The air gaps were all within production tolerances, but the impact it had on test data differences came as a surprise. Its impact was later examined by a FE analysis which confirmed the observation.

The data collection resulted in 20×2 frequency response function matrices of the five subframes. A system identification was made on a SIMO subset of data from one subframe using the N4SID state-space sub-space system identification method as implemented in the System Identification toolbox v9.0 in Matlab. Test data from 100 to 550 Hz were used to identify a model of state order 44 using SIMO data from shaking at point #17 (see Fig. 2.4 for its location). Frequency response function data up to 400 Hz from shaking at this point and a selection of 11 responses (accelerometers 1, 5, 7, 10, 12, 13, 14, 15, 18, 19 and 20) were used for calibration of the model. Data from the remaining 9 responses were used for a hold-out model validation.

2.3.3 Calibration and Validation Results

For the calibration and validation, test data in the frequency range 100–400 Hz have been used. The calibration and cross-validation were made using 11 transfer functions resulting from the SIMO testing with shaking at accelerometer #17. The remaining nine transfer functions obtained from the SIMO shaking at the same location were used for standard (hold-out) validation. The calibration was made using a damping equalization at 1 % relative damping and discrete frequency evaluations at a density of three samples per half-bandwidth. This relatively strong damping was set to regularize the calibration metric. Typical damping values for the identified modes below 400 Hz were in the order of 0.1 %. The calibration minimization was started from 20 randomized parameter settings in addition to the start from the nominal parameter setting.

The calibration resulted in a reduction of the square deviation for the calibration data from 1.748 to 1.156[−], thus a reduction by 33 %. The square deviation for validation data was reduced from 1.654 to 1.247[−], thus a reduction by 25 %. The cross-validation was done using 20 Monte Carlo realizations. In each such realization the data were split at random in two sets of equal size. While one set was used for calibration the other was used for calculation of the squared deviation. The *RSE* was calculated to be 1.036[−].

Resulting data for the calibration parameters are shown in Table 2.1. For the calibrated FE model, the mapping of experimentally found modal damping could be mapped to FE modes by mode correlation analysis. After such mapping, the FE model was used to synthesize frequency response functions corresponding to the experimental ditto. Typical frequency response functions from the calibration set are compared in Fig. 2.5. The comparison is made for the ensemble of tested subframes that showed similar behavior and for the FE model before and after calibration. Typical frequency response functions from the validation set are compared in Fig. 2.6.

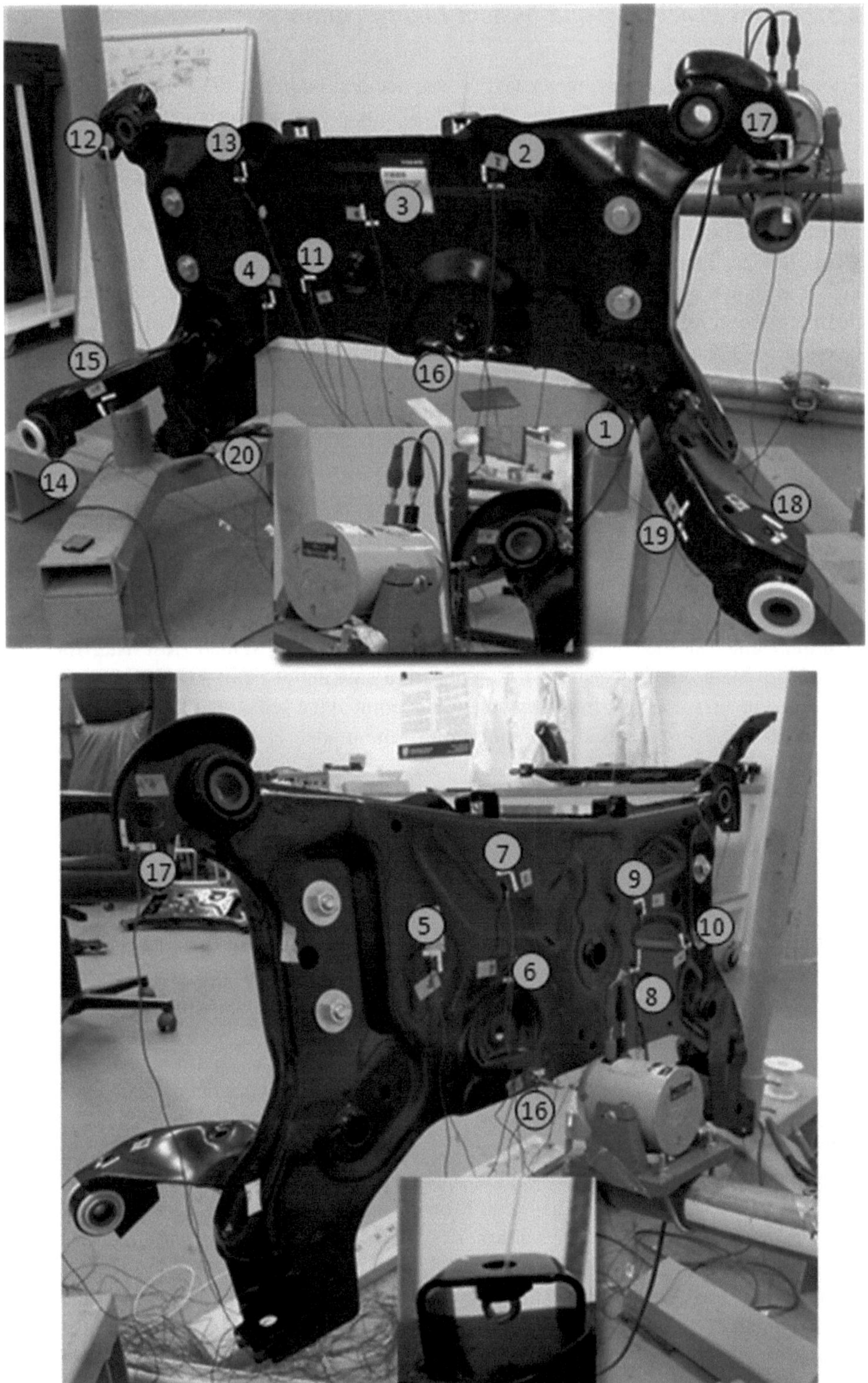

Fig. 2.4 Subframe from two different views. *Top view* shows position of 14 accelerometers visible from this viewpoint. The insert shows the position of the shaker with 5 mm long stinger and force sensor that sits just opposite of accelerometer #17 in one shaker configuration. *Bottom view* shows opposite side of subframe with position of other six accelerometers plus the two (16 and 17) that are at locations of shaker positions. Shaker is just opposite accelerometer #16 in *bottom view*. *Bottom insert* shows the supporting doubled fishing line that is threaded through a locking ring

2.3.4 Computational Efficiency

The calibration was made from 20 random starts and from the nominal parameter setting. The randomized starts were obtained from a latin hypercube sampling scheme. The calculations were made in 8 min 2 s to establish the nominal FE model, form the reduction basis and calculate the surrogate model gradients by parallel computation. It took 58 s to do the

Table 2.1 Parameter statistics with nominal and calibrated values. Mean values and coefficient of variation (CV) from cross-validation

#	Parameter	Nominal setting	Calibrated setting	Mean	CV [%]
1	E main	210 GPa	217	217	0.04
2	ρ main	7,850 kg/m^3	7, 750	7, 750	0.0
3	t bracket RHS	2.5 mm	2.46	2.46	0.012
4	t bracket LHS	2.5 mm	2.52	2.53	0.018
5	t top skin	2.3 mm	2.21	2.20	0.072
6	t bottom skin	2.0 mm	1.96	1.95	0.151
7	t arm base RHS	2.3 mm	2.33	2.33	0.0
8	t arm base LHS	2.3 mm	2.22	2.22	0.0
9	t arm RHS	3.0 mm	3.02	3.02	0.060
10	t arm LHS	3.0 mm	3.04	3.05	0.071
11	t arm extended base RHS	2.3 mm	2.46	2.47	0.080
12	t arm extended base LHS	2.3 mm	2.35	2.35	0.036
13	Bushing mass M_1	0.136 kg	0.135	0.135	0.031
14	Bushing mass M_2	0.136 kg	0.138	0.138	0.030
15	Bushing mass M_3	0.215 kg	0.211	0.211	0.015
16	Bushing mass M_4	0.215 kg	0.227	0.227	0.023
17	Weld elasticity E	210 GPa	207	207	0.0

evaluation of 200 latin hypercube sampling evaluations in parallel and it took 5 min 50 s to do the calibration calculations in parallel. This result in 14 min 50 s total calculation time for the calibration. The hold-out validation took 5 min 16 s which includes the calculation of an updated surrogate model at the calibration parameter configuration and a mapping of the experimentally found modal dampings. The cross-validation with 10 Monte Carlo samples took 17 min 8 s with the majority of time spent on the 10 recalibrations from the calibrated parameter configuration as a starting guess. The overall time spent by the computer was 37 min 14 s.

The calculations were made in Matlab on a desktop PC with an AMD FX-8350 eight-core CPU running at 4 GHz and having 32 GB memory. The FE model was established in MSC.Nastran. A DMAP alter was introduced in a SOL103 solution sequence to get Nastran to release mass and stiffness matrix data as soon as it had been created.

2.4 Concluding Remarks

The work reported here is a continuation of the work reported in [22]. The results that were reported in [22] led to further physical insight that was used to modify the subframe FEM into one that is in better agreement with test observations. The starting point for this work was that refined FE model.

Although not fully reported here, a significant effort was made to get proper test data. The test campaign started with five subframe units. Preliminary testing showed a non-linear behavior also at low excitation levels. Two units were disassembled and the cause of the nonlinearity was found to be internal parts that were connected by loose joints. These parts are tightly bounded in the operation condition in the car by bolt connections. A surrogate of the true bolt connections was introduced by four bolts to the remaining three subframes which are visible in Fig. 2.4 as the blank bolt heads. When properly mounted, these surrogate bolts significantly diminished the nonlinear behavior.

It has been demonstrated that the calibration of a large-scale FE model is feasible and with reasonable outcome. With respect to validation data, the model has been made more trustworthy and the square deviation in the validation data set has decreased by about 25 %. The full calibration and validation calculations can be made in little more than ½ hour on a modern desktop computer anno 2014.

Since some FE model parameters are used as surrogates for the true physics, still some interpretation of the calibration results is needed to get more insight into the true physical phenomena that cause the deviation between analysis and test. However, it is strongly believed that such interpretation of the calibration results is of the essence for gaining physical insight to the benefit of future modeling.

The calibration and validation has been made using the open source Matlab application FEMcali, which can be downloaded from Mathwork's web at www.mathworks.com.

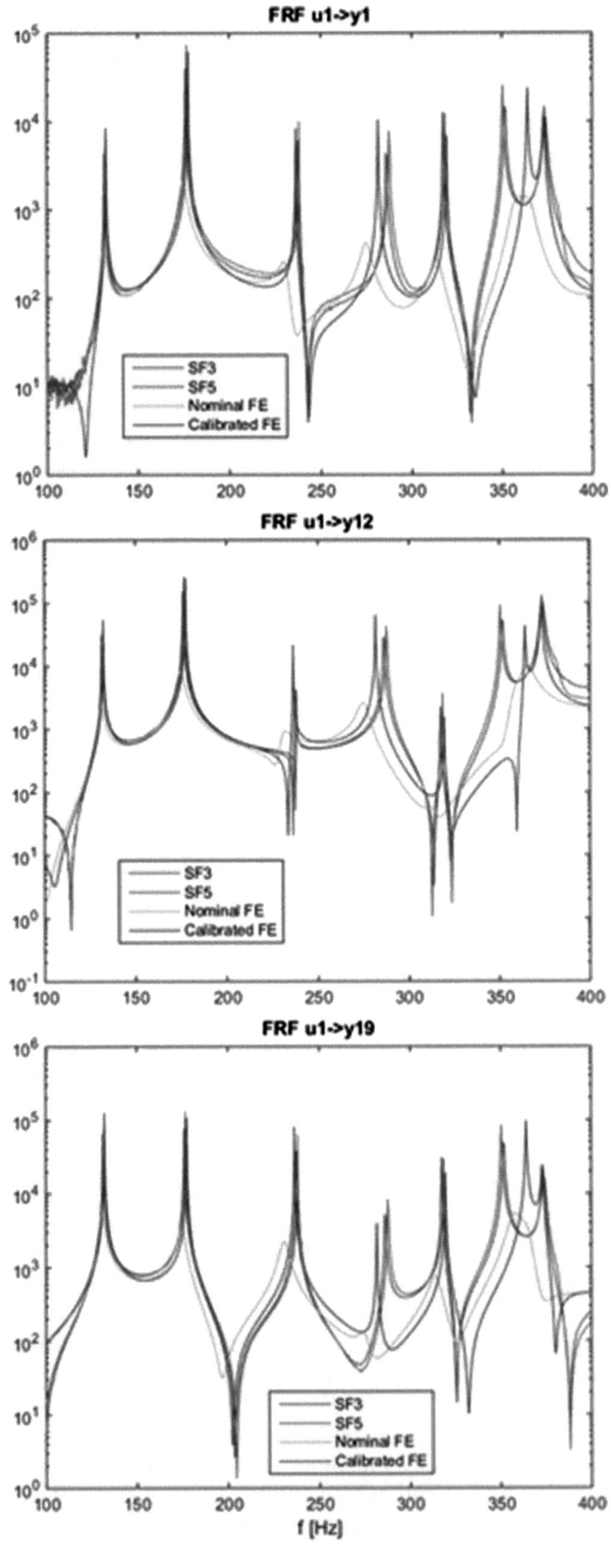

Fig. 2.5 Typical frequency response functions from calibration set. Note that the nominal FE model damping was set to 1 % while the calibrated FE model has mapped damping from test

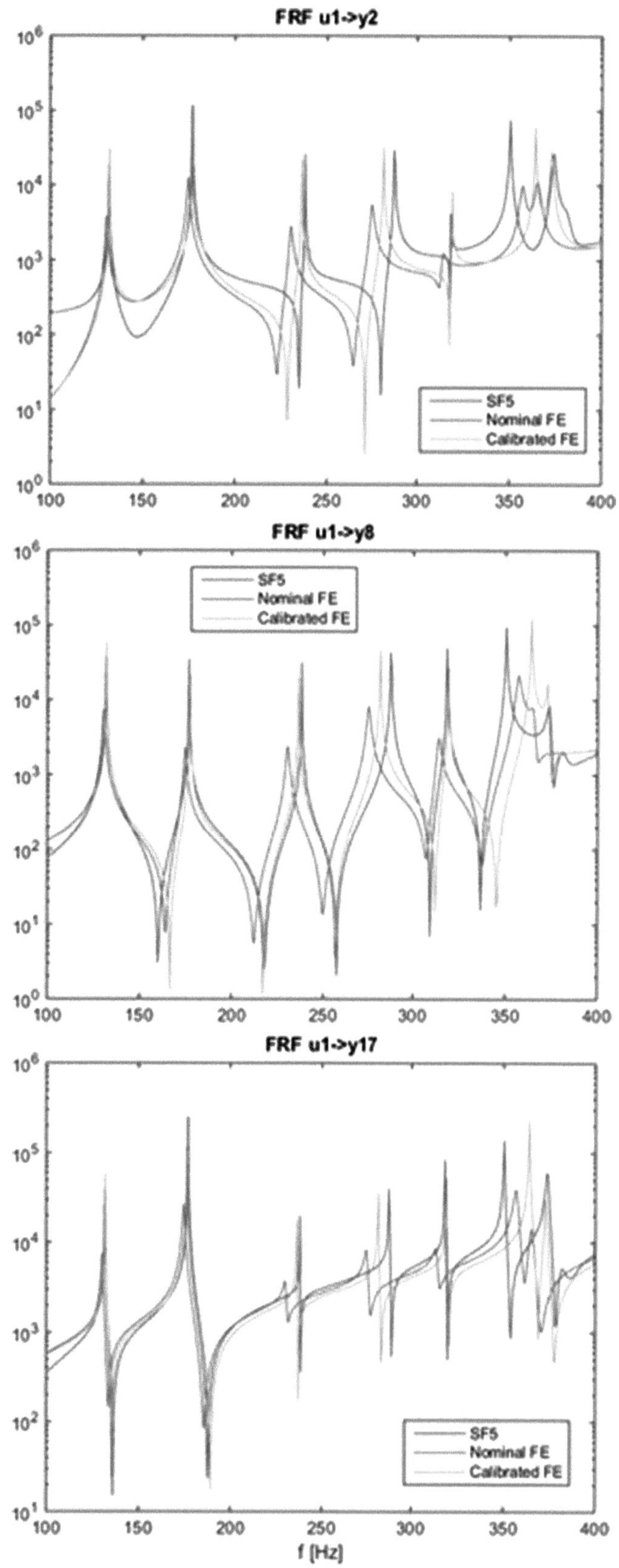

Fig. 2.6 Typical frequency response functions from validation set

References

1. Hasselman TK, Coppolino RN, Zimmerman DC (2000) Criteria for modeling accuracy – a state-of-the-practice survey. In: Proceedings of IMAC XIIX, San Antonio
2. Sarkar A, Ghanem R (2002) Mid-frequency structural dynamics with parameter uncertainty. Comput Meth Appl Mech Eng 191:5499–5513
3. Sarkar A, Ghanem R (2003) A substructure approach for the midfrequency vibration of stochastic systems. J Acoust Soc Am 113(4):1922–1934
4. Hasselman T, Anderson M, Lai YC (1998) Linking FEA and SEA by principal component analysis. In: Proceedings of IMAC XVI, Santa Barbara
5. Grafe H (1998) Model updating of large structural dynamics models using measured response functions. Ph.D. thesis, Imperial College, London
6. Larsson P-O, Sas P (1992) Model updating based on forced vibration testing using numerically stable formulations. In: Proceedings of IMAC X, San Diego
7. Gordis JH (1993) Spatial, frequency domain updating of linear structural dynamic models. In: Proceedings of 34th AIAA/ASME/ASCE/AHS/ASC structures, structural dynamics, and materials conference
8. Schultz MJ, Pai PF, Abdelnaser AS (1996) Frequency response function assignment technique for structural damage identification. In: Proceedings of IMAC XIV, Dearborn
9. Rad SZ (1997) Methods for updating numerical models in structural dynamics. Ph.D. thesis, University of London, London
10. Thyagarajan SK, Schultz MJ, Pai PF (1998) Detecting structural damage using frequency response functions. J Sound Vib 210:162–170
11. Zimmerman DC, Simmermacher T, Kaouk M (1995) Structural damage detection using frequency response functions. In: Proceedings of IMAC XIII, Nashville
12. Nauerz A, Fritzen C (2001) Model based damage identification using output spectral densities. ASME J Dyn Syst Meas Control 123:691–698
13. Balmès E (1993) A finite element updating procedure using frequency response functions – application to the MIT/SERC interferometer testbed. In: Proceedings of IMAC XI, Kissimmee
14. Herendeen DL, Woo L, Hasselman TK, Zimmerman DC (1998) Analysis-test correlation and model updating of dynamic systems using MDO software tools. In: Proceedings of 7th AIAA/USAF/NASA/ISSMO symposium on multidisciplinary analysis and optimization, St. Louis
15. Dascotte E, Strobbe J (1999) Updating finite element models using FRF correlation functions. In: Proceedings of IMAC XVII, Kissimmee
16. Lin RM, Ewins DJ (1994) Analytical model improvement using frequency response functions. Mech Syst Sign Process 8(4):437–458
17. Lammens S (1995) Frequency based validation of structural finite element models. Ph.D. thesis, Katholieke Universiteit Leuven, Leuven
18. Babuska V, Carter D, Lane S, Lacy S (2005) FRF correlation and error metrics for plant identification. In: Proceedings of 46th AIAA/ASME/ASCE/AHS/ASC structures, structural dynamics, and materials conference, Austin
19. Abrahamsson TJS, Kammer DC (2014) FEM calibration with FRF damping equalization. In: Proceedings of IMAC XXXII, Orlando
20. Craig RR Jr, Kurdila AJ (2006) Fundamentals of structural dynamics. Wiley, Hoboken
21. Kammer DC (1991) Sensor placement for on-orbit modal identification and correlation of large space structures. J Guid Control Dyn 14(2):251–259
22. Abrahamsson TJS, Bartholdsson F, Hallqvist M, Olsson KHA, Olsson M, Sällström Å (2014) Calibration and cross-validation of a car component using frequency response functions and a damping equalization technique. In: Proceedings of ISMA 2014, Leuven

Chapter 3
Subspace Identification of a 5 MW Reference Wind Turbine

Jennifer M. Rinker and Henri P. Gavin

Abstract Nonlinear systems, a category into which essentially all real-world engineering systems fall, are challenging in their design, response prediction, and control. Simulating their dynamic response requires numerical solvers, and the use of frequency-domain techniques is prohibited, which eliminates many commonly used engineering techniques. Subspace identification is a convenient method to approximate a complex, nonlinear system with a linear model, significantly enhancing simulation time and allowing the use of traditional control schemes. This paper investigates the utility of subspace identification for the National Renewable Energy Laboratory (NREL) 5 MW reference model, which runs on their open-source wind turbine simulator, FAST. The ability of the identified model to predict a variety of outputs is compared. The linear models found with subspace identification are used to create a simple gain-scheduled controller, and its performance is compared with the standard NREL 5 MW controller.

Keywords Subspace identification • Wind turbine • Control • Gain scheduling • Output feedback

3.1 Introduction

All real engineering systems are nonlinear in some way, whether it be by inherently nonlinear dynamics, the truncation of inputs or outputs, or some other characteristic that prevents the application of superposition. This is extremely inconvenient for the prediction of the system's behavior, as the only way to know with certainty how the system will respond to a specific input is to model the dynamics perfectly (essentially impossible) or to test the system with that exact input. Linear, time-invariant (LTI) systems do not suffer such a complication, as knowing the system's response to a particular input automatically yields insight into the system's response to any scaled version of that input.

Designing controllers for LTI systems is also much easier than for nonlinear systems. The control signal for an LTI system is most commonly assumed to be a function of the system state or output, known as feedback control; well-established techniques such as pole placement [1], proportional-integral-derivative (PID) control [1], or optimal control [2] can then be used to determine the feedback gains. There are a variety of techniques that have been developed for controlling nonlinear systems, the simplest of which is called *gain scheduling*. With gain scheduling, the system is assumed to behave linearly around a certain operating point, and a linear feedback controller is used in a range near the operating point. As the system shifts from one operating point to another, the controller feedback gains are changed accordingly. The benefit of gain scheduling techniques is that they are simple to implement and take advantage of well-established linear control theory; their drawback is that the stability of the global system is not easy to verify, and identifying a system with offsets can yield poor results.

To design a gain-scheduled controller, the nonlinear system must be linearized about an operating point. One method to linearize a nonlinear system is subspace identification, a system identification technique that uses input and output data to estimate a linear, state-space realization of the system. The technique can easily be used on multi-input, multi-output (MIMO) systems, and there are a variety of algorithms that can be used to estimate the state space matrices [3].

This paper uses subspace identification to create a linearized model of a wind turbine, which is then used to design a gain-scheduled controller. The data used for the identification are simulation data from an open-source wind turbine simulator called FAST that is produced by the National Renewable Energy Laboratory (NREL), and the wind turbine model used for

J.M. Rinker (✉)
Mechanical Engineering & Materials Science, Duke University, Box 90300 Hudson Hall, Durham, NC 27708-0300, USA
e-mail: jennifer.rinker@duke.edu

H.P. Gavin
Civil & Environmental Engineering, Duke University, Box 90287 121 Hudson Hall, Durham, NC 27708-0287, USA

M. Mains (ed.), *Topics in Modal Analysis, Volume 10*, Conference Proceedings of the Society for Experimental Mechanics Series, DOI 10.1007/978-3-319-15251-6_3

the program is their 5 MW reference model. Section 3.2 details the system identification, including specifics on the turbine model and the identification process. The process for determining the gains for the output feedback controller is given in Sect. 3.3. Lastly, a comparison of the turbine performance with the standard controller that is shipped with the 5 MW model and the gain-scheduled output feedback controller is presented in Sect. 3.4, and conclusions are drawn in Sect. 3.5.

3.2 Wind Turbine Identification

The controller to be designed is an output feedback controller that is gain-scheduled on the wind speed. The system to be identified is the turbine, where the control inputs are the blade pitch angle and generator torque and the outputs are the power generated and rotor speed. A diagram of the system is shown in Fig. 3.1. The identification is done for a variety of wind speeds ranging from the cut-in to cut-out wind speeds, and the wind excitation is prescribed a steady value. The system is excited by prescribing filtered noise time histories to the blade pitch angle and generator torque. With this formulation, the identified system will not feature wind speed as a disturbance, as it is a gain-scheduling parameter.

3.2.1 Turbine Model

The system to be identified is the 5 MW reference turbine model developed by the National Renewable Energy Laboratory (NREL) [4]. The model is constructed in the open-source turbine simulator also produced by NREL, called FAST (Fatigue, Aerodynamics, Structures, Turbulence). FAST allows a user to prescribe a variety of parameters for a turbine model of interest, and the system can either be linearized to return modal information or run dynamically with wind input, where Kane's method is used to form the governing equations and the nonlinear dynamics are solved using a constant-time-step Adams-Bashforth-Adams-Moulten predictor-corrector integration scheme [5, 6]. The code provides a variety of turbine outputs, including tip deflection, generator speed, tower deflections, generated power, blade pitch angles, and generator torque. The 5 MW model is based off of a variety of design information, with much of the data being used from the REpower 5M turbine. The model is a three-bladed, upwind horizontal-axis wind turbine (HAWT) with a 126-m rotor, 90-m hub height, and cut-in, cut-out, and rated wind speeds of 3 m/s, 25 m/s, and 11.4 m/s, respectively.

The standard 5 MW model uses the "user-defined controller" FAST option, with a controller designed by the researchers at NREL [4, Sec. 7]. The standard controller consists of a variable-speed torque controller for operation in below-rated winds and a blade-pitch-angle controller for above-rated winds. Two linear regions are used to smooth the transition between control regions. The torque controller uses the optimal quadratic law with the filtered generator speed (single-pole, lowpass filter with exponential smoothing) and a constant of proportionality determined by the optimal power coefficient. The blade pitch controller uses a proportional-integral (PI) controller developed with a first-order Taylor expansion of the drivetrain and is gain-scheduled on the pitch angle. The generator torque, generator torque rate, blade pitch angle, and blade pitch rate are all saturated to ensure the bounds of the equipment are not exceeded.

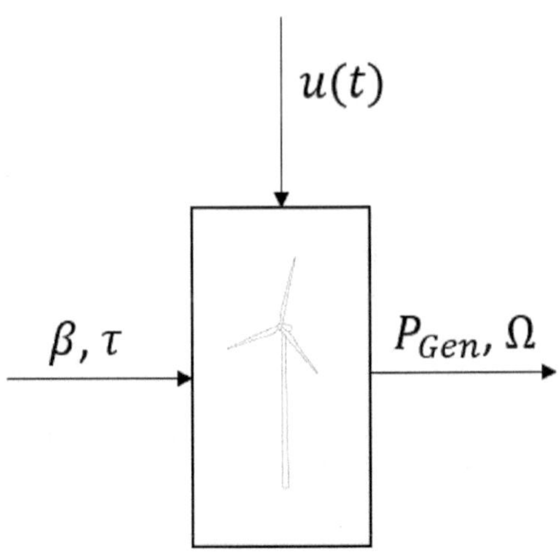

Fig. 3.1 Diagram of wind turbine system for identification. The control inputs are the blade pitch angle, β, and the generator torque τ. The system outputs are the power generated, P_{Gen}, and rotor speed, Ω. The wind input $u(t)$, although shown in this diagram, is steady and not included during identification

3.2.2 Subspace Identification

The subspace identification method used in this paper is the "robust combined algorithm" proposed by Overschee and Moor [7] and summarized by Fig. 4.8 in the same reference. The algorithm uses the singular value decomposition (SVD) of a weighted projection matrix to either determine the model order or isolate a prescribed number of modes. The corresponding state space matrices and covariance matrices are then determined from sets of linear minimization equations. The inputs to the system are the blade pitch angle and generator torque; the outputs are the rotor speed and generated power. Both the inputs and the outputs are detrended to remove any slope and mean offset, as the quality of the results was found to be extremely sensitive to mean offsets in the inputs and outputs.

The identification was done with steady wind (i.e., zero turbulence) at a variety of values throughout the range between the cut-in and cut-out speeds. The total FAST simulation time was 300 s, but only the period from 100 to 150 s was used for identification. To reveal the relationships between the inputs and outputs, the blade pitch angle and generator torque were prescribed a mean value plus a stochastic excitation. The mean value, hereafter referred to as "setpoints," for both of the inputs were prescribed as the steady-state value from the standard 5 MW model as a function of wind speed. The stochastic variation was filtered white noise (single-pole, lowpass with exponential smoothing). The standard deviation of the variation for the generator torque was either 1 % of the setpoint value or 1,500 N-m, whichever was smaller, and the standard deviation of the variation for the blade pitch angle was 1°.

Six states were found to characterize the system with acceptable accuracy at all wind speeds, and the identified systems at all wind speeds were stable. Comparisons of the outputs of the identified system and the FAST simulations for two different wind speeds are shown in Fig. 3.2. The subspace identification returns discrete system matrices, which can be transformed to continuous matrices assuming a zero-order hold. The identified system is then

$$\dot{x} = \mathrm{A}(U)x + \mathrm{B}(U)\delta u, \tag{3.1}$$

$$\delta y = \mathrm{C}(U)x + \mathrm{D}(U)\delta u, \tag{3.2}$$

where U is the wind speed. Here, δu represents the deviation of the control input from the mean; i.e., $\delta u = u - \bar{u}(U)$, where u is the total control input and $\bar{u}(U)$ is the means of the control signal at wind speed U. Similarly, δy is the deviation of the outputs from the mean: $\delta y = y - \bar{y}(U)$.

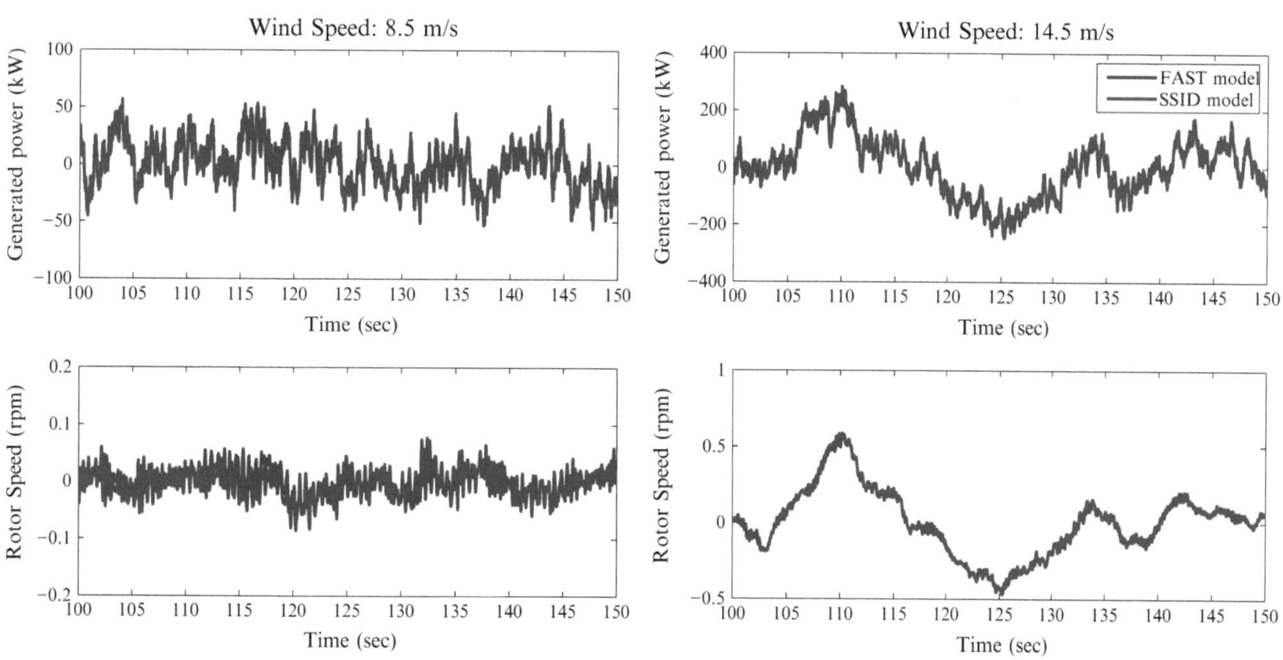

Fig. 3.2 Comparison of detrended output of FAST simulation and identified LTI system for a constant wind speed of 8.5 m/s (*left column*) and 14.5 m/s (*right column*)

3.3 Controller Design

This paper uses an output-feedback controller on the linearized system to control its behavior:

$$\delta u = -\mathrm{K}\delta y. \tag{3.3}$$

This controller is responsible for controlling the dynamics of the system; the setpoints must be determined from the wind speed (or an estimate of the wind speed). Substituting Eq. (3.3) into (3.1) and (3.2) and rearranging yields the closed-loop dynamical system:

$$\dot{x} = [\mathrm{A} - \mathrm{BK}(\mathrm{I} + \mathrm{DK})^{-1}\mathrm{C}]x \tag{3.4}$$

$$= \mathrm{A}_C x, \tag{3.5}$$

$$y = (\mathrm{I} + \mathrm{DK})^{-1}\mathrm{C}x \tag{3.6}$$

$$= \mathrm{C}_C x, \tag{3.7}$$

where I is the identity matrix and K is chosen such that I + DK is invertible.

The objective of the controller is to cause the system to respond as quickly as possible; however, if the controlled dynamics are too fast, the system overshoot will be physically unrealistic. Thus, the problem can be stated as

$$\min_{\mathrm{K}} \left[\max \mathcal{R}\{\lambda_{A_c}\} \right], \tag{3.8}$$

where λ_{A_C} are the eigenvalues of A_C, subject to

$$\mathcal{R}\{\lambda_{A_c}\} < 0, \tag{3.9}$$

$$\det(I + DK) \neq 0, \tag{3.10}$$

$$\max y(t) \leq 1.1 y_0. \tag{3.11}$$

The objective stated in Eq. (3.8) will push the eigenvalues of the controlled system as far into the left-half plane as possible, thus ensuring a fast system response. The constraints in Eqs. (3.9) and (3.10) will ensure that the controlled system is stable and than I + DK is invertible. The constraint shown in Eq. (3.11) will ensure that, when started from some initial state that will produce an output y_0, the system will not overshoot to more than 10 % of the initial output value. For the purposes of this paper, y_0 was assumed to be the standard deviation of the FAST output at each wind speed, and x_0 was calculated using the pseudoinverse of C_C.

The objective and cost functions were coded in Matlab, and the optimization was completed using `fmincon`. An optimal solution was found for each identified system corresponding to a different wind speed, though the ability of the solver to find a solution was found to depend upon the initial conditions. The constraints were verified in all cases; the linearized, controlled system was found to be stable and to have acceptable overshoot.

The final controller then operates as follows. The wind speed is used to determine the setpoints for the blade pitch angle and generator torque and the output feedback gain. The deviation of the outputs away from the setpoints is then scaled by the optimal feedback gain to produce δu, and the input setpoint values are added to create the total control signal. This controller is gain-scheduled on the wind speed, which is not always known in real wind turbine applications. Thus, some estimation of the wind speed should be done to determine the gain-scheduling parameter.

3.4 Results

The gain-scheduled controller was tested using turbulent wind records with different mean wind speeds, with $\sigma_u = 0.5\,\mathrm{m/s}$, and with a Kaimal spectrum with $L \approx 297\,\mathrm{m}$. The results with the designed controller for select wind speeds are shown in Figs. 3.3 and 3.4. Both results are in the above-rated region, as the controller was found to prescribe negative blade-pitch angles in the below-rated region that produced system instabilities.

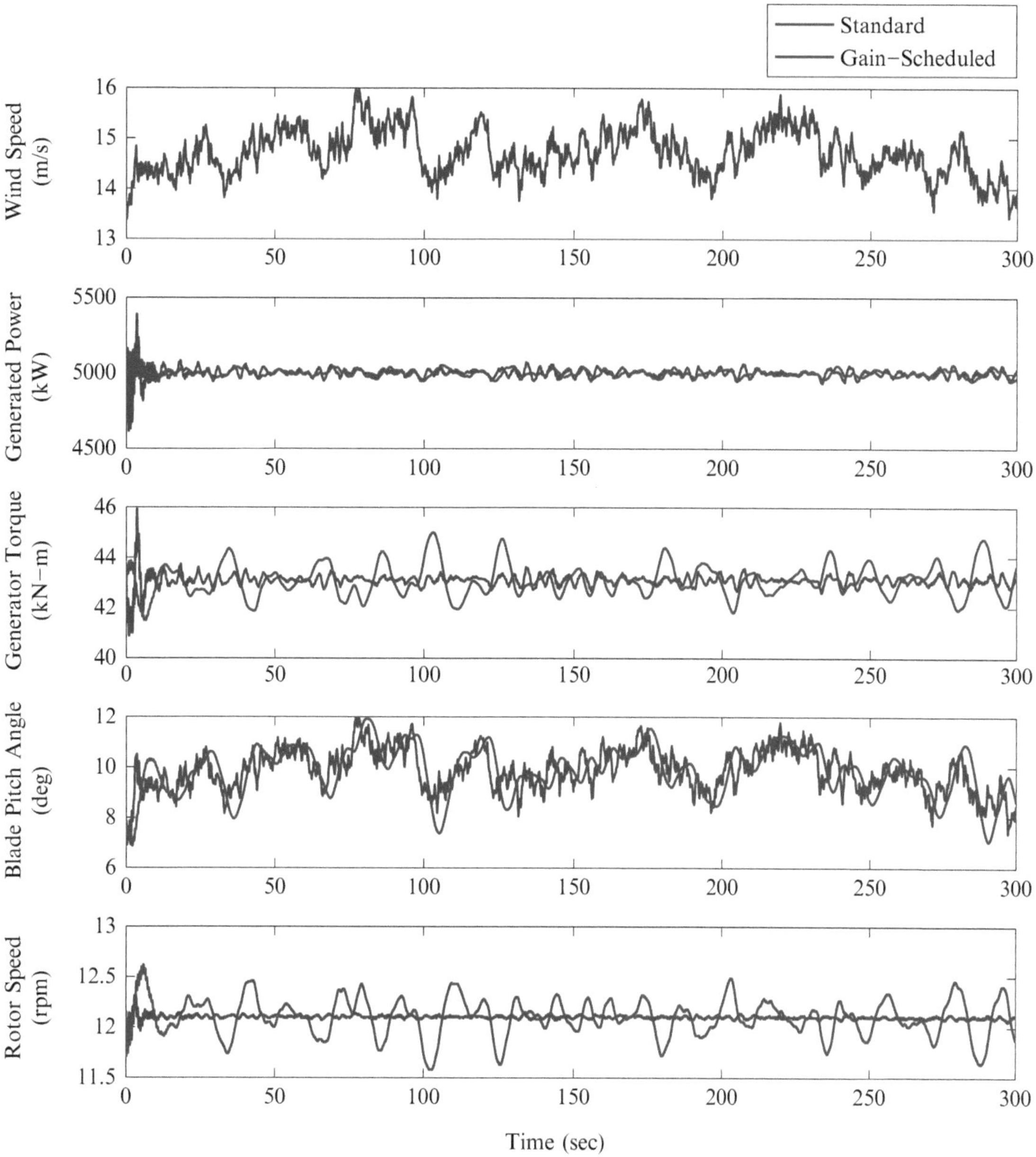

Fig. 3.3 Comparison of FAST simulations with standard controller (*blue*) and gain-scheduled controller (*green*) for $U = 14.5\,\mathrm{m/s}$

As can be seen in the figures, the designed controller is capable of producing a power output that is near the desired rated power, with some oscillations. In general, the power oscillations for the designed controller appear independent of wind speed, but the oscillations in the power output with the standard controller tend to decrease as the wind speed increases. The proposed gain-scheduled controller features oscillations in the rotor speed and generator torque that are not present in the standard controller, which will lead to increased dynamical loading in the system. The general trends of the pitch angle with the gain-scheduled controller match those of the pitch angle with the standard controller, though the standard controller is able to react to changes in the wind speed more quickly, which leads to the steadier rotor speed. However, this will produce much more fatigue on the pitch actuator.

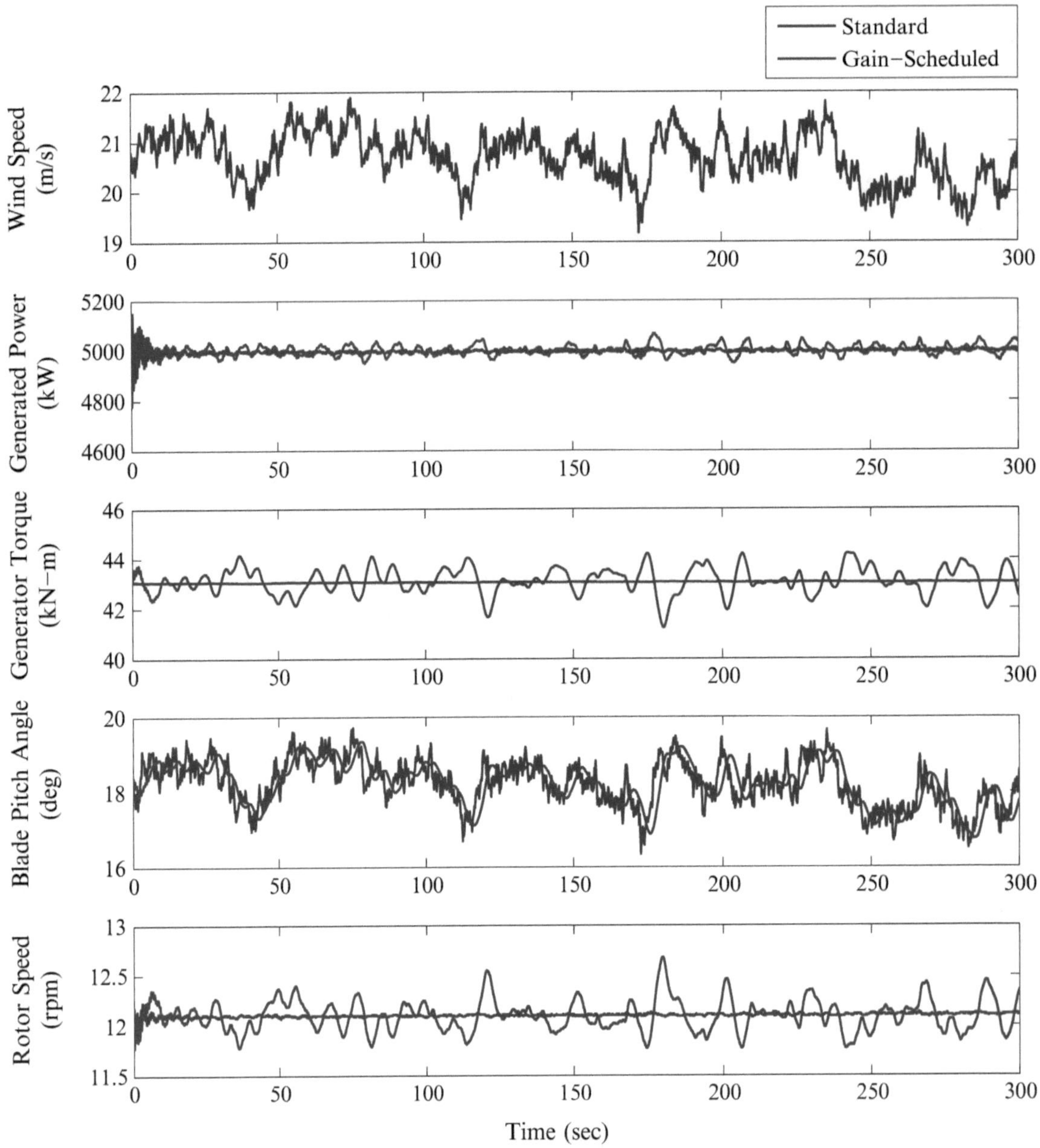

Fig. 3.4 Comparison of FAST simulations with standard controller (*blue*) and gain-scheduled controller (*green*) for $U = 20.5\,\text{m/s}$

3.5 Conclusions

This paper uses a subspace identification algorithm to create linearized 6-state state space models that approximate the behavior of the FAST 5 MW reference model at different wind speeds. The algorithm is found to perform well on detrended data and the linearized system is able to mimic the behavior of the nonlinear system very well. These identified state-space models were used to design an output feedback controller that is gain-scheduled on the wind speed. This feedback controller mapped the deviations of the outputs from prescribed setpoints to the controller deviations from their prescribed setpoints. The feedback gains are determined by a nonlinear optimization that optimizes the speed of the system while ensuring the system overshoot and stability remain within specified bounds. The performance of the gain-scheduled controller is compared with that of the standard controller that is shipped with the 5 MW model for turbulent wind. The developed controller is found to produce the correct power output fairly well, with slightly increased oscillations around the rated power. However, the generator torque and rotor speed have oscillations that are not present with the standard 5 MW controller and could lead to increased wear and tear on the generator. On the other hand, the pitch angle control signal for the standard controller is quite jagged and will produce increased wear and tear on the pitch actuator.

References

1. Franklin GF, Powell JD, Emami-Naeini A, Powell JD (2014) Feedback control of dynamic systems, 7th edn. Addison-Wesley, Reading
2. Lewis F, Syrmos V (1995) Optimal control, 2nd edn. Wiley, New York
3. Qin SJ (2006) An overview of subspace identification. Comput Chem Eng 30(10):1502–1513
4. Jonkman JM, Butterfield S, Musial W, Scott G (2009) Definition of a 5-MW reference wind turbine for offshore system development. National Renewable Energy Laboratory, Golden, CO. Technical Report, NREL/TP-500-38060
5. Jonkman JM (2003) Modeling of the uae wind turbine for refinement of fast_ad. Master's thesis, Colorado State University
6. Jonkman JM, Buhl ML Jr (2005) Fast user's guide. National Renewable Energy Laboratory, Golden, CO. Technical Report, NREL/EL-500-38230
7. Overschee PV, Moor BD Subspace identification for linear systems: theory, implementation, applications. Kluwer Academic Publishers, Dordrecht (1996)

Chapter 4
Dynamic Analysis of Complex Mechanical Structures Using a Combination of Numerical and Experimental Techniques

Dimitrios Giagopoulos and Sotirios Natsiavas

Abstract A systematic methodology is applied, leading to an accurate prediction of the dynamic response of a large and geometrically complex mechanical structures (e.g., a vehicle superstructure supported on a given chassis). The basic idea is to first measure the acceleration time histories at the connection points of the vehicle superstructure with its suspension system and use them subsequently as a base excitation in a finite element model of the superstructure. The reliability of the methodology applied was tested in a small scale nonlinear laboratory vehicle model. In this model, first the study is purely numerical and the emphasis is placed in demonstrating and verifying the accuracy and validity of the methodology applied. Then, the method is applied and examined, using real measurements. Next, the method applied in the superstructure of a real large military vehicle. The vehicle superstructure is first discretized by finite elements. The model is then updated through an experimental modal analysis procedure in a support-free state. Then, a series of experimental trials is performed in real operating conditions, aimed at recording the acceleration time histories at the connection points of the superstructure with the chassis. These time histories are used as a ground excitation for the FE model of the superstructure and the stresses developed are evaluated. In this way, the critical points of the superstructure can be identified by numerical means. The reliability of the methodology applied was tested by placing strain gauges at the critical points of the superstructure and performing a new set of measurements for the vehicle under similar loading conditions. Direct comparison of the numerical and experimental data obtained in this manner verified that the hybrid methodology applied is quite reliable.

Keywords System identification • Nonlinear dynamics • Substructuring • Model updating • Structural dynamics

4.1 Introduction

The main objective of the present work is to demonstrate the advantages of applying an appropriate hybrid modelling method in order to study the dynamics of a class of mechanical structures in an accurate and effective way. The method applied in the present work can be used to determine the dynamic behavior of vehicle superstructures. This is a necessary step in order to capture, explain and improve the dynamic behavior of many contemporary complex mechanical systems. At the same time, this leads to an accurate identification of points where the critical (largest) stresses appear in the structural part of interest. This is done by applying a numerical method for determining the equations of motion for the superstructure of the vehicle, while the dynamic characteristics of the remaining components are taken into account through the application of appropriate experimental measurements.

More specifically, the above procedure can be easily extended to systems with more than two structural components. In particular, the procedure followed in this work is applicable to all cases of vehicles carrying superstructures, where the supporting system (suspension and wheels) is not modeled, since its dynamic characteristics are not known. Essentially, it involves the following basic steps. First, the vehicle superstructure is divided in a number of substructures, which are discretized geometrically by application of the finite element method. The resulting model for each substructure is then updated through an experimental investigation of its dynamic response in a support-free state. In this way, all the elements of the matrix with the frequency response functions (FRFs) required for determining the response of the substructure are determined by imposing impulsive loading. Based on the measured FRFs, the natural frequencies and damping ratios of each substructure are then estimated. The natural frequencies provide a basis for checking the accuracy of similar results obtained

D. Giagopoulos (✉)
Department of Mechanical Engineering, University of Western Macedonia, Kozani, Greece
e-mail: dgiagopoulos@uowm.gr

S. Natsiavas
Department of Mechanical Engineering, Aristotle University, Thessaloniki, Greece

M. Mains (ed.), *Topics in Modal Analysis, Volume 10*, Conference Proceedings of the Society for Experimental Mechanics Series, DOI 10.1007/978-3-319-15251-6_4

by the corresponding finite element models, while the damping ratios are used for updating the corresponding damping matrices. Next, a series of experimental trials are performed in real operating conditions, aimed at recording the acceleration histories at the connection points of the superstructure with the chassis. These histories are subsequently used as a ground excitation for the finite element model resulting by performing a synthesis of all the components of the superstructure, as explained in the first part of this section. In this way, the stresses developed under the specific loading conditions are evaluated. From these numerical results, the critical points of the superstructure are finally selected, based on the level of the larger magnitude stresses observed.

The organization of this paper is as follows. In the following section, a brief but complete outline of the methodology applied is presented. Then, the effectiveness and accuracy of this methodology is first demonstrated in the third section, by presenting numerical and experimental results obtained for an experimental small scale vehicle model, involving suspensions with nonlinear characteristics. In the fourth section, results for a real military vehicle are presented. Here, the reliability of the methodology was tested by placing strain gauges at the critical points of the superstructure and performing a new set of measurements under similar dynamic load conditions, in order to experimentally verify the stress levels developed. The study concludes by presenting a summary of the results obtained.

4.2 Class of Mechanical Systems Examined: Equations of Motion

The equations of motion of mechanical systems with complex geometry are commonly set up by applying finite element techniques. Quite frequently, a systematic investigation of the dynamics of large scale mechanical structures leads to models involving an excessive number of degrees of freedom. Therefore, a computationally efficient solution requires application of methodologies reducing the numerical dimension of the original model [1–11]. Next, the basic philosophy of a time domain reduction method, which was part of the hybrid methodology applied, is presented briefly.

For simplicity, consider a mechanical system consisting of two subsystems, say A and B. Moreover, let the equations of motion for subsystem A be derived in the following classical form

$$\widehat{M}_A \ddot{\underline{x}}_A + \widehat{C}_A \dot{\underline{x}}_A + \widehat{K}_A \underline{x}_A = \widehat{\underline{f}}_A(t), \tag{4.1}$$

where $\widehat{M}_A$, $\hat{C}_A$ and $\widehat{K}_A$ are the mass, damping and stiffness matrix of the subsystem, respectively, while the vector $\widehat{\underline{f}}_A(t)$ represents the external forcing. For a typical model, the number of these equations may be quite large. However, for a given level of forcing frequencies it is possible to reduce significantly the number of the original degrees of freedom, without sacrificing the accuracy in the numerical results, by applying standard component mode synthesis methods [1, 5, 12]. This can be achieved through an approximate coordinate transformation with form

$$\underline{x}_A = \Psi_A \underline{q}_A. \tag{4.2}$$

The transformation matrix Ψ_A includes an appropriately chosen set of the lowest frequency normal modes of component A, corresponding to support-free conditions [1]. The number of these modes depends on the accuracy required in the response frequency range examined. Consequently, matrix Ψ_A is completed by a set of static correction modes of component A [4, 5, 13]. Employing this transformation, the original set of Eq. 4.1 can be replaced by a considerably smaller set of equations, expressed in terms of a new generalized coordinates $\underline{q}_A$. More specifically, application of the Ritz transformation Eq. 4.2 into the original set of Eq. 4.1 yields the much smaller dimension set

$$M_A \ddot{\underline{q}}_A + C_A \dot{\underline{q}}_A + K_A \underline{q}_A = \underline{f}_A(t), \tag{4.3}$$

where

$$M_A = \Psi_A^T \widehat{M}_A \Psi_A, \; C_A = \Psi_A^T \widehat{C}_A \Psi_A, \; K_A = \Psi_A^T \widehat{K}_A \Psi_A \text{ and } \underline{f}_A = \Psi_A^T \widehat{\underline{f}}_A.$$

Moreover, the set of unknowns can be split in the form

$$\underset{\sim}{q}_A = \left(\underset{\sim}{p}_A^T\ \underset{\sim}{x}_b^T\right)^T,$$

where $\underset{\sim}{p}_A$ includes coordinates related to the response of internal degrees of freedom of component A, while $\underset{\sim}{x}_b$ includes the boundary points of component A with component B. Next, similar sets of equations of motion are obtained for component B. Namely, the equations of motion are first set up in the form

$$M_B\ddot{\underset{\sim}{q}}_B + C_B\dot{\underset{\sim}{q}}_B + K_B\underset{\sim}{q}_B = \underset{\sim}{f}_B(t), \tag{4.4}$$

with coordinates

$$\underset{\sim}{q}_B = \left(\underset{\sim}{p}_B^T\ \underset{\sim}{x}_b^T\right)^T.$$

Then, a proper combination of Eqs. 4.3 and 4.4 leads to the equations of motion of the composite system in the classical form

$$M\ddot{\underset{\sim}{q}} + C\dot{\underset{\sim}{q}} + K\underset{\sim}{q} = \underset{\sim}{f}(t), \tag{4.5}$$

with coordinates

$$\underset{\sim}{q} = \left(\underset{\sim}{p}_A^T\ \underset{\sim}{p}_B^T\ \underset{\sim}{x}_b^T\right)^T.$$

As usual, the stiffness matrix of the composite system can be obtained by considering the total potential energy of the system. Likewise, the mass matrix of the composite system is obtained by considering the corresponding kinetic energy, while the forcing vector is determined by considering the virtual work [1, 5].

4.3 Application to Laboratory Vehicle Model

In this section, the emphasis is placed on applying the methodology proposed to a laboratory vehicle model. First in Sect. 4.3.1 the study is purely numerical and the emphasis is placed in demonstrating and verifying the accuracy and validity of the methodology applied. Then, in Sect. 4.3.2, the method is applied and examined in the experimental device of the vehicle, using real measurements.

4.3.1 Numerical Application to a Small Scale Vehicle-Like Frame Structure

The accuracy and effectiveness of the methodology demonstrated by applying it to a complex mechanical system, shown in Fig. 4.1. Specifically, the example system is a laboratory structure, which was designed in order to simulate the frame structure of a vehicle in a small scale. Details of material and geometrical dimensions of the frame can be found in previous studies [14, 15]. In brief, the selected frame structure comprises a frame structure with predominantly linear response and high modal density plus four supporting systems with strongly nonlinear action. These supporting systems consist of a lower set of linear discrete spring-damper units, connected to a concentrated mass, simulating the wheel subsystems, as well as of an upper set of a nonlinear discrete spring-damper (bushings) units connected to the frame and simulating the action of the vehicle suspension. The measurement points, indicated by 1–4, correspond to connection points between the frame and its supporting structures, while the measurement points 5–7 shown were chosen on the frame.

Based on the geometric details and the material properties of the structure, a detailed finite element model of the vehicle frame was first created. This model was built using mainly quadrilateral shell and hexahedral solid elements. The final model of the frame consists of 45,564 DOF. The wheel subsystems were simulated with linear spring, damper and mass elements,

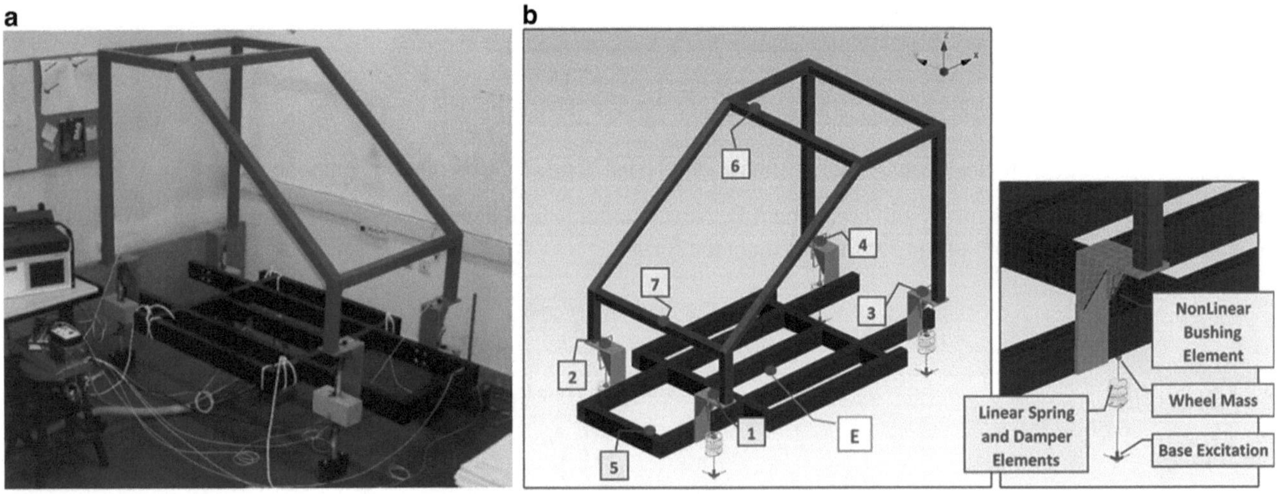

Fig. 4.1 (**a**) Experimental set up of the structure tested, (**b**) frame substructure nonlinear simulated supports and measurement locations

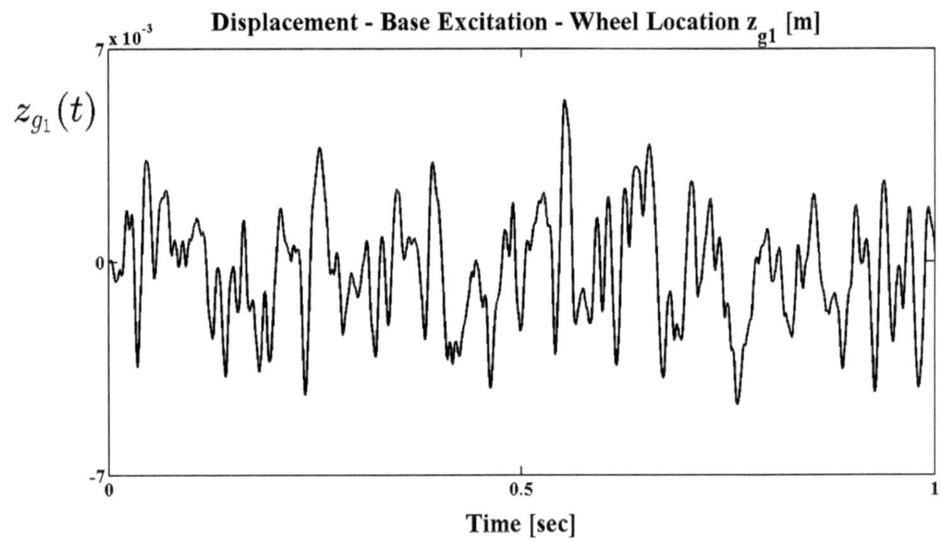

Fig. 4.2 Displacement history applied as base excitation in the vertical direction at first wheel

while the suspension subsystems were simulated with nonlinear bushing elements. More specifically, the nonlinear restoring and damping forces in the suspensions were selected to have the following specific forms

$$f_k = k_s x + \mu_k x^3 \text{ and } f_c = c_s \dot{x} + \mu_c \dot{x}^3, \tag{4.6}$$

respectively, with $x = x_1 - x_w$.

The basic idea of the applied methodology is to re-run the model by using the accelerations obtained at the boundary locations between the frame and the four supporting systems, as a base excitation to the frame, in order to calculate its dynamic response. More specifically, a nonlinear transient response analysis of the full vehicle model (frame and supports) was performed first, by applying four different transient displacement base excitation histories to the wheel subsystems in the vertical direction (Z). The model was then solved by using a numerical method belonging to the well-known class of Newmark's methods (instead of performing experiments) and the acceleration histories at the selected locations 1–7 were calculated. The form of the applied excitation in the first wheel is presented in Fig. 4.2, while the "measured" histories of the acceleration at boundary location 1, in the three directions (X-longitudinal, Y-transverse, Z-vertical), are presented in Fig. 4.3.

The "measured" acceleration histories at the boundary locations 1–4, in all three directions, were then used as base excitation to the frame of the vehicle. Next, the model for the frame was solved by performing a linear transient response analysis and the accelerations at the selected locations 1–7 were obtained. Results for the acceleration predictions obtained by this analysis are presented in Fig. 4.4 for the reference location 6 at the top of the vehicle frame, along the three directions.

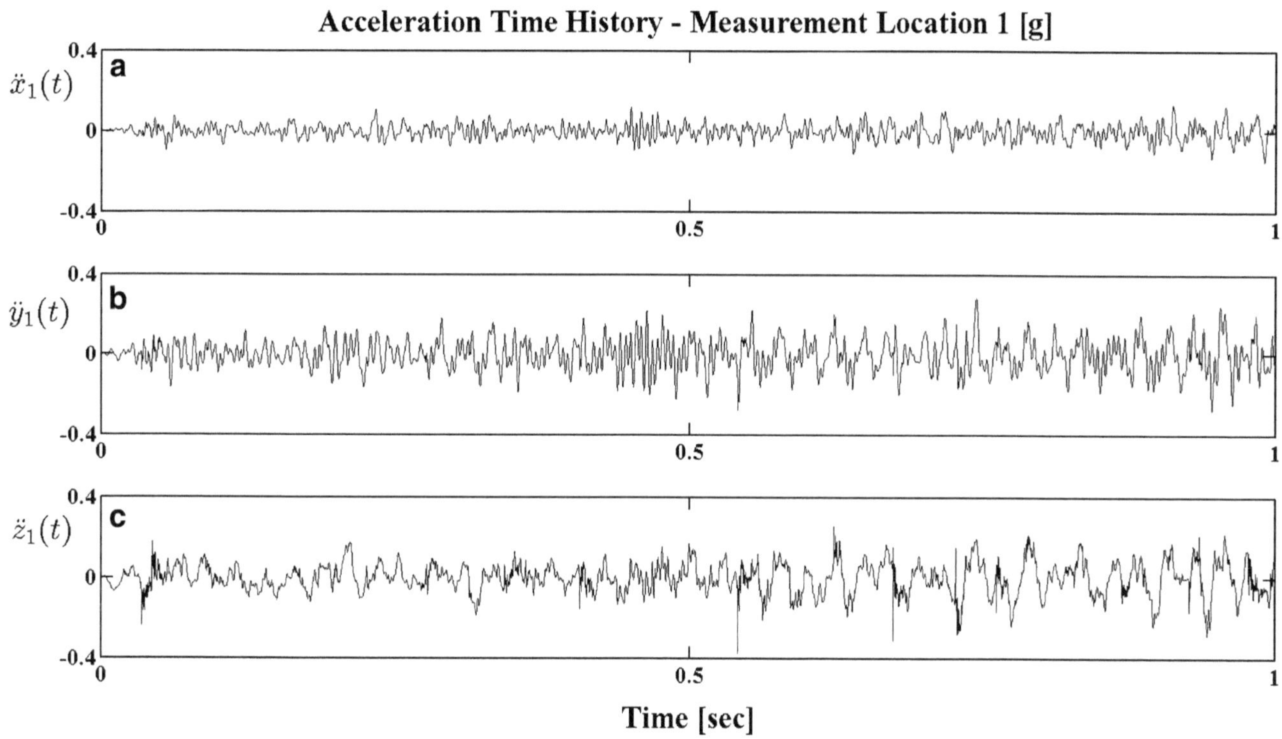

Fig. 4.3 Acceleration histories at boundary measurement location 1 in the: (**a**) X-longitudinal, (**b**) Y-transverse and (**c**) Z-vertical direction

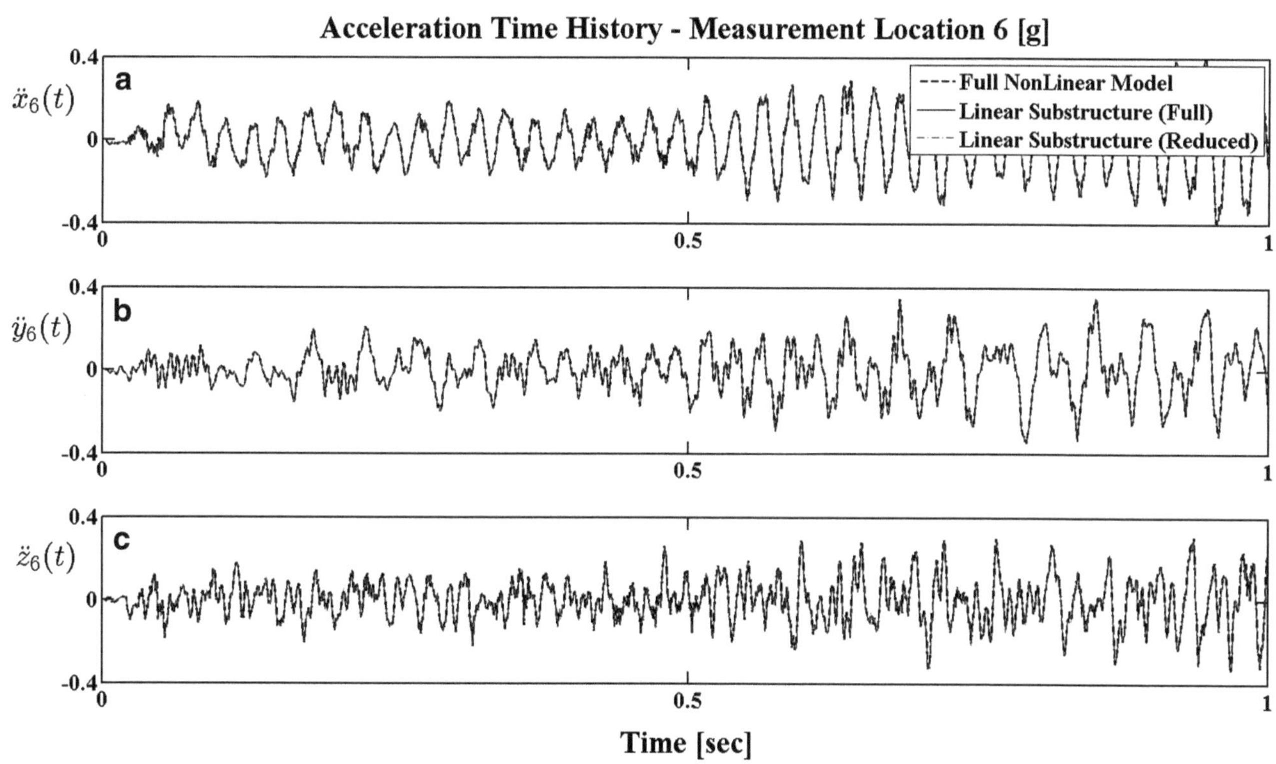

Fig. 4.4 Comparison of acceleration histories obtained at reference location 6 by the full nonlinear model (*broken lines*), the full linear model (*continuous lines*) and the reduced linear model (*dotted lines*) of the vehicle frame in the: (**a**) X-longitudinal, (**b**) Y-transverse and (**c**) Z-vertical direction

The predictions are compared with the simulated measurements of the full nonlinear model. In these figures, the results of the full nonlinear model are represented by the broken lines, while the results obtained for the linear substructure are represented by the black continuous lines. Again, direct comparison shows that the acceleration predictions are virtually the same as those obtained by the complete model.

Next, in order to check the accuracy of the coordinate reduction method presented in section 2, the same calculations were repeated with a model resulting by keeping only some of the lowest linear modes of the frame structure. It was found that by keeping modes with frequency up to about 500 Hz, leading to a new (reduced) model with only 50° of freedom, the results obtained were sufficiently accuracy. The corresponding results, shown in Fig. 4.4 by the dotted lines, are indistinguishable from those obtained by the full model.

4.3.2 Experimental Application to a Small Scale Vehicle-Like Frame Structure

Next the method applied and validated to the experimental device of the small scale vehicle which presented in Fig. 4.1a. More specifically, in the experimental process used the same setup with the numerical application (boundary and internal measurement locations), with one difference. We have in our disposal only one electromagnetic shaker and for this reason it is impossible to apply four different transient displacement base excitation histories to the wheel subsystems. Thus, the electromagnetic shaker placed in the location (E) which presented in Fig. 4.1b.

In this application, the acceleration time histories at the selected locations 1–7 were measured, by applying a force with the shaker in the vehicle frame. The measured acceleration histories at the boundary locations 1–4, in all three directions, were then used as base excitation to the FE model of the frame. This FE model was updated and validated, by applying a modal analysis procedure and Bayesian techniques in previous works [14, 16, 17]. Next, the model of the frame was solved by performing a linear transient response analysis and the accelerations at the selected locations 1–7 were obtained.

Results for the acceleration predictions obtained by this analysis are presented in Fig. 4.5 for the reference locations 5 and 6, in the vertical direction. More specifically in these figures presented results that corresponds to resonance region close to 3.4 Hz, which is dominated by local wheel body deflections. To achieve this, apply with the shaker a harmonic force with forcing frequency at 3.4 Hz. This case was selected because at this resonance region it is observed that is obtained a large uncertainty in the response spectra. The predictions are compared with the measurements of the full nonlinear model. In these figures, the experimental measurements of the full nonlinear model are represented by broken black line, while the results obtained for the linear substructure are represented by the red continuous lines. Direct comparison shows that the acceleration predictions are very close to the experimental measurements of the full nonlinear model.

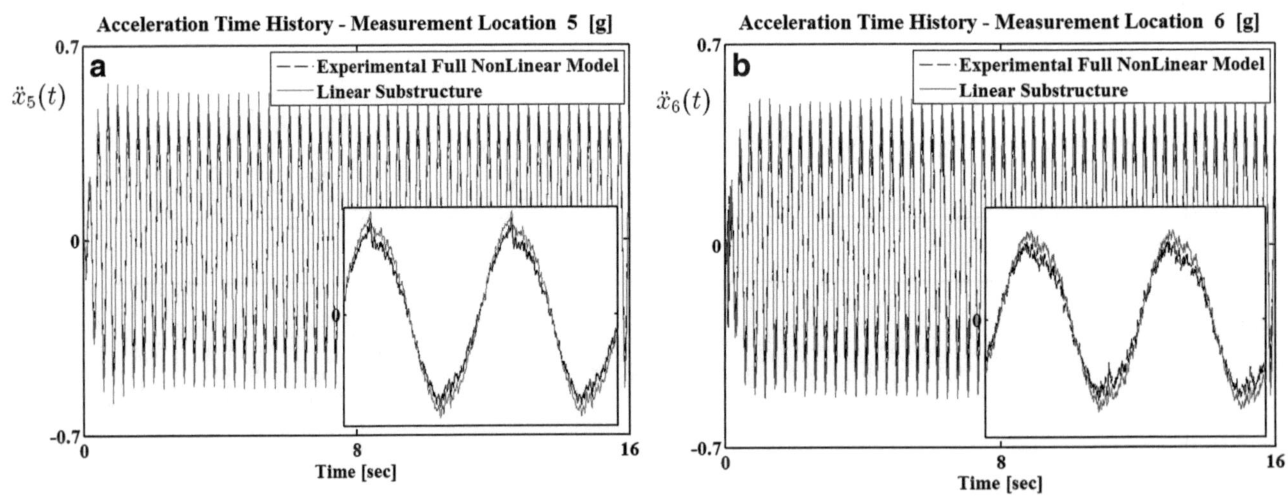

Fig. 4.5 Comparison of the experimental measurements of the full nonlinear model (*broken black line*) with the results obtained for the linear substructure (*red continuous lines*) of the vehicle frame at resonance frequency 3.4 Hz, in the: (**a**) location 5 and (**b**) location 6

4.4 Experimental Application to a Real Vehicle

The vehicle superstructure examined is shown in Fig. 4.6. In this application, the superstructure consisted of two main substructures. The first part includes the main body of the platform, while the second part includes the spacer. Both of these parts are presented in Fig. 4.7a, b, respectively. The geometry of these structures was discretized mainly by rectangular and triangular shell finite elements. In addition, some other elements like solid (hexahedral), bar and rigid body elements were also used. The total number of degrees of freedom of the resulting model was about 520,000. Due to the large size of this model, the development and solution of the finite element model was completed by using appropriate commercial software [18, 19].

After development of the overall finite element model, including the coordinate reduction and synthesis part, an experimental modal analysis survey of the vehicle superstructure was performed to quantify its dynamic characteristics and verify the dynamic predictions of the model. The experimental modal analysis procedure was applied separately for the two parts. All the necessary elements of the FRF matrix required for determining the response of the two substructures were determined by imposing impulsive loading [15–17, 20–23]. A schematic illustration of the measurement geometry of the test structures is presented in Fig. 4.7a, b. Based on the measured FRFs, the natural frequencies and the damping ratios of the frame substructure were estimated by applying the "Rational Fraction Polynomial Method" (RFPM).

Fig. 4.6 Vehicle superstructure and corresponding finite element model of its two substructures

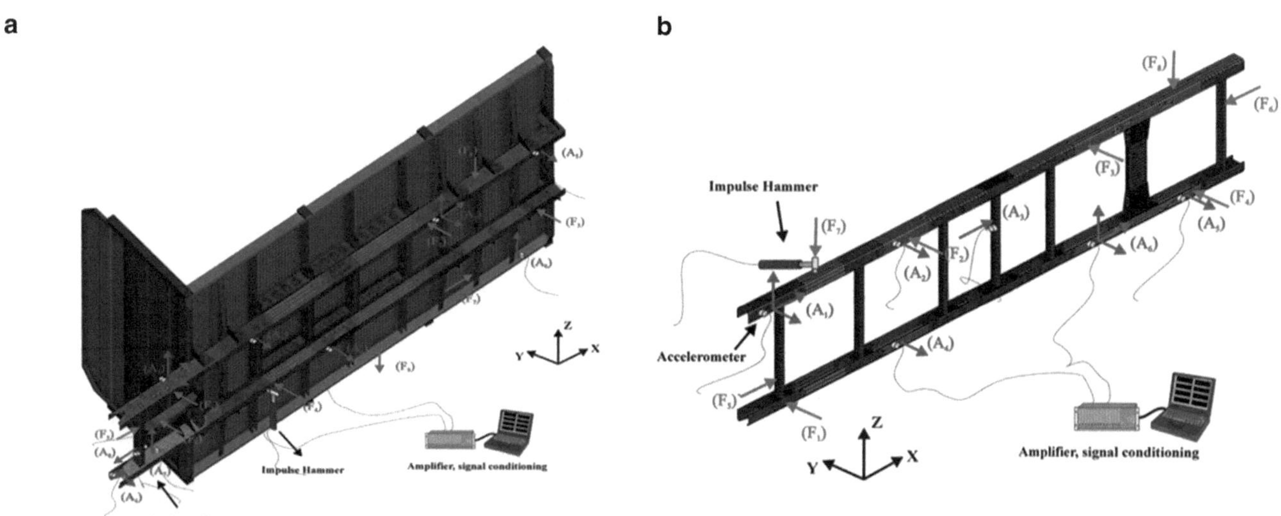

Fig. 4.7 Schematic illustration of the measurement geometry: (**a**) main body and (**b**) spacer

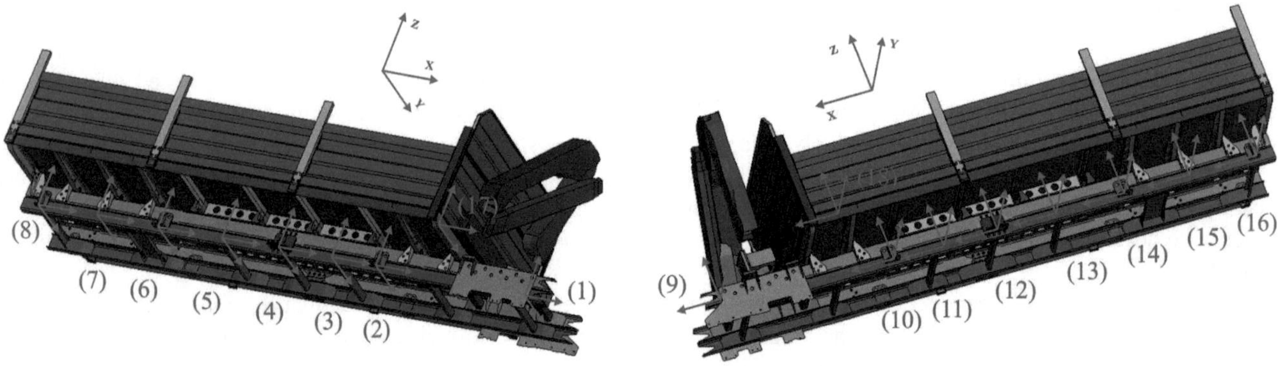

Fig. 4.8 Measurement locations of acceleration histories

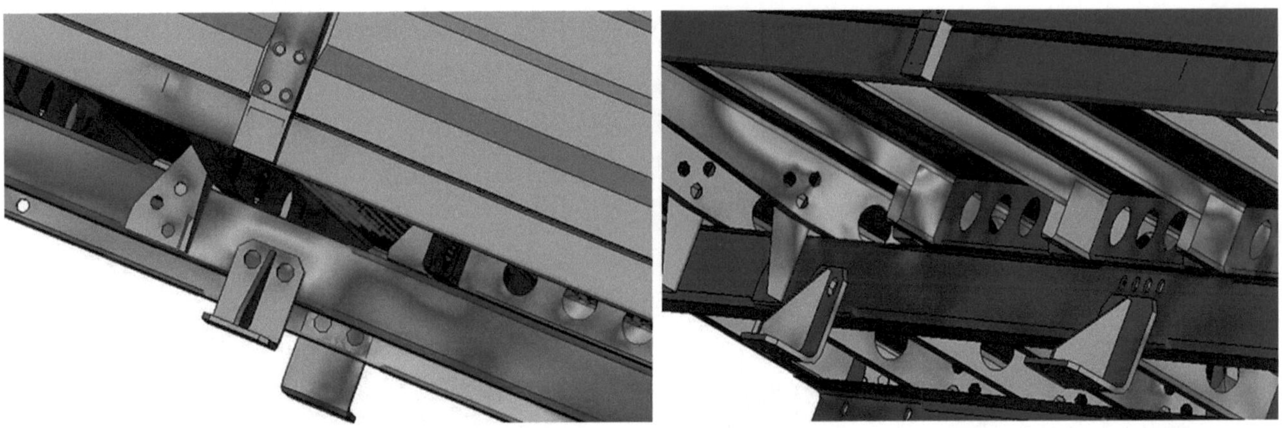

Fig. 4.9 Points of the superstructure where maximum stresses appear

After testing the accuracy of the superstructure finite element model, the ultimate goal was the identification of those areas where the larger stresses appear for the given loading. To achieve this, triaxial accelerometers were placed initially at 18 selected positions. These positions with the measurement directions are presented in Fig. 4.8 and include the connection points of the superstructure with the chassis of the vehicle as well as some other characteristic positions of the superstructure.

The measured acceleration histories corresponding to the biggest response amplitudes (worst case) were imported as a ground excitation in the finite element model of the superstructure. A reduction in the dimensions of the original system was also performed, so that the results are accurate in the frequency range 0–1,024 Hz. The total number of degrees of freedom of the reduced model was about 2,400, which is much smaller than the number of degrees of freedom of the original model. Then, the resulting model was solved numerically in order to calculate the maximum stresses developed. Figure 4.9, shows selected results, in which some indicative points of the superstructure with maximum stresses are presented.

Finally, in order to test the reliability of the method applied, strain gauges were placed at 16 selected critical points of the superstructure and a new set of measurements was carried out under similar dynamic loading conditions, to experimentally verify the stress levels developed. Some of the experimental and numerical results are summarized in Table 4.1. In this table are presented the maximum values of the von Mises stress obtained for five selected tests (indicated by MT1–MT5) and for all the points where measurements were taken (denoted by S1–S16). More specifically, in the penultimate column are presented the maximum values of all tests, for each measurement location, while in the last columns are presented the corresponding maximum values obtained by the finite element model and their percentage difference. Finally, in an effort to further illustrate the accuracy of the results, in Fig. 4.10 is presented a comparison of experimentally measured (continuous lines) and numerically determined (broken lines) acceleration time histories at two indicative locations of the vehicle superstructure.

Table 4.1 Types of base changes

Structural Tests / Strain Gage Positions	Maximum Value (peak) of Equivalent Stress Von Mises [MPa] MT1	MT2	MT3	MT4	MT5	Maximum Value of all Structural Tests	Maximum Value of FEM Solution
S1	181.45	152.51	178.39	112.29	175.12	**181.45**	174.39
S2	125.50	85.13	198.26	97.74	179.20	**198.26**	195.77
S3	161.61	135.29	201.56	167.61	198.82	**201.56**	193.85
S4	22.50	42.43	26.91	34.84	24.86	**42.43**	55.21
S5	141.85	89.98	141.75	90.63	127.96	**141.85**	135.15
S6	23.68	37.27	85.49	49.16	54.65	**85.49**	91.40
S7	58.05	72.20	76.83	67.62	80.88	**80.88**	83.58
S8	22.12	54.43	27.36	25.06	26.01	**54.43**	48.77
S9	196.43	147.56	181.78	210.15	135.80	**210.15**	198.40
S10	91.81	67.94	105.48	118.17	81.37	**118.17**	121.39
S11	147.62	143.51	144.60	159.99	109.01	**159.99**	152.59
S12	12.62	25.92	21.18	26.66	16.45	**26.66**	41.50
S13	32.88	32.67	38.24	54.39	29.18	**54.39**	48.07
S14	80.61	110.97	122.42	93.14	95.22	**122.42**	116.32
S15	43.39	74.83	84.36	68.35	43.94	**84.36**	79.32
S16	12.97	34.57	26.38	18.85	16.39	**34.57**	35.12

Yellow cells highlight similar frequencies after addition of ancillary subcomponent (AC) to the system

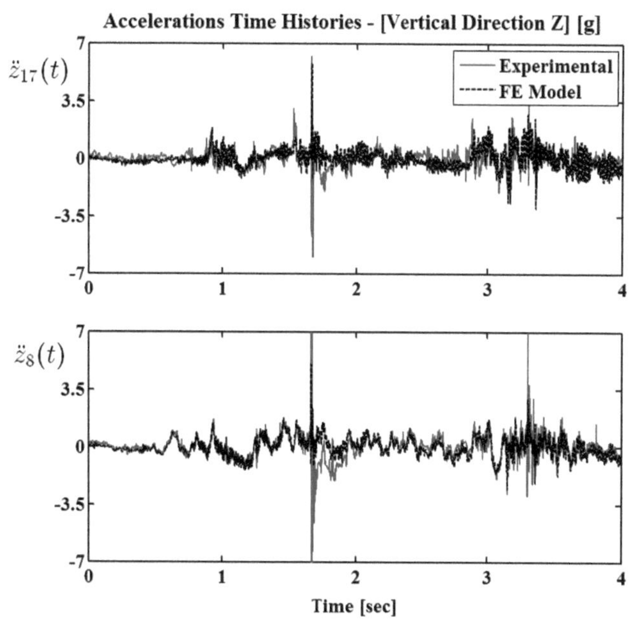

Fig. 4.10 Comparison of experimentally measured (*continuous lines*) and numerically determined (*broken lines*) accelerations at two locations

4.5 Conclusions

In this work, presented a systematic methodology for determining the dynamic response and identifying the critical points of the superstructure of a large real vehicle, without knowing the dynamic characteristics of the vehicle supporting structure. The basic idea was to start the solution process by first discretizing geometrically the superstructure with the finite element method. The damping parameters of the resulting model were then updated through an experimental investigation of its dynamic response in a support-free state. Then, a series of experiments was performed in real operating conditions, aimed at recording the acceleration histories at the connection points of the superstructure with the vehicle chassis. Next, a component mode synthesis method was applied in order to eliminate a substantial number of degrees of freedom of the original model. The measured acceleration histories were subsequently used as a ground excitation for the reduced finite element model and the stresses developed under specific loading conditions were evaluated. From these numerical results, the critical points of the superstructure were selected, based on the level of the largest stresses. Finally, a new set of measurements was carried out in order to experimentally verify the stress levels developed. Before application to the real vehicle, the validity and reliability of the method was also verified numerically in two simpler example structures. In all cases, comparison of numerical and experimental results indicated that the methodology applied gives accurate results and provides a useful tool in predicting the critical stress levels developed in the vehicle superstructure under given loading conditions.

Acknowledgements This research was supported by a grant from the Hellenic Vehicle Industry (ELVO S.A.).

References

1. Adams ML (1980) Nonlinear dynamics of flexible multi-bearing rotors. J Sound Vib 71:129–144
2. Craig RR Jr (1981) Structural dynamics – an introduction to computer methods. Wiley, New York
3. Craig RR Jr (1987) A review of time-domain and frequency domain component mode synthesis methods. Int J Anal Exp Modal Anal 2:59–72
4. Vermot des Roches G, Bianchi JP, Balmes E, Lemaire R, Pasquet T (2010) Using component mode in a system design process. In: Proceedings of the IMAC-XXVIII 2010, Jacksonville
5. Papalukopoulos C, Natsiavas S (2007) Dynamics of large scale mechanical models using multi-level substructuring. ASME J Comput Nonlinear Dyn 2:40–51
6. Verros G, Natsiavas S (2002) Ride dynamics of nonlinear vehicle models using component mode synthesis. ASME J Vib Acoust 124:427–434
7. Craig RR Jr (1977) Methods of component mode synthesis. Shock Vib Dig J 9:3–10
8. Klosterman A (1972) A combined experimental and analytical procedure for improving automotive system dynamics. SAE Technical Paper 720093
9. Craig RR Jr, Chang CJ (1977) Substructure coupling for dynamic analysis and testing. Technical report CR2781, NASA
10. MacNeal RH (1971) A hybrid method of component mode synthesis. J Comput Struct 1:581–601
11. Bennighof JK, Kaplan MF (1998) Frequency window implementation of adaptive multi-level substructuring. ASME J Vib Acoust 120: 409–418
12. Cuppens K, Sas P, Hermans L (2000) Evaluation of FRF based substructuring and modal synthesis technique applied to vehicle FE data, ISMA 2000. K.U. Leuven, Belgium, pp 1143–1150
13. Farhat C, Geradin M (1994) On a component mode synthesis method and its application to incompatible structures. Comput Struct 51:459–473
14. Giagopoulos D, Natsiavas S (2007) Hybrid (numerical-experimental) modeling of complex structures with linear and nonlinear components. Nonlinear Dyn 47:193–217
15. Ewins DJ (1984) Modal testing: theory and practice. Research Studies Press, Somerset
16. Papadimitriou C, Ntotsios E, Giagopoulos D, Natsiavas S (2012) Variability of updated finite element models and their predictions consistent with vibration measurements. Struct Control Health Monit 19:630–654
17. Giagopoulos D, Papadioti D-Ch, Papadimitriou C, Natsiavas S (2013) Bayesian uncertainty quantification and propagation in nonlinear structural dynamics. In: Proceedings of the IMAC-XXXI 2013, Garden Grove
18. DYNAMIS 3.1.1 (2013) Solver reference guide. DTECH, Thessaloniki
19. MSC.NASTRAN (2008) Quick reference guide. MSC.SOFTWARE
20. Mottershead JE, Friswell MI (1997) Model updating in structural dynamics: a survey. J Sound Vib 167:347–375
21. Mohanty P, Rixen DJ (2005) Identifying mode shapes and modal frequencies by operational modal analysis in the presence of harmonic excitation. Exp Mech 45:213–220
22. Spottswood SM, Allemang RJ (2007) On the investigation of some parameter identification and experimental modal filtering issues for nonlinear reduced order models. Exp Mech 47:511–521
23. Richardson MH, Formenti DL (1985) Global curve fitting of frequency response measurements using the rational fraction polynomial method. In: Third IMAC conference, Orlando
24. Huizinga ATMJM, van Campen DH, Kraker A (1997) Application of hybrid frequency domain substructuring for modelling an automotive engine suspension. ASME J Vib Acoust 119:304–310
25. Jetmundsen B, Bielawa RL, Flannelly WG (1988) Generalized frequency domain substructure synthesis. J Am Helicopter Soc 85:55–64
26. Ren Y, Beards CF (1995) On substructure synthesis with FRF data. J Sound Vib 185:845–866
27. Giagopoulos D, Natsiavas S (2013) Dynamic analysis and identification of critical points in the superstructure of a vehicle through FE modeling and mobility tests. In: Proceedings of the ASME IDETC/CIE 2013. Portland

Chapter 5
Predicting Approximate Governing Formula from Experimental Observations

Sagar Agarwal and R.P. Shimpi

Abstract Experiments help in understanding systems. Often experiments are conducted by varying some controllable parameters. The governing mathematical expressions involving the basic parameters may not always be known. This often causes a hindrance in the deep understanding of the experimental observations. In this paper, a method is suggested by which it may be possible to approximately predict the governing formula using the experimental data. Predicting the approximate formula is instrumental in developing appropriate theory and making sense out of the experimental data. The governing formula is predicted by suitable mathematical combination of basic parameters. The use of 'Simulated Annealing' optimization method is made to approximately predict the governing formula from the data obtained from experiments. The method seems to be robust and can accommodate various experimental errors (like small instrumental errors, human negligence in some readings). Illustrative examples are given. One of the examples is about beam experiments, wherein it is shown how to approximately predict deflection formula. The suggested methodology holds promise in understanding system behavior from experimental observations.

Keywords Predicting formula • Experimental data • Optimization • Simulated annealing

5.1 Introduction

5.1.1 Problem

5.1.1.1 Importance

Some experiments are conducted without knowing the relationship between the parameters. Predicting the relation is very important in building up of a new theory as it gives a lot of insight on the relation between the parameters and helps use understand the data properly. Sometimes trouble is faced trying to understand the data because of lack of theory or mathematical relation.

5.1.1.2 Presence of Experimental Errors

Due to some experimental or human error some sort of randomness often gets incorporated in the readings. Sometimes due to human negligence some of the readings are not as accurate as the experimenter might have hoped. It then can lead to false results for the nature of equation. In this paper, an attempt is made to develop a method to tackle the problem just stated.

5.2 Methodology

To predict the relation we have used Optimization techniques. All the relation possibilities are varied and compared with a hypothetical experimental data.

S. Agarwal • R.P. Shimpi (✉)
Department of Aerospace Engineering, IIT Bombay, Mumbai 400076, India
e-mail: rpshimpi@aero.iitb.ac.in

M. Mains (ed.), *Topics in Modal Analysis, Volume 10*, Conference Proceedings of the Society for Experimental Mechanics Series, DOI 10.1007/978-3-319-15251-6_5

5.2.1 Primary Parameter

One parameter is termed *as Primary parameter*, this parameter is chosen to be the parameter which is observed and noted while varying other parameters that are under our control. These controllable parameters are changed to get set of readings of the Primary Parameter in the experiment.

5.2.1.1 Calculating Error Between the Data and Random Relation

Once the Primary parameter is chosen, the error between the value of the Primary parameter obtained from the experimental data and value obtained from our random relation is noted. The sum of modulus of error for all the readings is calculated. This summed error is minimized to predict the relation using optimization technique.

In this paper, Simulated annealing optimization technique is used to predict the relation from the hypothetical experimental data. Simulated annealing is a probabilistic metaheuristic for global optimization problem to find the global optimum of a given function in a domain space. Guidelines of simulated annealing given by [16] (Vecci) are being followed in this method.

5.2.1.2 Choosing the Initial Temperature and Temperature Reduction Function

For our problem, Initial temperature is chosen to be sufficiently high that is 10,000 K so that it is able to search a larger domain and are able to search around freely all along the domain. The temperature reduction function is also chosen to be high that is 0.9. This was chosen to be a high number so that it is able to search better around the domain.

5.2.1.3 Developing the Function to be Optimized

The primary concept behind the method is to:

1. Generate a random relation keeping in mind the rules of dimensional analysis.
2. Calculate the error in Primary parameter between the data and value from the relation for each reading.
3. Take the average of mode of errors for all readings.
4. Minimize this average error obtained from step 3 using Simulated Annealing technique

To generate the random relation it is written in the following form:

$$f(x) = \sum_{i=1}^{l} \left(a_i \prod_{j=1}^{m} x_j^{bi,j} \cdot \prod_{k=m+1}^{n} (c_k + x_k)^{bi,k} \right)^d$$

Where,

a, b, c, d are constants

x_1 *to* x_m are dimensional parameters

x_{m+1} *to* x_n are dimensionless parameters

n > m and l is a sufficiently large number

The constants a, b, c are varied using simulated annealing to generate all possible relation. Dimensional analysis is taken care of by ensuring dimension of $\prod_{j=1}^{m} x_j^{b_{i,j}}$ is equal to that of primary parameter. This is done using equality constrain in simulated annealing.

5.2.2 Increasing Accuracy

Dimensional analysis is also incorporated in the problem to arrive at the relation faster and more accurately. Dimensional analysis is incorporated by substituting the coefficients and powers of some of the parameters by combination of coefficients and powers of other parameters obtained from equations of dimensional analysis. This also restricts the randomness of the coefficients such that the relation obtained is always dimensionally correct. Computational time of the solution is reduced and accuracy of the relation is increased especially when the number of experimental readings were less than 5. In general, minimum readings required to predict the relation was five times the total number to parameters varying in the experiment. However, results produced from less than this also seemed to give promising results.

5.2.3 *Testing of Method*

Various tests are performed on the robustness of the method to predict the relation. Small random 2–5 % error is incorporated in the hypothetical data. The relation seemed to be in the limits of 5 % error. The method is also tested for some completely absurd readings which depict human negligence in experiments. The method seemed to be robust to these types of errors. Several tests were performed to check dependence of variables as well for example an independent parameter was introduced in the data in of the test case, such that the primary parameter is not affected by this independent parameter. The method correctly points out its independence in its results.

5.3 Results

The code has been tested for several hypothetical experimental data for many experiments. The method is tested for three different sets of data:

1. Perfect Data
2. Data with Experimental Error of around 5 %
3. Some readings with significant error

Results are shown in this paper for case in which experimental error exists of around 5 % and 1 reading with huge error. Results are tabulated here for different experiments:

1. *Natural Frequency* (Table 5.1):

   Actual Formula: $f = .159\sqrt{\frac{k}{m}}$
2. *Deflection of Beam* (Table 5.2):

   Actual Formula: $\delta = \frac{4FL^3}{Ebt^3}$
3. *Shear Modulus* (Table 5.3):

   Actual Formula: $G = \frac{0.5E}{(1+\vartheta)}$

The results produced from the method are within the limit of 5 % error. It seems to be robust towards experimental error and large error in some of the readings. It can predict the formula and dependence of parameters in the experiment fairly accurately. It can help the experimenter to predict the theory from the experiment he has conducted. It can be an instrumental method to help to researchers validate and develop new theory and will help them analyze the experimental parameters better. They will be able to make sense of the experiment.

Table 5.1 Hypothetical experimental data for natural frequency experiment

S.no	k(N/m)	M(Rg)	f(Hz)
1	1	1	0.16
2	2	1	0.23
3	2	2	0.17
4	3	1	0.27
5	3	2	0.19
6	3	3	0.17
7	3.4	2.1	0.19
8	2.4	2.4	0.14
9	1.2	3	3.60

Note: This data contains small experimental error and last reading is triple the value it should have been formula generated by the method is $f = .1592\sqrt{\frac{k}{m}}$

Table 5.2 Hypothetical experimental data for deflection of a beam experiment

S.no.	F(N)	L(m)	E(Pa)	b(m)	t(m)	Deflection(m)
1	1	1	2	0.33	0.33	168.8
2	2	1	2	0.33	0.33	338.2
3	3	1	2	0.33	1	18.2
4	4	1	2	1	0.66	27.9
5	5	1	2	0.66	0.66	52.7
6	2	2	2	4	2	1.0
7	2	3	2	2	4	0.8
8	2	1	1	4	1.5	0.5
9	2	1	0.5	0.33	0.33	1349.1
10	2	1	0.5	0.66	0.33	224.4

Note: This data contains small experimental error and last reading is third of the value it should have been formula generated by this method is $\delta = \frac{4.04F^{0.98}L^{3.02}}{E^{0.98}b^{0.99}t^{2.99}}$

Table 5.3 Hypothetical experimental data for shear modulus

S.no.	E(m)	ν	G(Pa)
1	1	1	0.27
2	2	1	0.54
3	3	1	0.78
4	1	2	0.19
5	1	3	0.18
6	2	2	0.33
7	2	3	0.25
8	3	3	0.33
9	3	2	0.47
10	3	1	0.72
11	1.5	1	0.36
12	1	1.5	0.17
13	2.4	1.2	0.51
14	2	0.5	0.62
15	1	0.2	1.20

Note: This data contains small experimental error and last reading is triple of the value it should have been formula generated by this method is $G = \frac{0.495E}{(0.99+\vartheta)^{0.99}}$

References

1. Andresenz B, Nourani Y (1998) A comparison of simulated annealing cooling strategies. J Phys A Math Gen 31:8373–8385
2. Cochran WG (1957) Experimental design
3. Corte H (2011) Simulated annealing. Matlab, Oct 2011
4. Dunning T (2008) Analysis of experimental data
5. Frausto-Solís J, Sanvicente-Sánchez H (2004) A method to establish the cooling scheme in simulated. Springer
6. Box GEP, Hunter WG, Hunter JS, Hunter WG (1978) Statistics for experimenters: an introduction to design, data analysis, and model building
7. Field JE, Walley SM (2004) Review of experimental techniques for high rate deformation and shock studies. Int J Impact Eng 30:725–775
8. Kempthorne O (1952) Design and analysis of experiments
9. Keppel G (1991) Design and analysis: a researchers handbook
10. Langhaar HL (1951) Dimensional analysis and theory of models
11. Mandel J (2012) Statistical analysis of experimental data
12. Montgomery DC (1997) Design and analysis of experiments
13. Peters CA (2001) Statistics for analysis of experimental data. In: Environmental engineering process manual
14. Pharr GM, Oliver WC (2011) An improved technique for determining hardness and elastic modulus using load and displacement sensing indentation experiments. J Mater Res 7:1564–1583
15. Schwarz G (1978) Estimating the dimension of a model. Ann Stat 6:461–464
16. Vecchi MP (1983) Optimization using simulated annealing. Science 220:671–680
17. Vigeh A (2011) Investigation of a simulated annealing cooling schedule

Chapter 6
Experimental Characterization and Simulation of Vibration Environmental Test

Washington J. DeLima and Melanie N. Ambrose

Abstract The dynamic characterization of shakers as well their dynamic interaction with adapters and test fixture are fundamental to understand the response of an unit under environmental testing (random vibration, SRS shock, sine sweep). Numerical modeling of these systems provides support for their design and a good insight how the systems may respond during test. Critical inputs to achieve reliable predictive simulations require accurate input environments, coupling behavior, and test article definition. In this paper experimental methods such as modal analysis, operational deflection shape were used to characterize the dynamic behavior for a variety of vibration test setups, to define realistic input and model validation for the numerical simulation.

Keywords Virtual test • Electrodynamic shaker • Simulation • Characterization

6.1 Introduction

It is important to understand the interaction between the DUT (device under test) and the test equipment (shaker, shaker-DUT adapter and control system) during a closed-loop vibration test to establish the test robustness. Some factors that affect the test robustness include: weight of the DUT, weight of adapter, closed loop vibration control system, frequency range and vibration levels in the test specification.

Evaluating the robustness of a test procedure experimentally can be time consuming, creating delays and adding cost. One alternative is the creation of a model to simulate the test. Test simulation tools have been developed to support vibration test and evaluate the robustness of the test procedure. Rice et al. [1, 2] presented an approach for virtual shaker test where it was developed simulation model that would represent the actual vibration test. A lumped parameter electro-mechanical model of the shaker was used following the work in [3–5] together with a model of the vibration controller. Klenke et al. [6] developed VETO (Vibration Virtual Environment for Test Optimization) to help test engineer to select the optimal test parameters needed in the design test. Xiang et al. [7, 8] presented a sequential process to simulate vibration test with application to satellite test.

Simulation technology can bring forward solutions that offer predictive modeling to prevent accidental over-test of sensitive units or achieving challenging test requirements. Creating a calibrated shaker model allows the opportunity to design test fixtures that are robust and well-behaved under the testing frequencies. A complete model that captures the interactions and cumulative response of the shaker, adapter, and fixtured assembly provides the basis for predictive design of the test setup. This paper presents an experimental characterization of vibration test components and a reduced order model to simulate the vibration test setup.

W.J. DeLima (✉) • M.N. Ambrose
Honeywell Federal Manufacturing & Technologies, Kansas City, MO, USA
e-mail: wdelima@kcp.com

M. Mains (ed.), *Topics in Modal Analysis, Volume 10*, Conference Proceedings of the Society for Experimental Mechanics Series, DOI 10.1007/978-3-319-15251-6_6

6.2 Shaker Characterization of Closed Loop Vibration Test

A schematic diagram of closed-loop vibration test is presented in Fig. 6.1. In a closed-loop vibration test, the controller sends a signal (driver voltage) to an amplifier that sends the amplified signal to the shaker. The shaker sends the signal to the DUT that is attached to the shaker through a shaker-adapter and fixture. A feedback response is sent back to the vibration controller that compensates the signal to adjust it to the test specification.

The efficacy of the test simulation tool depends greatly on how accurate the elements on Fig. 6.1 are represented. One of the most import elements is the shaker. Electro-mechanical shakers are widely used in vibration testing. The structure of a conventional shaker resembles a loudspeaker where an AC electrical current passes through a coil wire that is in a fixed magnetic field [3–5]. That creates a dynamic axial force in the coil proportional to the current. The motion of the coil is transmitted to the table where the DUT, fixture and the shaker adapter are attached. Another approach to shaker design is IAR (Induct-A-ring Armature) developed by Unholtz-Dickie [9]. IAR used a solid metal cylinder as moving coil. The force generated in the coil was created by induction which eliminates the need for epoxy-bonded coil wires.

In order to better simulate the shaker and its influence to the closed-loop vibration test a series of tests were performed in a group of shakers. Four similar shakers where tested. The following tests were used to characterize the shaker:

1. Driver voltage of single point control on the center of an empty shaker. Test specification is random vibration 0.001 g^2/Hz from 10 to 4,000 Hz.
2. Operational deflection shape (ODS) on the surface of the table of an empty shaker with a single control point at the center of the shaker and test specification random vibration 0.001 g^2/Hz from 10 to 4,000 Hz
3. Operational deflection shape (ODS) on the top surface of a shaker adapter attached to the shaker table with a single control point at the center of the adapter and test specification random vibration 0.001 g^2/Hz from 10 to 4,000 Hz

6.2.1 Driver Voltage of Empty Shaker

The driver voltage of a single control point test was measured during a random vibration (0.001 g^2/Hz from 10 to 4,000 Hz). The control point is located at the center of the shaker table. Figure 6.2 shows the driver voltage for four shakers with their respective voltage RMS value. Remember that the valleys in the curve represent resonance frequencies and the peaks represent the nodal frequencies of the shakers. All shakers have similar behavior up to 2 kHz. That is somewhat expected since the manufacture of the four shakers guarantees the armature resonances of the shakers are above 2 kHz. However, there is a significant difference between the shakers above 2 kHz. In terms of overall driver voltage, Shaker 4 requires the highest driver voltage and Shaker 3 the lowest. The armature resonance of Shaker 4 is about 100 Hz less that the other three shakers. As the frequency goes up the difference among the shakers is more evident. Figure 6.3 shows a detail of Fig. 6.2 for frequency range from 3 to 4 kHz. Shakers 1, 2 and 3 have resonance and nodal frequency in these frequency ranges and Shaker 2 has one more nodal frequencies than Shaker 1 and 3. Shaker 4 has only one nodal frequencies. This difference in modal behavior of the shakers explains the difference in driver voltage RMS values since more nodal points means that the shaker requires more driver voltage and certain random vibration tests would not be able to run on this shaker

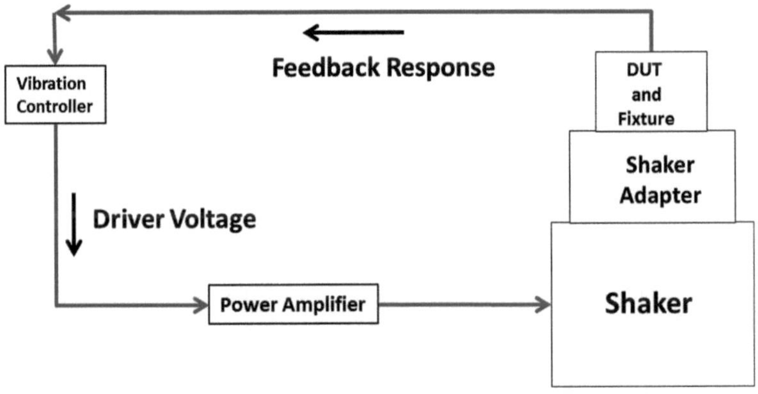

Fig. 6.1 A schematic diagram of closed-loop vibration test

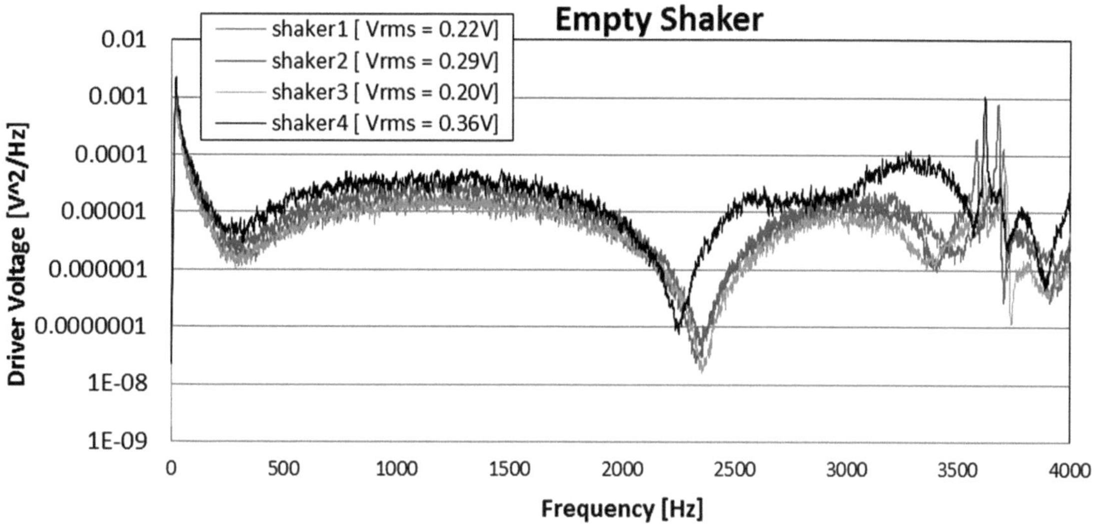

Fig. 6.2 Driver voltage empty shaker configuration

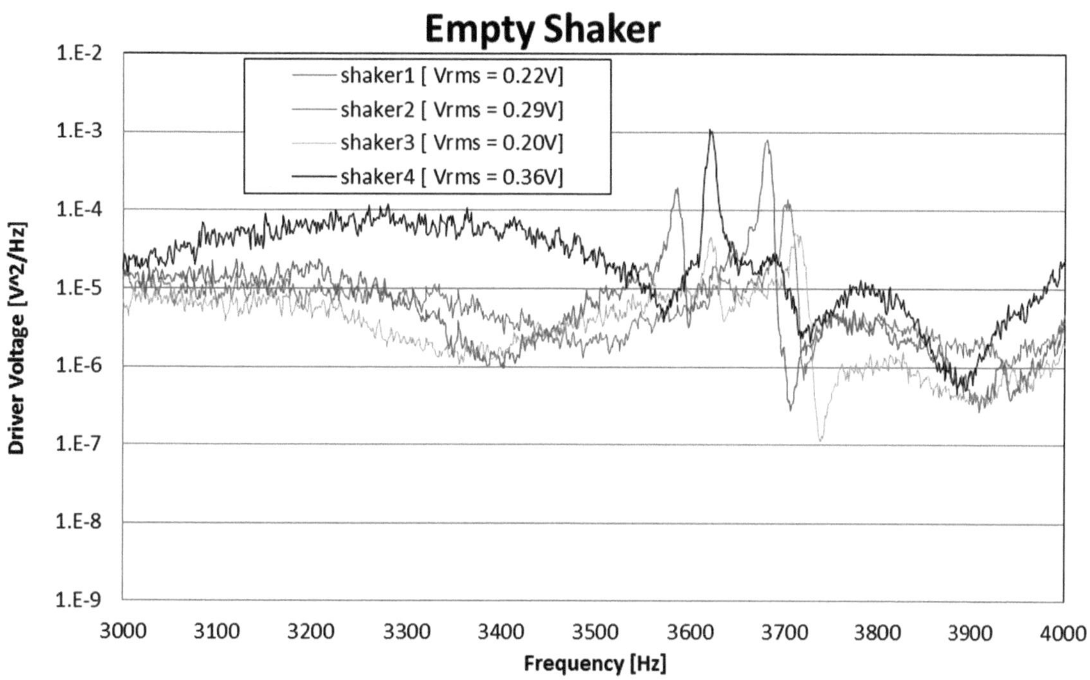

Fig. 6.3 Driver voltage empty shaker configuration between 3 and 4 kHz

6.2.2 Operational Deflection Shape of Empty Shakers

Operational deflection shape tests were performed on empty shakers using random vibration 0.001 g^2/Hz from 10 to 4,000 Hz. The control accelerometer was located in the center of the surface of the shaker tables (Fig. 6.4) and controls the vibration perpendicular to the table surface (z-direction in Fig. 6.4). LMS Testlab and LMS SCADA were used in this test. The ODS maps the surface of the shaker table into a mesh and calculates the FRF between the control point and all the response points to calculate the relative motion between the control points and the response points.

Twenty eight FRF resulted from the test. The envelope FRF is used to represent these 28 FRF. Figure 6.4 shows the envelope FRF for the shakers. Note that FRF = 1 implies that the response points moves in phase with the control point and FRF different from 1 implies that the motion of the response points is not in phase with the control location. All the response points are in phase with the control point up to around 2.8 kHz. Above 2.8 kHz the response points are not in phase with the control points and there is great variation in the behavior of the FRF from shaker to shaker (Fig. 6.5).

Fig. 6.4 ODS geometry for empty shaker configuration

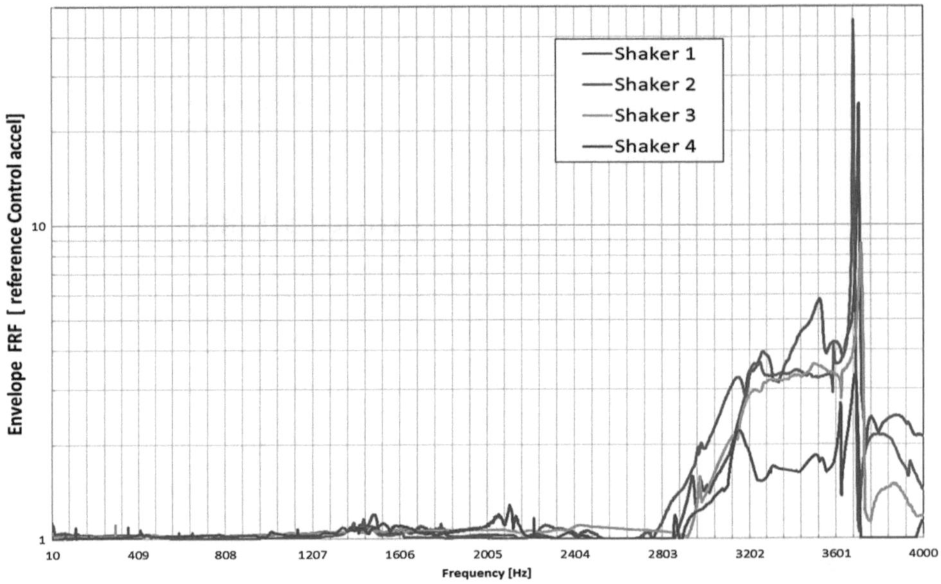

Fig. 6.5 Envelope FRF of ODS test for empty shakers configuration

The off-axis responses are another factor that affect the test. In this paper X and Y directions are the off-axis direction and tell us that during the test the shaker motion is not unidirectional and that it will transfer energy to the DUT in a direction not specified in the test requirement. Figure 6.6 shows the off-axis response at the control point. The off-axis response appears to increase as the frequency increases and approaches the control response in the frequency range 3.5 and 4 kHz. In this frequency range Shaker 1 shows off-axis responses above the control response. The off-axis response in a non-controlled point (point 37 in Fig. 6.7) is presented in Fig. 6.8. For all four shakers, the response at the control direction follows the response of the control point up to around 2 kHz. After 2 kHz the response varies from shaker to shaker. The off-axis response at the non-controlled point (point 37) increases as the frequency increases and reaches, or even surpasses, the response of the control direction after the 2 kHz. Again the off-axis behavior after 2 kHz varies from shaker to shaker.

6.2.3 Operational Deflection Shape of Adapter on Top of the Shaker

In order to gain more insight of how the elements in the diagram in Fig. 6.1 can be modeled, an Operational Deflection Shape test was performed on the top surface of an shaker adapter attached to the shakers. A 7×7 inch and 70 lb cube was used. Figure 6.9 show the actual test setup and Fig. 6.10 shows a LMS testlLab mesh of the top surface of the adapter and the outer points on the shaker table.

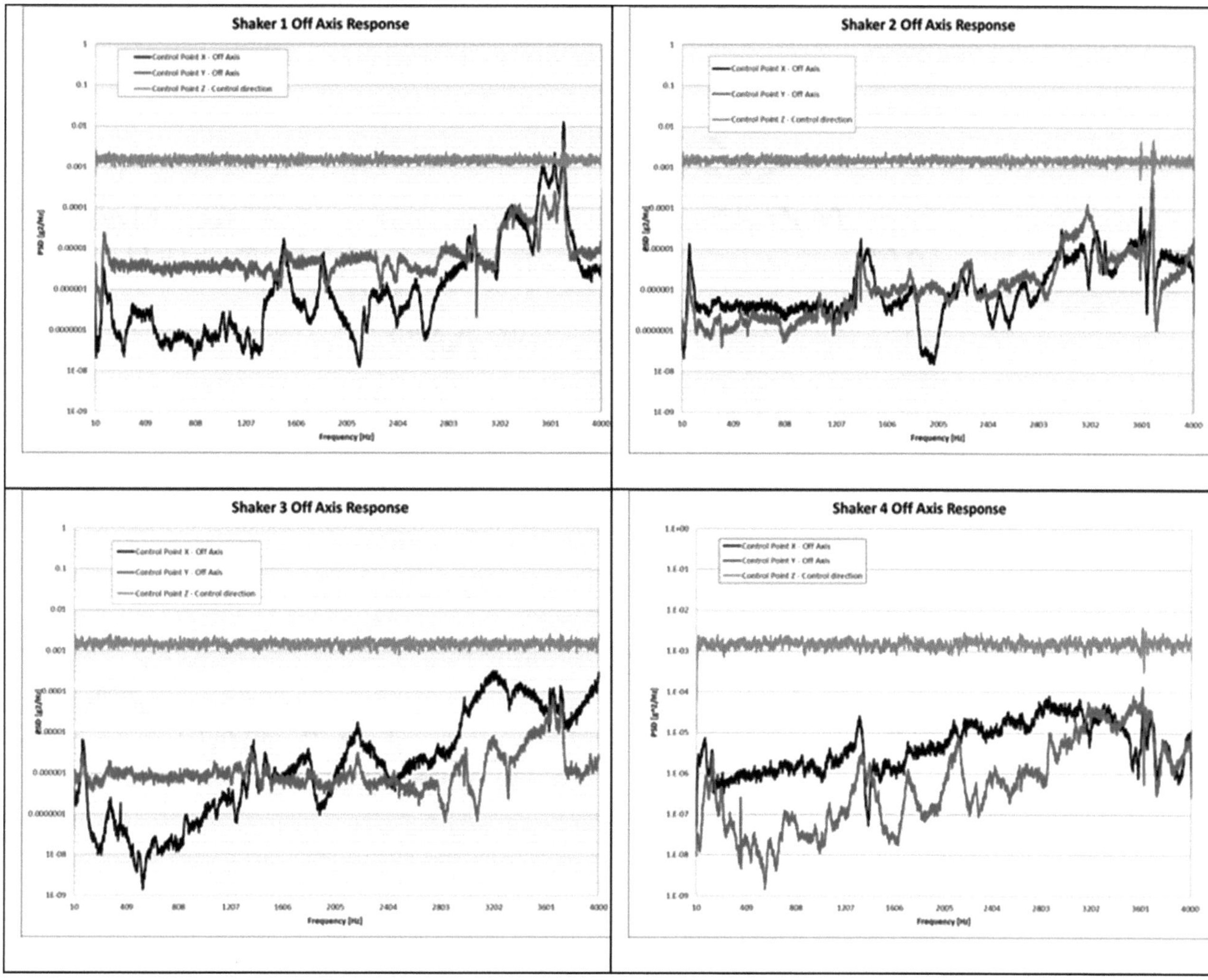

Fig. 6.6 Off axis response at control location during ODS test for empty shakers configuration

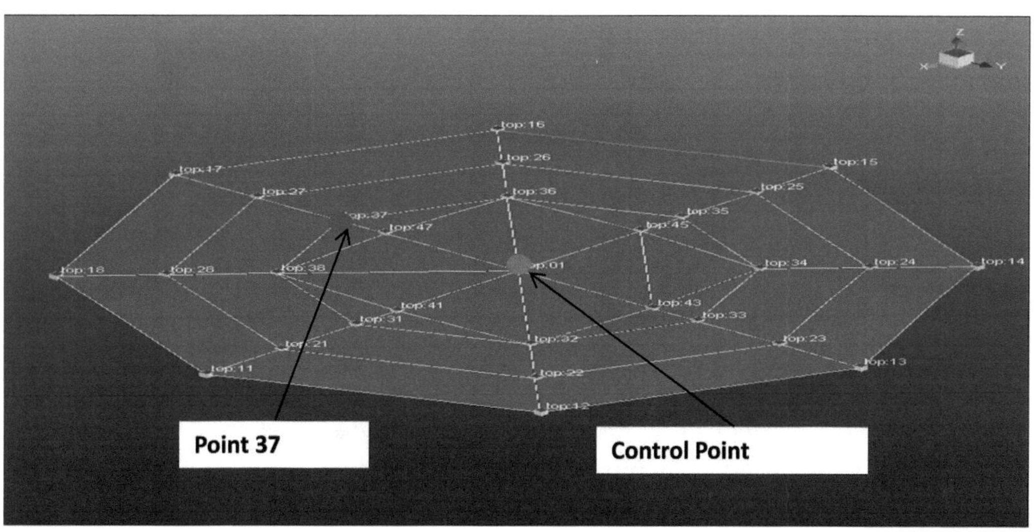

Fig. 6.7 Location of control point and point 37 in the ODS geometry during ODS test for empty shakers configuration

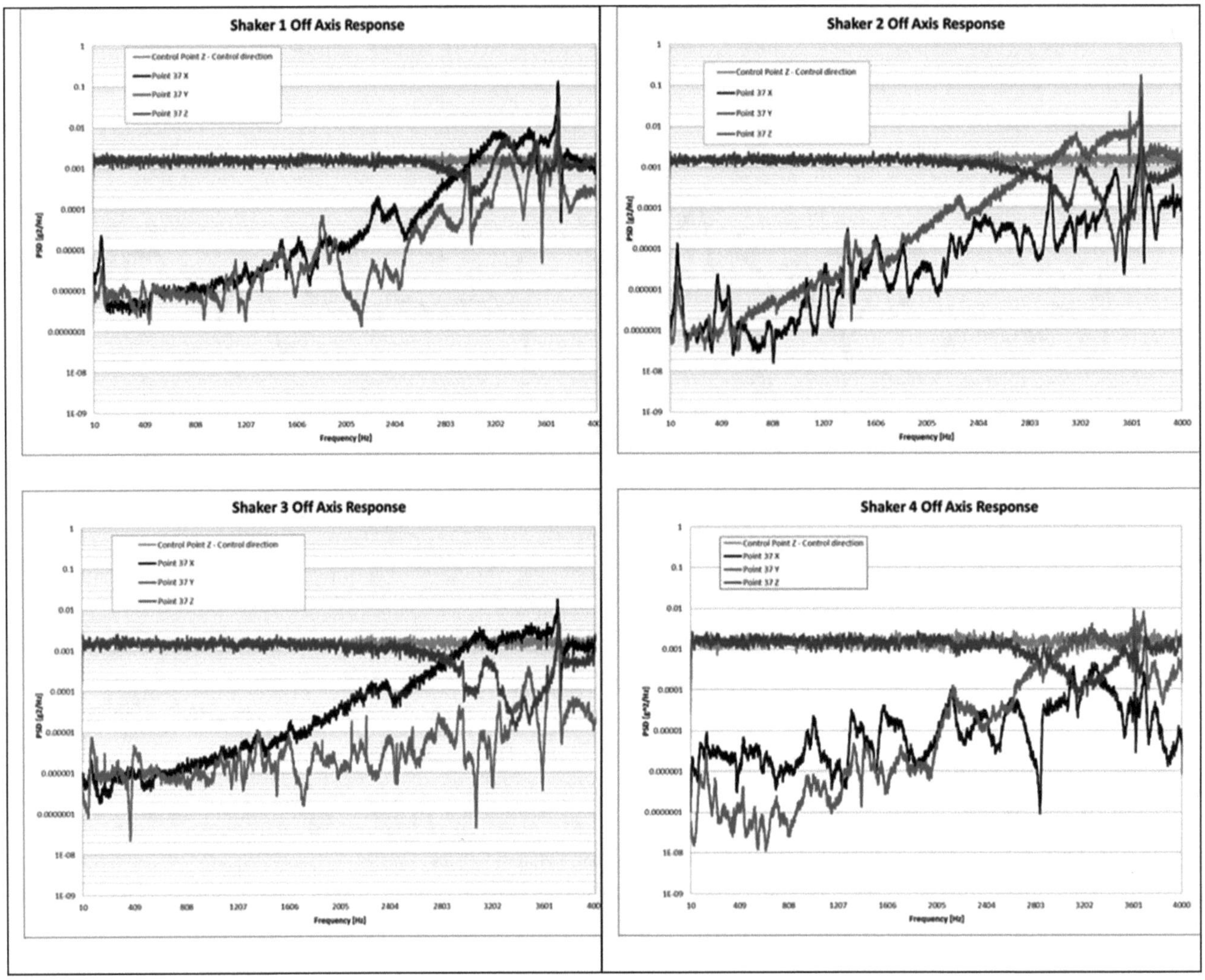

Fig. 6.8 Off Axis response at point 37 (see Fig. 6.7) during ODS test for empty shakers configuration

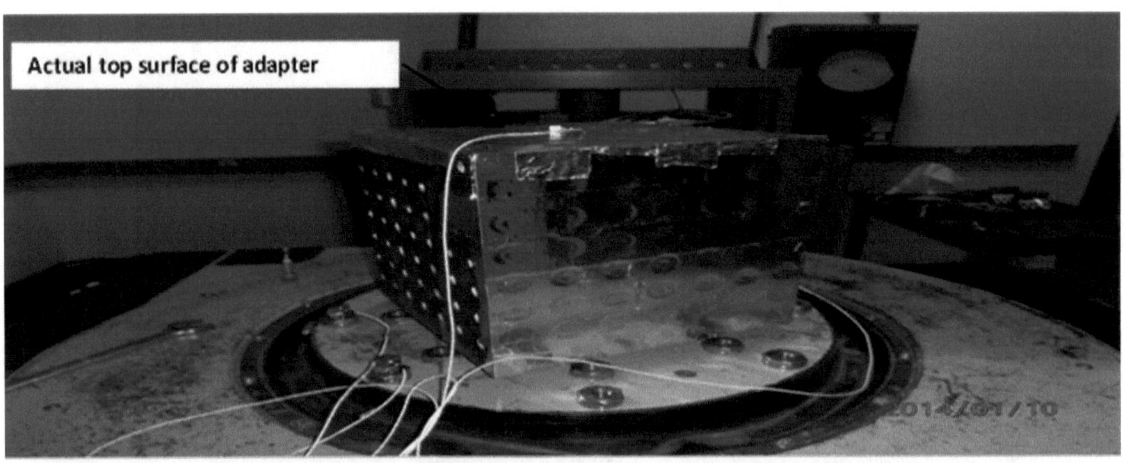

Fig. 6.9 Actual test set up of ODS on the top surface of shaker adapter

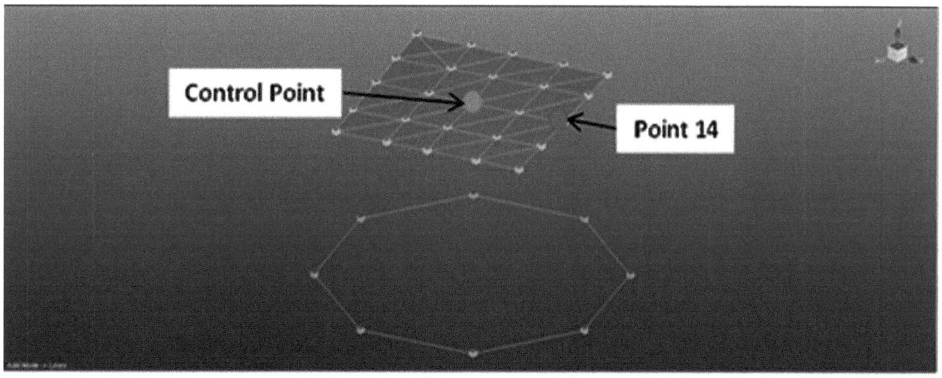

Fig. 6.10 LMS TestLab mesh of the top surface of shaker adapter and the shaker table

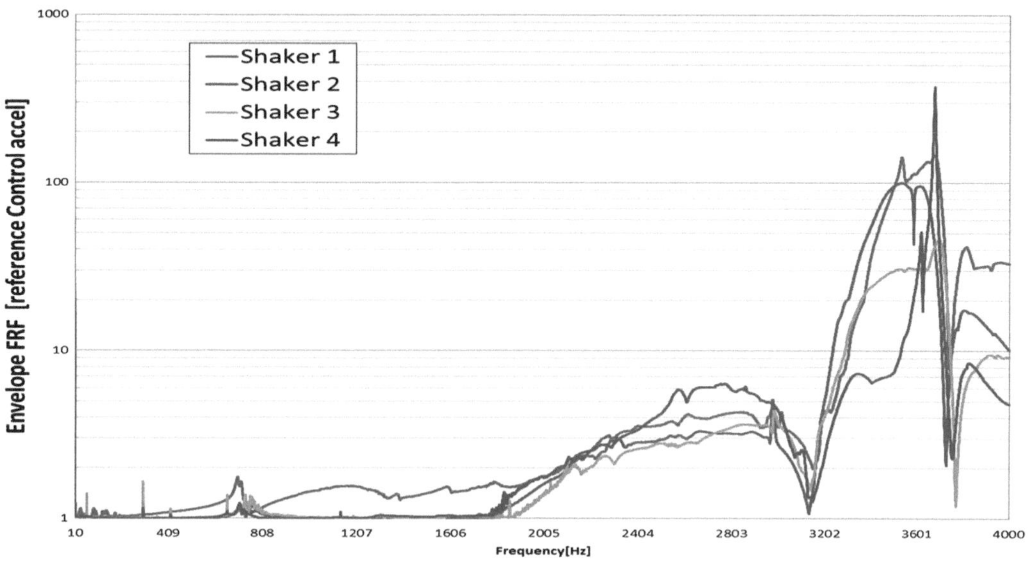

Fig. 6.11 Envelope FRF of ODS test on the top surface of an shaker adapter for all four shakers

The control point is located at the center of the cube top surface (Figs. 6.9 and 6.10). Figure 6.11 presents the envelope FRF for all shaker. All the points of the top surface of the cube move in phase with the control point up to about 1.8 kHz. Note the decreasing in frequency relative to the empty shaker results are due to the interaction between the cube and the shaker. As in the previous section, there is a shaker to shaker variation in the higher frequency range.

The off-axis responses at the control and non-control points (Fig. 6.10) are presented in Figs. 6.12 and 6.13 respectively. The off-axis response at the control and no-control points approach the control response at around 800 Hz frequency range. The 800 Hz is a resonance frequency of the cube. As in the empty shaker case, the off-axis responses tend to increase with frequency and can reaches value equal or higher than the control response at the frequency range above 3 kHz.

6.3 Numerical Simulations

Predictive modeling is an integral piece of successful environmental testing. To obtain a predictive model, verification and validation of the assumptions and results of the model are critical. Finite element analysis was used to create a model of the 7×7 adapter which was then combined with a reduced order model of the shaker. Both the adapter and system models were calibrated using the dynamic characterization data described previously.

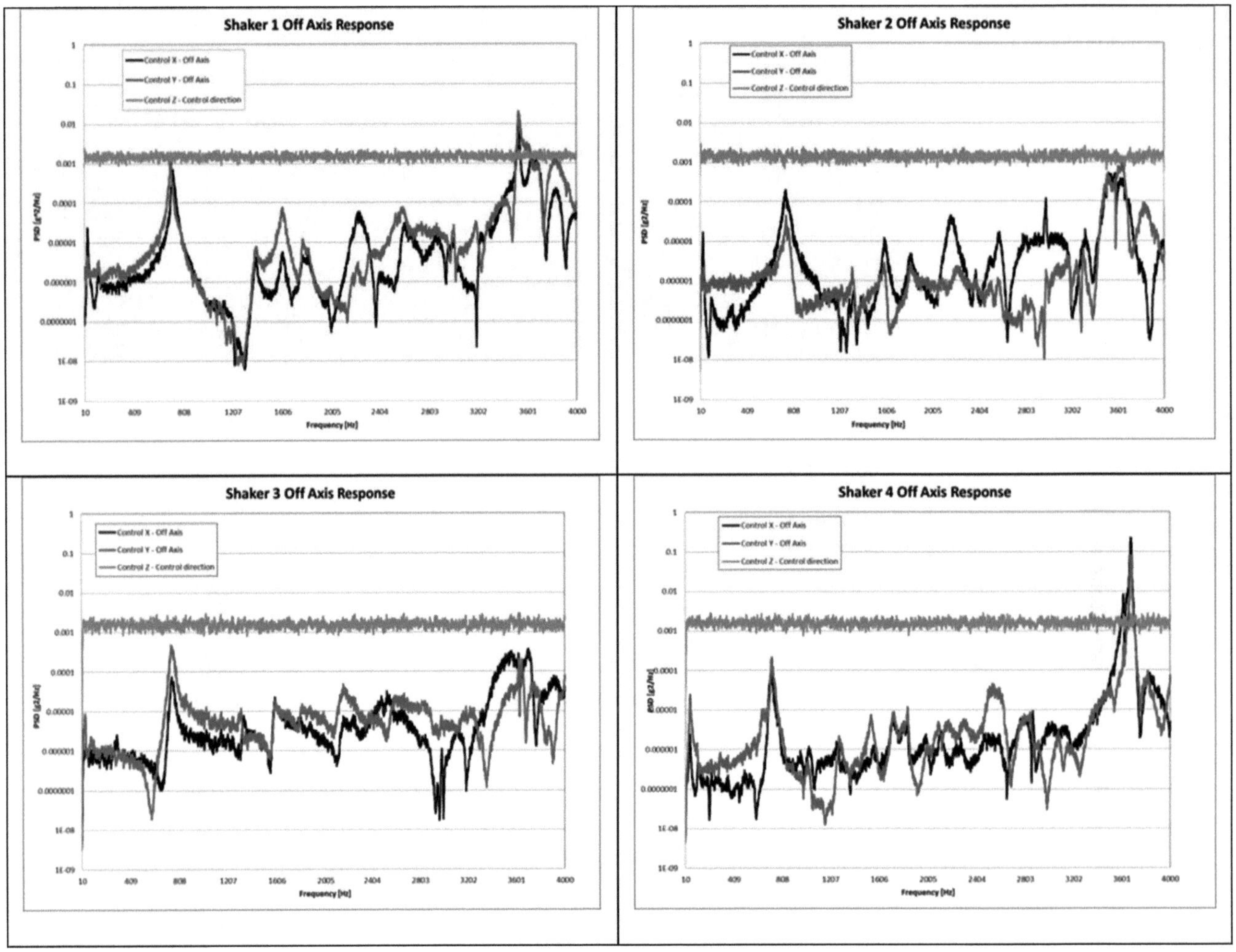

Fig. 6.12 Off Axis response at control point (Fig. 6.10) during ODS test on the top surface of an shaker adapter

A three-dimensional solid finite element model of a magnesium adapter was created to include the unique weight reduction features of the 7 × 7 adapter (Fig. 6.14). The adapter is physically mounted to the shaker by 13 shaker mounts. The boundary condition describing the interaction between the adapter and the shaker mounts was evaluated using three different scenarios: (1) fixing the bottom of the shaker mounts, (2) fixing the bottom of the cube, and (3) fixing a small patch area near the bolt circumference. It was found that the reduction in stiffness as a result of the actual mounting setup is best modeled as a small surface area representing a bolted interface that is fixed in all six degrees of freedom, modeled by scenario 3 (Fig. 6.15). This model assumption correlated best with the off-axis motion as shown in the ODS test results (Figs. 6.11, 6.12, and 6.13) and the axial compression/elongation modal response as shown by the shaker driver voltage (Fig. 6.16).

The next step was to model a system with a shaker and adapter. A simplified lumped parameter single-degree-of-freedom model was created to represent the shaker. The shaker adapter was modeled with FEA as presented in the previous paragraph. The axial armature resonance of the empty shaker was found to be approximately 2,300 Hz from the modal characterization of the four shakers. When a 7 × 7 adapter cube was placed on top of one of the shakers, the axial resonance shifted down to approximately 1,700 Hz (Fig. 6.16). The drive voltage indicates there is another prominent mode that is introduced to the system as a result of the adapter, occurring at 2,800 Hz, this is a resonance of the adapter. This reduced order system model of the shaker with the adapter (Fig. 6.17) was calibrated to the experimental response mentioned previously. This calibration successfully captured both the resonance of the adapter and that of the armature as shown in Fig. 6.16 by the normalized acceleration output. The calibrated modes of the system are shown in Figs. 6.17 and 6.18.

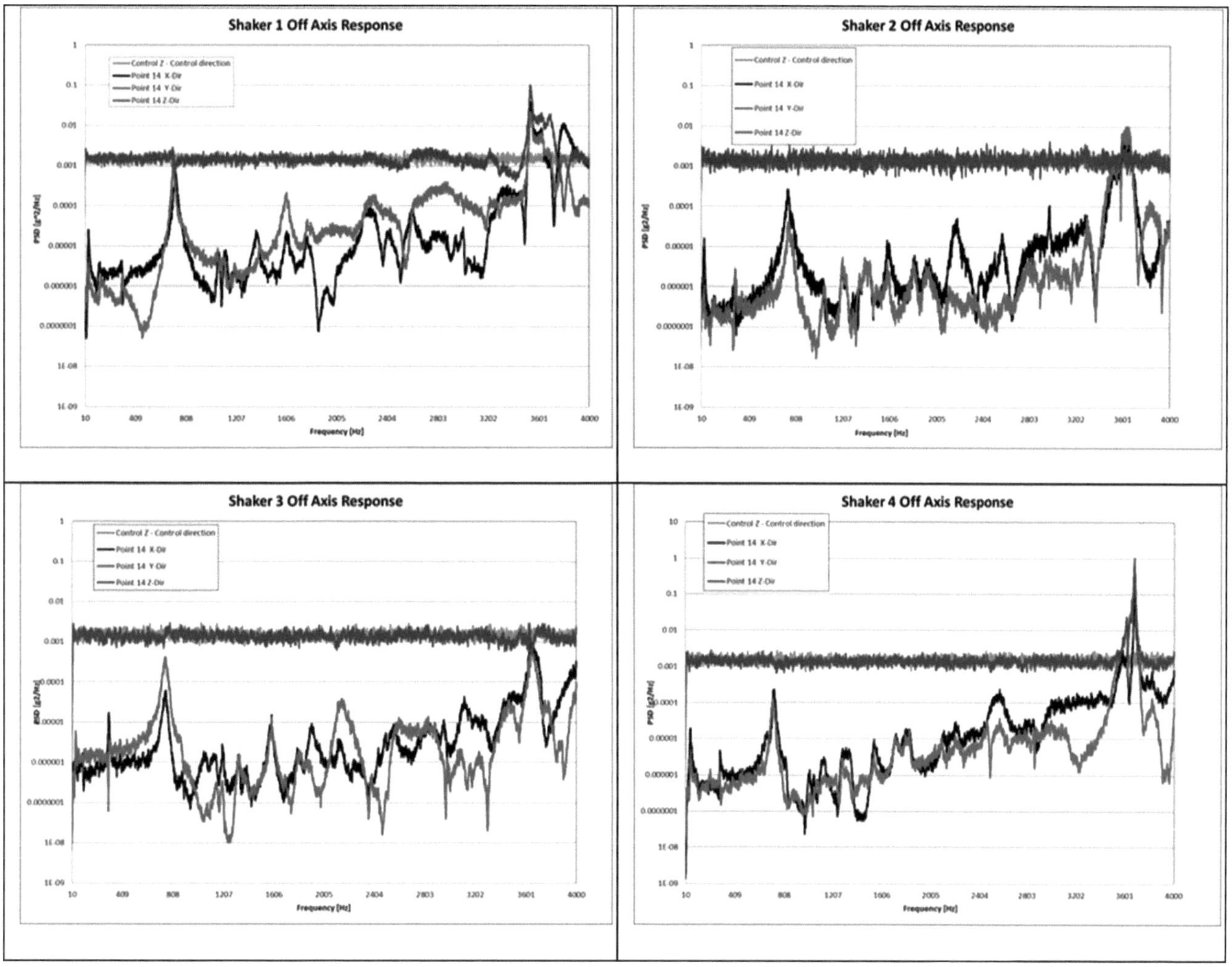

Fig. 6.13 Off Axis response at point 14 (Fig. 6.10) during ODS test on the top surface of an shaker adapter

Fig. 6.14 7 × 7 magnesium adapter

6.4 Conclusion

An experimental characterization of the vibration test setup was performed at four similar shakers using a flat random vibration spectrum from 10 Hz to 4 kHz. A substantial shaker to shaker variation and off-axis response were found between 3 k and 4 kHz which can greatly affect the testability of DUT. The experimental characterization was used to create a model to simulate the test setup. A simplified lumped parameter of the shaker was calibrated to match the shaker armature resonance. Additional working are being carried out to create a more detail lumped parameter model of the shaker to accommodate the full shaker response, its iteration with the adapter and the shaker to shaker variation.

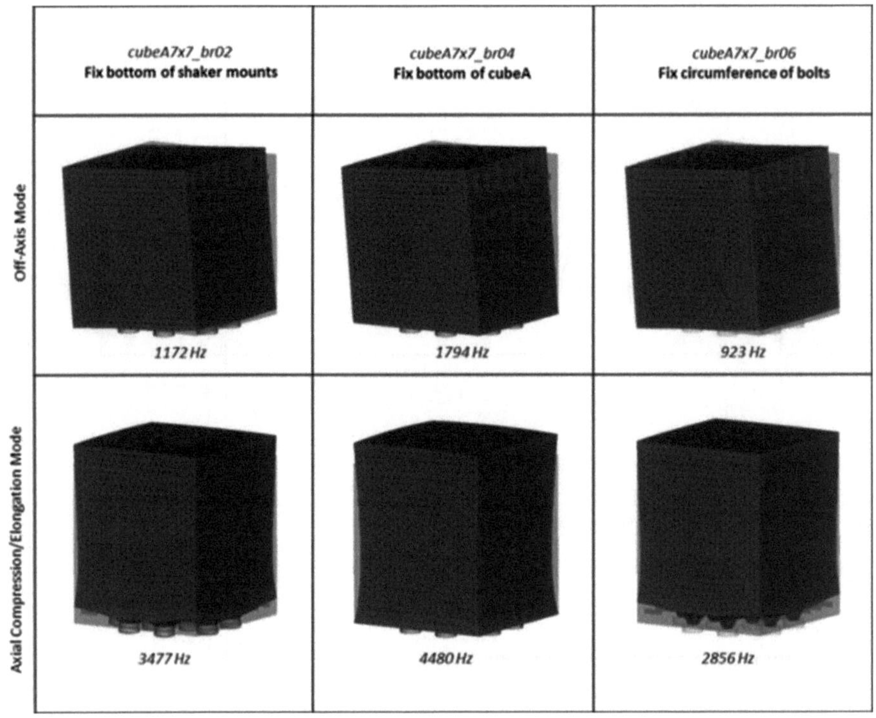

Fig. 6.15 Eigenvalue analysis of 7 × 7 adapter with various boundary conditions used for model calibration

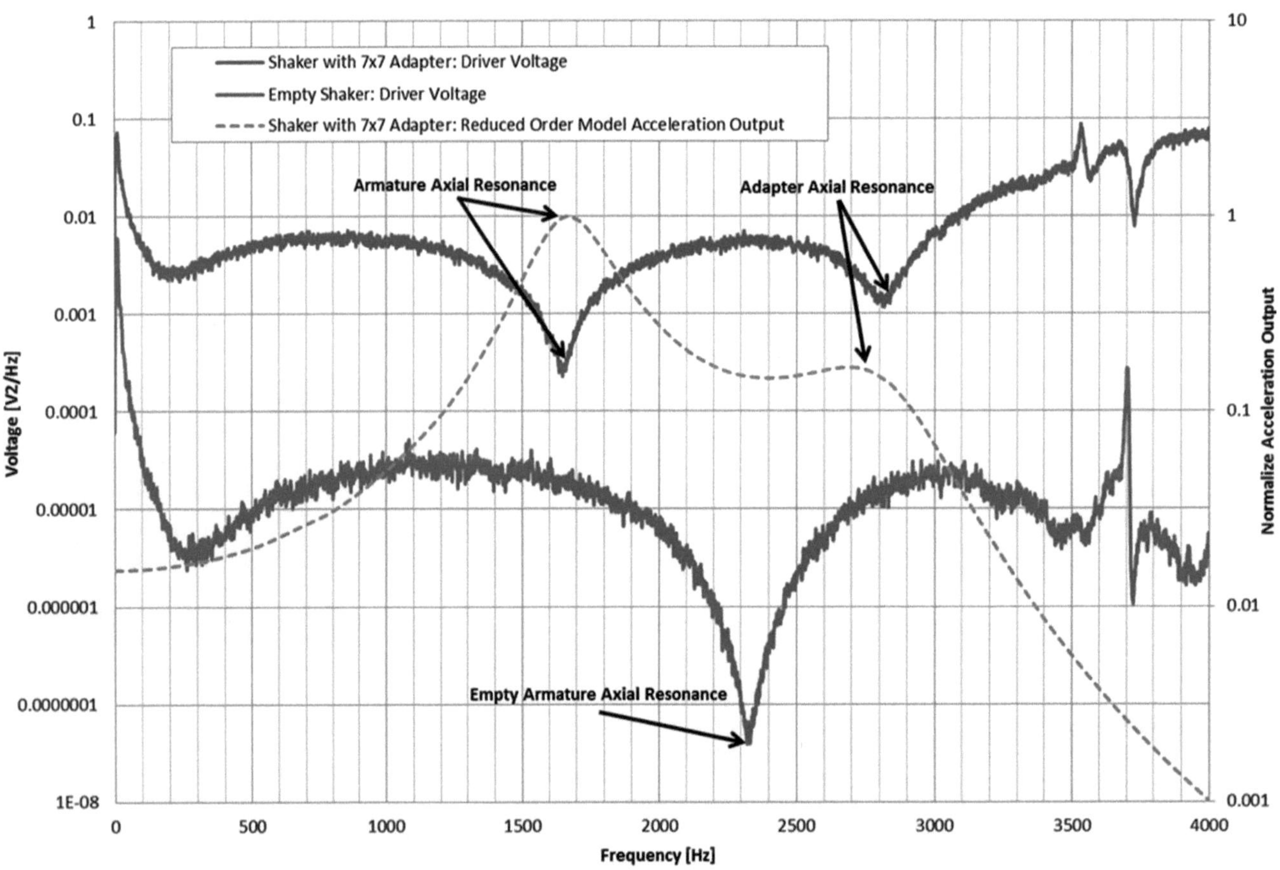

Fig. 6.16 Calibrated reduced order model modal response

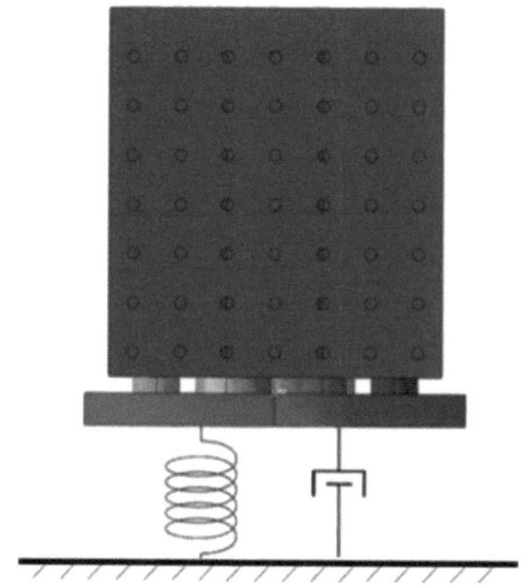

Fig. 6.17 Simplified reduced order model of the shaker armature and 7 × 7 adapter

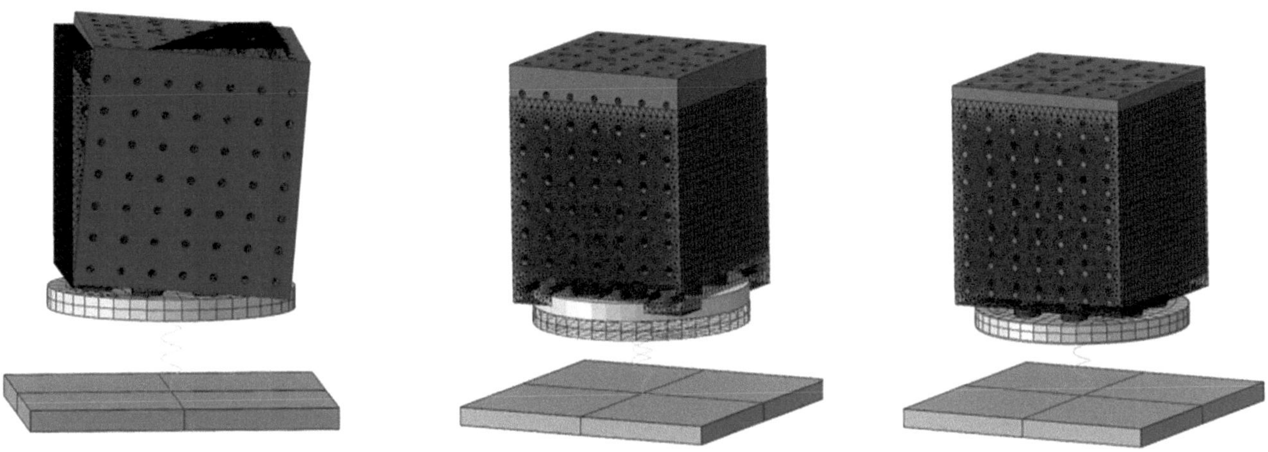

Fig. 6.18 Mode shapes of calibrated reduced order model

Acknowledgments Notice: This presentation has been authored by Honeywell Federal Manufacturing & Technologies under Contract No. DE-NA0000622 with the U.S. Department of Energy. The United States Government retains and the publisher, by accepting the article for publication, acknowledges that the United States Government retains a nonexclusive, paid up, irrevocable, world-wide license to publish or reproduce the published form of this manuscript, or allow others to do so, for the United States Government purposes.

References

1. Ricci S, Peeters B, Debille J, Britte L, Faignet E (2008) Virtual shaker testing: a novel approach for improving vibration test performance. In: Proceedings of the ISMA 2008 international conference on noise and vibration engineering, Leuven, 15–17 Sept 2008
2. Ricci S, Peeters B, Debille J, Britte L, Faignet E (2009) Virtual shaker testing for predicting and improving vibration test performance. In: Proceedings of the 27th international modal analysis conference, IMAC-XXVII, Orlando, 9–12 Feb 2009
3. Varoto PS, De Oliveira LPR (2002) Interaction between a vibration exciter and the structure under test. Sound & Vibration, Oct 2002
4. Fox Lang G, Snyder D (2001) Understanding the physics of electrodynamic shaker performance. Sound & Vibration, Oct 2001
5. Fox Lang G (1997) Electrodynamic shaker fundamentals. Sound & Vibration, Apr 1997
6. Klenke SE, Lauffer JP, Gregory DL, Togami TC (1996) The vibration virtual environment for test optimization (VETO), SAND-96-2217C
7. Chuang L, Shu-Hong Xiang, Feng Yao-Qi (2010) Research on virtual test system for the satellite. In: Proceedings of the 17th international congress on sound and vibration, ICSV 17, Cairo, 18–22 July 2010
8. Shu-Hong Xiang, Chuang L, Yan Tingfei (2007) Virtual vibration test and verification for the satellite. In: Proceedings of the 14th international congress on sound and vibration, ICSV 13, Cairns, 9–12 July 2007
9. Phillip Rogers, Michael Garofalo (2009) What is an induct-a-ring shaker? Sound & Vibration, Mar 2009

Chapter 7
System Identification of an MDOF Experimental Structure with a View Towards Validation and Verification

K. Worden, O.D. Tiboaca, I. Antoniadou, and R.J. Barthorpe

Abstract Validation approaches can determine the degree of accuracy of simulated models representing real structures. Therefore these approaches should deal with concepts concerning fidelity-to-data, the uncertainty quantification and the comparative metrics i.e. measures that quantify the level of agreement between simulation and experimental outcomes. In the context of nonlinear systems associated with nonlinear models additional care should be taken when attempting to perform a validation scheme, in order to evaluate the examined models, due to bifurcations that may occur.

In this paper, experimental datasets from a laboratory based three-story building structure are used in order to calibrate the parameters of a physics-based model using system identification methods. The structure described here is linear; however, the work is a necessary precursor to the investigation of the behaviour of the structure when nonlinearity is introduced in the form a cantilever beam impacting the highest story of the building for a specific range of excitations. In this study the linear model is identified using experimental data and the sensitivity of model predictions is examined when the parameters of the model are varied.

Keyword System identification

7.1 Introduction

This study focuses on the application of methods of Verification and Validation (V&V) to nonlinear structural dynamics models. V&V of numerical models is a pressing issue across science and engineering and one that is attracting increasing interest from both the industrial and academic research communities. Reliance upon the predictions of complex simulations has advanced to the point of dominating the test and analysis process in many applications. This is particularly the case in applications where experimental testing is either prohibitively expensive or simply not feasible, the study of structural response to extreme events being one such example. The aim of model validation is to provide a principled means of assessing the credibility that may be ascribed to numerical model predictions. In recent years a number of application-specific guidelines have been proposed for implementing this process. Among the first such frameworks to focus on physics-based engineering models was that produced in 1999 by the AIAA for computational fluid dynamics (CFD) problems [1]. These have been followed more recently by as series of standards introduced by the American Society of Mechanical Engineers (ASME), currently comprising the ASME Guide for V&V in Computational Solid Mechanics in 2006 [2] and the Standard for Verification and Validation in Computational Fluid Dynamics and Heat Transfer in 2009 [3]. These texts provide a solid foundation for the study of validation methods and many of the more general aspects of these frameworks are transferable to problems in structural dynamics. However, validation of nonlinear dynamical models presents additional challenges that are yet to be fully addressed in the research literature. An issue of particular interest is how to account for potential bifurcations in the frequency response of a non-linear system.

This paper presents some of the precursor analyses required as part of a broader framework for V&V of nonlinear systems. The layout of the paper is as follows. The experimental rig adopted for the study is introduced in Sect. 7.2. Sections 7.3 and 7.4 introduce tools for system identification and analysis and present results of their application to the bookshelf structure. Discussions of results and directions for future work are given in Sect. 7.5.

K. Worden (✉) • O.D. Tiboaca • I. Antoniadou • R.J. Barthorpe
Dynamics Research Group, Department of Mechanical Engineering, University of Sheffield, Mappin Street, Sheffield S1 3JD, UK
e-mail: k.worden@sheffield.ac.uk

M. Mains (ed.), *Topics in Modal Analysis, Volume 10*, Conference Proceedings of the Society for Experimental Mechanics Series, DOI 10.1007/978-3-319-15251-6_7

7.2 The Experimental Rig and Data Capture

For the purposes of the present work, a small-scale simulated shear building model was constructed. This was chosen to correspond closely with a structure previously designed and built at Los Alamos National Laboratories (LANL). Within LANL, the experimental rig was referred to informally as the 'bookshelf' rig and this informal nomenclature is also adopted here. The bookshelf structure, illustrated in Fig. 7.1 has four levels, floors or shelves, with the lower level being considered the base. Each shelf/floor is composed of a substantial rectilinear aluminium block with a mass of 5.2 kg and dimensions $35 \times 25.5x \times 0.5$ cm $(L \times w \times h)$. The shelves are joined by upright beams at each corner; each beam having a mass of 238 g and dimensions $55.5 \times 2.5 \times 0.6$ cm. The blocks used to connect the main plates and the upright beams have a mass of 18 g and dimensions $2.5 \times 2.5 \times 1.3$ cm. For each block, four bolts were used, each of Viraj A2-70 grade and with a mass of 10 g. The structure was mounted on a rail system which was securely clamped onto a substantial testing table. In order to introduce the excitation into the structure, an electrodynamic shaker with a force transducer was connected to the base. The output response of the structure was recorded using four accelerometers positioned, as shown in the pictures, at the middle of each main plate.

So far, the system is linear, and the intention of the first identification tests was that it remain so in order to simplify the identification problem. However, as the longer term goal of this work was to consider nonlinear systems, provision was made for nonlinearity. In the original LANL structure, nonlinearity was introduced via a comparatively complex series of bumpers connected between shelves; in the current structure, nonlinearity is introduced in a much more basic fashion in order to impose simpler equations of motion. The nonlinearity here will be introduced via a vertical cantilever beam which contacts (via a bolt) the topmost shelf of the structure at negative displacements of that shelf.

The experimental data were acquired using an LMS CADA system connected to a SCADAS-3 interface. A total of 93,184 points per channel were recorded at a sampling frequency of 1024 Hz. Lateral accelerations were recorded for each shelf from piezoelectric accelerometers fixed to the edges. 'FRFs' (transmissibilities) between the relative accelerations of the floor and the base acceleration are given in Fig. 7.2; the structures shown indicate that a three-DOF model of the rig is likely to capture the main dynamics.

Fig. 7.1 The 'bookshelf' experimental rig showing shaker attachment and guide rail system

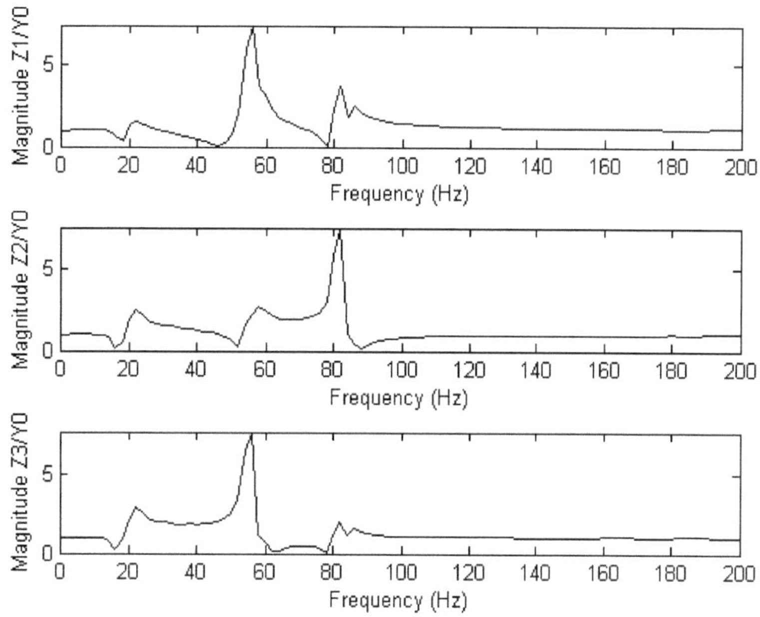

Fig. 7.2 'FRFs' between the base acceleration of the rig and the relative accelerations of the upper floors

7.3 System Identification Using Self-Adaptive Differential Evolution (SADE)

For the sake of completeness, a brief overview of the basic Differential Evolution (DE) and SADE algorithms will be given here; for more detail, the reader is referred to the original papers [11, 12]. As in all evolutionary optimisation procedures, a population of possible solutions (here, the vector of parameter estimates), is iterated in such a way that succeeding generations of the population contain better solutions to the problem in accordance with the Darwinian principle of survival of the fittest. The problem is framed here as a minimisation problem with the cost function defined as a normalised mean-square error between the measured data and that predicted using a given parameter estimate. Having established by FRF analysis that the base-excited system appears to correspond well to a three-DOF system, the model equations considered were,

$$\begin{aligned} m_1\ddot{z}_1 + c_1\dot{z}_1 + c_2(\dot{z}_1 - \dot{z}_2) + k_1 z_1 + k_2(z_1 - z_2) &= -m_*\ddot{y}_0 \\ m_2\ddot{z}_2 + c_2(\dot{z}_2 - \dot{z}_1) + c_3(\dot{z}_2 - \dot{z}_3) + k_2(z_2 - z_1) + k_3(z_2 - z_3) &= -m_*\ddot{y}_0 \\ m_3\ddot{z}_3 + c_3(\dot{z}_3 - \dot{z}_3) + k_3(z_3 - z_2) &= -m_*\ddot{y}_0 \end{aligned} \tag{7.1}$$

where the $\{z_i = y_i - y_0 \ : \ i = 1, \dots, 3\}$ are displacement coordinates relative to the base displacement. As discussed above, the rig was operated in its linear condition in order to acquire data for the identification. As it is not clear what the actual masses are prior to the identification, an estimate m_* is used for the RHS of the equations. The estimate is based on the physical masses of the shelves and associated fixings, Including m_1, m_2 and m_3 in the parameter vector $\underline{\theta} = (m_1, m_2, m_3, c_1, c_2, c_3, k_1, k_2, k_3)$ allows the identification algorithm to correct for the contribution of the vertical beams etc. Based on the design geometry and materials, m_* was taken here as 5.475 kg.

The cost function referred to above was defined here in terms of the prediction errors associated with each DOF. A set of Normalised Mean-Square-Errors (NMSEs) J_i were defined by,

$$J_i(\underline{\theta}) = \frac{100}{N\sigma^2_{\ddot{z}_i}} \sum_{i=1}^{N} (\ddot{z}_i - \hat{\ddot{z}}_i(\underline{\theta}))^2 \tag{7.2}$$

where $\sigma^2_{\ddot{z}_i}$ is the variance of the measured sequence of relative accelerations and the caret denotes a predicted quantity; N is the number of 'training' points used for identification. The total cost function J was then taken as the average of the J_i. Previous experience has shown that a cost value of less than 5.0 represents a good set of model predictions (or parameter estimates).

In order to generate the predictions $\hat{\ddot{z}}_i$, the coupled Eqs. (7.1) were integrated forward in time in Matlab [8] using a fixed-step fourth-order Runge–Kutta scheme for initial value problems [10]. A better solution could potentially be found by using an adaptive solver like the $(4, 5)$th-order Runge–Kutta method for the solution of non-stiff problems encapsulated in the Matlab function ode45; however, it was shown in [16] that the use of the adaptive scheme in the context of evolutionary system identification can lead to strange results. The excitations for the predictions were established by the measured base accelerations $\ddot{y}_0$ and the initial estimate m_*. Although a great deal of data were measured in the experiments, the SADE identification scheme is computationally expensive, with the main overhead associated with integrating trial equations forward in time. For this reason, the training set or identification set used here was composed of only $N = 700$ points. To avoid problems associated with transients, the cost function was only evaluated from the final 500 points of each predicted record.

Once the data were generated, the SADE algorithm was applied to the identification problem using a parameter vector $\underline{\theta}$. The standard DE algorithm of reference [11] attempts to transform a randomly generated initial population of parameter vectors into an optimal solution through repeated cycles of evolutionary operations, in this case: mutation, crossover and selection. In order to assess the suitability of a certain solution, the cost function referred to above was used; this casts the identification in the form of a minimisation problem.

Figure 7.3 shows a schematic for the procedure for evolving between populations. The following process is repeated with each vector within the current population being taken as a *target vector*; each of these vectors has an associated cost J defined above. Each target vector is pitted against a *trial vector* in a selection process which results in the vector with lowest cost advancing to the next generation. The process for constructing the trial vector involves variants of the standard evolutionary operators: mutation and crossover.

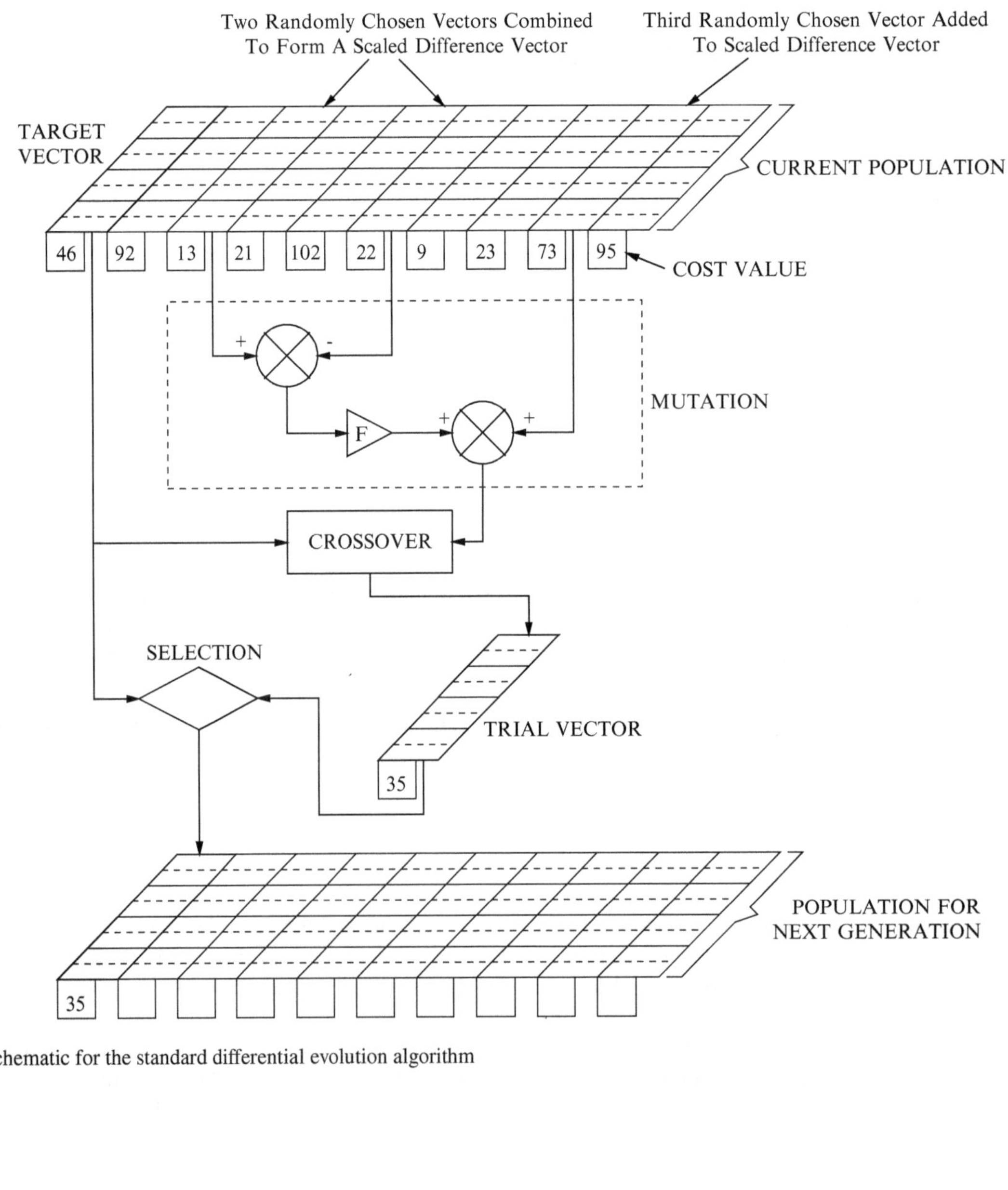

Fig. 7.3 Schematic for the standard differential evolution algorithm

The mutation procedure used in basic DE employs vector differentials. Two vectors A and B are randomly chosen from the current population to form a vector differential $A - B$. A *mutated vector* is then obtained by adding this differential, multiplied by a scaling factor, F, to a further randomly chosen vector C to give the overall expression for the mutated vector: $C + F(A - B)$. F is often found have an optimal value between 0.4 and 1.0.

The *trial vector* is the child of two vectors: the target vector and the mutated vector, and is obtained via a crossover process; in this work uniform crossover is used. Uniform crossover decides which of the two parent vectors contributes to each chromosome of the trial vector by a series of $D - 1$ binomial experiments, where D is the dimension of the problem i.e. the number of parameters in the model. Each experiment is mediated by a crossover probability C_p (where $0 \leq C_p \leq 1$). If a random number generated from the uniform distribution on [0,1] is greater than C_p, the trial vector takes its parameter from the target vector, otherwise the parameter comes from the mutated vector.

This process of evolving through the generations is repeated until the population becomes dominated by only a few low cost solutions, any of which would be suitable. The core of the SADE algorithm is essentially that of DE, the only real difference being that, in SADE, F and C_p are adapted generation by generation in order to arrive at optimal values. Furthermore, SADE can select between different mutation strategies in addition to the basic DE strategy described above in order to optimise its performance; the specific additional strategies applied here are the same as those described in [15].

In this case, the SADE algorithm was initialised with a population of randomly selected parameter vectors or individuals. The parameters were generated using uniform distributions on specified initial ranges. As in all iterative optimisation schemes, the initial estimates can prove critical; here, estimates based on engineering judgement were used. The masses in the model were not considered as a problem, as the inertia of the system was considered very likely to be dominated by the shelves and fixings, it was expected that the true values would be close to the estimate m_* given above. For this reason, a short range $[4.5, 6.0]$ was taken for the initial population. The situation with the stiffness parameters is a little more complicated as it is not clear what the appropriate boundary conditions are the upright beams connecting the floors. An approximate value of $k_* = 5.2 \times 10^5$ N/m can be obtained by assuming encastre conditions; however, the true value may vary substantially from k_* if the bolts do not impose a true fixed condition, for example. Because of the uncertainty in the initial physical estimates of the stiffness, the initial ranges for SADE were taken on roughly an order of magnitude below and above k_* i.e. $[5 \times 10^4, 5 \times 10^6]$. Taking into account the values of m_* and k_* and assuming damping ratios in the vicinity of 0.1 % for aluminium, the initial ranges for the damping parameters were estimated at $[0.1, 10.0]$, again an order or magnitude below and above a nominal value of $c_* = 1.0$ N s/m.

A population of 100 individuals was chosen for the SADE runs with a maximum number of generations of 200. In order to sample different random initial conditions for the DE algorithm, ten independent runs were made. The other parameters chosen used for SADE were a starting value for F of 0.9 and a starting value for C_p of 0.5 (these values proved to be effective in a number of previous studies; this completes the specification of SADE for the problem of this paper.

Each of the ten runs of the DE algorithm converged to a good solution to the problem in the sense that cost function values of around 2 % or below were obtained in all cases; the summary results are given in Table 7.1. The best solution gave a cost function value of 1.591. A visual comparison of the 'true' experimental responses and predicted responses for the best parameter set is given in Fig. 7.4. This comparison is based upon a set of testing data that was distinct from the training data used to fit the parameters.

The results are interesting. Although there are only small variations in the prediction errors, the coefficients of variation (standard deviation of estimate/mean of estimate) are quite high for a number of the parameters. This indicates that the errors are rather insensitive to the parameters. One of the questions usually investigated in the validation of models is this

Table 7.1 Parameter estimates from ten independent SADE runs

Parameter	Best	Maximum	Minimum	Mean	Standard deviation	Coefficient of variation
m_1	5.144	5.228	5.023	5.109	0.054	0.011
m_2	5.587	5.813	5.558	5.656	0.076	0.013
m_3	5.264	5.456	5.264	5, 348	0.068	0.013
c_1	6.681	10.00	4.725	5.918	3.290	0.556
c_2	0.476	0.973	0.100	0.312	0.304	0.975
c_3	3.411	4.045	0.100	1.423	1.493	1.049
k_1 ($\times 10^{-5}$)	1.156	1.219	0.500	0.856	0.333	0.390
k_2 ($\times 10^{-5}$)	5.874	6.920	5.333	6.113	0.429	0.070
k_3 ($\times 10^{-5}$)	3.126	4.374	0.500	2.661	1.095	0.411
J	1.591	2.021	1.591	1.773	0.156	0.088

Fig. 7.4 SADE model predictions on testing data: (**a**) $\ddot{z}_1$, (**b**) $\ddot{z}_2$, (**c**) $\ddot{z}_3$

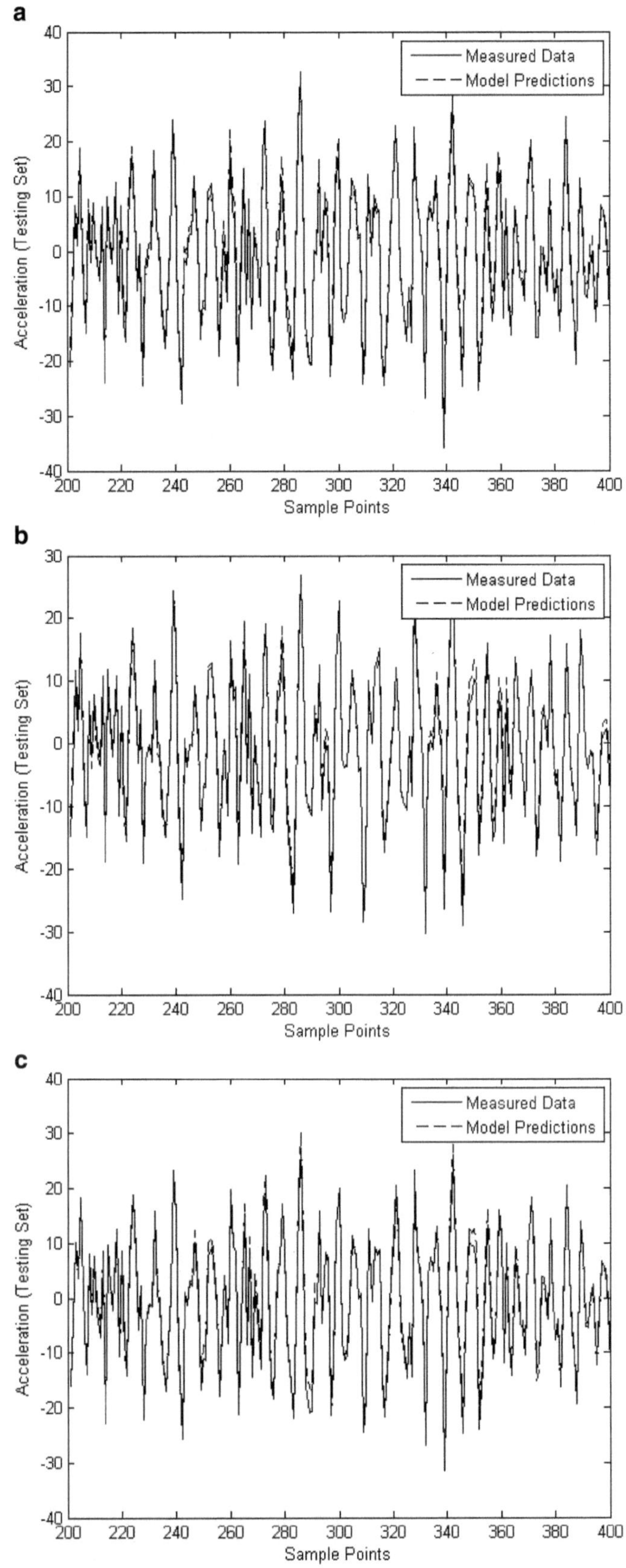

question of sensitivity; how is the uncertainty in an outcome (prediction error) related to the uncertainty in the inputs (model parameter estimates). This is considered in more detail in the following section, where a Bayesian sensitivity analysis is applied. The insensitivity of the errors in this case is most likely related to the fact that only a small training set is used; this would normally be a simple matter to deal with at the expense of a little more computational effort; however, this is not addressed here as the objective of the paper is to look at issues associated with sensitivity and fidelity and the current situation allows interesting discussion.

7.4 Bayesian Sensitivity Analysis

The Sensitivity Analysis (SA) technique used here is based on a non-parametric probabilistic approach as detailed in [9]. Each uncertain input parameter is represented as a probability distribution, and a Gaussian process emulator (fast running model or surrogate model) is fitted using multiple runs of the model as dictated by a Design-of-Experiments (DOE). From this emulator, statistical quantities relating to sensitivity and uncertainty can be inferred directly; for example, output uncertainty distributions and main effects. Importantly, this requires no additional runs of the original model for each sensitivity measure, unlike conventional SA methods detailed for example in [14]. Furthermore, the advantage of using Gaussian process regression is that the uncertainty of the emulator fit is itself quantified, giving the analyst a very pragmatic quantification of the uncertainty in the data. A very brief summary of the main points of the algorithm will be given, the reader is referred to the original papers for more details.

7.4.1 Gaussian Process Regression

Any computer model can be regarded as a function of its inputs: $f(\underline{x})$. Although this function is deterministic and governed by known mathematical relationships, it is often of such complexity as to be considered mathematically intractable. From a practical point of view, $f(\underline{x})$ could be regarded as an unknown function, given that one does not know the output for a given set of inputs until one has actually run the model. If, however, one samples the function (model) at a number of carefully chosen input points, it is possible to fit a response surface that can predict the output of the model for any point in input space without having to run the simulation itself. Although the idea of modelling a model (metamodelling) may seem a little abstract, for simulations that are computationally expensive it is a useful tool, since any approach for uncertainty analysis requires multiple runs of the model under investigation.

A particular approach to formulating the metamodel or emulator that has gathered interest in recent years is the use of Gaussian process regression [6, 7, 15]. Gaussian processes are an extension of a multivariate Gaussian probability distribution. Whereas most forms of regression return a crisp value $f(\underline{x})$ for any given $\underline{x}$, a Gaussian process returns a Gaussian probability distribution. Thus for a function, the Gaussian process can be considered as a multivariate Gaussian distribution, where the dimension of the multivariate distribution can be thought of as the resolution of the function, or the number of predictive points. In the case detailed here, the resolution need not be specified, since one is not interested in predicting particular output points, but rather quantities pertaining to the whole range of output space.

Gaussian processes adhere to the Bayesian paradigm, that is, a number of prior assumptions are made about the function being modelled, and then training data (samples from the model) are used to update and evaluate a posterior distribution over functions. A key assumption is that the model is a *smooth* function of its inputs; it is this that allows extra information concerning the response to be gained at reduced computational cost. No mathematical details will be given here: the reader is referred to [6, 7, 13]. The important point for the current context is that a model is formed which can estimate accurately what the cost function J is corresponding to a given set of parameter estimates $\underline{\theta}$ *without* integrating the equations of motion forward in time.

The dependence of the emulator on training data means that some model runs are always required. The advantage of the Bayesian sensitivity approach is that, typically far fewer runs are needed to train the emulator than would be needed for a full Monte Carlo sensitivity analysis. To deal with the sampling of training data in as principled manner as possible, ideas of experimental design are applied and a maximin Latin hypercube design (maximin LHD) generated using the GEM-SA software [5] is used here.

7.4.2 Inference for Sensitivity Analysis

Several quantities can be inferred from the GP regression model described above, that are relevant to sensitivity analysis. Fundamental quantities such as the mean and variance of the output distribution can be evaluated, as well as main effects, interactions and sensitivity measures for input parameters based on their contribution to output variance [ref].

The results presented here will be for variance and sensitivity indices. Variance-based methods are widely used in sensitivity analysis. This involves quantifying the proportion of output variance for which individual input parameters are responsible. In particular, sensitivity can be measured by conditional variance,

$$V_i = \mathrm{Var}\{E(J\,|\theta_i)\} \tag{7.3}$$

which is the expected value of the contribution of the input variable θ_i to the output variance i.e. the uncertainty in J. (Note that this is also commonly known as the main effect index (MEI)). This can be extended to measure conditional variance of interactions of inputs, i.e. the output variance resulting from co-varying of two parameters θ_i and θ_j, and so on for higher order interactions. Although this approach allows detailed insight into the effects of combinations of inputs on output uncertainty, it can be time-consuming to examine all possible interaction permutations for models with many input dimensions. An alternative sensitivity measure [4], describes the output variance that would remain if one were to learn the true values of all inputs except θ_i,

$$V_{Ti} = \mathrm{Var}(J) - \mathrm{Var}\{E(J\,|\underline{\theta}_{-i})\} \tag{7.4}$$

where the notation $\underline{\theta}_{-i}$ means the vector of all parameters *except* θ_i. This measure, called the total sensitivity index (TSI), measures the variance caused by an input θ_i and *any* interaction of any order including θ_i. It allows a more holistic view of the uncertainty attributed to each input, but does not give any details as to how it is distributed between main effects and interactions. Between the MEIs and TSIs, a detailed view of the sensitivities of inputs and their interactions can be gained. The advantage of the GP regression emulator is that the functional form of the emulator allows the variance and sensitivity measures to be evaluated as closed form integrals.

7.4.3 Results

As mentioned above the analysis carried out here used the code GEM-SA [5]. In total 200 design points for the Gaussian process emulator were generated as a Latin Hypercube Maximin design. The input parameter distributions were taken as uniform over the ranges assuming for the initial populations used for the SADE analysis. Following the regression stage, the variances and TSIs were computed and the results are given in Table 7.2.

It appears that the main contributors to output variance are stiffnesses, note by far the biggest proportion of output uncertainty comes from interactions; this makes sense because the things that determine the physics of the response i.e. natural frequencies and damping ratios are actually combinations of the basic physical parameters: masses, dampings and stiffnesses.

Table 7.2 Results of Bayesian sensitivity analysis

Parameter	Main effect index	Total sensitivity index
m_1	1.66	14.49
m_2	1.59	25.34
m_3	0.11	0.11
c_1	0.14	0.15
c_2	0.80	20.49
c_3	0.46	6.80
k_1	15.32	52.04
k_2	12.46	58.98
k_3	2.59	35.78
Interactions	69.83	–

7.5 Discussion and Conclusions

The contribution of this study is to demonstrate two of the tools required for model development and validation in structural dynamics. The focus of the SADE system identification exercise detailed in Sect. 7.3 is on improving fidelity-to-data through test-analysis correlation. The method was applied here to the parameters of a linear model using data acquired from the experimental structure, with the results indicating that it is possible to identify a physically feasible set of parameters with a good level of consistency between independent executions of the algorithm. The prediction errors of the resulting model were in this case evaluated using testing data distinct from the training data, although drawn from the same experiment. The derived model was found to perform well in this case although—as expected—the error metric was observed to be higher for the testing data than for the training data. The next step for this work is to evaluate the validity of the derived model form for more challenging scenarios, discussed below.

The second part of the study focused on developing a better understanding the model through application of sensitivity analysis. In this case a global method for sensitivity analysis was adopted based upon Bayesian inference. A particular feature of the structure under study is that a major component of its dynamic behaviour is driven by the interaction between it's physical parameters (masses, dampings and stiffnesses), which is effectively revealed by the global sensitivity analysis applied here. In one sense the application of SA methods allows an evaluation of the robustness-to-uncertainty of the model as it establishes what the expected impact of variability on the model parameters would be on the model predictions. It could be argued that such an analysis would be more usefully used as a precursor to a calibration study as it would allow those parameters and interactions to which the predictions are most sensitive to be identified. However, this would be contingent upon feasible ranges for the parameters having already been specified. In this study those ranges arose from the system identification step.

Ultimately the study serves as a necessary step towards developing validation methods for nonlinear systems using the introduced experimental testbed. The next stage for the work is to complete the specification of the validation exercise for the structure in its linear state, then explicitly address the complications that arise once nonlinearity is introduced. Completing the validation exercise will involve specifying a criteria for accepting the model as 'valid', then evaluating the model performance against this criteria for a series of more demanding validation experiments, for example under structural modification. For the nonlinear system, there is additional interest in investigating whether validating a model based upon its ability to accurately predict time domain responses is a sufficient basis for ascribing credibility to its ability to predict responses in the frequency domain.

References

1. AIAA (1998) Guide for the verification and validation of computational fluid dynamics simulations (G-077-1998). American Institute of Aeronautics and Astronautics
2. ASME (2006) V&V 10 Guide for verification and validation in computational solid mechanics. American Society of Mechanical Engineers
3. ASME (2009) V&V 20 Standard for verification and validation in computational fluid dynamics and heat transfer. American Society of Mechanical Engineers
4. Homma T, Saltelli AK (1996) Importance measures in global sensitivity analysis of model outputs. Reliab Eng Syst Saf 52:1–17
5. Kennedy MC Gem-SA Homepage. Available from: http://ctcd.group.shef.ac.uk/gem.html
6. Kennedy MC, O'Hagan A (2001) Bayesian calibration of computer models. J R Stat Soc Ser B 63:425–464
7. Kennedy MC, Anderson CW, Conti S, OHagan A (2006) Case studies in Gaussian process modelling of computer codes. Reliab Eng Syst Saf 91:1301–1309
8. Matlab v7 (2004) The Mathworks
9. Oakley JE, O'Hagan A (2004) Probabilistic sensitivity analysis of complex models: a Bayesian approach. J R Stat Soc Ser B 66:751–769
10. Press WH, Teukolsky SA, Vetterling WT, Flannery BP (2007) Numerical recipes: the art of scientific computing, 3rd edn. Cambridge University Press
11. Price K, Storn R (1997) Differential evolution—a simple and efficient heuristic for global optimization over continuous spaces. J Glob Optim 11:341–359
12. Qin AK, Suganthan PN (2005) Self-adaptive differential evolution algorithm for numerical optimization. In: Proceedings of IEEE congress on evolutionary computation (CEC 2005), Edinburgh, Scotland
13. Sacks J, Welch WJ, Mitchell TJ, Wynn HP (1989) Design and analysis of computer experiments. Stat Sci 4:409–435
14. Saltelli AK, Chan A, Scott EM (2000) Sensitivity analysis. Wiley, New York
15. Worden K, Becker WE (2012) On the identification of hysteretic systems, part II: Bayesian sensitivity analysis and parameter confidence. Mech Syst Signal Process 29:213–227
16. Worden K, Manson G (2012) On the identification of hysteretic systems, part I: fitness landscapes and evolutionary identification. Mech Syst Signal Process 29:201–212

Chapter 8
Some Non-conventional Boundary Conditions (From Marshmallows to Plungers: Who Would Have Guessed)

Tina Dardeno, Patrick Logan, and Peter Avitabile

Abstract When performing a modal test, a number of mechanisms are available to approximate the free-free boundary condition. Some test setups may be more difficult to implement than others, but ultimately the effect of the boundary condition on the actual modes of the system are of concern. An expensive setup may not be any better than an equivalent, inexpensive "Rube Goldberg" approach. In lieu of an expensive setup, a low cost, easily deployed configuration may be improvised from common objects not native to the test lab.

In this paper, several different boundary conditions are employed with two being very, very unconventional. The effects of the supports on the flexible modes are explored through testing with multiple support configurations. Natural frequencies and mode shapes of the modes (flexible and rigid body modes) are presented. Drive point frequency response functions for the different support configurations are compared. Some other considerations are discussed relative to the general items of concern when performing these types of free-free modal tests.

Keywords Modal testing techniques • Boundary conditions effects

8.1 Introduction

Often times, there is a need to test a structure in a free-free configuration. This is usually one of the more simple boundary conditions to achieve in a test lab. Typically a truly built-in condition is very hard to achieve due to the large, massive seismic type anchor that is needed. Free-free conditions are also used many times when correlating the test results with a finite element model; again this is a very easy boundary condition to simulate.

Now it is very easy to say "free-free" but many times how this is achieved in a practical lab setting may actually have an effect on the flexible modes of the system as well as the rigid body modes themselves [1]. There have been many, many different approaches to achieve this free-free condition and some have been very expensive. But the bottom line is how do the test boundary conditions affect the measured frequency, damping and mode shapes for the structure under test – that is the most critical question to answer [2, 3].

In some work involving correlation of the test data to the finite element model, analysts will often just use the analytical truly free condition. While this is very easy to do analytically, the practical reality is that the structure under test will actually have some type of spring support system which really needs to be modeled in order to be compared to the actual test data which has an actual "real" support system. If this is ignored then there can be improper characterization of the analytical frequencies and mode shapes identified from the model.

In order to illustrate some of the problems with the boundary conditions, three structures are evaluated. One is a simple frame type structure with very closely spaced frequencies for the first bending and first torsion of the frame which is a very common test scenario seen in many structures. The second structure is a cantilevered plate attached to a larger mass used to mimic some turbine blade qualification tests. The third structure is a shock response plate fixture. All of these structures use some very non-traditional test support systems as described in each test case.

T. Dardeno (✉) • P. Logan • P. Avitabile
Structural Dynamics and Acoustic Systems Laboratory, University of Massachusetts Lowell,
One University Avenue, Lowell, MA 01854, USA
e-mail: tina_dardeno@student.uml.edu

M. Mains (ed.), *Topics in Modal Analysis, Volume 10*, Conference Proceedings of the Society for Experimental Mechanics Series, DOI 10.1007/978-3-319-15251-6_8

8.2 Frame Structure with Marshmallow Supports

The dynamic characterization of the small aluminum frame was determined with the frame placed on four large marshmallows and again with the frame placed on ten small marshmallows. The marshmallows provide a reasonably good free-free condition but there is some effect on the type and location of the marshmallows used. Data gathering was performed with the frame located on a seismic anchor (Fig. 8.1).

For the four large marshmallows test, the marshmallows were placed in the center of each side of the frame, as shown in Fig. 8.1. For the ten small marshmallows test, four marshmallows were placed beneath the long end of the frame and one marshmallow was placed beneath the short end of the frame as shown in Fig. 8.1.

Two teardrop accelerometers were placed on the frame: one beneath point one and another beneath point seven. A modally tuned impact hammer with a white plastic tip was used to excite the frame at 16 locations. All measurements and impacts were performed in the +Z direction (Fig. 8.2).

Nine flexible modes were evaluated over a bandwidth of 2,000 Hz using Photon software via LDS Dactron. Data was then processed in LMS (Fig. 8.3).

Figure 8.3 shows that the first and second modes were swapped depending on the boundary conditions. Specifically, the first and second modes for the four large marshmallow test were bending and torsional modes, respectively. Conversely, for the ten small marshmallow test, the first mode was a torsional mode and the second was a bending mode.

The natural frequencies and damping of the two tests are compared in Table 8.1.

Table 8.1 shows that the natural frequencies associated with the flexible modes remained relatively constant with both boundary conditions. The frequency difference between the first and second modes was greater for the ten small marshmallows test (1.72 Hz) compared to the four large marshmallows test (0.275 Hz), indicating greater overlap of modes for the four large marshmallows test. Furthermore, the large marshmallow test produced frequencies slightly lower than those obtained with the small marshmallow test. This frequency shift implies that the small marshmallow configuration produced greater stiffness than the large marshmallows. However, the large marshmallows produced greater damping for all but the second and fifth modes compared to the small marshmallow configuration.

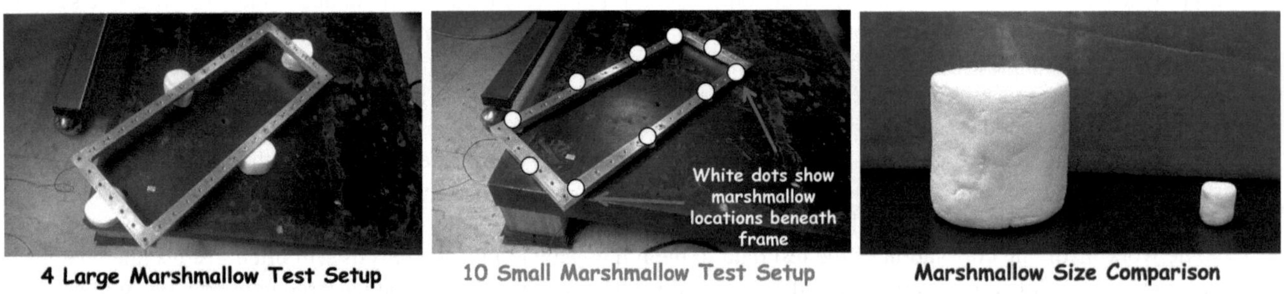

Fig. 8.1 Testing configuration for aluminum frame on four large marshmallows and on ten small marshmallows

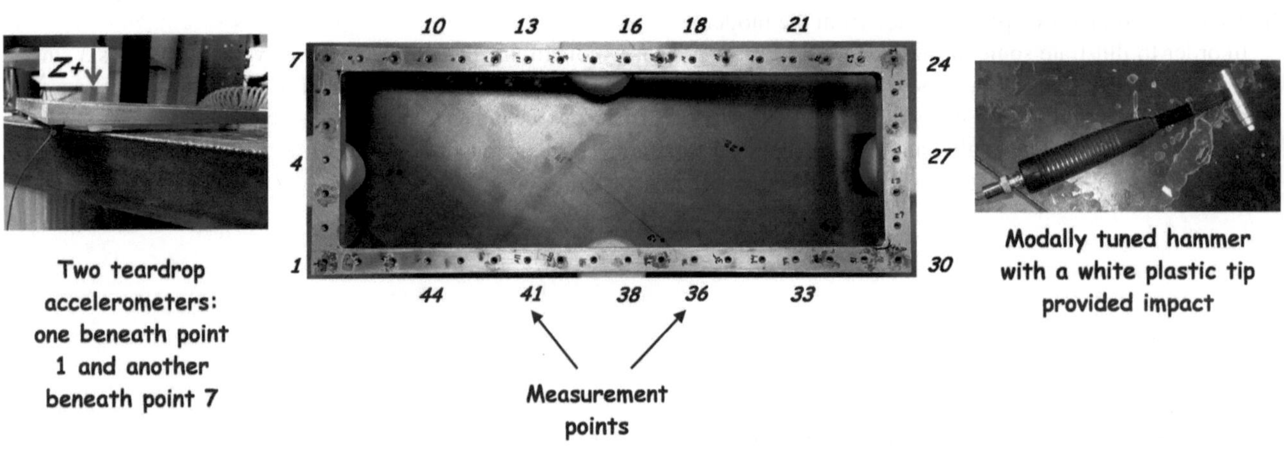

Fig. 8.2 Measurement points and instrumentation for four large marshmallow and ten small marshmallow tests

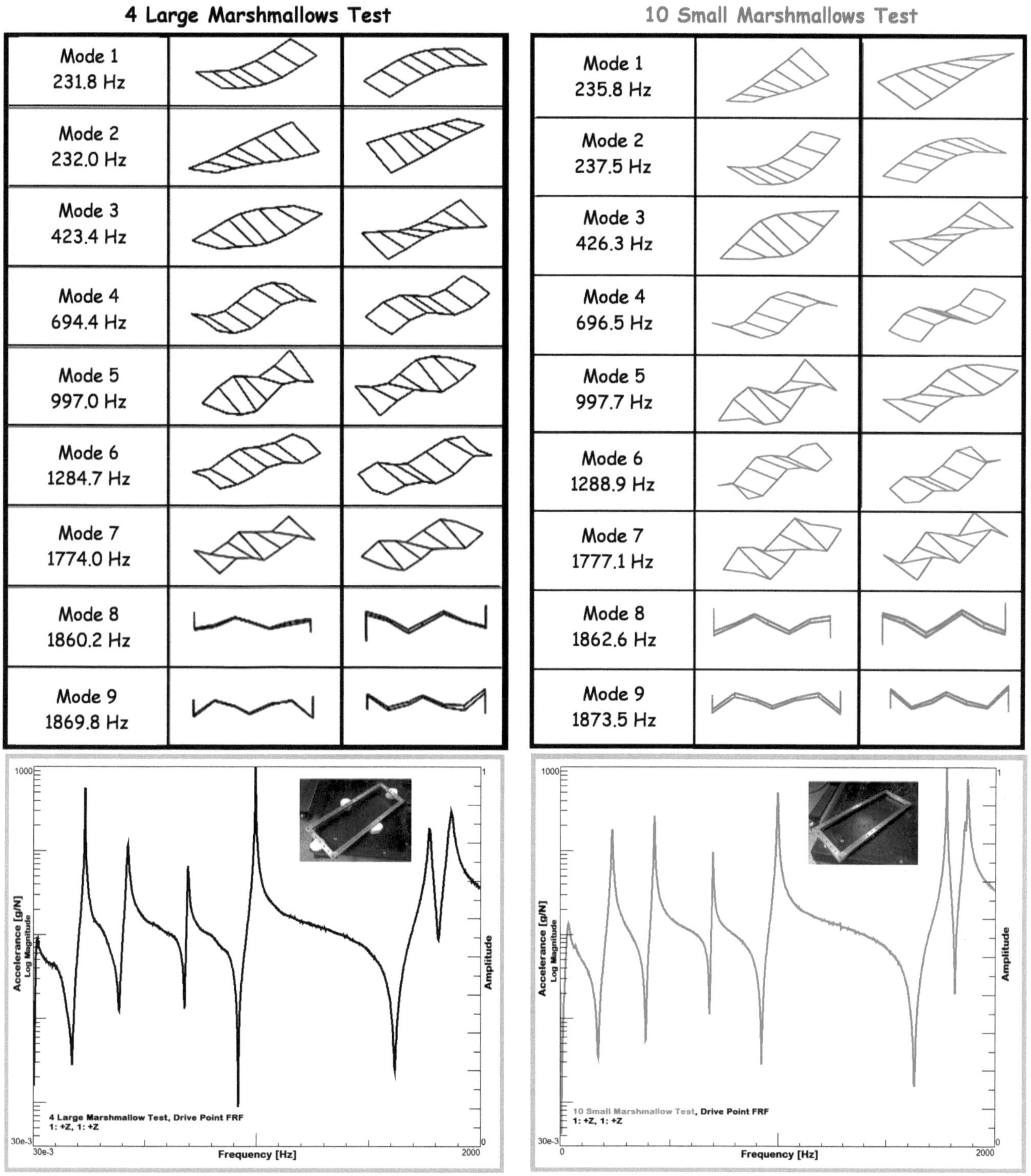

Fig. 8.3 Results of four large marshmallow test and ten small marshmallow test

From Fig. 8.3, the large marshmallows were located at the node points of the first torsion mode. Different marshmallow locations were selected to create a boundary condition with more of an effect for the first torsion mode. The experiment was repeated with the large marshmallows located at the corners of the frame (Fig. 8.4). The same equipment and testing parameters were used.

Table 8.1 Natural frequencies and damping for four large marshmallow test and ten small marshmallow test

	Frequency (Hz)			Damping (% critical)		
Mode	Four large marshmallows	Ten small marshmallows	% difference	Four large marshmallows	Ten small marshmallows	% difference
1	231.766	235.784	1.72 %	2.31	1.14	67.83 %
2	232.041	237.502	2.33 %	0.16	1.12	150.00 %
3	423.397	426.309	0.69 %	0.96	0.43	76.26 %
4	694.349	696.520	0.31 %	0.38	0.29	26.87 %
5	996.980	997.680	0.07 %	0.07	0.27	117.65 %
6	1,284.748	1,288.930	0.32 %	0.57	0.11	135.29 %
7	1,774.023	1,777.070	0.17 %	0.33	0.05	147.37 %
8	1,860.231	1,862.640	0.13 %	0.37	0.18	69.09 %
9	1,869.836	1,873.460	0.19 %	0.42	0.13	105.45 %

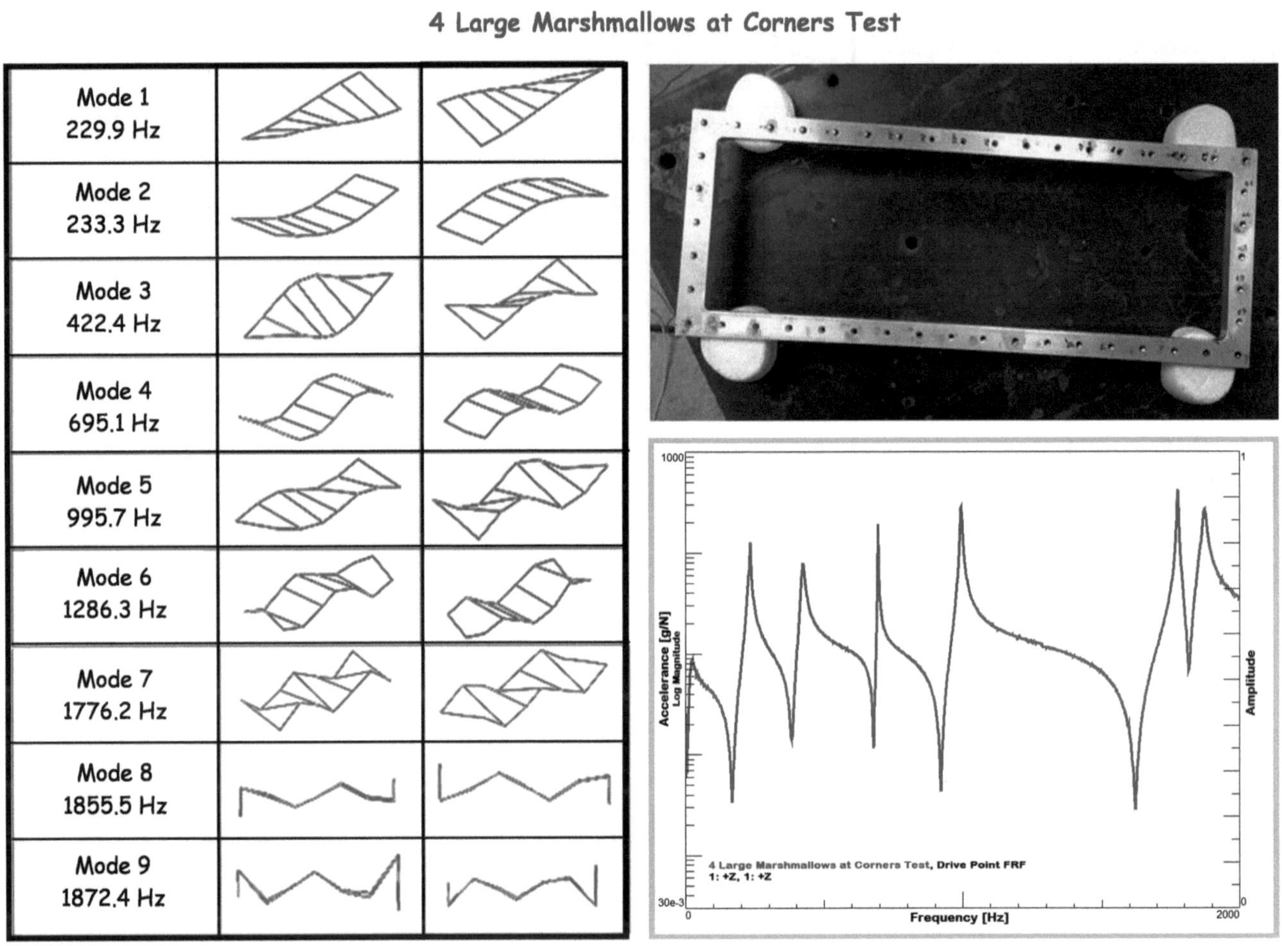

Fig. 8.4 Results of four large marshmallows at corners test

Figure 8.4 shows that moving the four large marshmallows to the corners of the frame resulted in the first torsional mode occurring at a lower frequency than the first bending mode, as was seen with the ten small marshmallows test. Figure 8.5 compares the first two modes of all three tests.

The natural frequencies and damping of the four large marshmallows at the corners test are compared to that of the ten small marshmallows test in Table 8.2. Compared to the ten small marshmallows test, the four large marshmallows at the corners test produced lower frequencies. However, the large marshmallows at the corners test produced greater damping for all but the second and fifth modes compared to the small marshmallow configuration.

The natural frequencies and damping of the four large marshmallows at the corners test are compared to that of the original four large marshmallows test in Table 8.3. Compared to the original large marshmallows test, the four large marshmallows

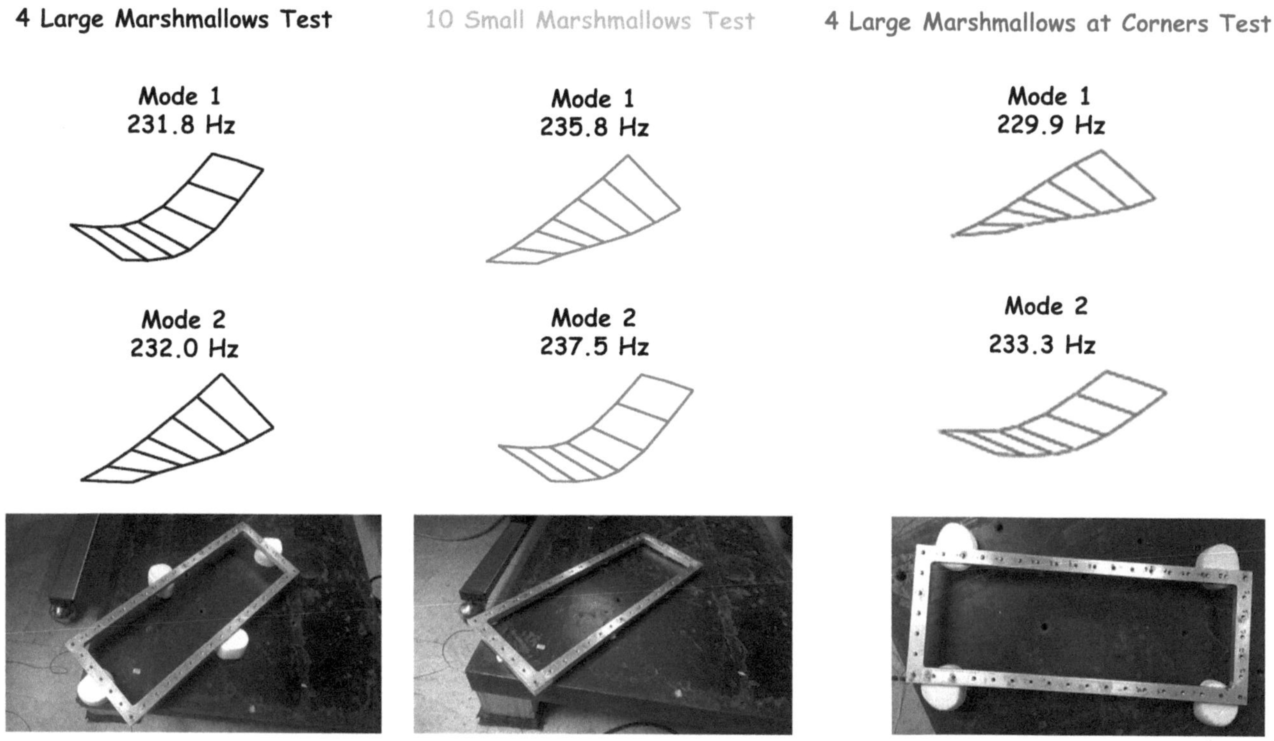

Fig. 8.5 Results comparison for different marshmallow boundary conditions

Table 8.2 Natural frequencies and damping for four large marshmallows at corners test and ten small marshmallows test

	Frequency (Hz)			Damping		
Mode	Four large marshmallows at corners	Ten small marshmallows	% difference	Four large marshmallows at corners	Ten small marshmallows	% difference
1	229.910	235.784	2.52 %	3.11	1.14	92.71 %
2	233.318	237.502	1.78 %	0.76	1.12	38.30 %
3	422.352	426.309	0.93 %	1.55	0.43	113.13 %
4	695.133	696.520	0.20 %	0.15	0.29	63.64 %
5	995.690	997.680	0.20 %	0.35	0.27	25.81 %
6	1,286.290	1,288.930	0.21 %	0.51	0.11	129.03 %
7	1,776.203	1,777.070	0.05 %	0.15	0.05	100.00 %
8	1,855.454	1,862.640	0.39 %	0.60	0.18	107.69 %
9	1,872.439	1,873.460	0.05 %	0.38	0.13	98.04 %

Table 8.3 Natural frequencies and damping for four large marshmallows at corners test and four large marshmallows test

	Frequency (Hz)			Damping		
Mode	Four large marshmallows at corners	Four large marshmallows	% difference	Four large marshmallows at corners	Four large marshmallows	% difference
1	229.910	231.766	0.80 %	3.11	2.31	29.52 %
2	233.318	232.041	0.55 %	0.76	0.16	130.43 %
3	422.352	423.397	0.25 %	1.55	0.96	47.01 %
4	695.133	694.349	0.11 %	0.15	0.38	86.79 %
5	995.690	996.980	0.13 %	0.35	0.07	133.33 %
6	1,286.290	1,284.748	0.12 %	0.51	0.57	11.11 %
7	1,776.203	1,774.023	0.12 %	0.15	0.33	75.00 %
8	1,855.454	1,860.231	0.26 %	0.60	0.37	47.42 %
9	1,872.439	1,869.836	0.14 %	0.38	0.42	10.00 %

Table 8.4 MAC for all marshmallow tests

		4 Large Marshmallows Test									10 Small Marshmallows Test									4 Large Marshmallows at Corners Test								
		1	2	3	4	5	6	7	8	9	1	2	3	4	5	6	7	8	9	1	2	3	4	5	6	7	8	9
4 Large Marshmallows Test	1	100	3.26	0.01	0.33	0.02	9.9	0	0.8	3.19	7.73	92.91	0.14	0.42	0.02	8.92	0.01	0.84	4.17	0.17	97.07	0.03	0.35	0.01	8.49	0.02	0.42	3.04
	2	3.26	100	0.19	0.02	0.35	1.67	0.01	0.05	0.21	92.25	6.18	0.3	0.14	0.14	0.28	0.1	0.01	0.03	96.25	0.02	0.28	0.04	0.22	0.18	0.07	0.08	0.12
	3	0.01	0.19	100	0	0.38	0.14	0.23	0.06	0	0.19	0.17	99.66	0	0.46	0.21	0.06	0.02	0	0.18	0.03	99.74	0	0.38	0.15	0.09	0.01	0
	4	0.33	0.02	0	100	0.13	0.32	0.07	5.55	0.39	0.03	0.35	0	99.8	0.15	0.15	0.01	6.59	0.16	0.04	0.37	0	99.83	0.09	0.36	0.04	6.37	0.27
	5	0.02	0.35	0.38	0.13	100	0.49	0.25	0.04	0.01	0.48	0.03	0.24	0.24	99	0.5	0.13	0.01	0.04	0.48	0.03	0.32	0.11	98.9	0.03	0.37	0.18	0.02
	6	9.9	1.67	0.14	0.32	0.49	100	0.18	4.63	1.21	0.11	9.82	0.33	0.31	0.75	88.59	0.11	0.25	2.28	1.07	9.06	0.16	0.38	0.89	86.9	0.23	0.89	2.2
	7	0	0.01	0.23	0.07	0.25	0.18	100	0.09	0.1	0.01	0.01	0.11	0.05	0.23	0.04	99.61	0.03	0.07	0.03	0	0.19	0.01	0.31	0.28	99.56	0.23	0.1
	8	0.8	0.05	0.06	5.55	0.04	4.63	0.09	100	1.69	0.07	0.8	0.11	4.58	0.03	2.94	0.06	93.88	0.73	0.13	0.82	0.09	4.89	0.01	1.6	0.05	86.83	4.92
	9	3.19	0.21	0	0.39	0.01	1.21	0.1	1.69	100	0.88	2.39	0	0.29	0	0.58	0.01	0.87	97.36	0.23	2.98	0.01	0.4	0	0.6	0.08	16.68	97.07
10 Small Marshmallows Test	1	7.73	92.25	0.19	0.03	0.48	0.11	0.01	0.07	0.88	100	0.05	0.24	0.15	0.03	0.18	0.04	0.06	0.33	93.95	7.11	0.27	0.06	0.06	0.22	0.02	0.1	0.61
	2	92.91	6.18	0.17	0.35	0.03	9.82	0.01	0.8	2.39	0.05	100	0.54	0.51	0.01	8.01	0.07	0.69	3.61	4.92	93.66	0.27	0.37	0.01	7.63	0.06	0.28	2.38
	3	0.14	0.3	99.66	0	0.24	0.33	0.11	0.11	0	0.24	0.54	100	0	0.3	0.46	0.01	0	0	0.28	0.21	99.76	0	0.23	0.33	0.02	0	0
	4	0.42	0.14	0	99.8	0.24	0.31	0.05	4.58	0.29	0.15	0.51	0	100	0.27	0.15	0.01	5.51	0.09	0.15	0.47	0.02	99.87	0.19	0.36	0.03	5.26	0.16
	5	0.02	0.14	0.46	0.15	99	0.75	0.23	0.03	0	0.03	0.01	0.3	0.27	100	0.41	0.11	0	0.01	0.14	0.01	0.38	0.14	99.91	0.02	0.34	0.12	0
	6	8.92	0.28	0.21	0.15	0.5	88.59	0.04	2.94	0.58	0.18	8.01	0.46	0.15	0.41	100	0.02	0.09	0.03	0.16	8.55	0.23	0.21	0.66	97.41	0.11	0.16	0.1
	7	0.01	0.1	0.06	0.01	0.13	0.11	99.61	0.06	0.01	0.04	0.07	0.01	0.01	0.11	0.02	100	0.01	0	0.14	0.01	0.04	0.01	0.16	0.24	99.65	0.22	0.01
	8	0.84	0.01	0.02	6.59	0.01	0.25	0.03	93.88	0.87	0.06	0.69	0	5.51	0	0.09	0.01	100	0.32	0.02	0.85	0	5.79	0	0.37	0	87.12	0.89
	9	4.17	0.03	0	0.16	0.04	2.28	0.07	0.73	97.36	0.33	3.61	0	0.09	0.01	0.03	0	0.32	100	0.02	4.07	0	0.16	0.01	0.04	0.05	10.17	97.33
4 Large Marshmallows at Corners Test	1	0.17	96.25	0.18	0.04	0.48	1.07	0.03	0.13	0.23	93.95	4.92	0.28	0.15	0.14	0.16	0.14	0.02	0.02	100	2.93	0.26	0.04	0.22	0.26	0.09	0.06	0.11
	2	97.07	0.02	0.03	0.37	0.03	9.06	0	0.82	2.98	7.11	93.66	0.21	0.47	0.01	8.55	0.01	0.85	4.07	2.93	100	0.06	0.37	0	8.24	0.01	0.37	2.9
	3	0.03	0.28	99.74	0	0.32	0.16	0.19	0.09	0.01	0.27	0.27	99.76	0.02	0.38	0.23	0.04	0	0	0.26	0.06	100	0.01	0.3	0.16	0.07	0	0
	4	0.35	0.04	0	99.83	0.11	0.38	0.01	4.89	0.4	0.06	0.37	0	99.87	0.14	0.21	0.01	5.79	0.16	0.04	0.37	0.01	100	0.08	0.41	0	5.64	0.25
	5	0.01	0.22	0.38	0.09	98.9	0.89	0.31	0.01	0	0.06	0.01	0.23	0.19	99.91	0.66	0.16	0	0.01	0.22	0	0.3	0.08	100	0.01	0.42	0.1	0
	6	8.49	0.18	0.15	0.36	0.03	86.9	0.28	1.6	0.6	0.22	7.63	0.33	0.36	0.02	97.41	0.24	0.37	0.04	0.26	8.24	0.16	0.41	0.01	100	0.44	0.21	0.08
	7	0.02	0.07	0.09	0.04	0.37	0.23	99.56	0.05	0.08	0.02	0.06	0.02	0.03	0.34	0.11	99.65	0	0.05	0.09	0.01	0.07	0	0.42	0.44	100	0.17	0.11
	8	0.42	0.08	0.01	6.37	0.18	0.89	0.23	86.83	16.68	0.1	0.28	0	5.26	0.12	0.16	0.22	87.12	10.17	0.06	0.37	0	5.64	0.1	0.21	0.17	100	17.99
	9	3.04	0.12	0	0.27	0.02	2.2	0.1	4.92	97.07	0.61	2.38	0	0.16	0	0.1	0.01	0.89	97.33	0.11	2.9	0	0.25	0	0.08	0.11	17.99	100

at the corners test produced similar frequencies. The frequency difference between the first and second modes, however, increased from 0.275 to 3.41 Hz, better aligning with that of the small marshmallows test. The large marshmallows at the corners test produced greater damping at the first, second, third, fifth, and eighth modes. The increased damping at the first torsional mode is expected as the marshmallows were moved from the node line to the corners where maximum displacement occurred. Discrepancies in damping at the other modes are likely due to similar movement of the supports either closer to or further from node lines or differences in data processing. In addition, moving the large marshmallows inward towards the center of the frame produced greater damping. Therefore, differences in how the marshmallows were arranged with respect to the center of the frame may have caused some discrepancy. However, an effort was made to be consistent between the tests. Additional mode comparison for the three tests can be found in Table 8.4.

These three tests show that the boundary condition may have an important effect on the frequencies of the modes as well as the organization and damping of the modes. Therefore, when performing a test on a "free-free" object, one must not only be cautious about the shifting of frequencies but also about the organization of the modes due to the test setup. Overall the marshmallows provided a very useful free-free boundary condition support.

8.3 Upright Cantilever Calibration Structure on Pads

The upright cantilever calibration structure is a plate bolted into a block with a much larger mass; this is similar to a test configuration for small turbine blade qualification tests. The structure is generally tested on padded material to minimize the effects of any incompatibilities between the bottom of the structure and the work surface. Experiments with the calibration structure had previously been performed, but with different types of padding. In attempting to compare results, the question arose as to the effects of the different types of padded material on the flexible modes. The structure is shown in Fig. 8.6.

Experimental mode shapes for the structure, obtained from prior testing at 25 points, are shown in Fig. 8.7aa. A drive point FRF, and input and coherence spectra from point 45 are shown in Fig. 8.7b. For this supplemental investigation, six points were selected for study based on the locations of maximum responsiveness of the various flexible modes.

Impact testing was performed with a mini impact hammer at points 11, 15, 21, 25, 41, 45. The impact hammer was mounted to a tripod (with a Dunkin Donuts® straw) to maintain a very consistent impact location on the structure. In addition,

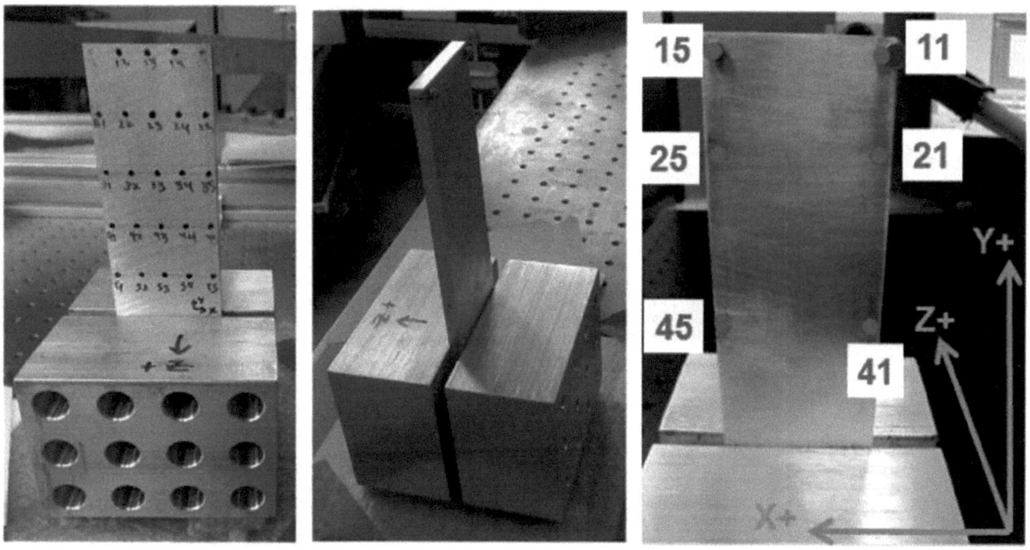

Fig. 8.6 Calibration structure, from the front, side and rear, with impact points and coordinate axes

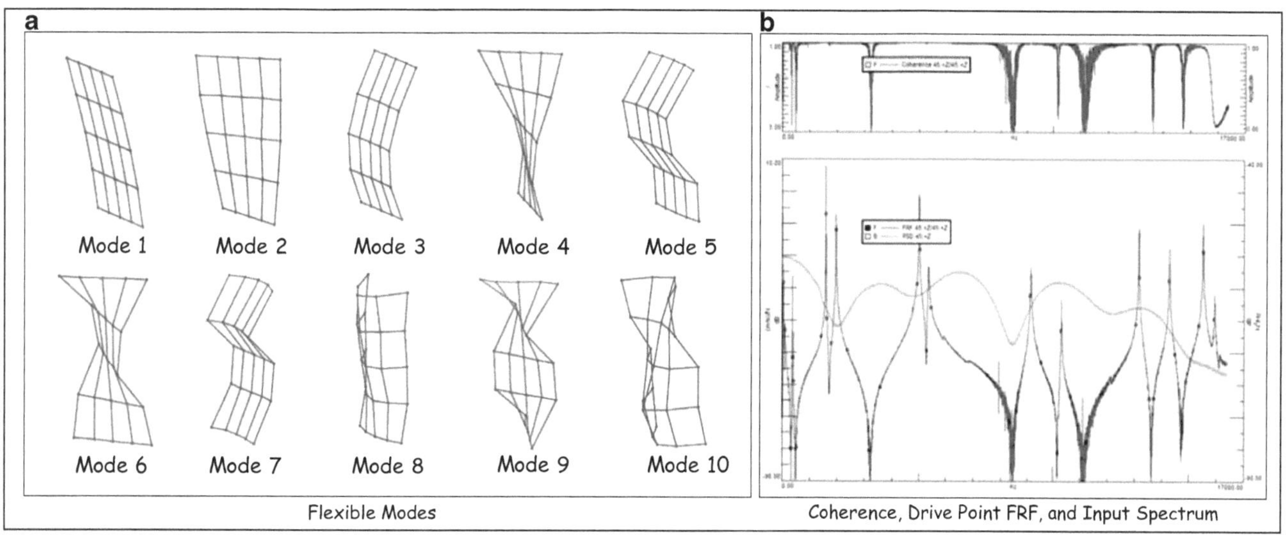

Fig. 8.7 (**a**) Flexible mode shapes of upright cantilever structure obtained from prior testing and (**b**) typical coherence, drive point FRF, and input spectrum obtained during prior testing

impacts at points 11 and 15 were made through ball bearings previously glued to the structure for testing with larger impact hammers; this further controls the accuracy of the impact location. Response was measured at point 11 with a laser. For each point, 10 averages were taken over a 20 kHz bandwidth with a Data Physics Quattro Dynamic Signal Analyzer. Acquisition time was 1.28 s for each measurement. A force-exponential window was employed to reduce the effects of the structure ringing. Data processing was performed in LMS. The impact hammer and laser setup are shown in Fig. 8.8. The structure was tested on 4¾ in. diameter suction cups, bubble wrap and blue packing foam, shown in Fig. 8.9. While bubble wrap and packing foam might routinely be employed, the suction cups were studied as a slightly different mechanism to support the structure. Drive point FRFs and input spectra were consistent with earlier results.

The frequencies and damping of the identified modes are summarized in Table 8.5. The identified flexible mode shapes are shown in Fig. 8.10. The flexible modes from prior testing are shown for reference.

Table 8.5 shows little variation in the natural frequencies and modal damping between the suction cups, bubble wrap and foam for the flexible modes. This implies that this boundary condition does not affect the modes of the structure in a significant way which is largely due to the large attached mass. The large attached mass has the effect of simulating a seismic anchor and therefore the boundary condition plays a much smaller role in affecting the frequencies and mode shapes. But overall the suction cups were determined to be the best of the configurations investigated.

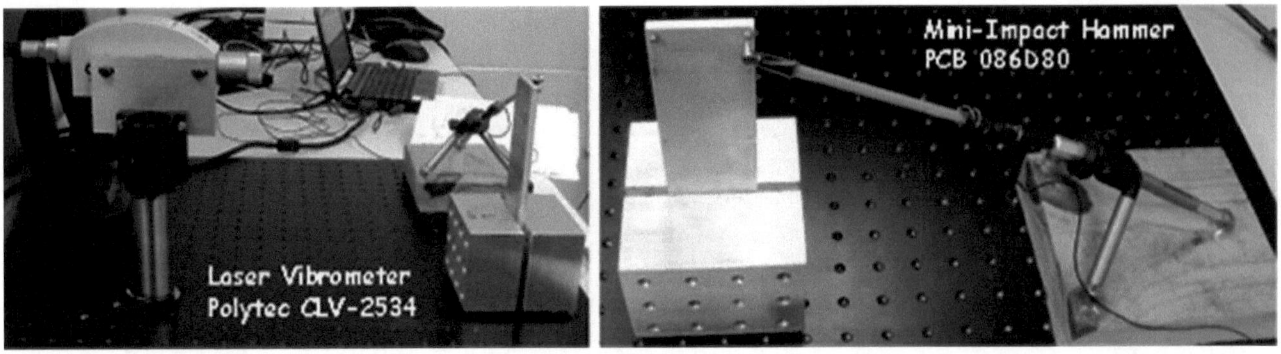

Fig. 8.8 Mini impact hammer and laser setup

Fig. 8.9 Suction cup, bubble wrap and packing foam used as support padding

Table 8.5 Natural frequencies and modal damping for the upright cantilever plate, by padding type

	Natural frequency (Hz)			Damping		
Mode	Suction cups	Bubble wrap	Foam	Suction cups	Bubble wrap	Foam
1	370.0	371.2	369.5	0.41 %	0.18 %	0.21 %
2	1,565.6	1,565.4	1,565.4	0.02 %	0.02 %	0.02 %
3	1,958.2	1,958.5	1,957.9	0.12 %	0.11 %	0.11 %
4	4,924.3	4,923.6	4,923.7	0.06 %	0.06 %	0.06 %
5	5,344.8	5,343.8	5,344.4	0.16 %	0.14 %	0.15 %
6	9,007.4	9,009.7	9,008.8	0.06 %	0.07 %	0.09 %
7	10,277.0	10,275.6	10,276.1	0.05 %	0.04 %	0.05 %
8	12,793.3	12,791.6	12,792.5	0.06 %	0.06 %	0.06 %
9	13,989.2	13,987.1	13,988.6	0.09 %	0.09 %	0.08 %
10	15,248.2	15,245.8	15,245.4	0.07 %	0.06 %	0.04 %

8.4 Shock Response Plate Fixture with Plunger Supports

Supporting larger, heavier structures can sometimes be more difficult. Expensive air ride systems or other complicated support mechanisms have often been deployed. Before mounting a new shock plate on an air piston system, a modal test was necessary for some preliminary shock calculations. A good support mechanism that was economical, easy to set up, and would provide useful results was needed. After some brainstorming, a very different configuration was used – toilet plungers! The dynamic characterization of the aluminum shock response plate was determined with the plate placed on three plungers and again with the plate placed on six plungers (Fig. 8.11). To prevent interference with the accelerometer, for the six plunger test, the plungers were placed at the edges of the plate. For the three plunger test, the plungers were positioned directly beneath measurement points.

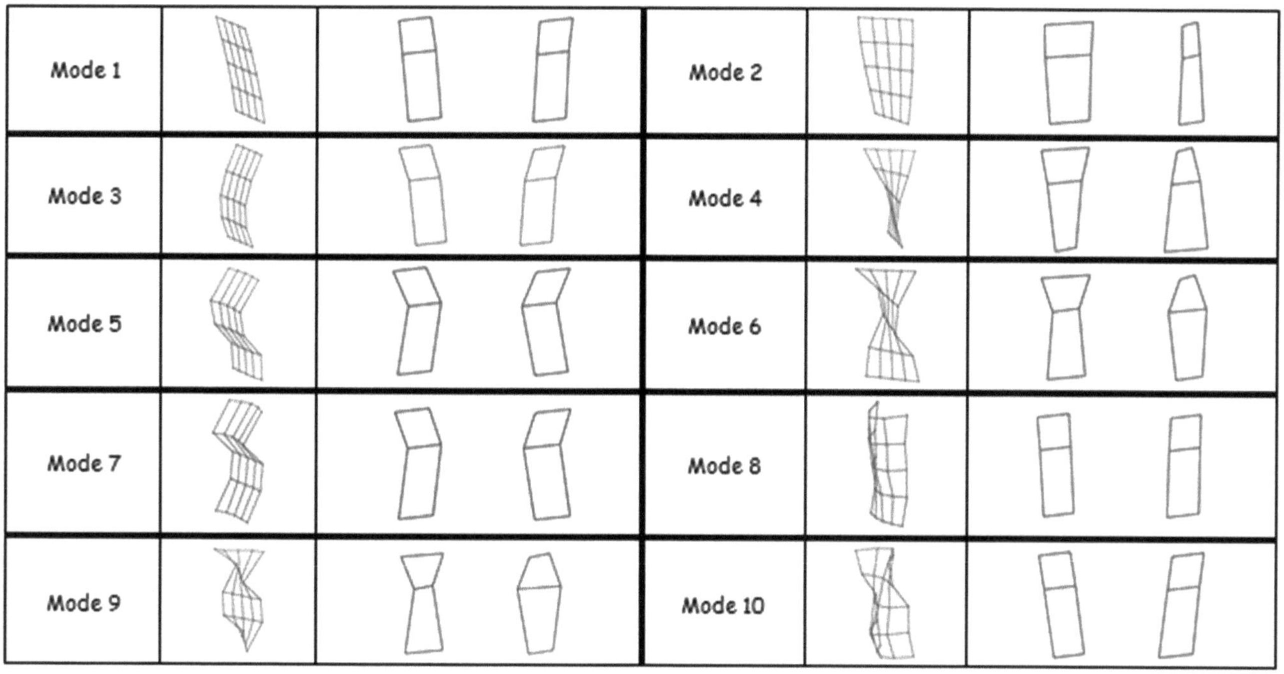

Fig. 8.10 Mode shapes of identified flexible modes, with prior determined flexible modes as reference

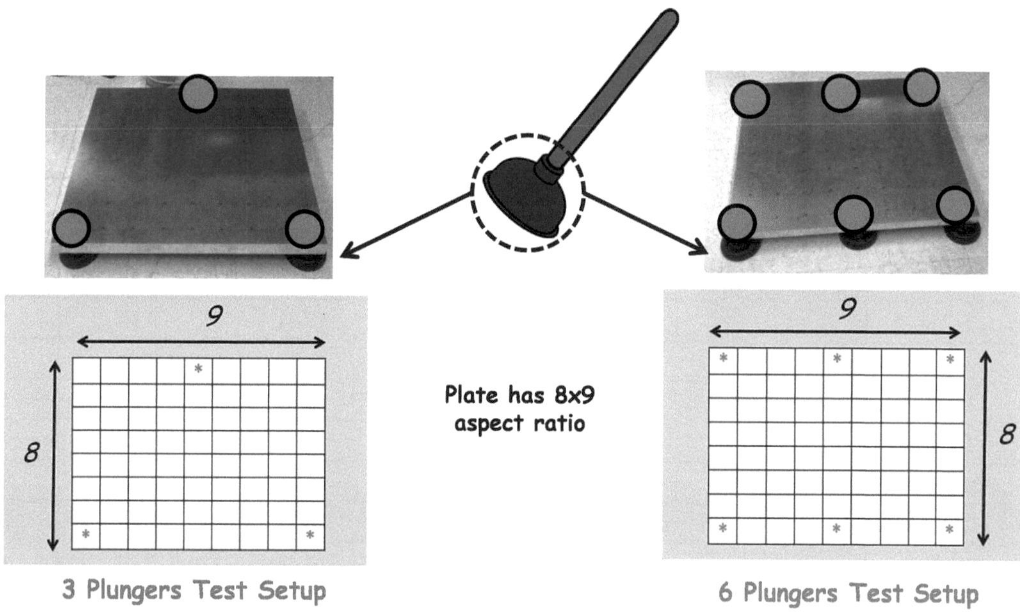

Fig. 8.11 Test setup for three plungers and six plungers tests

For the three plungers test, two plungers were placed at the corners along one of the long sides of the plate and one plunger was placed in the center of the opposite side, as shown in Fig. 8.11. For the six plungers test, three plungers were placed along each of the long sides of the plate as shown in Fig. 8.11.

Measurement points and a coordinate system were defined for both tests. A tri-axial accelerometer was placed on the plate beneath point nine. For the three plunger test, a small sledge hammer with a black plastic tip was used to impact the plate. For the six plunger test, a large sledge hammer with a black plastic tip was used. Measurements were acquired in the X, Y, and Z directions as appropriate. The measurement point locations and testing equipment is shown in Fig. 8.12.

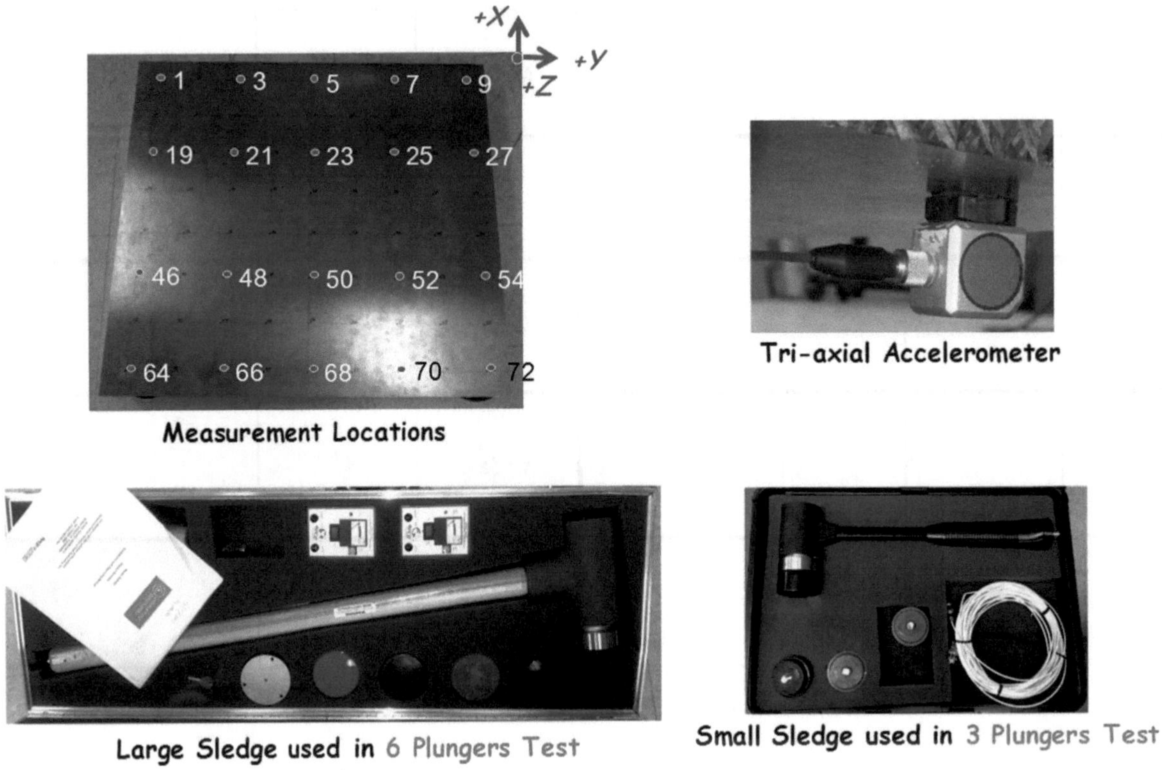

Fig. 8.12 Measurement point locations, coordinate system, and equipment used for three plungers and six plungers tests

Table 8.6 Natural frequencies and damping for three plunger test and six plunger test

	Natural Frequency (Hz)				Damping		
Mode	3 Plungers	MAC	6 Plungers	% Difference	3 Plungers	6 Plungers	% Difference
1	4.167	77.183	4.519	8.10%	6.45	6.39	0.93%
2	4.339	93.340	4.602	5.88%	6.41	6.30	1.73%
3	5.756	91.744	6.816	16.86%	6.81	7.34	7.49%
4	9.013	94.797	9.350	3.67%	5.36	8.21	42.00%
5	10.278	90.015	12.359	18.39%	5.62	10.23	58.17%
6	13.989	91.585	15.967	13.21%	6.16	9.20	39.58%
7	219.399	94.612	217.166	1.02%	0.18	1.36	153.25%
8	315.940	88.481	316.362	0.13%	0.22	0.78	112.00%
9	439.26 3	80.409	439.762	0.11%	0.48	1.20	85.71%

The six rigid body modes and first three flexible modes were evaluated over a bandwidth of 500 Hz using Photon software via LDS Dactron. Data was then processed in LMS. The natural frequencies and damping of the two tests are compared in Table 8.6. The mode shapes from the plunger tests are shown in Fig. 8.13.

Table 8.6 shows that using six plungers rather than three plungers increased the frequencies of the six rigid body modes and the second two flexible modes; decreased the frequency of the first flexible mode; and either increased or decreased the damping depending on the mode. Differences in damping were most noticeable after the first three rigid body modes. Overall, the toilet plungers were seen to be a very good support configuration for the heavy shock plate and provided very good vertical and lateral support (at a cost which is far cheaper than any of the more elaborate air floating systems used).

3 Plungers Test

Mode 1 4.167 Hz	Translational X	X
Mode 2 4.339 Hz	Translational Y	Y
Mode 3 5.756 Hz	Rotational Z	Z
Mode 4 9.013 Hz	Translational Z	Z
Mode 5 10.278 Hz	Rotational X	X
Mode 6 13.989 Hz	Rotational Y	Y
Mode 7 219.399 Hz	Torsional	
Mode 8 315.940 Hz	Bending	
Mode 9 439.263 Hz	Bending	

6 Plungers Test

Mode 1 4.519 Hz	Translational X	X
Mode 2 4.602 Hz	Translational Y	Y
Mode 3 6.816 Hz	Rotational Z	Z
Mode 4 9.350 Hz	Translational Z	Z
Mode 5 12.359 Hz	Rotational X	X
Mode 6 15.967 Hz	Rotational Y	Y
Mode 7 217.166 Hz	Torsional	
Mode 8 316.362 Hz	Bending	
Mode 9 439.762 Hz	Bending	

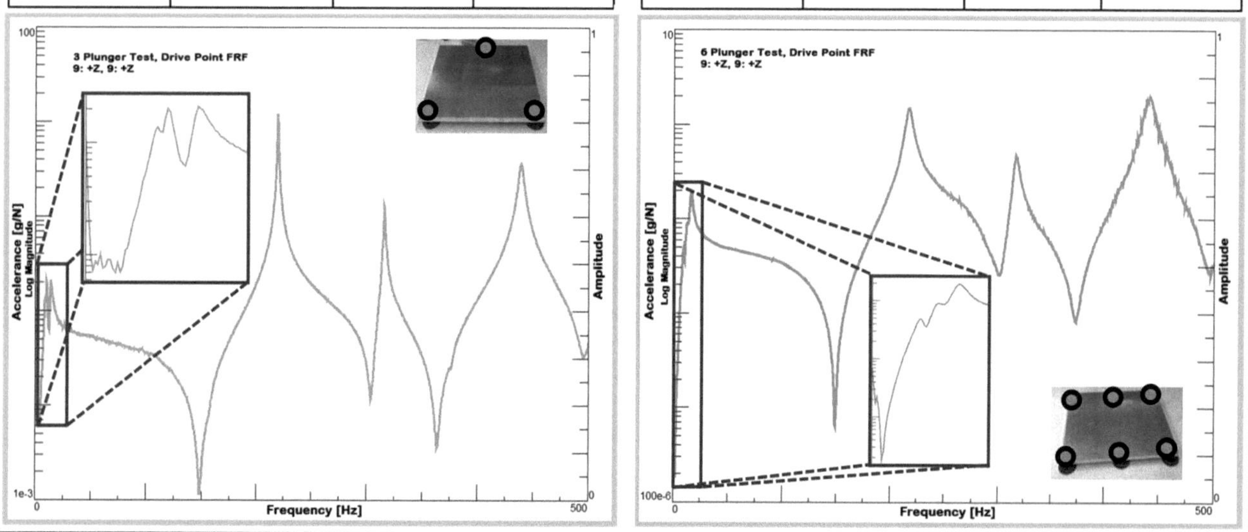

Fig. 8.13 Results of three plungers and six plungers tests

8.5 Conclusion

The test boundary conditions can have an effect on the measured frequencies for an experimental modal test. A few examples were shown to illustrate some non-conventional boundary conditions that could be employed – from marshmallows to plungers. Analysis of the various cases showed that the frequencies might have slight shifting but, more importantly, the organization of the mode shapes might change; this was seen in the organization of the closely spaced bending and torsion modes of one structure. Expensive solutions were not necessary to obtain suitable support scenarios. Care must be exercised when setting up the boundary condition mechanism and variations may need to be considered to assure that proper data is collected to identify the modes of the structure under test.

References

1. Avitabile P (2012) How free does a test need to be? Does it really matter that much? SEM Experimental Techniques, Aug 2012
2. Avitabile P (2001) Can the test setup have an effect on the measured modal data? SEM Experimental Techniques, Dec 2001
3. Avitabile P (2010) Will the support mechanism have any effect on FRFs? Does bungie cord vs. fishing line make any difference? SEM Experimental Techniques, Dec 2010

Chapter 9
Robustness of Disc Brake Systems Regarding Squeal

Philippe Stegmann, Sebastian Kruse, and Klaus Augsburg

Abstract Despite the efforts in research and development of brake systems there still is a lack of procedures to evaluate the robustness of these systems with respect to friction induced vibrations during the development process. Hence, the identification of the modal properties and their characteristics using experimental methods becomes even more important. Experience shows that these modal properties can change easily due to variations in operating conditions or due to manufacturing tolerances.

This work compares existing approaches to define a brake system's robustness regarding squeal. Moreover it shows experimentally the strong dependence of the system's resonance frequencies and modal damping values to slight changes in operating conditions. Consequently, a new approach is developed, in order to allow a modal analysis of a complete brake system with its' inherent robustness.

The results of this study investigating the influence of different operating conditions on the modal properties are shown and these are discussed with respect to the system's robustness.

Keywords Brake squeal • Modal-analysis • System-excitation • Modal parameters

9.1 Motivation

In today's development of brake systems the noise behavior plays a significant role. Depending on their in the frequency range the different types of brake noises can be classified for example as shown in [1]. This work concentrates on the most discussed issue of brake squeal [2] that typically occurs in a frequency range from 1 to 20 kHz, even though the approach is suitable for a lower frequency range as well. Different approaches explaining the origin of brake squeal are presented in [2–4]. A widespread view is the presence of coupled vibration-modes and eigenfrequencies [5] of the brake system that result in an unstable equilibrium state and lead to limit cycle oscillations [6]. An enabler for mode-coupling is the dependency of the system's modal properties on operational parameters as for example the brake pressure, the temperature or the friction level [7].

Despite simulation and vehicle tests dynamometer testing plays the most important role in the brake's NVH-development. The evaluation of brake systems in dynamometer-testing is based on procedures that consist of a matrix of varying operating conditions. Typically occurring vibrational states are assumed to be critical if noise above a threshold of 70 dB is detected in the cabin of the dynamometer. The evaluation of a full test matrix according to [8] consists of the absolute occurrence of noise events, their frequencies, the sound pressure levels, the length of a noise and the operating conditions leading to each noise event.

Using these test approaches repeatedly specific noise events that occur in later vehicle tests might not be detected. That can be due to specific load conditions in the car or due to special environmental conditions. Also the variance of parameters during the production process might cause this phenomenon. This obviously lengthens the development process and possibly causes customer complaints.

Hence, there is a need for more efficient and reliable procedures to evaluate the robustness of brake systems. While a system's robustness is influenced by sensitivities to tolerances and changing system properties over time, these influences are not taken into account during today's dynamometer tests. In general only one system of one specification is tested.

P. Stegmann (✉) • S. Kruse
Audi AG, Development Foundation Brake, Ingolstadt, Germany
e-mail: philippe.stegmann@audi.de

K. Augsburg
Ilmenau University of Technology, Ilmenau, Germany

M. Mains (ed.), *Topics in Modal Analysis, Volume 10*, Conference Proceedings of the Society for Experimental Mechanics Series, DOI 10.1007/978-3-319-15251-6_9

The work in process tries to overcome this issue by characterizing the vibrational behavior of a brake system in a wider sense than today's development approaches. This is based on the assumption that already in states near instabilities the dynamics of these states are inherent in the system and can be detected. This assumption is verified in the remainder of this work.

This contribution starts out with a description of the setup for measuring the modal properties of a brake system during braking conditions. Subsequently it describes analysis techniques for the signal processing. Eventually it presents first results for the development of modal properties under varying operational conditions employing the introduced setup. This setup will be further developed in future work to fully describe the robustness of a brake system with respect to its vibrational behavior.

9.2 Experimental Setup

In order to obtain a more detailed evaluation of the NVH behavior of brake systems the approach of a modal analysis of the full system under operating conditions is chosen. Thereby the identification of the resonance-frequencies in the frequency-response-function enables the determination of potential squeal-frequencies also when no noise is detected. In addition the modal damping of the resonance-frequencies is calculated and analyzed. For this analysis the test setup is described starting out with the setup of the brake system before introducing the equipment for the measurement.

9.2.1 Brake System

The system for the ongoing investigation consists of brake-disc, brake-pads, caliper and knuckle. In order to gain access to the brake pad's backplate for the installation of sensors a floating caliper is chosen. The brake system is manipulated in order to show squeal reproducibly by removing damping elements. This system's basic squeal behaviour is determined from a standard brake matrix-test according to the SAE-J 2521 [8] performed on a brake dynamometer. The result is shown in Fig. 9.1. Based on this test procedure three squeal frequencies with multiple noise occurrences are identified at around 2, 3 and 6 kHz.

In the following the investigations focus on the modal properties at approximately 6 kHz. This noise shows to be easy to reproduce for brake-pressures from 11 to 30 bar and temperatures from 20 °C to 120 °C.

9.2.1.1 Excitation and Measurement Concept

For the modal analysis an external excitation of the system is chosen. This allows a selective excitation of chosen frequencies as well as the investigation of nonlinearities by a variation of the excitation amplitude. For the excitation the brake piston is

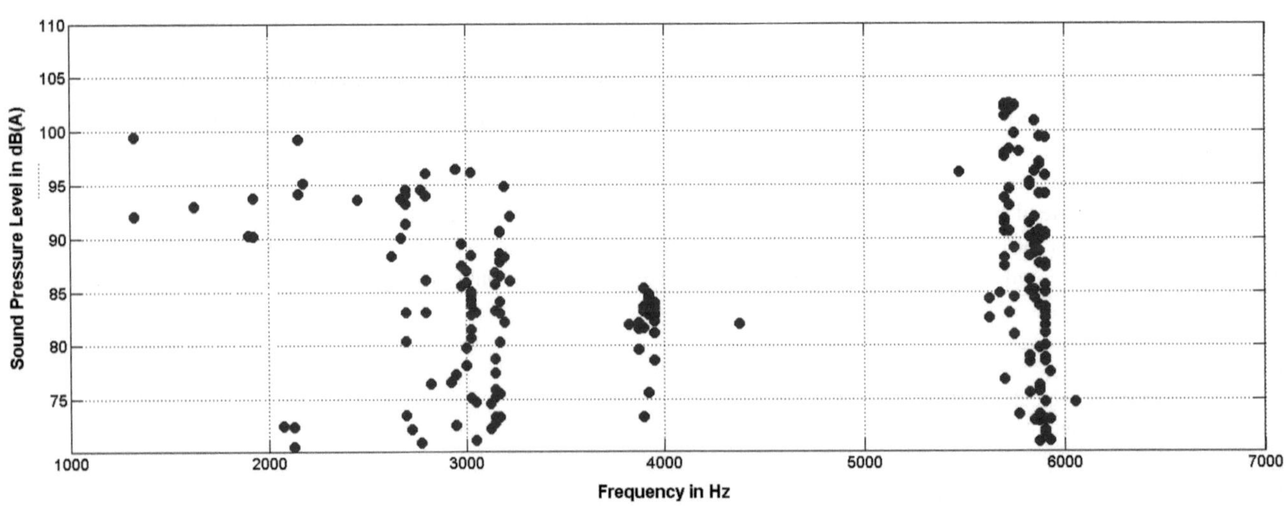

Fig. 9.1 Results of the SAE J 2521 Matrix test

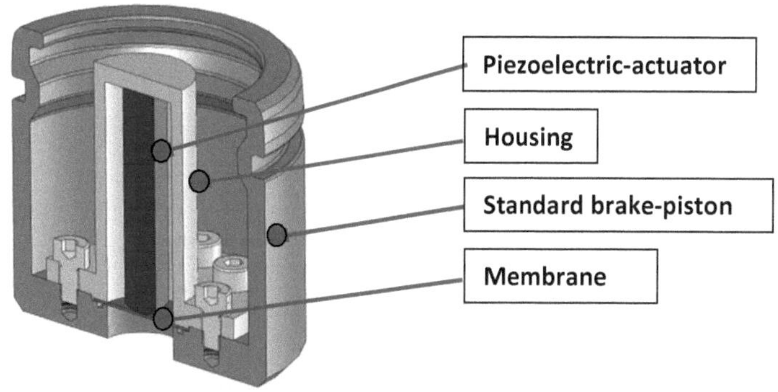

Fig. 9.2 Piezo equipped brake-piston

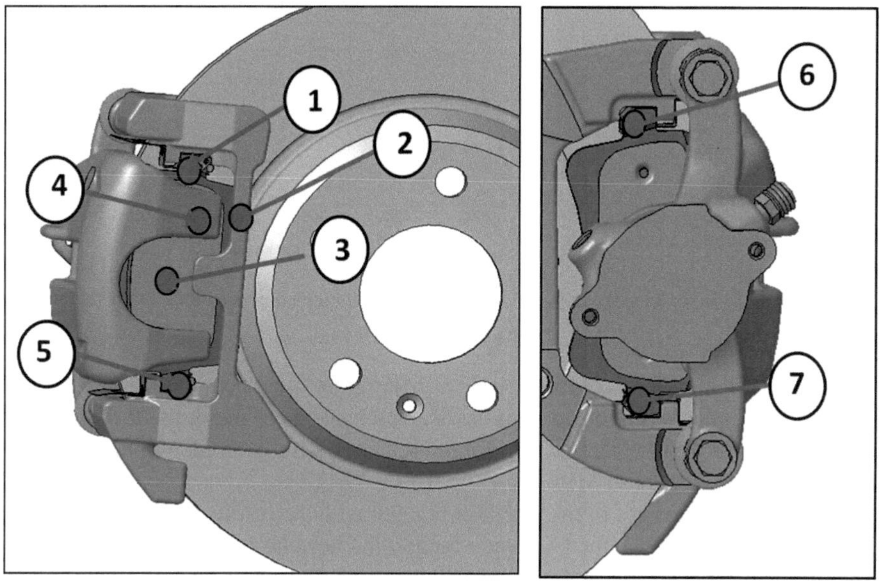

Fig. 9.3 Positions of the acceleration sensors

equipped with a piezoelectric actuator as depicted in Fig. 9.2. Similar concepts of an integrated piezoelectric-actuator in the brake piston have already been introduced in [9–11] and have proven their ability for brake system's actuation. Through the application of an electric potential the piezoceramic actuator is distended whereby the deflection of the membrane is attained. Due to the fact that the membrane is in contact with the brake fluid, pressure oscillations are generated. For the examination a sinusoidal voltage with constant amplitude is applied to the piezoceramic actuator which results in the system's excitation due to pressure oscillations. Using this concept the influence of the additional installation of a membrane and its dynamic behavior needs to be taken into account. The resonance frequencies of the modified piston have been identified. Especially the first resonance-frequencies of the membrane strongly influence the system and therefore have to be avoided during excitation. For the presented piston this means that the usable frequency span has to be set to a maximum of 7 kHz.

The system's response is recorded with PCB Type 356A01 triaxial acceleration sensors. Sensor-positions on the backplate of the brake-pads and on the caliper are depicted in Fig. 9.3. The acceleration-sensors are attached on the structure using a HBM X60 two component adhesive. By using this special type of adhesive the attachment of the sensors is ensured for temperatures for the complete temperature range of the sensor ranging from 20 °C to 120 °C.

The applied test method consists of a stepped-sine actuation during a drag-application of the brake with constant pressure and velocity. As illustrated in [12] the stepped-sine allows the most accurate analysis of the nonlinear system. In order to avoid an increase of the temperature during the measurement, the time for one run with the stepped-sine should be minimized. In the present case a frequency span from 1 to 7 kHz is chosen with a step-size of 5 Hz. That leads to a cycle time of approximately 15 s for a full system analysis.

The stepped-sine is performed with the application of constant voltage amplitude on the piezoelectric-actuator. Concurrently the voltage amplitude is used as the reference for the calculation of the frequency-response-function (FRF). The FRFs are calculated from response and actuation signals. Resonance-frequencies and modal damping values are calculated from the FRF. For the identification of the resonance-frequencies two approaches have been applied. The first one is to directly identify the resonance-frequencies from the peaks of the different sensor signals in frequency domain. That implies the need to analyse all FRFs for the 21 different sensors. The second practice is the calculation of the complex mode indicator

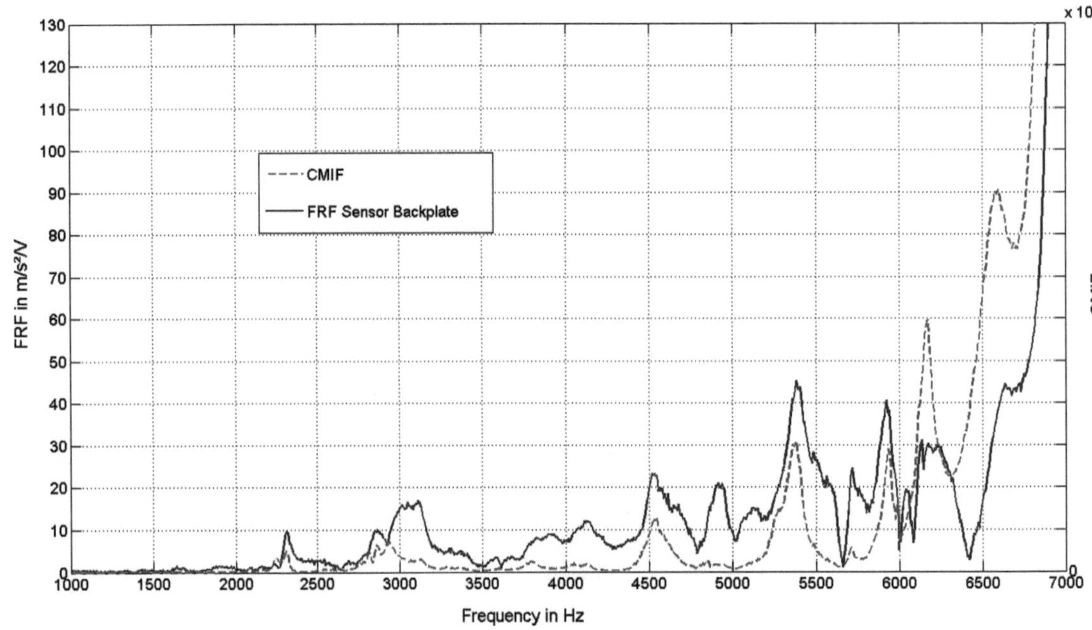

Fig. 9.4 Results of the modal analysis at 10 bar, 100 °C disc temperature with 3 km/h velocity

function from the FRFs according to [13, 14]. While the calculation of the CMIF provides a quick overview of the main resonance-frequencies of the system the detailed analysis of the individual signals allows to evaluate the participation of the components on the different modes. Figure 9.4 shows for example the FRF of one sensor on the backplate of the brakepad as well as the calculated CMIF.

Comparing the results of the CMIF in Fig. 9.4 with the results in Fig. 9.1 it becomes apparent that it is possible to identify resonance-frequencies in the frequency-ranges that have high noise occurrence in the SAE-J 2521 matrix test. It is important to note the present operational conditions. For the ongoing investigation the focus is set on the frequency range around 6 kHz. This is chosen because the 6 kHz noise was steadily reproducible for brake-pressures from 11 to 30 bar and temperatures from 20 °C to 110 °C.

9.3 Special Requirements for the Modal Analysis of Brake Systems

Under operational conditions there are aspects that need to be taken into account in terms of the accuracy of the modal analysis. The friction between brake-pad and disc generates vibrational noise that influences the energy of the additional excitation that is needed as well as quality of the result. Figure 9.5 shows the increase of the measured vibrations on the backplate when brake-pressure is applied (left) as well as the spectrum of the vibration signal under braking conditions (right). At the same time the measured acceleration-amplitudes also show an increase with increasing rotational speed of the brake disc and increasing brake pressure. In order to improve the quality of the results the rotational speed for the investigation is set to a value of 3 km/h corresponding 0.4 rev/s.

In order to expose the impact of the frictional noise on the results of the modal analysis the measurement has been performed first on a brake system without rotation of the disc and subsequently on a system with rotation of the disc. The results depicted in Fig. 9.6 are meant to illustrate exemplarily this influence of this friction induced vibration. It has to be mentioned that this investigation has been performed on another type of brake system so that the identified eigenfrequencies differ from the other results.

Another particularity of brake systems that leads to low-frequency-vibration phenomena lies in the occurrence of imperfections like disc thickness variations, side run out of the brake disc or local variations of the coefficient of friction on the disc that result in significant changes of the modal properties during one revolution. Therefore in particular the reproducibility of the results has to be evaluated and if necessary components like the brake-disc need to be changed to ensure the quality of the results. The detailed investigation of the these effects are not the focus of this publication but will be explored in future work.

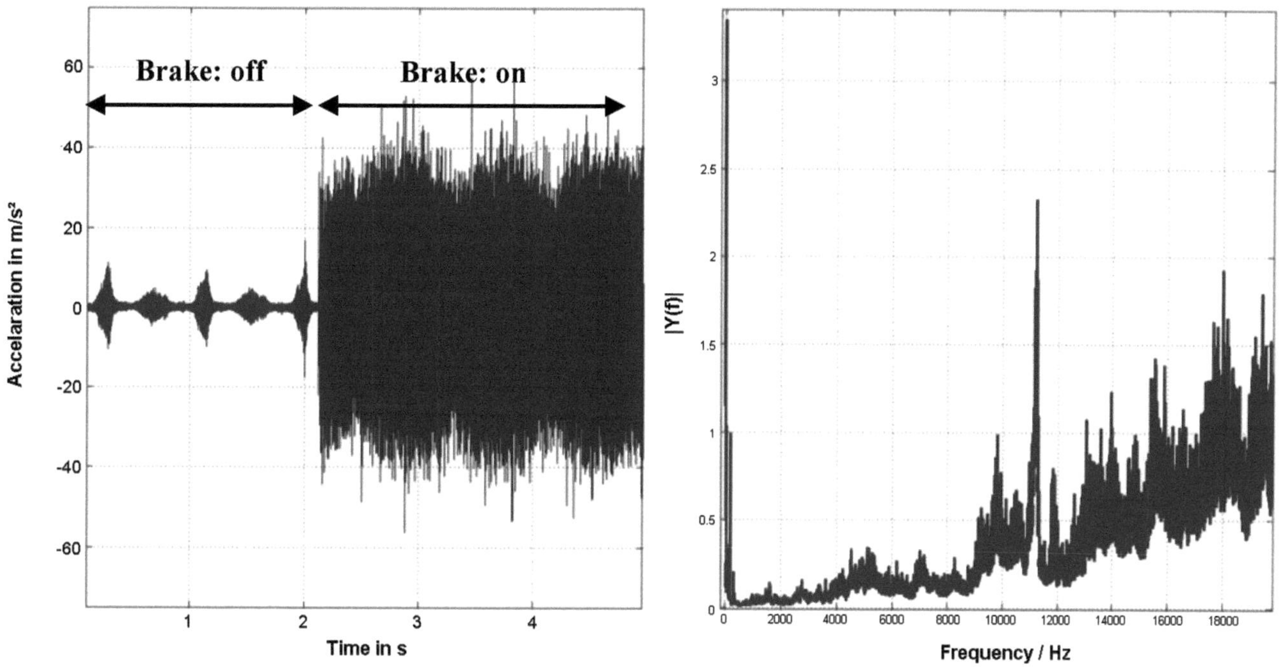

Fig. 9.5 Comparison of acceleration-signal in off and on-brake situation (*left*). Frequency spectrum of "on-brake""signal (*right*)

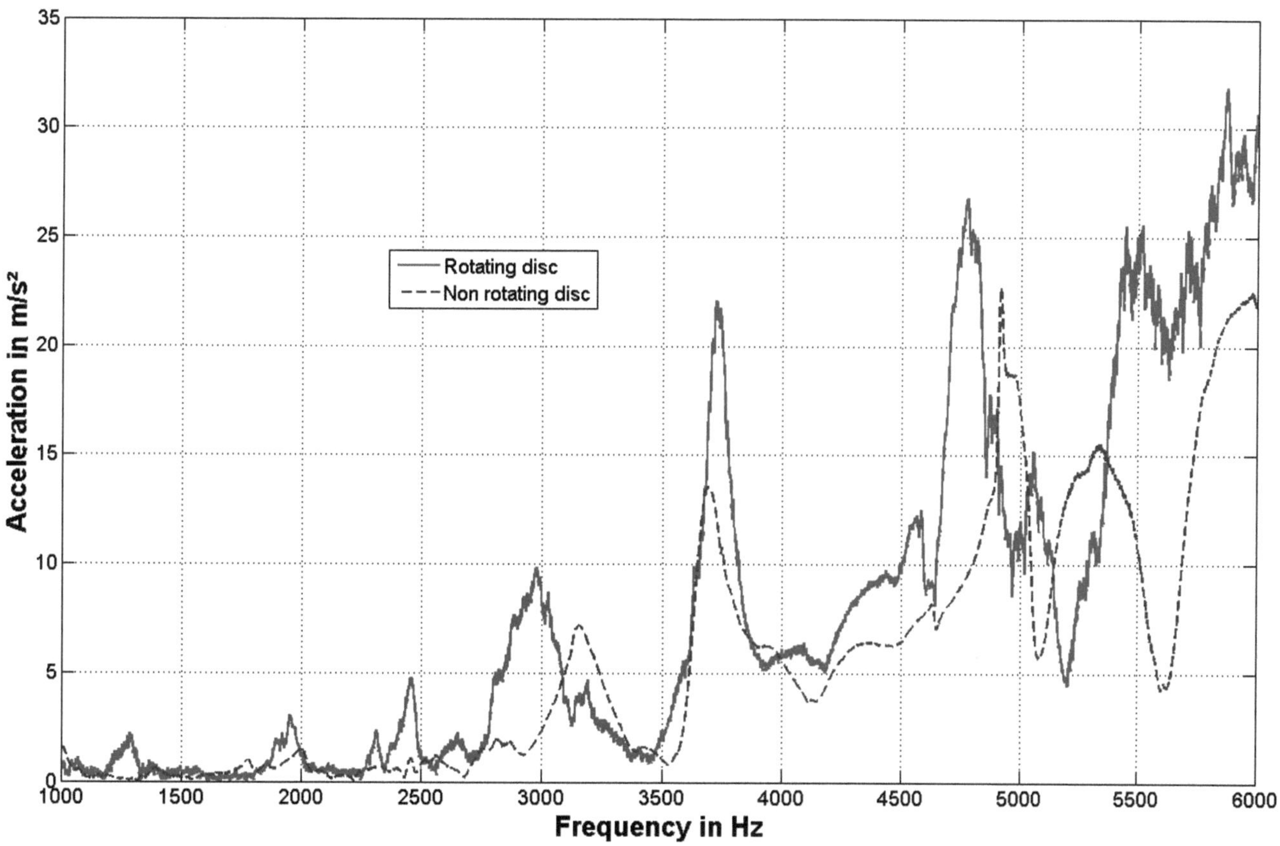

Fig. 9.6 Results of the modal analysis for rotating and non-rotating disc

Considered these previously mentioned particularities of the brake system the focus of the further development has to be on appropriate data-analysis methods in order to obtain reproducible and reliable results. Subsequently methods to calculate the damping from the FRFs have to be implemented and evaluated for this specific application. This includes for example the possibility to filter out all interfering signals that reduce the quality of the results. The influence of the interfering signals on the results will become clear in the following section as it complicates the calculation of the modal damping values.

9.4 Results

In order to be able to use a modal analysis of the brake system for the evaluation of its' robustness with respect to squeal two important requirements for the mechanism of brake squeal in a real brake system need to hold. As a part of the following investigation these two hypotheses upon the relevance of the modal damping are formulated and verified. The first one states that the modal damping is changing with a variation of parameters like the brake pressure or the temperature. Subsequently the second hypothesis says that the modal damping decreases when parameters are change in direction of a critical state.

While using the modal-analysis-technique, variations of the operational conditions are performed. This allows the verification of the hypothesis building the basis for the robustness approach: Modal damping changes when system parameters are changed as the system is approaching an unstable state. In the present case the variation of the brake pressure has the largest influence on the noise occurrence while the system proved to have a lower sensitivity towards changes of the temperature and velocity. The used system shows an instability when pressure is increase to 11 bar. Figure 9.7 shows the results of the modal analysis in the considered frequency range when brake-pressure is increased towards 11 bar. As hypothesized the modal damping of the mode around about 6 kHz decreases when pressure is approaches the critical brake-pressure of 11 bar.

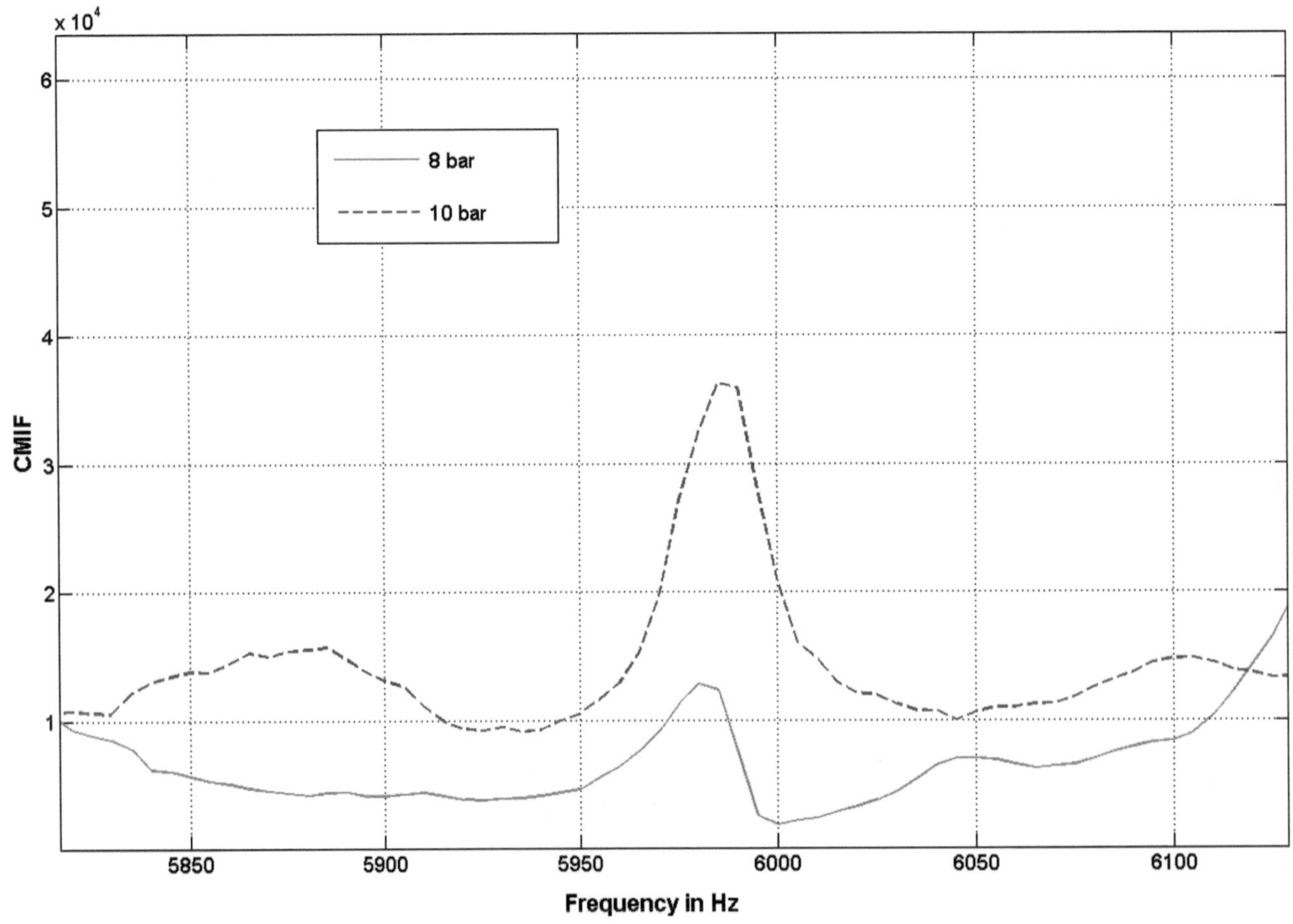

Fig. 9.7 Results for different brake-pressure CMIF and modal damping (D)

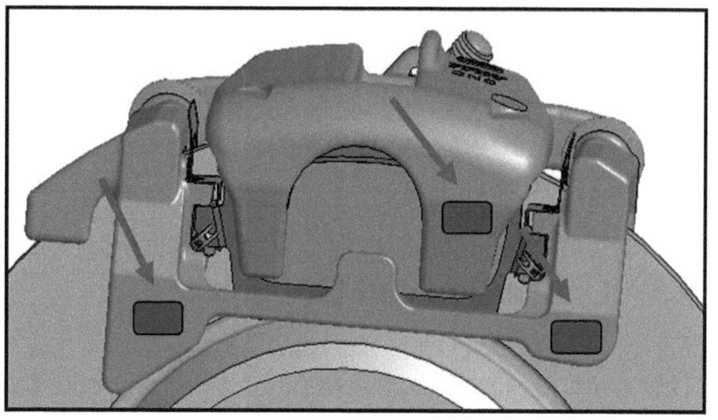

Fig. 9.8 Positions of the additional masses

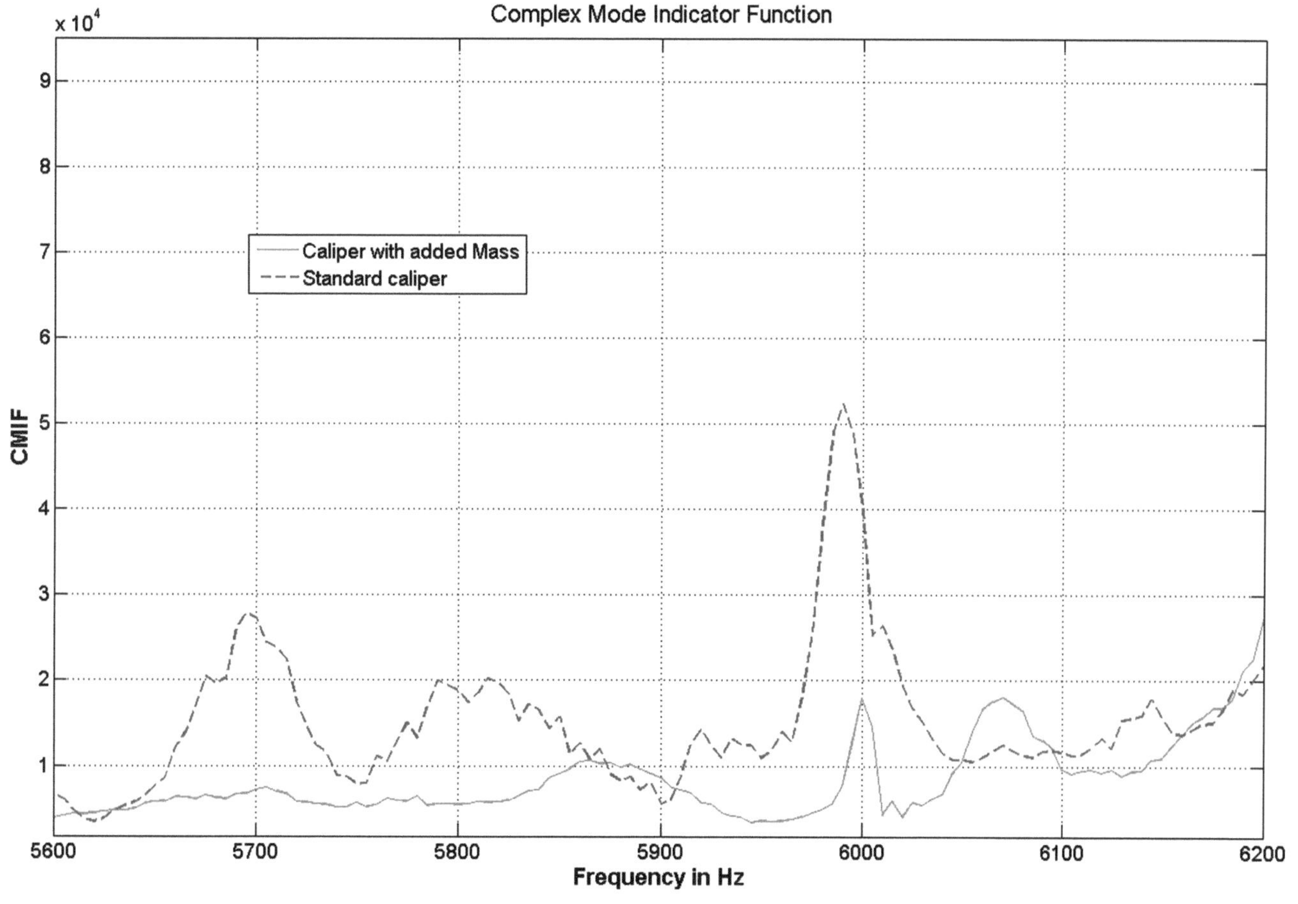

Fig. 9.9 Results of the modal-analysis for the standard and the modified caliper

To demonstrate the potential of this method for the development process of a brake system an additional approach is chosen. This approach may serve to illustrate the influence of structural modifications. For this purpose the structure of the caliper is modified on basis of the results of a deflection-shape-analysis using the laser-scanning-vibrometer as described in [15]. In order to verify the influence of modifications additional masses of 50 g are placed on points of the caliper that show high deflection amplitudes during a 6 kHz squeal-phenomenon.

Figure 9.8 depicts this modification. Subsequently the modal analysis is performed for operational conditions near an instability of the unmodified caliper.

Figure 9.9 illustrates the comparison of the CMIF of the brake system with the standard caliper and the modified caliper. An influence of the modification on the modal characteristics at about 6 kHz is apparent. However, the aforementioned difficulties with the quality of the results complicate especially the accurate estimation of the modal damping. This has to be considered interpreting these results and needs to be tackle during future development. The need of a better method for data analysis in order to have a more reliable estimation for the modal damping has to be outlined at this point of the development. Nonetheless the results demonstrate the potential of the method to evaluate and compare different noise countermeasure during the development process.

9.5 Conclusion and Outlook

This work presents the investigation of a full brake system on a dynamometer using the experimental modal analysis with an additional concept for systems excitation. The comparison with a standard test method and the accordance of the identified resonance-frequencies with the noise-frequencies confirm the validity and potential of this approach.

The influence of operational parameters and structural modifications on the resonance-frequencies and on modal damping is shown for two selected examples. As assumed the results show that the modal damping of the brake system decreases continuously once the system parameters approach an unstable state. Hence, the potential of the introduced methodology for a more detailed evaluation of the robustness of brake-systems has been outlined.

As the work is ongoing and this contribution presents a first step towards an approach estimating the robustness of a full brake system with respect to brake noise, there is a need for further development. Several details of the method will have to be improved. This includes an optimization of the signal and data processing especially to reduce the influence of interfering signals on the quality of the results. Focusing on the actuation concept a normalization of the actuation force will allow the direct comparison of different modes. Moreover extended parameter variations of all operational conditions will need to be performed and several different brake-systems have to be investigated.

This future work shall lead to the long-term aim of a method that on the one hand is able to determine systems' robustness through localization of instable areas in the full parameter space. On the other hand this method will be able to provide a fast evaluation of the efficiency of different countermeasures in the development process.

References

1. Breuer B, Bill KH (2012) Bremsenhandbuch, 4.Auflage; Springer Vieweg
2. Kinkaid NM, Reilly OM, Papadopoulos P (2003) Automotive disc brake squeal. J Sound Vib 267:105–166
3. Allgaier R (2001) Experimentelle und numerische Untersuchungen zum Bremsenquietschen. Dissertation Universität Stuttgart, Chapter 4
4. Shi X (1996) Enstehung des Bremsenquietschens. Dissertation Technische Universität Braunschweig, pp 8–13
5. Hoffmann N, Gaul L (2003) Effects of damping on mode-coupling instability in friction induced oscillations. ZAMM Z Angew Math Mech 83(8):524–534
6. Sinou J-J, Jezequel L (2007) The influence of damping on the limit cycles for a self-exciting mechanism. J Sound Vib 304:875–893
7. Hetzler H (2008) Zur Stabilität von Systemen bewegter Kontinua mit Reibkontakten am Beispiel des Bremsenquietschens. Dissertation Universität Karlsruhe, Chapter 4
8. SAE International (2006) SAE-J 2521 Surface vehicle recommended practice, disc and drum brake dynamometer squeal noise matrix, Rev. 2006
9. Marschner H (2012) Mit Sicherheit leise – Neue Wege zur geräuscharmen Bremse. μ-symposium 2012
10. Waldschmidt A (2007) Active car brake system for squeal suppression based on piezohydraulic actuation. Adaptronic Congress Göttingen
11. Stegmann P, Kruse S, Augsburg K (2014) Comparison of excitation concepts for the characterization of brake systems robustness regarding squeal. Eurobrake Lille
12. Gygax PE Dr (1986) Experimentelle Testprozeduren: Übersicht über Probleme und Verfahren bei Frequenzgangmessungen. Symposium Nichtlineare Strukturdynamik ETH-Zürich, Oktober 1986
13. Allemang R-J, Brown D-L (2006) A complete review of the complex mode indicator function (CMIF) with applications. In: International conference on noise and vibration engineering ISMA, Leuven Belgium
14. Ewins DJ (2000) Modal testing: theory, practice and application, 2nd edn. Research Studies Press, Baldock, pp 300–301
15. Marschner et al (2008) Innovative Schwingungsmesstechnik in der Bremsenentwicklung. Automobiltechnische Zeitschrift (ATZ) 02/2008

Chapter 10
Measurement of Vibration Resulting from Non-contact Ultrasound Radiation Force

Thomas M. Huber, Spencer M. Batalden, and William J. Doebler

Abstract In modal testing, the most common excitation method is a transducer in mechanical contact with the object under test. While this method is effective, there are delicate structures where it is desirable to excite vibrations without physical contact. The ultrasound radiation force provides a noncontact excitation method resulting from the nonlinear interaction of sound waves scattering from an object. The incident ultrasound consists of sine waves with frequencies of f_1 and f_2; the resulting radiation force has a component at the difference frequency $f_1 - f_2$. By combining the difference frequency radiation force with a scanning vibrometer, previous studies have demonstrated completely non-contact measurements of resonance frequencies and operating deflection shapes of structures ranging from microcantilevers to classical guitars. In the current study, a 19.6 by 8.1 by 0.37 mm clamped-free brass cantilever, with a resonance frequency of 610 Hz, was excited using the radiation force from a focused ultrasound transducer. By mounting the transducer on a computer-controlled translation stage, it enabled measurements of the edge-spread function for the transducer; measuring this distribution is an important first step towards quantifying the applied radiation force.

Keywords Ultrasound • Radiation force • Edge-spread function • Vibrometer • Cantilever

10.1 Introduction

In modal testing of small objects, one challenge can be to excite the vibrational modes without distortions due to mass loading. Using a laser Doppler vibrometer, it is relatively straightforward to measure the vibration in a non-contact manner; however, it is not always possible to excite these vibrations without contact. The ultrasound radiation force is a non-contact method of excitation resulting from ensonifying a structure with a pair of ultrasound frequencies with the frequency difference adjusted to the resonance frequency of the structure. Previous studies have demonstrated that this technique can be used in modal testing in air for structures as small as microcantilevers [1] to as large as the face of a classical guitar [2]. These previous studies demonstrated the feasibility of determining resonance frequencies of structures, and also operating deflection shapes using a laser Doppler vibrometer. They, did not, however, quantify the magnitude or distribution of the applied radiation force.

The current study focused on using the ultrasound radiation force applied to a cantilever to directly measure the edge-spread function for an ultrasound transducer. Similar to its definition in optics, the edge-spread function is a measure of the integrated force distribution for a sharp-edged section of the circular ultrasound focus point. In the current study, this was determined by translating the ultrasound focus point across the edge of a cantilever, and monitoring the resulting radiation force using a single-point vibrometer.

10.2 Theory

Previous papers have described in detail the mechanism for ultrasound stimulated audio-range excitation, both in air [3] and in water [4–6]. If an object is ensonified with a pair of ultrasound frequencies, f_1 and f_2, interference between the two frequencies produces a radiation force that results in a vibration of the object at the difference frequency $\Delta f = f_2 - f_1$. Both frequency components were emitted from a single transducer using a double-sideband suppressed-carrier amplitude

T.M. Huber (✉) • S.M. Batalden • W.J. Doebler
Department of Physics, Gustavus Adolphus College, 800 College Avenue, Saint Peter, MN 56082, USA
e-mail: huber@gac.edu

M. Mains (ed.), *Topics in Modal Analysis, Volume 10*, Conference Proceedings of the Society for Experimental Mechanics Series, DOI 10.1007/978-3-319-15251-6_10

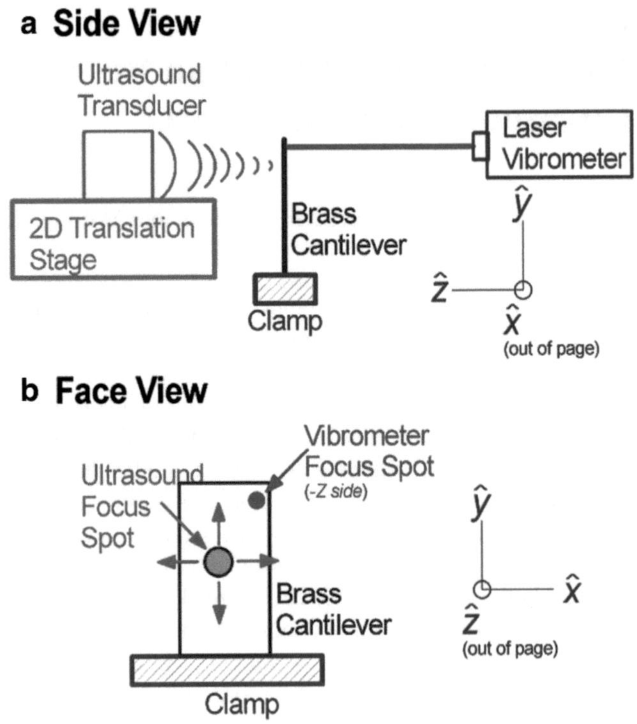

Fig. 10.1 Schematic diagram of apparatus used for measuring edge-spread functions for ultrasound radiation force excitation

modulated (AM) waveform [7, 8]. As shown in Fig. 10.1, an object was excited by a transducer emitting two different ultrasound frequencies $f_1 = f_c - \Delta f/2$ and $f_2 = f_c + \Delta f/2$ where f_1 and f_2 are ultrasound frequencies which are symmetrical about a central frequency f_c.

The radiation force [9, 10] is caused by changes in the energy density of an acoustic field. In the following derivation, it is assumed that the total ultrasound pressure field $P(\boldsymbol{r})$ at a point $\boldsymbol{r}$ will be the same at both frequencies f_1 and f_2 that are emitted by the transducer. However, as the waves of different frequencies traverse the distance between the transducer and the arrival point $\boldsymbol{r}$, they will arrive with different phases $\varphi_1(\boldsymbol{r})$ and $\varphi_2(\boldsymbol{r})$, thus the total pressure field due to the two frequency components may be written as

$$p(\boldsymbol{r}, t) = P(\boldsymbol{r}) \cos[2\pi f_1 t + \varphi_1(\boldsymbol{r})] + P(\boldsymbol{r}) \cos[2\pi f_2 t + \varphi_2(\boldsymbol{r})]. \tag{10.1}$$

This causes an instantaneous energy density given by $e(r,t) = p(r,t)^2/\rho c^2$; this energy density will have a time-independent component, a component at the difference frequency Δf, and high-frequency components at multiples of f_1 and f_2. The radiation force of interest for the current technique is the energy density component at the difference frequency, which can be written as

$$e_{\Delta f}(\boldsymbol{r}, t) = P(\boldsymbol{r})^2 \cos\left[\left(2\pi\Delta f\right)t + \Delta\varphi(\boldsymbol{r})\right]/\rho c^2. \tag{10.2}$$

Assuming that $P(\boldsymbol{r})$ is a plane wave, this will impart a force in the beam direction on an object of area dS with drag coefficient $d_r(\boldsymbol{r})$ given by [9, 10]

$$F_{\Delta f}(\boldsymbol{r}, t)\, dS = e_{\Delta f}(\boldsymbol{r}, t)\, d_r(\boldsymbol{r})\, dS = P(\boldsymbol{r})^2 \cos\left[\left(2\pi\Delta f\right)t + \Delta\varphi(\boldsymbol{r})\right]/\rho c^2\; d_r(\boldsymbol{r})\, dS. \tag{10.3}$$

The total radiation force $F_{Tot,\Delta f}(r,t)$ as a function of time is the integral of Eq. 10.3 over the ensonified surface of the object; this radiation force can induce a vibration of the object at a frequency Δf. Object vibration due to this radiation force is a function of the size, shape and mechanical impedance of the object. Previous studies have shown that this radiation force can be used for modal analysis of a variety of systems [11–13] including hard-drive suspensions [14–16]. To eliminate standing waves between the transducer and surface, the frequencies f_1 and f_2 were rapidly varied using a random carrier packet algorithm that maintained the difference frequency Δf [17].

The ultrasound radiation force distribution $F_{\Delta f}(r,t)$ from a circular focused transducer will have a radial distribution function with a central circular maximum, surrounded by successive rings of minima and maxima. If this circular beam focus, centered at (x_s,y_s), overlaps a section at the edge of a cantilever with known drag function $d_r(\boldsymbol{r})$ that is centered at $x=0$ and parallel to the y axis, the total radiation force would then be a fraction of the circular beam distribution given by

$$F_{\mathrm{Tot},\Delta \mathrm{f}}\left(x_s, y_s, \mathrm{t}\right) = \iint_{x>0,y} F_{\Delta \mathrm{f}}\left(x - x_s, y - y_s, \mathrm{t}\right) d_r\left(x, y\right) dxdy \tag{10.4}$$

A plot of $F_{\mathrm{Tot},\Delta \mathrm{f}}(x_s, y_s, \mathrm{t})$ versus x_s as the transducer is moved across the edge at $x=0$ is called the edge-spread function. By determining this edge-spread function, it should be possible to infer the radial distribution function $F_{\Delta f}(r,t)$ which is useful in modeling the response of ultrasound on an object with a more complicated shape or unknown drag function.

For the current study, a brass cantilever was clamped at one end as shown in Fig. 10.1. According to Euler beam theory [18], the displacement of the cantilever is independent of position in the x direction. The measured deflection shape $z(y)$ along the y direction was fit to the functional form expected for a clamped-free cantilever,

$$z(y) = a\left\{\left[\sin\left(\lambda\left(y - y_0\right)\right) - \sinh\left(\lambda\left(y - y_0\right)\right)\right] + b\left[\cosh\left(\lambda\left(y - y_0\right)\right) - \cos\left(\lambda\left(y - y_0\right)\right)\right]\right\} \tag{10.5}$$

where y_0 is the position of the clamped edge of the cantilever, a is an amplitude parameter, the product of the parameter λ and cantilever length ℓ has the value $\lambda\ell = 1.875104$ for the fundamental mode, and b is given by

$$b = \left[\frac{\sinh\left(\lambda\ell\right) + \sin\left(\lambda\ell\right)}{\cosh\left(\lambda\ell\right) + \cos\left(\lambda\ell\right)}\right]. \tag{10.6}$$

10.3 Experimental Setup and Procedure

The apparatus used is shown schematically in Fig. 10.1. The cantilever was a strip of brass with dimensions of 19.6 by 8.1 by 0.37 mm clamped at one end in a machinist's vise. The vise had a mass of about 2 kg and was bolted to a Newport optical vibration isolation table to make a stable support.

The transducer used in this portion of the testing was a custom-made ultrasound transducer for operation in air (MicroAcoustics Instruments, Gatineau, Canada). This transducer has a focal length of 7 cm and produces a focused ultrasound spot with a beam profile roughly 2 mm in diameter. The transducer's central maximum is located near 700 kHz with a bandwidth of over 200 kHz. The transducer was attached to an orthogonal pair of Newport 423 translation stages with computer-controlled Zaber Technologies T-NA08A25 μm with 0.05 μm resolution and 25.4 mm travel. Thus, the location of the transducer focus point could be raster scanned over a range of positions extending beyond the edges of the cantilever.

The transducer's waveforms were output from a 4-channel Strategic Test UF2e-6022 60 MSamples/s Arbitrary Waveform Generator PCI express board (Stockholm, Sweden). This board generated the Double-Sideband, Suppressed Carrier (DSB-SC) waveform of Eq. 10.1 with a carrier frequency in the vicinity of 700 kHz, and a pair of sidebands separated by 610 Hz. This waveform was amplified using an ENI-240 L RF amplifier to about 250Vpp. Another DAC output channel continuously cycled a simple 610 Hz sine wave that was used as a reference signal.

To determine the vibration of the cantilever, a Polytec PSV-400 Scanning Laser Doppler Vibrometer (Waldbronn, Germany) was focused near the free end of the cantilever. The vibrometer analog output signal was routed into a Zurich Instruments HF2LI Lock-In Amplifier (Zurich, Switzerland) with the reference signal being the 610 Hz sine wave produced by the Strategic Test card.

10.4 Results

The ultrasound transducer produced a DSB-SC waveform that had a difference frequency of 610 Hz, which is the resonance frequency of the brass cantilever used. By moving the transducer using the computer-controlled translation stage system, it was possible to perform horizontal scans where the transducer was kept at a fixed position in the y direction and multiple points were taken at different x positions, as shown in Fig. 10.2. The transducer was scanned such that the ultrasound focus

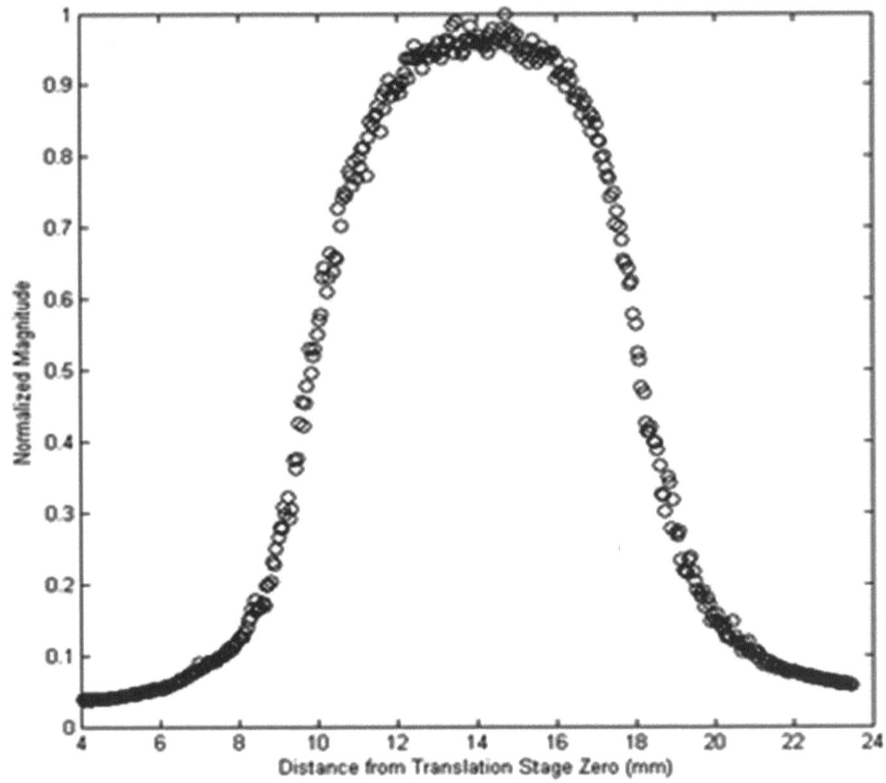

Fig. 10.2 Normalized response as ultrasound focus point is scanned across horizontal width of cantilever

point was directed beyond one edge of the cantilever, moved across the cantilever until it moved beyond the other side. The vibrometer measured the velocity amplitude corresponding to each position of the ultrasound focus point; each of these velocities were divided by the maximum velocity to produce the normalized magnitude. Since the response of a cantilever to a point force is independent of the horizontal position, the response function would have square edges that would go from 0 to 1 if the ultrasound focus point was infinitesimally small. However, because of the finite size of the ultrasound focus spot, there will be a distribution as shown in Fig. 10.2 since the cantilever's response will be reduced if a fraction of the ultrasound focus is not striking the cantilever. The normalized magnitude would be 0.5 when the center of the ultrasound focus spot was centered on the edge of the cantilever since half of the focus spot would be striking the cantilever and half would miss the cantilever off of the edge. The distance between these half-response points of 8.2 ± 0.1 mm is consistent with the 8.1 mm width of the cantilever.

Alternately, vertical scans consisted of positioning the transducer at the midpoint of the cantilever in the x direction and taking multiple measurements with varying positions in the y direction, as shown in that data points on Fig. 10.3. Even though the center of the ultrasound transducer began this scan below the clamped edge of the cantilever, the response did not go to zero because a fraction of the ultrasound focus spot was still striking the cantilever. As the transducer was moved in the positive y direction towards the free end of the cantilever, the cantilever's response would increase as expected if the driving force was concentrated at a single location. However, when the ultrasound transducers focus spot extended beyond the end of the cantilever, the applied force on the cantilever decreased, which leads to the decreasing amplitude of data points towards the right side of Fig. 10.3.

The solid line in Fig. 10.3 is a least-squares fit of a section of the data set illustrated to Eq. 10.5 which is the response expected for a point-source applied to a cantilever at a location y. The free parameters in this fit were y_0, the position of the clamped edge of the cantilever, and the amplitude parameter a. The section of the data set used for the fit was selected to insure that the majority of the ultrasound focus spot was incident on the transducer and not falling beyond the clamped or free ends of the cantilever. This fit demonstrates that, as long as the ultrasound focus point is entirely striking the cantilever, it causes a response that is essentially equivalent to a point source at that location. However, once the ultrasound focus spot extends beyond the edge of the cantilever, it is necessary to take into account the integrated response of the fraction of ultrasound incident on the surface.

The edge-spread function is defined as the integrated fraction of the radial distribution function incident on a sharp edge. Alternately, it can be modeled as a 2-d convolution between the square edge of the cantilever and the radial distribution function. Figure 10.4 shows two independent measurements of the edge-spread function determined using this

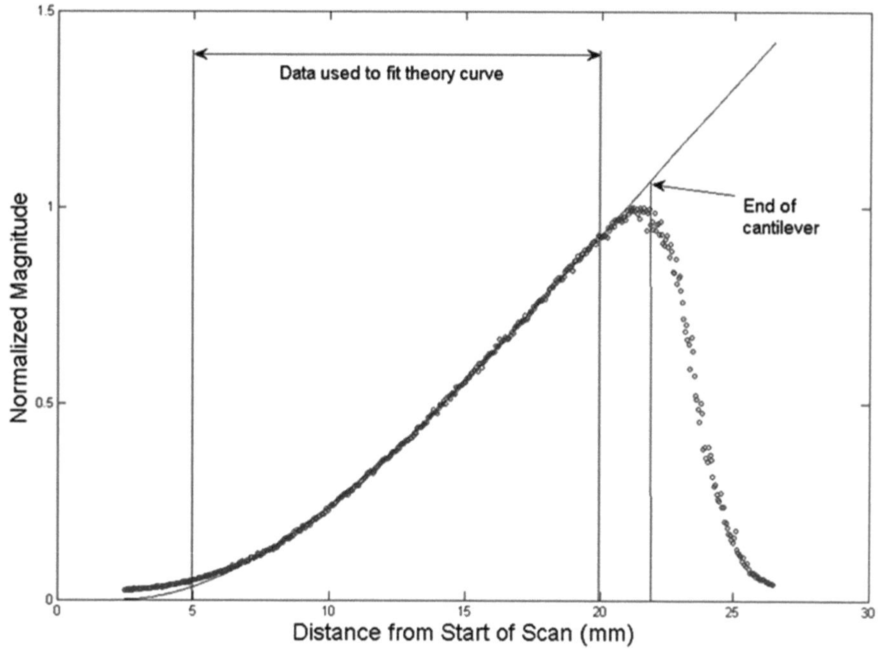

Fig. 10.3 Comparison of measurements of normalized response of cantilever as ultrasound focus point is moved along vertical axis of cantilever (*open circles*) and theoretical fit for response of cantilever to a point-source driving force (*solid line*). The reduction of response, particularly at distances above about 22 mm in this graph, is because only a fraction of the ultrasound focus point strikes the cantilever

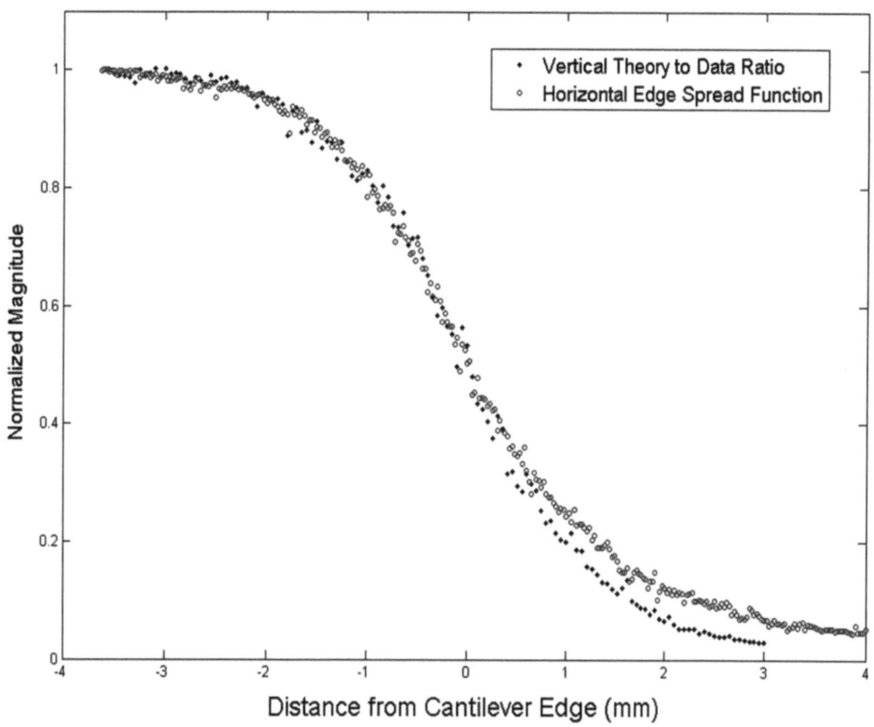

Fig. 10.4 Measurement of edge-spread function using horizontal scan (*open circles*) and vertical scan (*solid points*)

roving-transducer experiment. For the open circles, this plot shows the data set from the right-hand edge of Fig. 10.2, which is a horizontal scan across the width of the cantilever with the 50 % location shifted to correspond to $x = 0$. For negative positions less than -3 mm, essentially the entire ultrasound focus spot was striking the cantilever, thus the normalized response was essentially 1. For positive positions greater than about $+3$ mm, essentially the entire ultrasound focus spot was beyond the edge of the cantilever, leading to a very small normalized response. The shape of this curve between these limits is the measured edge-spread function for this transducer.

As an independent measurement of the edge-spread function, the vertical scan shown in Fig. 10.3 was used to determine the solid dots shown in Fig. 10.4. To create these points, the ratio between the measured response of Fig. 10.3 was divided by the theoretical prediction from a point source (solid line in Fig. 10.3). This ratio essentially measures how much the response decreases because the actual response is an integral of a finite-sized radial distribution that might extend beyond the edge of the cantilever. The x values for this plot were the distance of the center of the ultrasound focus spot from the free end of the cantilever. Again, when the transducer was more than about 3 mm below the free end of the cantilever (−3 mm on this plot), the normalized response was about 1, and this fell to near zero once the ultrasound focus point was more than about 3 mm above the free end of the cantilever. These two independent measurements of the edge-spread function are in very good agreement. The slight variation may be due to uncertainties in determining the location where the center of the ultrasound focus is exactly at the edge of the cantilever.

10.5 Conclusions

The experiments described demonstrate that it is possible to use the ultrasound radiation force of a roving transducer moving across a fixed-free cantilever to measure the edge-spread function of an ultrasound transducer. Future studies will involve utilizing this edge-spread function to determine the radial-distribution of the ultrasound radiation force. The combination of knowing the ultrasound radiation force applied to a surface and the measured operating deflection shapes using a scanning laser Doppler vibrometer, the goal is determination of frequency response functions using a fully non-contact method for both excitation and measurement.

Acknowledgements This material is based upon work supported by the National Science Foundation under Grant Nos. 0959858, and 1300591. Any opinions, findings and conclusions or recommendations expressed in this material are those of the author(s) and do not necessarily reflect the views of the National Science Foundation (NSF). The authors would like to thank B. Bjork for preliminary measurements and assistance with some of the analysis.

References

1. Huber TM, Abell BC, Mellema DC, Spletzer M, Raman A (2010) Mode-selective noncontact excitation of microcantilevers and microcantilever arrays in air using the ultrasound radiation force. Appl Phys Lett 97:214101
2. Huber TM, Beaver NM, Helps JR (2013) Noncontact modal excitation of a classical guitar using ultrasound radiation force. Exp Tech 37(4):38
3. Huber TM, Fatemi M, Kinnick RR, Greenleaf JF (2006) Noncontact modal analysis of a pipe organ reed using airborne ultrasound stimulated vibrometry. J Acoust Soc Am 119:2476
4. Fatemi M, Greenleaf JF (1998) Ultrasound stimulated vibro-acoustic spectography. Science 28:82
5. Fatemi M, Greenleaf JF (1999) Vibro-acoustography: an imaging modality based on ultrasound-stimulated acoustic emission. Proc Natl Acad Sci 96:6603
6. Greenleaf JF, Fatemi M (1999) Acoustic force generator for detection, imaging and information transmission using the beat signal of multiple intersecting sonic beams. US Patent 5,903,516
7. Chen S, Fatemi M, Kinnick RR, Greenleaf JF (2004) Comparison of stress field forming methods for vibro-acoustography. IEEE Trans Ultrason Ferroelect Freq Control 51(3):313
8. Fatemi M, Greenleaf JF (1999) Acoustic force generation by amplitude modulating a sonic beam. US Patent 5,921,928
9. Westervelt PJ (1951) Theory of steady force caused by sound waves. J Acoust Soc Am 23:312
10. Borgnis FE (1953) Acoustical radiation pressure of plane compressional waves. Rev Mod Phys 25:653
11. Fatemi M, Greenleaf JF (2000) A novel method for modal analysis of fine structures. In: Proceedings 2000 IEEE international ultrasonics symposium short courses, p 252, Oct 2000
12. Fatemi M, Greenleaf JF (2002) Mode excitation and imaging by the radiation force of ultrasound. J Acoust Soc Am 111:2472
13. Mitri FG, Trompette P, Chapelon J-Y (2003) Detection of object resonances by vibro-acoustography and numerical vibrational mode identification. J Acoust Soc Am 114:2648
14. Huber TM, Calhoun D, Fatemi M, Kinnick RR, Greenleaf JF (2006) Noncontact modal testing of hard-drive suspensions using ultrasound radiation force. In: Proceedings of international modal analysis conference (IMAC XXIV), 2 Feb 2006, paper 363; see http://physics.gac.edu/~huber/Presentations/imac_2006_february
15. Huber TM, Calhoun D, Fatemi M, Kinnick RR, Greenleaf JF (2005) Noncontact modal testing of hard-drive suspensions using ultrasound radiation force. J Acoust Soc Am 118:1928; see http://physics.gustavus.edu/~huber/presentations/asa_2005_october/
16. Huber TM, Purdham JC, Fatemi M, Kinnick RR, Greenleaf JF (2005) Noncontact mode excitation of small structures in air using ultrasound radiation force. J Acoust Soc Am 117:2455; see http://physics.gustavus.edu/~huber/presentations/asa_2005_may/
17. Huber TM, Beaver NM, Helps JR (2011) Elimination of standing wave effects in ultrasound radiation force excitation in air using random carrier frequency packets. J Acoust Soc Am 130:1838
18. de Silva CW (2007) Vibration: fundamentals and practice, 2nd edn. CRC Press/Taylor & Francis Group, Boca Raton, pp 346–351

Chapter 11
The Use of Fiber Bragg Grating Sensors for Strain Modal Analysis

Fábio Luis Marques dos Santos, Bart Peeters, Ludo Gielen, Wim Desmet, and Luiz Carlos Sandoval Góes

Abstract This paper discusses the use optical fiber Bragg grating (FBG) strain sensors for structural dynamics measurements and modal analysis. For some industrial applications, the use of strain sensors (combined or not with accelerometers) can bring benefits such as reduced size and weight. In many of these applications, FBG sensors lead the class of new sensor technologies that make dynamic strain measurements more attractive, with additional qualities such as the reduction of cabling, immunity to electromagnetic interference and higher sensor robustness. On the other hand, the main difficulty in the use of this technology is their integration and synchronization with other types of sensors, since their acquisition usually requires a separate specialized measurement unit. This is an important requirement in modal analysis, where synchronization between input and output measurements is a key issue that can directly affect the quality of the data. In this paper, FBG sensors are used in an experimental modal analysis, where their analogue signal is digitalized on the same way as the electrical sensors, guaranteeing synchronization.

Keywords Strain modal analysis • Dynamic strain • Strain field • Fiber bragg grating • Strain mode shape

11.1 Introduction

Strain gauges have been commonly used for static load testing, durability analysis and lifetime prediction in the mechanical industry [1–3]. Moreover, there are applications in which dynamic strain measurements are more usual, such as in load identification in Transfer Path Analysis, or sometimes the sensor size makes it for a more suitable choice, which is the case for gas turbine blade testing or helicopter and wind turbine blade measurements. Another application that can use dynamic strain measurements is strain modal analysis. Usually modal testing has been associated with the use of displacement responses (or their derivatives with respect to time), but the use of strain sensors for modal testing has been already discussed [4–7]. Recently there has been growing interest from both industry [8] and academia [9] on the topic—assessing and evaluating structural integrity on design prototype stages and also monitoring in real-time (with structural health monitoring systems (SHM)) has led to an increase in the number of dynamic strain applications, to the development of improved identification and measurement techniques, as well as to improved sensor technology.

F.L.M. dos Santos (✉)
Siemens Industry Software, Interleuvenlaan 68, 3001 Leuven, Belgium

Katholieke Universiteit Leuven (KUL), Division PMA, Celestijnenlaan 300B, 3001 Heverlee, Belgium
e-mail: fabio.m.santos@siemens.com

Institute Tecnológico de Aeronáutica (ITA), Praça Marechal Eduardo Gomes,
50-Vila das Acácias CEP 12.228-900 São José dos Campos SP, Brazil

B. Peeters • L. Gielen
Siemens Industry Software, Interleuvenlaan 68, 3001 Leuven, Belgium
e-mail: bart.peeters@siemens.com; ludo.gielen@siemens.com

W. Desmet
Katholieke Universiteit Leuven (KUL), Division PMA, Celestijnenlaan 300B, 3001 Heverlee, Belgium
e-mail: wim.desmet@kuleuven.be

L.C.S. Góes
Institute Tecnológico de Aeronáutica (ITA), Praça Marechal Eduardo Gomes,
50-Vila das Acácias CEP 12.228-900 São José dos Campos SP, Brazil
e-mail: goes@ita.br

M. Mains (ed.), *Topics in Modal Analysis, Volume 10*, Conference Proceedings of the Society for Experimental Mechanics Series, DOI 10.1007/978-3-319-15251-6_11

A recent and important contribution on the field of dynamic strain measurements are the fiber optic sensors, of which the most popular types are the Fiber Bragg Grating (FBG) sensors [10, 11]. Some of their advantages are robustness to magnetic interference, small sensor dimensions and weight, reduced cabling, possibility to embed on composites plus the possibility of embedding these sensors in composite structures, makes for an attractive solution for use in SHM systems. The availability of such an array of sensors, ready to be used and adequate for modal testing, is another incentive to carrying out a strain modal analysis, saving up on time and instrumentation.

Another application of dynamic strain measurements is related to the strain displacement relations [12]. In many systems, strain gauges are used as the standard vibration sensor, especially when size or sensor location is an issue. Such is the case in aerospace applications, like gas turbines, wind turbines and helicopters [13], where size and weight are very restricted, and any sensor place on a blade should affect its aerodynamic properties as little as possible. One particular use of the strain measurements and strain to displacement relations is the strain pattern analysis (SPA), where strain measurements are used to predict blade displacements.

11.2 Strain Modal Analysis Theory

To obtain the strain modal formulation, one can start with the fundamental theory of modal analysis. Modal theory states that the displacement on a given coordinate can be approximated by the summation of a n number of modes:

$$u(t) = \sum_{i=1}^{n} \phi_i q_i(t) \tag{11.1}$$

where u is the displacement response in x direction, ϕ_i is the ith (displacement) vibration mode, and q_i is the generalized modal coordinate and t is time. For small displacements, given the theory of elasticity, the strain/displacement relation is:

$$\varepsilon_x = \frac{\partial}{\partial x} u \tag{11.2}$$

And similarly, the same relationship exists between the strain vibration modes and the displacement modes:

$$\psi_i = \frac{\partial}{\partial x} \phi_i \tag{11.3}$$

This way, by the relations on Eqs. (11.2) and (11.3), the expression on (11.1) can be rewritten as:

$$\varepsilon(t) = \sum_{i=1}^{n} \psi_i q_i(t) \tag{11.4}$$

Moreover, the relationship between the generalized modal coordinate q and an input force F is:

$$q_i = \Lambda_i^{-1} \phi_i F, \mathrm{with} \Lambda_i = (-\omega^2 m_i + j\omega c_i + k_i) \tag{11.5}$$

where m_i, c_i and k_i are the ith modal mass, modal damping and modal stiffness, and ω is the excitation frequency.

Substituting (11.5) into (11.4), the relation between a force input and a strain output, in terms of displacement and strain modes is represented as:

$$\varepsilon_i = \sum_{i=1}^{n} \psi_i \Lambda_i^{-1} \phi_i F \tag{11.6}$$

And finally, the strain frequency response function (SFRF) can be obtained, in matrix form:

$$[H^{\varepsilon}] = \sum_{i=1}^{n} \Lambda_i^{-1} \{\psi_i\} \{\phi_i\} = [\psi] [\Lambda]^{-1} [\phi]^T \tag{11.7}$$

The expansion of (11.7) is:

$$\begin{bmatrix} H_{11}^{\varepsilon} & H_{12}^{\varepsilon} & \cdots & H_{1N_i}^{\varepsilon} \\ H_{21}^{\varepsilon} & H_{22}^{\varepsilon} & \cdots & H_{2N_i}^{\varepsilon} \\ \vdots & \vdots & \vdots & \vdots \\ H_{N_o1}^{\varepsilon} & H_{N_o2}^{\varepsilon} & \cdots & H_{N_oN_i}^{\varepsilon} \end{bmatrix} = \sum_{i=1}^{n} \Lambda_i^{-1} \cdot \begin{bmatrix} \psi_{1i}\phi_{1i} & \psi_{1i}\phi_{2i} & \cdots & \psi_{1i}\phi_{N_i i} \\ \psi_{2i}\phi_{1i} & \psi_{2i}\phi_{2i} & \cdots & \psi_{2i}\phi_{N_i i} \\ \vdots & \vdots & \vdots & \vdots \\ \psi_{N_o i}\phi_{1i} & \psi_{N_o i}\phi_{2i} & \cdots & \psi_{N_o i}\phi_{N_i i} \end{bmatrix} \tag{11.8}$$

where N_o represents the number of strain gauge measurement stations (or the number of output measurements) and N_i represents the number of excitation points (or the number of inputs).

The columns of the matrix correspond to the strain responses due to the excitation points along the rows of the matrix. Some important characteristics can be inferred from Eq. (11.8). First of all, differently from displacement FRFs, the SFRF matrix is not symmetric, that is, for instance, $H_{12}^{\varepsilon} \neq H_{21}^{\varepsilon}$. This means that reciprocity is not guaranteed for strain modal analysis—exciting point a and measuring point b will not yield the same FRF as if exciting point b and measuring point a. Moreover, any column of the SFRF matrix contains all the information regarding the strain modes (ψ), while any row of the SFRF matrix contains information about the displacement modes (ϕ). This particular property leads to practical applications—to obtain the strain mode shapes, one must use a fixed excitation point and measure the strain responses. On the other hand, by using a strain gauge as a fixed reference sensor and moving the excitation point (as with impact testing), the displacement mode shapes can be obtained.

Due to the similarity of the strain modal formulation and the displacement modal formulation, the same modal identification methods can be used in both cases, as long as the appropriate caution is taken. In this article, the PolyMAX identification method [14] was used without any modifications.

11.3 FBG Measurement Principle

The fiber Bragg grating (FBG) is a periodic microstructure that acts as a wavelength selective mirror. This means that if light from a broadband source is injected in the optical fiber, only light within a very narrow spectral width, centered at the Bragg wavelength, will be back-reflected by the grating. The remaining light will continue its way through the optical fiber without experiencing any loss. The fiber Bragg grating is a symmetric structure, so it will always reflect light at the Bragg wavelength no matter which side the light is coming from.

The Bragg wavelength (λ_B) is essentially defined by the period of the microstructure (Λ) and the index of refraction of the core (n_{eff}):

$$\lambda_B = 2n_{\text{eff}}\Lambda \tag{11.9}$$

A fiber Bragg grating has unique characteristics to perform as a sensor. For example, when the fiber is stretched or compressed, the FBG will measure strain. This happens essentially because the deformation of the optical fiber leads to a change in the period of the microstructure and, consequently, of the Bragg wavelength. There is also some contribution from the variation of the index of refraction, through the photo-elastic effect. Sensitivity to temperature is also intrinsic to a fiber Bragg grating. In this case, the main contributor to Bragg wavelength change is the variation of the silica refraction index, induced by the thermo-optic effect. There is also a contribution from the thermal expansion, which alters the period of the microstructure. This effect is, however, marginal given the low coefficient of thermal expansion of silica.

The strain dependence of a fiber Bragg grating can be determined by differentiating the wavelength:

$$\frac{\Delta\lambda_B}{\lambda_B} = \frac{\Delta(n_{\text{eff}}\Lambda)}{n_{\text{eff}}\Lambda} = \left(1 + \frac{1}{n_{\text{eff}}} \cdot \frac{\partial n_{\text{eff}}}{\partial \varepsilon}\right)\Delta\varepsilon = (1 + p_e)\Delta\varepsilon = \beta_\varepsilon \Delta_\varepsilon \tag{11.10}$$

where β_ε is the strain sensitivity of the Bragg grating, p_e is the photo-elastic constant (variation of the index of refraction with axial tension), which for the optical fiber is equal to: $p_e \approx -0.212$, so the strain sensitivity of a FBG from (11.10) becomes:

$$\frac{\Delta\lambda_B}{\Delta_\varepsilon} = \beta_\varepsilon \cdot \lambda_B = 0.788\lambda_B \tag{11.11}$$

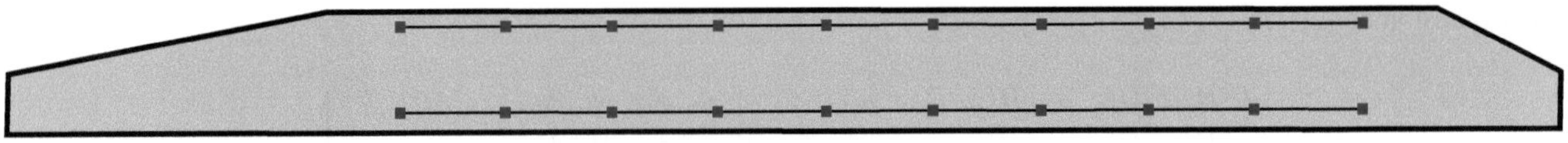

Fig. 11.1 Helicopter blade: location of FBG sensors and strain gauges

and therefore an FBG with wavelength $\lambda_B = 1,550$ nm, will have the following sensitivity:

$$\frac{\Delta\lambda_B}{\Delta_\varepsilon} = 1.2\frac{pm}{\mu\varepsilon} \tag{11.12}$$

Multiple gratings can be used within a single fiber, as long as each one has a different wavelength. There are two methods of interrogating multiple wavelengths - a broadband light source that covers all the wavelengths, or a tunable light source that sweeps through all the wavelengths.

11.4 Experimental Analysis

Two analysis cases will be presented—initially, an experimental modal analysis on a helicopter main rotor blade will be shown. For this case, a standalone acquisition unit will be used, and the synchronization procedure with the rest of the electrical sensors will be described. Then a synchronized heterogeneous acquisition on the same structure will be carried out, with impact tests and the impulse responses from collocated strain gauges and FBG sensors will be shown.

11.4.1 Unsynchronized FBG Measurements

The main rotor blade of a PZL SW-3 helicopter was used for the strain modal analysis with the FBG sensors. For this experiment, the blade was suspended with elastic cords to obtain a free-free boundary condition and fiber bragg grating sensors were used to measure the dynamic strain on the surface of the blade. In total, 20 FBG sensors were instrumented on the surface of the blade, following two straight fiber lines of ten sensors each, and two strain gauges (in quarter bridge configuration) were used on one of the fiber lines to help with the synchronization. Figure 11.1 shows the position of the strain gauges and the FBG sensors on the two fiber lines on the surface of the blade.

The blade was excited using an electrodynamic shaker, and the driving point was chosen close to the tip of the blade, near the trailing edge, with a sine sweep excitation. Two acquisition units were used—an LMS Scadas Mobile with VB8 modules for the strain gauges and a separate acquisition unit for the FBG sensors—the interrogation method for this unit is based on a tunable light source that sweeps through all the FBG wavelengths. The excitation frequency ranged from 2 to 80 Hz, with the sampling rate from the FBG acquisition system being 200 Hz. In the specified bandwidth, 11 vibration modes were identified—six bending modes, three in-plane modes and two torsional modes. The natural frequencies and the mode types for these modes are shown on Table 11.1. The displacement modes of the blade (and their types) were already known from previous experiments [15].

The next step for the strain modal analysis using the FBG sensors is to synchronize their signals with the measured force signal. Since the resistive strain gauges, accelerometers and force cell signals were all acquired with the same acquisition unit and are therefore synchronized, the FBG sensors have only to be synchronize them with one of these sensors—hence, the use of the collocated strain gauges. This synchronization procedure is carried out in an offline manner, or practically speaking, after the data acquisition has already been done. The steps for the data synchronization are as follows:

- Selection of the best suitable strain gauge to be used for the synchronization
- Division of the data by blocks equal to number of averages
- Synchronization (alignment) of the data, block-by-block
- Reassembly of all the blocks in one data signal
- FRF calculation
- Data import into LMS Test.Lab for modal analysis

Table 11.1 Natural frequencies and mode types for the helicopter main rotor blade

Mode number	Natural frequency (Hz)	Damping ratio (%)	Mode type
1	3.6	1.72	Bending
2	10.3	0.50	Bending
3	14.0	0.84	In-plane
4	20.3	0.37	Bending
5	30.4	0.98	Torsional
6	33.8	0.4	Bending
7	37.6	0.66	In-plane
8	49.5	0.52	Bending
9	61.2	0.93	Torsional
10	67.2	0.43	Bending
11	75.9	0.74	In-plane

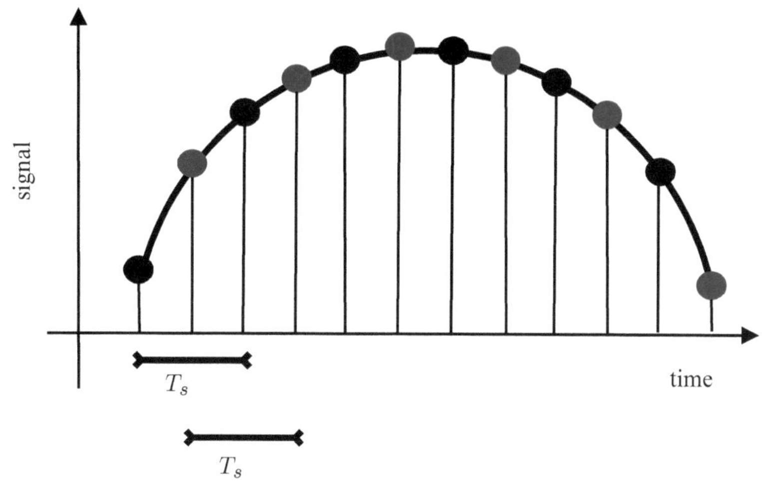

Fig. 11.2 Sampling time error example

After an initial analysis, the strain gauge and FBG sensor on point 4 (from Fig. 11.1) were chosen for the synchronization procedure—overall, all sensor pairs were suitable to be used, but one pair had to be chosen. Next, all signals were upsampled to improve the offline synchronization efficiency. This step can help to reduce errors—even if both signals would have the same sampling frequency, it is still possible that the samples are taken at a different time and therefore unsynchronized. Figure 11.2 shows an example of how this can happen—even though the sampling time T_s shown in the Figure is the same for both signals (black and grey), they can still be shifted in time. By reducing the sampling period T_s, one can reduce how big this error will be. The resampling factor for the strain gauge was of 4, and the resampling factor for the FBG sensor was of 16, bringing both sensors' sampling frequency up to 3,200 Hz.

Furthermore, the signals were divided by blocks—in total, they were divided in 20 blocks representing the 20 excitation cycles for the sine sweep. As a standard procedure, the first and last blocks are also discarded, so in the end 18 blocks were available. The next step is to align each block individually. This is carried out by using the cross-correlation function. For 2 very similar signals, synchronized in time, the cross correlation function should have its peak value exactly on the 0 lag position in the x-axis. If the signals are misaligned (which is our case), then the peak value will occur outside of the 0 lag position, but will represent how many lags (or sample differences) one of the signals should be shifted to be aligned. The result, after realigning all 18 blocks and putting them back together into one signal, is shown in Fig. 11.3, where one of the realigned blocks is shown.

Consequently, the FRFs can be computed by using the H1 estimator. To calculate the crosspower and autopower functions to be used in the H1 estimator calculations, a rectangular window was used and no overlap was performed (each block consisted of a full sweep, starting and ending with 0 excitation, so leakage was not a problem). The resulting FRFs for the one of the FBG and strain gauge sensor pairs and their respective coherence functions are shown in Fig. 11.4, where a comparison between the strain gauge and FBG signal quality can be made.

Finally, the FRFs can be imported in LMS Test.Lab so that the modal analysis can be carried out. For this purpose, only the FBG sensor FRFs are really needed, since the strain gauge measurements were only used initially for the synchronization. The procedure to identify the modes is the same as in the classic displacement modal analysis. The PolyMAX identification

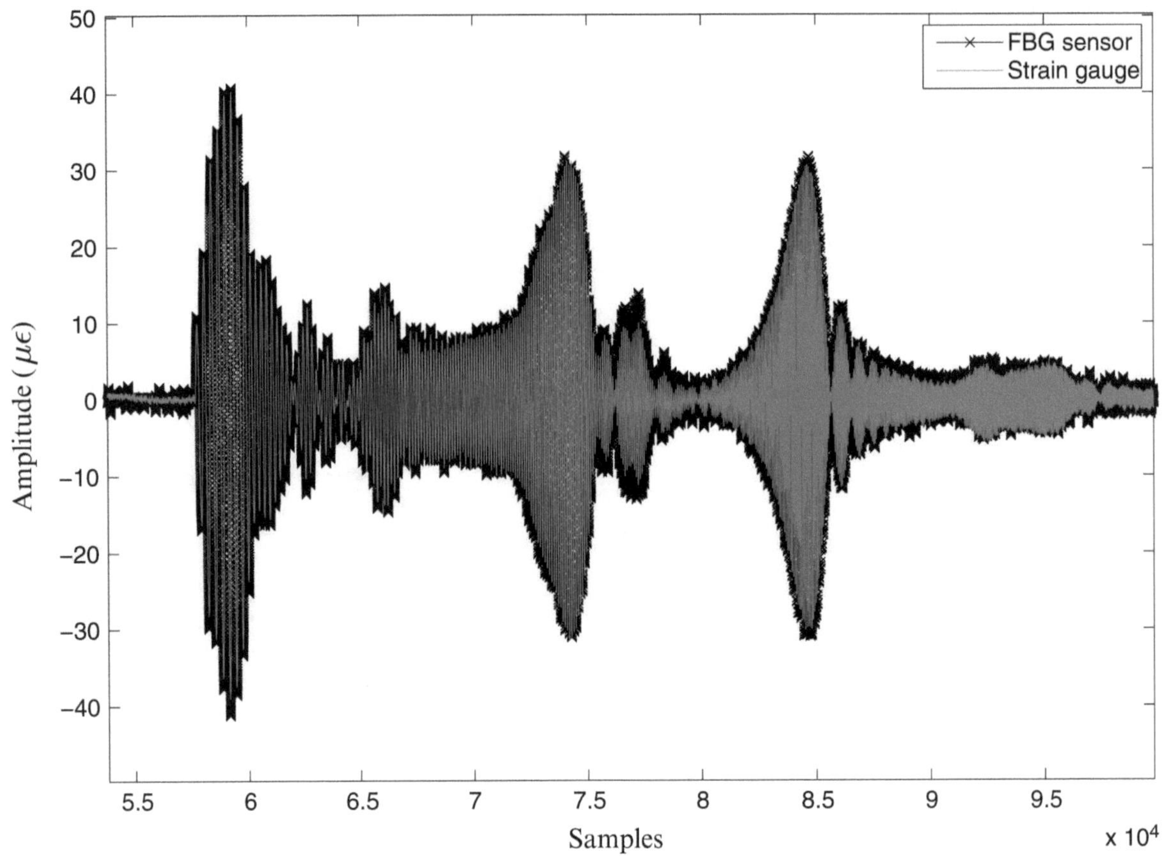

Fig. 11.3 FBG and strain gauge time signal alignment: zoomed in one of the blocks

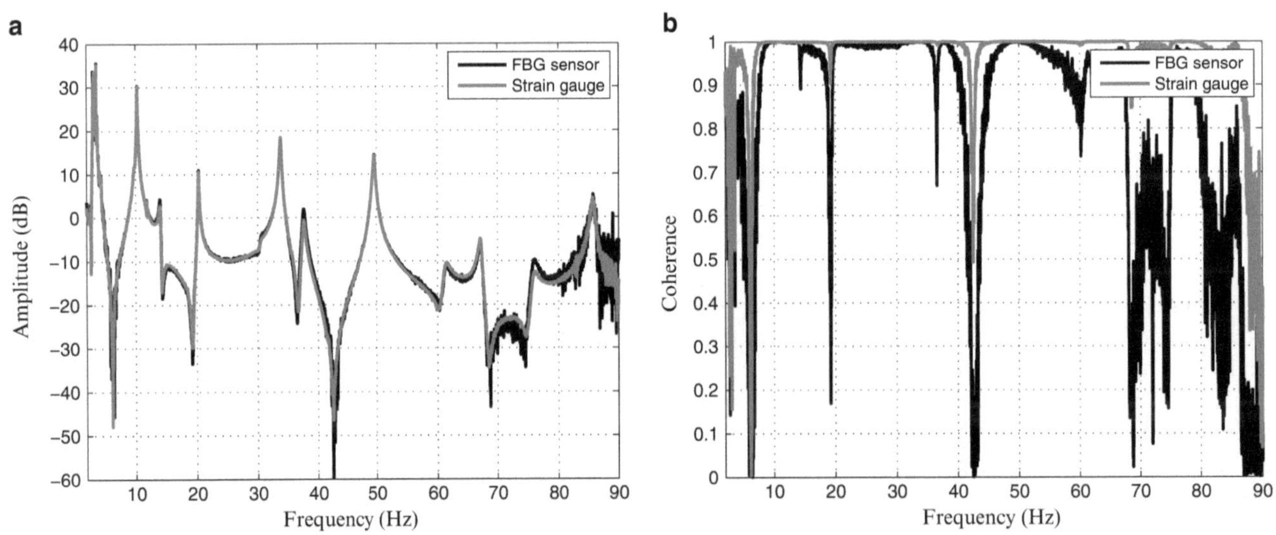

Fig. 11.4 Comparison of FRFs and coherence for one of the sensor pairs. (**a**) FRF comparison for sensor pair on point 4. (**b**) Coherence comparison for sensor pair on point 4

algorithm was used and all the 11 strain mode shapes shown in Table 11.1 were identified. The first five of these mode shapes are shown in Figs. 11.5, 11.6, 11.7, 11.8, and 11.9.

Analyzing the acquired data from the strain gauges and FBG sensors, it is visible that the FRFs match very well up to 65 Hz. After this frequency, there is more noise present on the FBG FRF (Fig. 11.4a), which is also visible in the coherence function Fig. 11.4b. This decrease in quality can be attributed to two facts—as the frequency gets closer to the maximum

Fig. 11.5 Helicopter blade strain mode—first bending mode at 3.58 Hz

Fig. 11.6 Helicopter blade strain mode—second bending mode at 10.27 Hz

Fig. 11.7 Helicopter blade strain mode—first in-plane mode at 13.9 Hz

Fig. 11.8 Helicopter blade strain mode—third bending mode at 20.30 Hz

sensor bandwidth (100 Hz), it is expected that the quality might decrease. Moreover and most importantly, the described offline synchronization procedure works better for lower frequencies, while in higher frequencies it is still expected that the synchronization will not be perfect, which is the case for the FBG sensors.

11.4.2 Synchronized Measurements

The way that the measurements were carried out with the FBG sensors meant that the quality was limited by the offline synchronization procedure. To overcome this problem, a different methodology for interrogating the FBG wavelengths was used by using a broadband light source that covers all the FBG wavelengths. In this way, it is possible to obtain an analogue signal from the FBG sensors, which can be connected directly to the same acquisition system as the other sensors, guaranteeing synchronous measurements with higher sampling rates. The downside is that less sensors are supported per fiber, as the broadband light source has a much more limited bandwidth.

To check for synchronicity between the sensors, impact tests were carried out, measuring the signals from two FBG sensors and collocated resistive strain gauges, their position number (4 and 7) shown in Fig. 11.1 an both sensor pairs can be seen in Fig. 11.10.

The time trace for the impulse response of the sensor pairs can be seen in Fig. 11.11a and b. In this case, no synchronization procedure was necessary, as the sensor acquisition is done using the same acquisition unit. Furthermore, the noise levels for both sensor types was investigated. Figure 11.12 shows the noise levels for both types of sensors. The FBG sensor has lower noise level and additionally, there is a high amplitude harmonic component at 50 Hz on the quarter bridge strain gauge, due to the noise from the electrical network.

11.5 Results Analysis and Conclusion

In this paper, the theory for strain modal analysis was presented, followed by an introduction on the measurement principle for Fiber Bragg Grating sensors. For this purpose, a brief explanation of the basic properties of the sensor were given and how it can be used to measure strain.

Moreover, it was shown how FBG sensors can be used to carry out a strain modal analysis and visualize the strain mode shapes. To be able to properly do the modal identification procedure, an offline synchronization procedure had to be carried out, as a means of synchronizing the signals from the FBG sensors to the signal from the force sensor. The results show that

Fig. 11.9 Helicopter blade strain mode—first torsional mode at 30.4 Hz

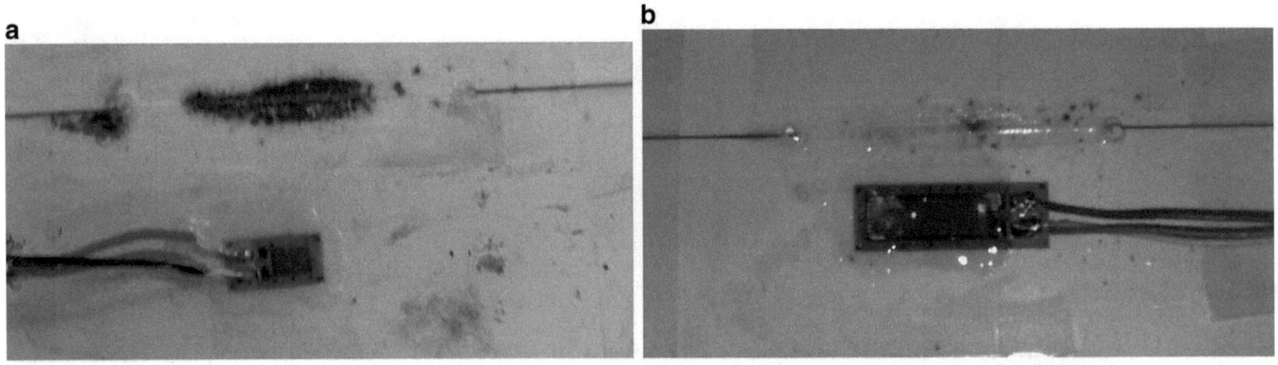

Fig. 11.10 Collocated FBG sensors and strain gauges on helicopter blade. (**a**) FBG sensor and strain gauge collocated on point 4. (**b**) FBG sensor and strain gauge collocated on point 7

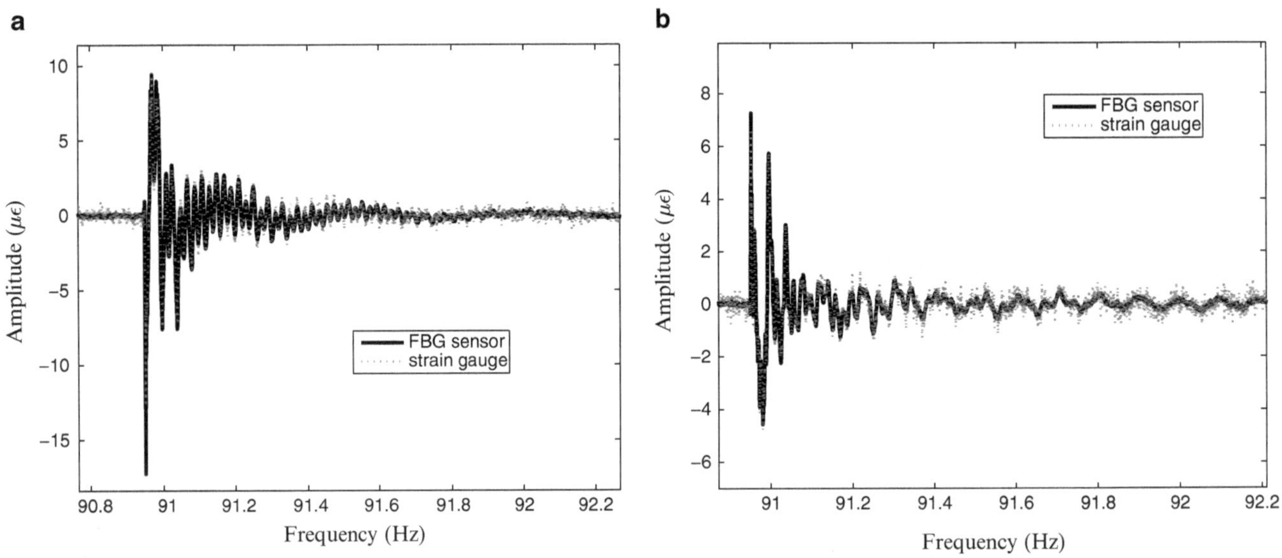

Fig. 11.11 Impulse response for sensor pairs on locations 4 and 7 on blade. (**a**) FBG and strain gauge impulse response: synchronized signal acquisition on point 4. (**b**) FBG and strain gauge impulse response: synchronized signal acquisition on point 7

the strain mode shapes can be identified, but there is a limitation on the fact that the offline synchronization procedure is less efficient as the frequency gets higher. To overcome this limitation, a different acquisition principle for the FBG sensors was used, which could yield as an output an analogue signal that could be directly acquired by the same acquisition unit as the electrical sensors, guaranteeing synchronization. A noise analysis of both strain gauges and FBG sensors was carried out, and it was seen that the FBG could yield lower levels of noise.

Future studies include more complex measurements and a full modal analysis with the synchronized FBG sensors and use of the sensors in other strain applications.

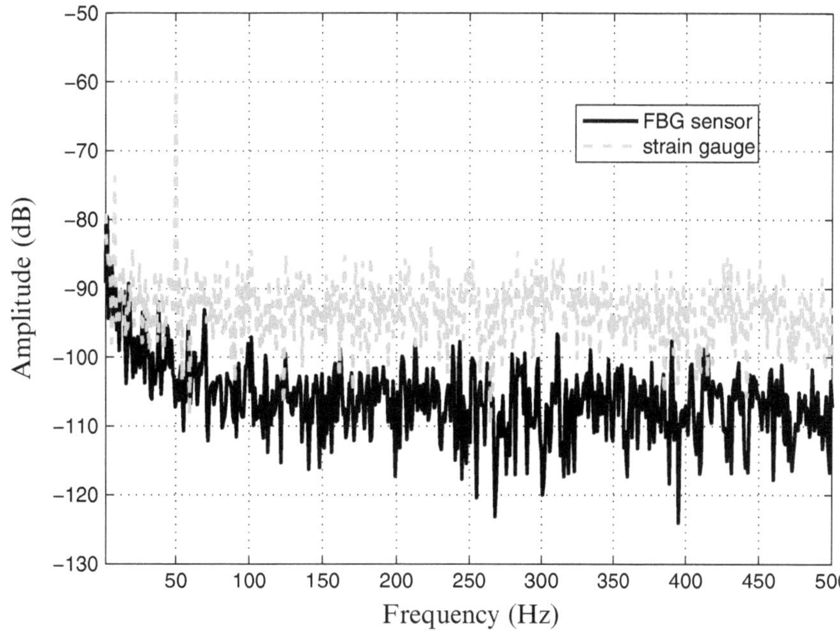

Fig. 11.12 FBG sensor and strain gauge noise levels

Acknowledgements Fábio Luis Marques dos Santos, first author of this paper, is an Early Stage Researcher at Siemens Simulation & Test Solutions, under the FP7 Marie Curie ITN project "IMESCON" (FP7-PEOPLE-2010-ITN, Grant Agreement No. 264672). This research was also carried out in the Framework of FP7 ICT Collaborative project "WiBRATE" (FP7-ICT-2011-7, Grant Agreement No. 289041). The authors of this work gratefully acknowledge the European Commission for the support.

References

1. Vandepitte D, Sas, P (1990) Case study of fracture of a mechanical component due to resonance fatigue. Mech Syst Sig Process 4(2):131–143.
2. Liefooghe C, Van der Auweraer H, Janssen P, Sas P (1992) Validation of modal filtering/editing approach to dynamic fatigue analysis. In: Proceedings of ISMA 17, Tokyo
3. Wentzel H (2013) Fatigue test load identification using weighted modal filtering based on stress. Mech Syst Sig Process 40:618–627
4. Bernasconi O, Ewins DJ (1989) Application of strain modal testing to real structures. In: Proceedings of the 7th international modal analysis conference, vol 2, pp 1453–1464
5. Vári LM, Heyns PS (1994) Using strain modal testing. In: Proceedings of the 12th international conference on modal analysis, vol. 2251, p 1264
6. Li S, Wu Z (2005) Structural identification using static macro-strain measurements from long-gage fiber optic sensors. J Appl Mech 8:943–948
7. Reich GW, Park KC (2001) A theory for strain-based structural system identification. J Appl Mech 68(4):521–527
8. Peeters B, Luis Marques dos Santos F, Pereira A, Araujo F (2014) On the use of optical fiber bragg grating (fbg) sensor technology for strain modal analysis. In: 11th international conference on vibration measurements by laser and noncontact techniques-AIVELA 2014: Advances and applications, vol 1600. AIP Publishing, New York, pp 39–49
9. Kranjc T, Slavič J, Boltežar M (2014) A comparison of strain and classic experimental modal analysis. J Vib Control. doi: 1077546314533137
10. Kang L-H, Kim D-K, Han J-H (2007) Estimation of dynamic structural displacements using fiber bragg grating strain sensors. J Sound Vib 305(3):534–542
11. Ling H-Y, Lau K-T, Cheng L (2004) Determination of dynamic strain profile and delamination detection of composite structures using embedded multiplexed fibre-optic sensors. Compos Struct 66(1):317–326
12. Pisoni AC, Santolini C, Hauf DE, Dubowsky S (1995) Displacements in a vibrating body by strain gage measurements. In: Proceedings of the 13th international conference on modal analysis
13. Tourjansky N, Edmond S (1992) The measurement of blade deflections - a new implementation of the strain pattern analysis. ONERA Technical Paper
14. Peeters B, Van der Auweraer H, Guillaume P, Leuridan J (2004) The polymax frequency-domain method: a new standard for modal parameter estimation? Shock Vib 11:395–409
15. Luis Marques dos Santos F, Peeters B, Van der Auweraer H, Carlos Sandoval Góes L (2013) Modal-based damage detection of a composite helicopter main rotor blade. 54th AIAA/ASME/ASCE/AHS/ASC structures, structural dynamics, and materials conference

Chapter 12
Using Mode Shapes for Real Time ODS Animation

Brian Schwarz, Shawn Richardson, and Mark Richardson

Abstract In a rotating machine, the dominant forces are applied at multiples of the machine running speed, called orders. An ***order-tracked ODS*** is assembled from the peaks at one of the order frequencies in a set of response frequency spectra of a machine. An order-tracked ODS is a convenient way to visualize and monitor the health of the machine.

In this paper, it is shown how ***modes participate*** in an order-tracked ODS of a rotating machine, and how they participate differently at different operating speeds. It is also shown how the ***modal participation*** can be used to ***expand*** an order-tracked ODS so that it is suitable for display on a model of the machine. With an animated ODS display, changing vibration levels and vibration ***hot spots*** can be observed while the machine is running.

Keywords Operating deflection shapes (ODS's) • Experimental modal analysis (EMA) mode shapes • Finite element analysis (FEA) mode shapes • Modal assurance criterion (MAC)

12.1 Introduction

It is well known that most rotating machines will exhibit different vibration levels under different loads and speeds. A convenient way to troubleshoot these problems is to visualize the operating deflection shapes (ODSs) of the machine in ***"real time"***. This is conveniently done by attaching multiple accelerometers to the machine surfaces, and acquiring vibration data from the machine while it is running. In addition to visualizing the deflection of the machine in real time, an ODS can be used to diagnose machine faults by numerically comparing its current values with previously archived values.

All machines and mechanical structures have resonances, also called modes of vibration. Each resonance has a ***unique natural frequency***, ***damping value,*** and ***mode shape***. It is well known that if a resonance is excited, the response of the machine or structure can be excessive. In this sense, modes are referred to a ***"mechanical amplifiers"***. Over time, sustained excessive vibration levels will cause a variety of failures.

When excited, ***modes participate in*** or ***contribute to*** the overall vibration response of a machine. In an ideal sense, a mechanical structure has an infinite number of modes, but in a practical sense only a few modes participate significantly in its response.

In this paper, it is first shown how ***modal participation*** in an ODS is calculated. Once the ***modal participation*** is known, mode shapes with many components in them can be used to ***expand*** the ODS, thus providing a more accurate and realistic description of the machine's vibration. When displayed in animation, an expanded ODS is useful for observing overall vibration levels and for quickly spotting ***areas of excessive vibration (hot spots)*** under different operating conditions.

12.1.1 Operating Deflection Shape (ODS)

If two or more sensors (located at different points and/or in different directions) are used to measure the response of a machine while it is running, this data is called an **Operating Deflection Shape (ODS)**. An ODS is a ***vector of complex values***, each component of which has a magnitude & phase, or real & imaginary parts. When an ODS is displayed in animation on a model of the machine, it shows how each test point is deflected with a magnitude & phase ***relative*** to the deflection of all other test points.

B. Schwarz • S. Richardson • M. Richardson (✉)
Vibrant Technology, Inc., Scotts Valley, CA, USA
e-mail: mark.richardson@vibetech.com

M. Mains (ed.), *Topics in Modal Analysis, Volume 10*, Conference Proceedings of the Society for Experimental Mechanics Series, DOI 10.1007/978-3-319-15251-6_12

In order to display it in animation, an ODS must have the correct ***relative magnitude & phase*** between all its components. This relative magnitude & phase requirement among its components is the reason why it is called a deflection ***"shape"***.

12.1.2 Ways to Acquire ODS Data

There are two ways to acquire an ODS so that the relative magnitudes & phases of all its components are valid;

1. Acquire ***data simultaneously from all sensors***
   A multi-channel simultaneous acquisition system is required in order to acquire data from all sensors at once. An ODS can be assembled either from the ***same sample*** of data in simultaneously acquired time records, or from the ***same sample*** of data in multiple frequency (Fourier) spectra which were calculated from the acquired time records.
2. Acquire ***data simultaneously from at least two sensors***, with one sensor remaining at the ***same fixed reference sensor*** location throughout the acquisition process.
   In this case, data is acquired sequentially in ***multiple measurement sets***, with each set containing data from ***one or more roving sensors*** and the ***same (fixed) reference sensor***. In order to preserve the correct relative magnitude & phase in all shape components, a ***cross channel*** function must be calculated between all roving responses and the reference response. A Cross spectrum between the roving and reference responses is commonly calculated. Another function called an ODS FRF can be calculated. An ODS FRF is the Auto spectrum of a roving response combined with the phase of the Cross spectrum between the roving and reference responses.

12.2 Order-Tracked ODS

In a rotating machine, excitation forces are caused by unbalances, shaft misalignments, broken or misaligned gear teeth, reciprocating parts, and many other faults. All of these faults involve rotating parts, and therefore create excitation forces that are ***cyclic*** in nature. These forces are functions of the running speed of the machine and are called ***"orders"***. The first order is equal to the running speed, the second order equal to twice the running speed, the third order equal to three times the running speed, and so on. In a rotating machine, the dominant forces are applied only at multiples of the machine running speed. In other words, forces are applied only in ***discrete narrow frequency bands.***

An ODS assembled from the peaks at one of the order frequencies in a set of response frequency spectra is called an ***order-tracked ODS***. The machine speed should remain unchanged during data acquisition, but could change between acquisitions due to varying operating conditions. Therefore, in order to obtain an order-tracked ODS, the machine speed must also be measured (using a tachometer), and order peaks located in the spectra from which an order-tracked ODS can be assembled.

12.3 Variable Speed Rotating Machine

Figure 12.1 shows a variable speed rotating machine, instrumented with eight tri-axial accelerometers. An accelerometer is attached to the top of each bearing block, and six accelerometers are attached to the base plate; three on the front edge and three on the back edge. This test setup was used to measure order-tracked ODSs under different machine speeds.

A laser tachometer with its beam pointed at the outer wheel of the machine was used to measure the machine speed, as shown in Fig. 12.1. The outer wheel had reflective tape on it, so the laser measured the once-per-revolution speed of the machine.

A spectrum of the tachometer signal is shown in Fig. 12.2. The running speed of the machine is obtained from the ***lowest frequency peak*** in the tachometer spectrum.

Figure 12.3 contains a model of the machine that was used to display ODSs in animation. Each of the test points is displayed as a cube icon and is numbered.

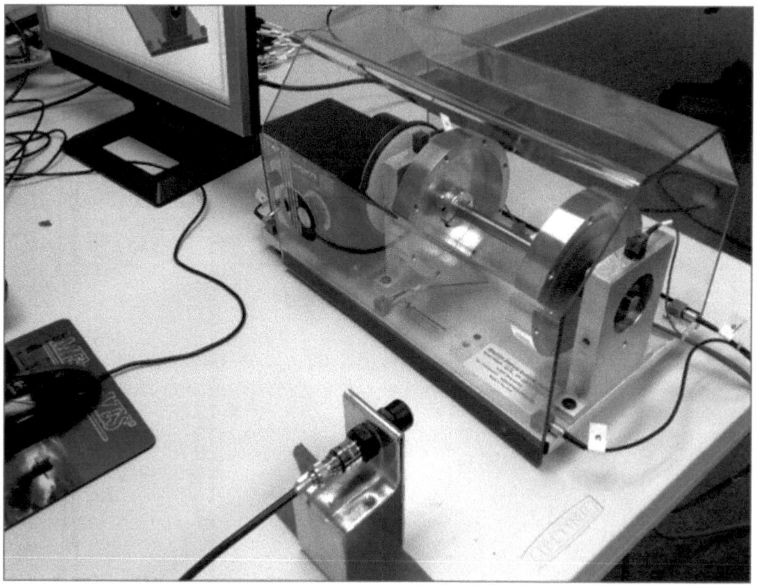

Fig. 12.1 Variable speed rotating machine

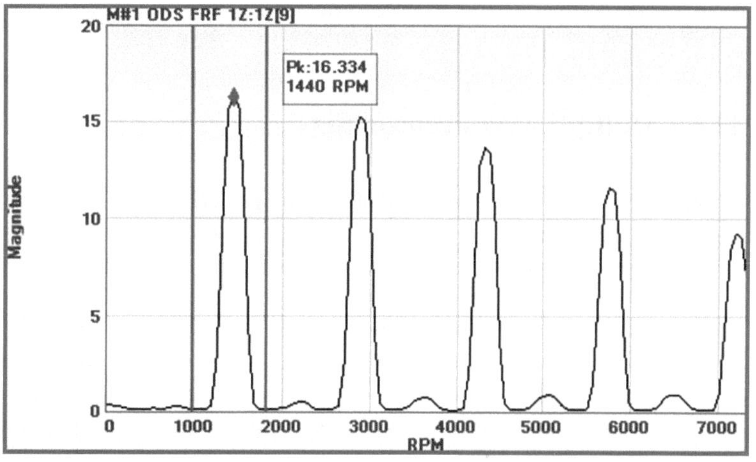

Fig. 12.2 Laser tachometer spectrum

Fig. 12.3 Rotating machine model

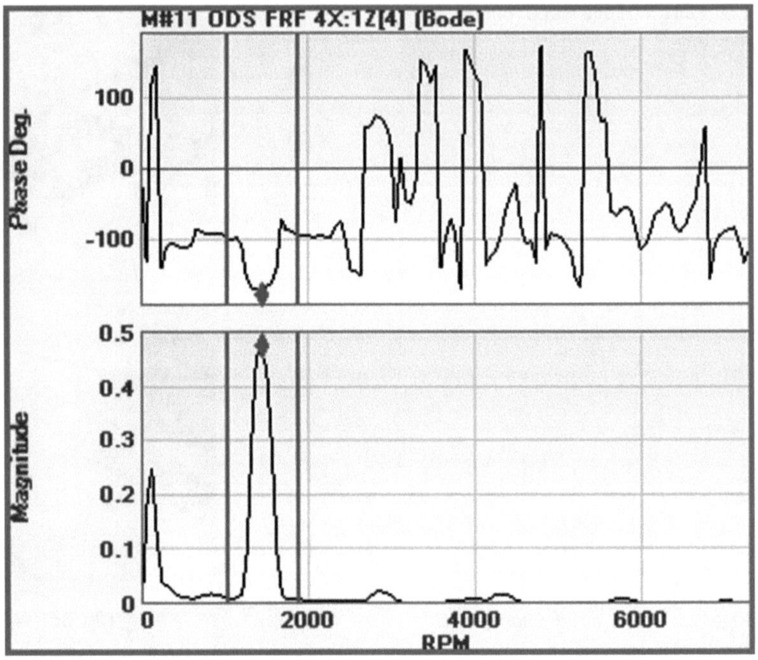

Fig. 12.4 Typical ODS FRF

12.4 Multiple Measurement Sets

A total of 24 ODS FRFs were calculated from data acquired sequentially from the machine in ***eight measurement sets***. Each measurement set was acquired with a 4-channel analyzer that simultaneously acquired four signals of accelerometer data. In each measurement set, three time waveforms were acquired from one of the tri-axial accelerometers, together with a time waveform from the reference accelerometer. Accelerometer 1Z, which measured the Z direction acceleration on the outside bearing block, was used as the reference.

Figure 12.4 shows a typical ODS FRF acquired from the rotating machine. Its magnitude is the Auto spectrum of the acceleration response, and its phase is the phase of the Cross spectrum between the response and the reference response.

Notice that the units of the ODS FRF are velocity units, indicating that it has been integrated from acceleration to velocity. Notice also that the magnitude peak appears to be somewhat ***"widened"***. This is because the time waveforms were windowed with a ***"flat top"*** window. When a waveform is a ***narrow band***, this window makes the peak values at the order frequencies more accurate [1].

An order-tracked ODS is assembled by taking the peak values of the ODS FRFs at one of the order frequencies. This is done by using a ***"peak cursor"***, that finds the peak value in a small band surrounding each order peak.

Figure 12.5 shows two order-tracked ODSs displayed side by side on the machine model, one at 985 RPM and the other at 2280 RPM. Of course, an ***animated display*** of the ODSs makes it easier to see the difference between them. The deflections at the eight test points are the actual values acquired from the accelerometers. The deflections at all of the other ***un-measured*** points are being ***"geometrically interpolated"*** from the deflections of nearby test points. The MAC value [2] of the two shapes is displayed on the right. Its low value **(0.10)** indicates that the two shapes are very different.

12.5 Mode Shapes of the Machine

It is apparent from the display in Fig. 12.5 that the ODS probably contains participation of both the ***"rigid body"*** and the ***"flexible"*** mode shapes of the base plate and bearing blocks. Since the machine is resting on four rubber mounts (one under each corner), its rigid body modes will participate in its ODS. The machine has ***six rigid body*** mode shapes. These mode shapes describe the ***free-free motion*** of the machine in space, but it will be shown that they also dominate its ODSs since it is resting on ***four soft springs***.

The rigid body and flexible body mode shapes of the machine were obtained from a finite element model of the base plate and bearing blocks. The easiest way to do this is to treat each of the three parts separately.

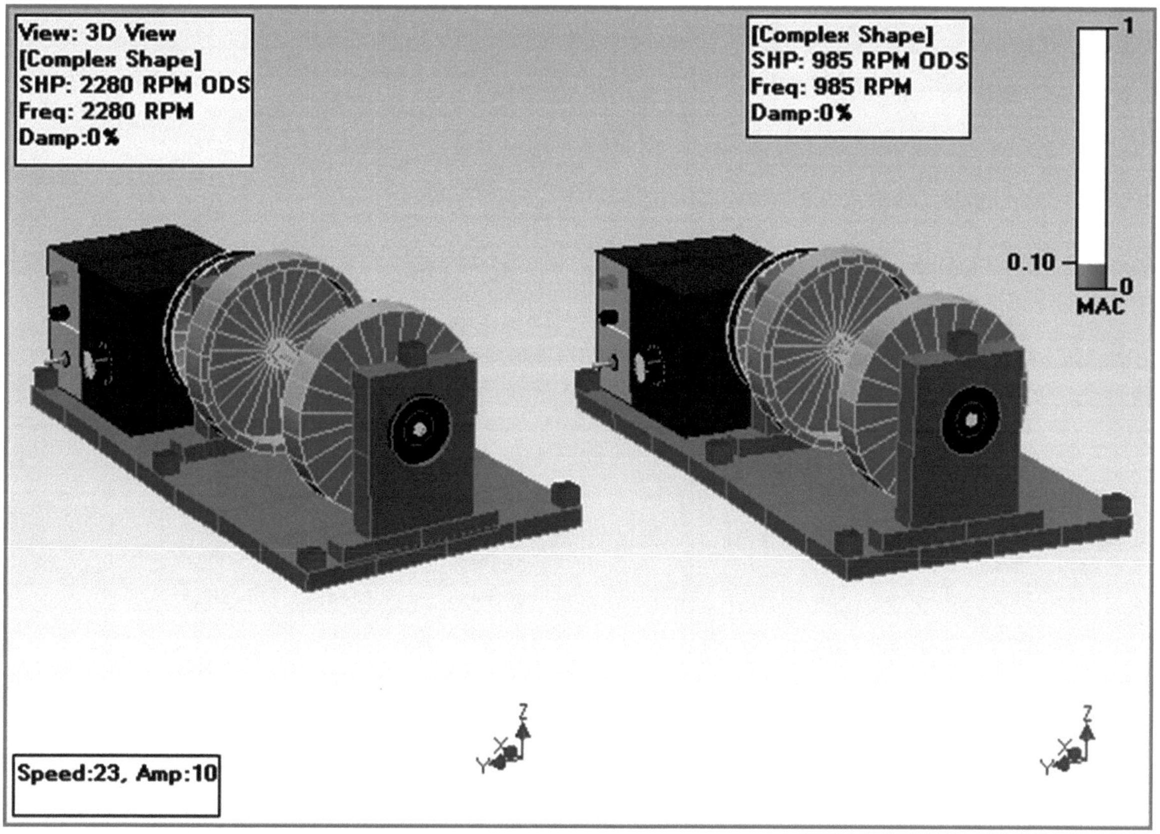

Fig. 12.5 ODSs at 985 and 2280 RPM

FEA models of the base plate and one of the bearing blocks were built first. Then the FEA modes of the base plate and the bearing block were calculated separately.

Figure 12.6 lists the FEA mode shapes of the base plate. It was modeled using 112 FEA brick elements. Figure 12.7 lists the FEA modes of the bearing block. It was modeled using 71 FEA bricks and 6 FEA prisms. Notice that both substructures have 6 rigid body modes.

The Structural Dynamics Modification (SDM) method [3] was used to ***"attach"*** the two bearing blocks to the base plate using stiff springs. This is called ***"substructuring". FEA springs*** were used to model the connections of the blocks to the base plate. Figure 12.8 shows one of the bearing block models connected to the base plate using ***18 very stiff FEA springs*** (with 1E6 lbf/in stiffness). Each spring applies stiffness between a point on a bearing block and a point on the base plate in one (X, Y, or Z) direction.

The mode shapes of the base plate and bearing blocks attached together are listed in Fig. 12.9. Notice that there are still ***six rigid body*** mode shapes, as expected.

12.6 Modal Participation Matrix

The mode shapes of the base plate and bearing blocks will be used to expand the order-tracked ODSs acquired from the eight accelerometers on the rotating machine. As a first step, a ***modal participation matrix*** that contains the contribution of each mode shape to the ODSs is calculated. The ***modal participation matrix***, introduced in a previous paper [4], is defined by Eq. 12.1 below,

$$[U][P] = [V] \tag{12.1}$$

$[V]$ = matrix of ***ODSs*** (n by o)
$[U]$ = matrix of mode shapes (e by m)
$[P]$ = ***modal participation*** matrix (m by o)

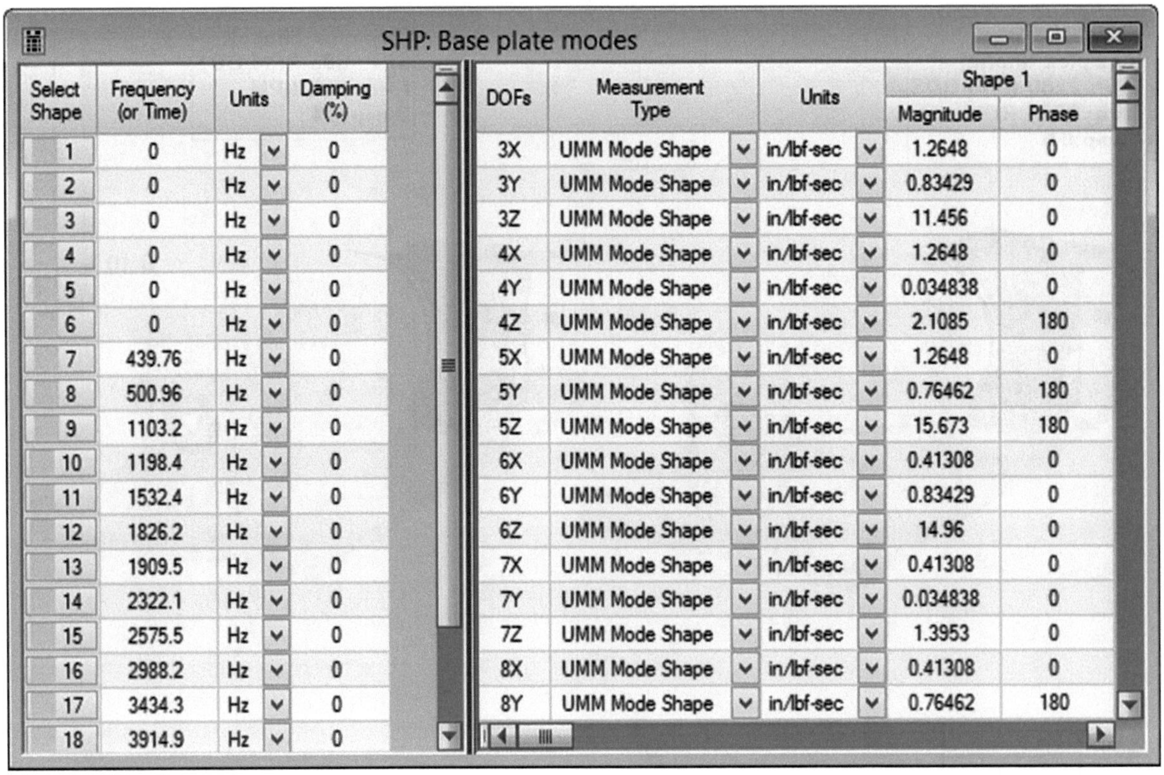

SHP: Base plate modes

Select Shape	Frequency (or Time)	Units	Damping (%)
1	0	Hz	0
2	0	Hz	0
3	0	Hz	0
4	0	Hz	0
5	0	Hz	0
6	0	Hz	0
7	439.76	Hz	0
8	500.96	Hz	0
9	1103.2	Hz	0
10	1198.4	Hz	0
11	1532.4	Hz	0
12	1826.2	Hz	0
13	1909.5	Hz	0
14	2322.1	Hz	0
15	2575.5	Hz	0
16	2988.2	Hz	0
17	3434.3	Hz	0
18	3914.9	Hz	0

DOFs	Measurement Type	Units	Shape 1	
			Magnitude	Phase
3X	UMM Mode Shape	in/lbf-sec	1.2648	0
3Y	UMM Mode Shape	in/lbf-sec	0.83429	0
3Z	UMM Mode Shape	in/lbf-sec	11.456	0
4X	UMM Mode Shape	in/lbf-sec	1.2648	0
4Y	UMM Mode Shape	in/lbf-sec	0.034838	0
4Z	UMM Mode Shape	in/lbf-sec	2.1085	180
5X	UMM Mode Shape	in/lbf-sec	1.2648	0
5Y	UMM Mode Shape	in/lbf-sec	0.76462	180
5Z	UMM Mode Shape	in/lbf-sec	15.673	180
6X	UMM Mode Shape	in/lbf-sec	0.41308	0
6Y	UMM Mode Shape	in/lbf-sec	0.83429	0
6Z	UMM Mode Shape	in/lbf-sec	14.96	0
7X	UMM Mode Shape	in/lbf-sec	0.41308	0
7Y	UMM Mode Shape	in/lbf-sec	0.034838	0
7Z	UMM Mode Shape	in/lbf-sec	1.3953	0
8X	UMM Mode Shape	in/lbf-sec	0.41308	0
8Y	UMM Mode Shape	in/lbf-sec	0.76462	180

Fig. 12.6 Mode shapes of the base plate

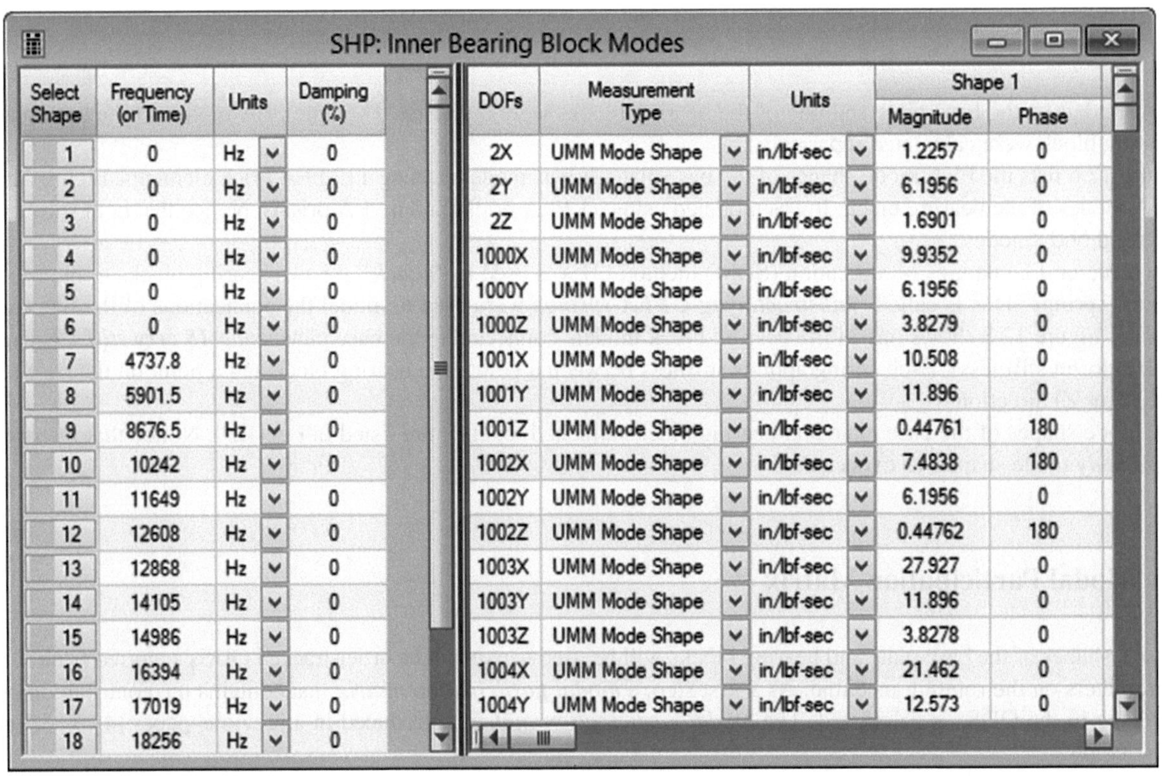

SHP: Inner Bearing Block Modes

Select Shape	Frequency (or Time)	Units	Damping (%)
1	0	Hz	0
2	0	Hz	0
3	0	Hz	0
4	0	Hz	0
5	0	Hz	0
6	0	Hz	0
7	4737.8	Hz	0
8	5901.5	Hz	0
9	8676.5	Hz	0
10	10242	Hz	0
11	11649	Hz	0
12	12608	Hz	0
13	12868	Hz	0
14	14105	Hz	0
15	14986	Hz	0
16	16394	Hz	0
17	17019	Hz	0
18	18256	Hz	0

DOFs	Measurement Type	Units	Shape 1	
			Magnitude	Phase
2X	UMM Mode Shape	in/lbf-sec	1.2257	0
2Y	UMM Mode Shape	in/lbf-sec	6.1956	0
2Z	UMM Mode Shape	in/lbf-sec	1.6901	0
1000X	UMM Mode Shape	in/lbf-sec	9.9352	0
1000Y	UMM Mode Shape	in/lbf-sec	6.1956	0
1000Z	UMM Mode Shape	in/lbf-sec	3.8279	0
1001X	UMM Mode Shape	in/lbf-sec	10.508	0
1001Y	UMM Mode Shape	in/lbf-sec	11.896	0
1001Z	UMM Mode Shape	in/lbf-sec	0.44761	180
1002X	UMM Mode Shape	in/lbf-sec	7.4838	180
1002Y	UMM Mode Shape	in/lbf-sec	6.1956	0
1002Z	UMM Mode Shape	in/lbf-sec	0.44762	180
1003X	UMM Mode Shape	in/lbf-sec	27.927	0
1003Y	UMM Mode Shape	in/lbf-sec	11.896	0
1003Z	UMM Mode Shape	in/lbf-sec	3.8278	0
1004X	UMM Mode Shape	in/lbf-sec	21.462	0
1004Y	UMM Mode Shape	in/lbf-sec	12.573	0

Fig. 12.7 Mode shapes of a bearing block

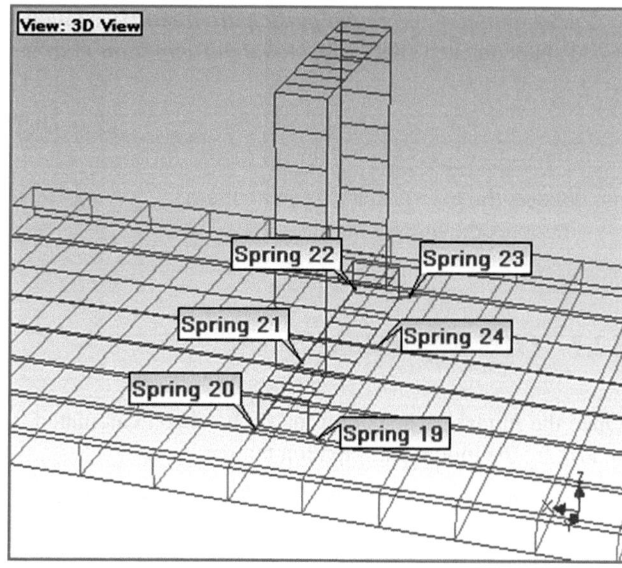

Fig. 12.8 Springs connecting base plate and bearing blocks

SHP: Unmodified Modes

Select Shape	Frequency (or Time)	Units	Damping (%)
1	0	Hz	0
2	0	Hz	0
3	0	Hz	0
4	0	Hz	0
5	0	Hz	0
6	0	Hz	0
7	439.76	Hz	0
8	500.96	Hz	0
9	1103.2	Hz	0
10	1198.4	Hz	0
11	1532.4	Hz	0
12	1826.2	Hz	0
13	1909.5	Hz	0
14	2322.1	Hz	0
15	2575.5	Hz	0
16	2988.2	Hz	0
17	3434.3	Hz	0
18	3914.9	Hz	0

DOFs	Measurement Type	Units	Shape 1	
			Magnitude	Phase
3X	UMM Mode Shape	in/lbf-sec	1.2648	0
3Y	UMM Mode Shape	in/lbf-sec	0.83429	0
3Z	UMM Mode Shape	in/lbf-sec	11.456	0
4X	UMM Mode Shape	in/lbf-sec	1.2648	0
4Y	UMM Mode Shape	in/lbf-sec	0.034838	0
4Z	UMM Mode Shape	in/lbf-sec	2.1085	180
5X	UMM Mode Shape	in/lbf-sec	1.2648	0
5Y	UMM Mode Shape	in/lbf-sec	0.76462	180
5Z	UMM Mode Shape	in/lbf-sec	15.673	180
6X	UMM Mode Shape	in/lbf-sec	0.41308	0
6Y	UMM Mode Shape	in/lbf-sec	0.83429	0
6Z	UMM Mode Shape	in/lbf-sec	14.96	0
7X	UMM Mode Shape	in/lbf-sec	0.41308	0
7Y	UMM Mode Shape	in/lbf-sec	0.034838	0
7Z	UMM Mode Shape	in/lbf-sec	1.3953	0
8X	UMM Mode Shape	in/lbf-sec	0.41308	0
8Y	UMM Mode Shape	in/lbf-sec	0.76462	180

Fig. 12.9 FEA mode shapes of combined substructures

Writing out Eq. 12.1 for ***matching*** shape components,

$$\underset{\text{(n by m)}}{\begin{bmatrix} u_{1,1} & \cdots & u_{1,m} \\ \vdots & \ddots & \vdots \\ u_{n,1} & \cdots & u_{n,m} \end{bmatrix}} \underset{\text{(m by o)}}{\begin{bmatrix} p_{1,1} & \cdots & p_{1,o} \\ \vdots & \ddots & \vdots \\ p_{m,1} & \cdots & p_{m,o} \end{bmatrix}} = \underset{\text{(n by o)}}{\begin{bmatrix} v_{1,1} & \cdots & v_{1,o} \\ \vdots & \ddots & \vdots \\ v_{n,1} & \cdots & v_{n,o} \end{bmatrix}}$$

n = number of ***matching*** mode shape & ODS components
m = number of mode shapes
o = number of ODSs
e = number of mode shape components

Each column of the ***modal participation matrix*** contains ***complex*** valued scale factors that define the contribution of each mode shape to each ODS. The ***modal participation matrix*** is calculated as a least squared solution to Eq. 12.1,

$$[P] = \left[[U]^h [U]\right]^{-1} [U]^h [V] \tag{12.2}$$

h – denotes the *transposed conjugate* matrix
−1 – denotes the *inverse* matrix

12.7 ODS Expansion

Once the modal participation matrix has been calculated, the ODS expansion is done by post-multiplying the mode shape matrix by the modal participation matrix,

$$\left[\overline{\overline{V}}\right] = [U]\,[P] \tag{12.3}$$

$\left[\overline{\overline{V}}\right]$ = expanded ODS matrix

12.8 Modal Participation at Two Speeds

ODS data was acquired from the rotating machine in Fig. 12.1 at two different speeds. The modal participation of the first 10 FEA modes in the 985 RPM ODS was calculated using Eq. 12.2, and is shown in Fig. 12.10.

The modal participation shows that the ***first three rigid body modes*** are the ***dominant contributors*** to the 985 RPM ODS. These modes are being excited and are contributing the most to the ODS at this speed. At this speed, the machine is simply ***"bouncing"*** on its rubber mounts.

The modal participation of the first 10 FEA modes in the 2280 RPM ODS is shown in Fig. 12.11. The participation of modes 2, 5, and 6 shows that they are ***dominating*** the 2280 ODS. At this higher speed, the machine is ***"rocking"*** on its rubber mounts with more deflection at the outer bearing location.

Notice also that the participation scale factors of the dominate modes at both speeds have ***significant real and imaginary parts***. This is because the ODSs are complex valued and the FEA mode shapes are real valued, also called ***normal*** modes. In order to match the normal mode shapes to complex ODSs, the complex modal participation factors, when multiplied by the normal mode shapes, create expanded ODSs that are complex valued.

12.9 ODS Expansion

Finally, Eq. 12.3 was used to expand the 985 and 2280 RPM ODSs. The MAC values for each expanded and un-expanded ODS pair are shown in Fig. 12.12. The MAC value for the **acquired and expanded 985 RPM ODSs** is **0.87**, indicating a strong correlation between this experimental and its expanded ODS. The MAC value between the **experimental and expanded 2280 RPM ODSs** is **0.94**, indicating an even stronger correlation between this experimental and its expanded ODS.

The two off-diagonal MAC values are between the **expanded 985 RPM ODS** and the **experimental 2280 RPM ODS**, and between the **expanded 2280 RPM ODS** and the **experimental 985 RPM ODS**. Both of these values are ***very low*** (**<0.10**), again indicating that the ODSs of the machine at these two speeds are quite different from one another.

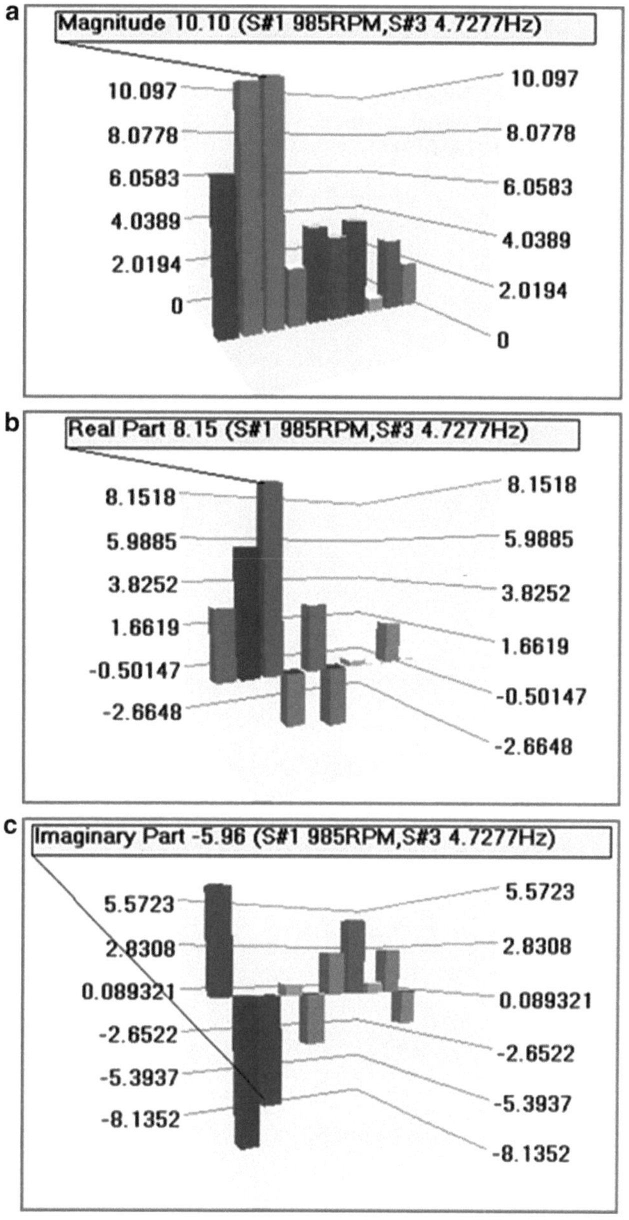

Fig. 12.10 (**a**) Magnitude of modal participation at 985 RPM. (**b**) Real part of modal participation at 985 RPM. (**c**) Imaginary part of modal participation at 985 RPM

12.10 Conclusions

A linear relationship between matrices of mode shapes and matrices of ODSs was presented in a previous paper [4], and was used here to calculate a ***modal participation matrix***. Equation 12.1 merely expresses the well known ***superposition property*** of mode shapes. That is, the overall vibration of a machine or structure contained in its ODS, can be represented as a ***linear combination*** of its mode shapes.

In a wide variety of practical cases, modes can been used to characterize and further understand resonant vibration. Whenever the dynamics of a structure can be represented by its modes, this ***superposition*** property can be utilized, and the participation of each mode in the ODS can be calculated.

This modal participation matrix shows which modes are ***dominating*** an ODS, and therefore which modes are being excited the most under a specific operating condition.

A key advantage of this technique is that **only mode shapes** are required to calculate the modal participation and expand an ODS for display in animation. Modal frequency and damping are not required.

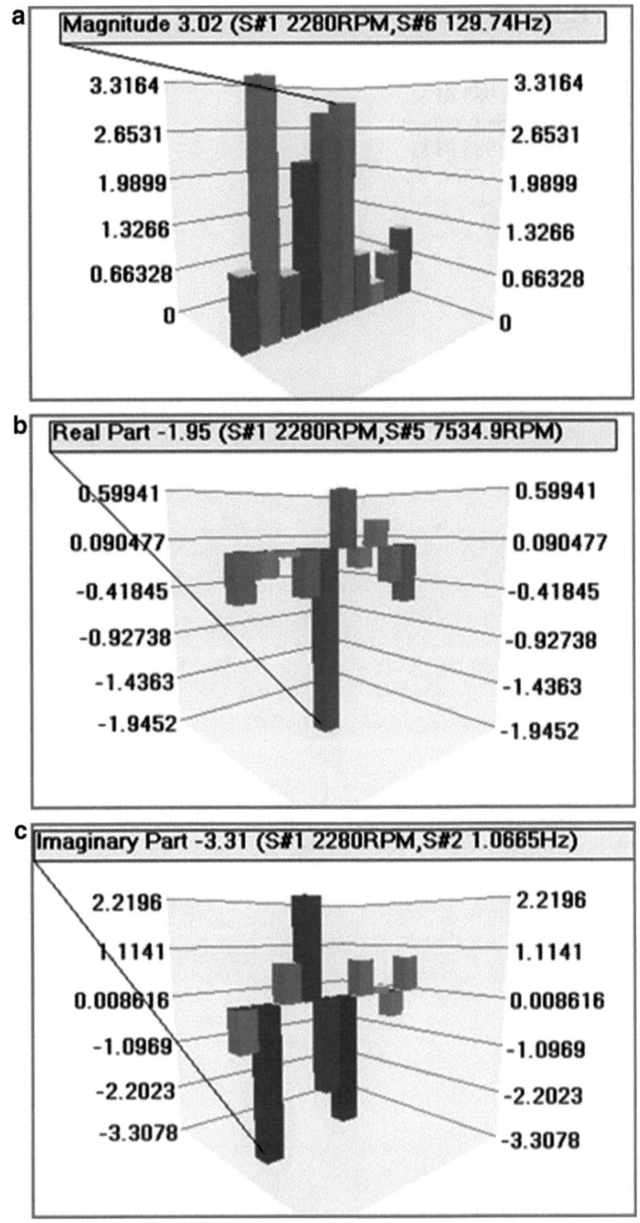

Fig. 12.11 (**a**) Magnitude of modal participation at 2280 RPM. (**b**) Real part of modal participation at 2280 RPM. (**c**) Imaginary part of modal participation at 2280 RPM

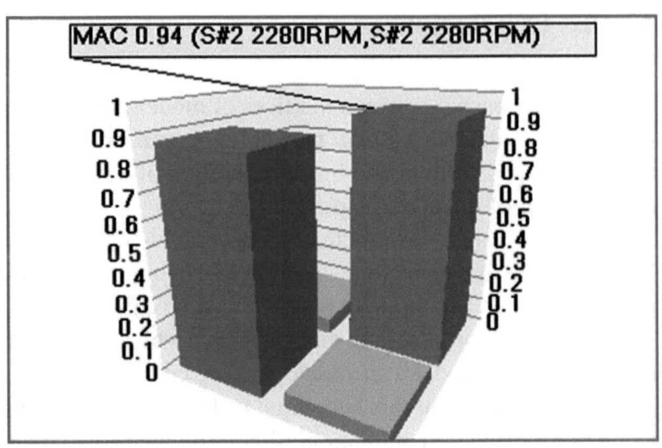

Fig. 12.12 MAC – expanded vs. un-expanded ODSs

In order to construct an analytical modal mode, it was also shown how substructuring and the SDM method can be used to ***"tie together"*** several parts of a machine using stiff FEA springs. SDM calculates the overall mode shapes using the mode shapes of each part, and these mode shapes are easy to obtain by modeling each part in a free-free condition.

ODSs of most rotating machines are ***complex*** valued because they are caused by ***cyclic*** forces within the machine. What makes this technique very appealing is that a set of FEA ***normal*** mode shapes can be ***"curve fit"*** to complex ODS data, resulting in ***complex*** modal participation factors. These participation factors can then be used to expand the ODS, thus providing a realistic animated display of machine deflections on a 3D model of the machine.

References

1. Potter RW Compilation of time windows and line shapes for fourier analysis. Hewlett Packard Technical Note
2. Allemang RJ (2002) The modal assurance criterion (MAC): twenty years of use and abuse. In: Proceedings of the international modal analysis conference
3. Schwarz BJ, Richardson MH (1997) Structural modifications using higher order elements. In: 15th IMAC conference, Feb 1997
4. Schwarz B, Richardson M (2014) Linear superposition and modal participation. IMAC XXXII, 3–6 Feb 2014

Chapter 13
Removing Unwanted Noise from Operational Modal Analysis Data

William K. Bonness and David M. Jenkins

Abstract Operational modal analysis data includes the measurement of dynamic signals such as structural vibration data and the corresponding excitation force or pressure. In addition to the desired information, measured structural vibration data can include unwanted electrical noise and vibration energy from adjacent structures. Measured dynamic pressures can contain unwanted signals such as acoustic and vibration induced pressures. In this paper, a noise removal technique is presented in which an unlimited number of unwanted correlated signals can be removed from a set of measured data. In its simplest form, this technique is related to coherent output power (COP). However, unlike COP noise removal, multiple signals can be removed from measured data while retaining the magnitude and phase of the original data required for modal analysis processing. This technique is demonstrated using vibration data and dynamic wall pressure measurements from a thin-walled aluminum cylinder filled with water flowing at 20 ft/s.

Keywords Operational • Modal • Noise • Removal • Processing

13.1 Introduction

Operational modal analysis involves measuring the vibration of a system during operation either because the desired operational excitation is not appropriately represented by traditional impulsive modal impact testing or because the system cannot be conveniently taken off-line. Transfer function data between vibration and input force are traditionally measured and used as input for an experimental modal analysis. However, when the input force cannot be measured, cross-spectrum measurements between accelerometers can be used in place of the transfer function data. Resonance frequency, damping, and operational modes shapes can all be obtained from cross-spectrum measurements between accelerometers placed at locations required to appropriately resolve the structural mode shapes. Occasionally, measurements of dynamic force or pressure can also be made to help determine modal mass.

Measurements of structural vibration or pressure often include unwanted signals not associated with the signals of interest -especially in an operational setting where other machinery or processes are often running. The unwanted signals may be associated with electrical noise, vibration energy from adjacent structures, vibration induced pressures, and acoustic pressures. While various noise removal techniques are available to "clean up" measured data, a conditioned spectral density technique [1] is presented which can remove multiple signals of unwanted noise from measurements of cross-spectra. This technique can be applied to any dynamic signal measurements which meet necessary criteria.

Lauchle and Daniels [2] employed a related subtraction technique using multiple sensors measured simultaneously to remove noise signals from their measurements of turbulent boundary layer (TBL) wall pressure. Naguib, Gravante, and Wark [3] discuss the advantages of an optimal filtering approach compared to a difference approach. In their work, the optimal filtering provides a better estimate of the noise than the difference approach. The optimal filtering approach yields a frequency domain transfer function which is equivalent to the estimated transfer function from the conditioned spectral density for one sensor of interest and one noise sensor. For offline data analysis, the conditioned spectral density allows one to iteratively remove the effects of several additive noise processes. Additionally, these frequency domain transfer functions can be inverse fast Fourier transformed to obtain a finite impulse response estimate without computing the inverse of a large matrix.

W.K. Bonness (✉) • D.M. Jenkins
Applied Research Laboratory, The Pennsylvania State University, P.O. Box 30, State College, PA 16804, USA
e-mail: wkb3@arl.psu.edu

M. Mains (ed.), *Topics in Modal Analysis, Volume 10*, Conference Proceedings of the Society for Experimental Mechanics Series, DOI 10.1007/978-3-319-15251-6_13

13.2 Approach

13.2.1 Noise Removal Scheme

The single-sided cross spectral density function between two measured signals, $x(t)$ and $y(t)$, is defined as

$$G_{xy}(f) = \frac{2}{T}\overline{X^*(f)Y(f)}. \tag{13.1}$$

where T is the time record length, $X(f)$ and $Y(f)$ are the Fourier Transforms of $x(t)$ and $y(t)$, and the asterisk and overbar denote complex conjugate and ensemble average, respectively. The auto spectrum for a single measured signal $x(t)$ is defined as

$$G_{xx}(f) = \frac{2}{T}\overline{X^*(f)X(f)}, \tag{13.2}$$

and the transfer function, $H_{xy}(f)$, between signals $x(t)$ and $y(t)$ is defined as

$$H_{xy}(f) = \frac{G_{xy}(f)}{G_{xx}(f)}. \tag{13.3}$$

The coherence, $\gamma^2_{xy}(f)$, between $x(t)$ and $y(t)$ is defined as

$$\gamma^2_{xy}(f) = \frac{|G_{xy}(f)|^2}{G_{xx}(f)G_{yy}(f)}, \tag{13.4}$$

and the Coherent Output Power (COP) is defined as

$$COP = \gamma^2_{xy}(f)G_{xx}(f). \tag{13.5}$$

When noise does not exist in a system, it is straightforward to measure the noise-free cross-spectrum between the desired signals $x(t)$ and $y(t)$. However when noise, common to both signals, is added to $x(t)$ and $y(t)$, the desired noise-free cross-spectrum can no longer be measured directly.

Using a three sensor model example illustrated in Fig. 13.1, the noise-free cross-spectrum between signals $x(t)$ and $y(t)$ can be obtained when correlated unwanted noise, represented by a third signal $n(t)$, is added to both signals. The conditioned spectral density technique outlined by Bendat and Persol [1] rejects the portion of the measured $x(t)$ and $y(t)$ signals which

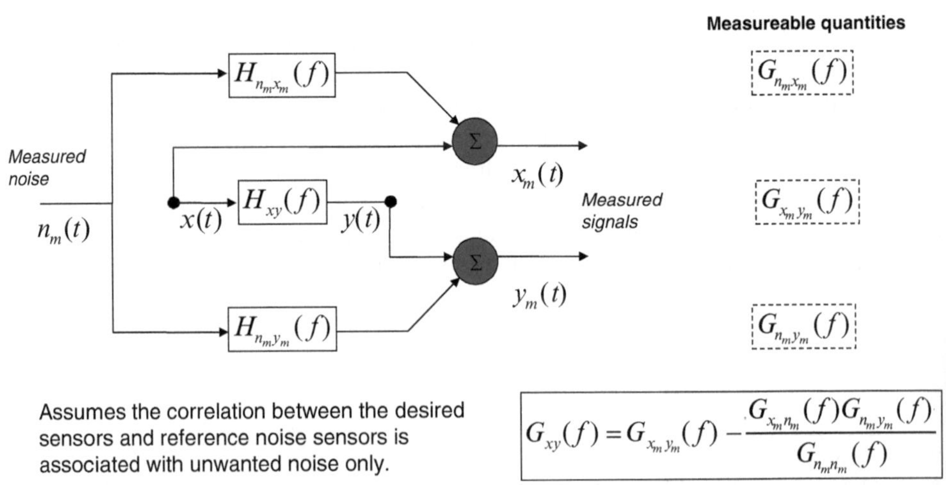

Fig. 13.1 Three sensor model of coherent noise removal process

correlate with the measured noise, $n(t)$. The noise-free cross-spectrum, $G_{xy}(f)$, can be obtained from the measured noisy cross-spectrum, $G_{x_m y_m}(f)$, and spectral measurements involving the noise sensor using

$$G_{xy}(f) = G_{x_m y_m}(f) - \frac{G_{x_m n_m}(f) G_{n_m y_m}(f)}{G_{n_m n_m}(f)}. \tag{13.6}$$

This requires that the cross-spectra between the sensors of interest and reference noise sensor be measured simultaneously. It also assumes that the correlation between the desired sensors and reference noise sensor is associated with unwanted noise only.

For a two sensor model, this noise removal technique reduces to the form,

$$G_{xx}(f) = G_{x_m x_m}(f) - \frac{|G_{x_m n_m}(f)|^2}{G_{n_m n_m}(f)}, \text{ or} \tag{13.7}$$

$$G_{xx}(f) = \left(1 - \gamma^2_{x_m n_m}(f)\right) G_{x_m x_m}(f). \tag{13.8}$$

This form is closely related to the Coherent Output Power (COP) from Eq. (13.5). This form yields a noise free autospectrum, $G_{xx}(f)$, obtained from a noisy autospectrum, $G_{x_m x_m}(f)$, and the coherence, $\gamma^2_{x_m n_m}(f)$, between signal, $x(t)$, and a noise signal, $n(t)$.

This technique can be generalized to include as many reference noise sensors as desired and still retain the cross spectrum between any pair of sensors of interest,

$$G_{xy\cdot(n)!}(f) = G_{xy\cdot(n-1)!}(f) - \frac{G_{xn\cdot(n-1)!}(f) G_{ny\cdot(n-1)!}(f)}{G_{nn\cdot(n-1)!}(f)}, \tag{13.9}$$

where conditioned spectral densities of order n! can be computed from previously known conditioned spectral quantities of order (n−1)!.

To implement this process, one must compute the entire cross-spectral density matrix using all signals including both signals of interest and measured noise signals. It is easiest to implement if the noise signals are at the end of the matrix. One must choose a reference noise signal and remove the correlated portion of this reference signal from all remaining cross-spectra in accordance with Eq. (13.9). This creates a new cross-spectral density matrix and reduces the dimensions by one. Then, one repeats this process for all reference noise signals. The final cross-spectral density matrix includes only signals of interest after removing the correlated portion of all reference noise signals.

13.2.2 Measured Data

The conditioned spectral density noise removal technique is illustrated using data obtained by Bonness, Capone, and Hambric [4]. They measured cylinder vibrations and TBL dynamic wall pressures from a thin-walled 0.15 m diameter aluminum cylinder internally filled with water (pipe-flow) flowing at 6.1 m/s. The gravity-fed test facility (no moving parts) was designed to minimize background vibration and noise from interfering with TBL induced measured quantities. However, because the TBL levels were so low relative to other sources, the measured data still contained a degree of unwanted noise.

A ring array of 12 accelerometers (PCB W532, sens ~ 100 mV/g) was attached around the shell at a single axial location corresponding to an evenly spaced circumferential grid as shown in Fig. 13.2. Two line arrays of flush mounted wall pressure sensors (PCB-105 M147, sens. ~50 mV/psi) were installed downstream of the 0.61 m long test-section also shown in Fig. 13.2. One array was aligned with the flow while the second was perpendicular to the flow. In addition to the sensors installed to measure TBL induced vibration and pressure, additional accelerometers and dynamic pressure sensors were installed at various locations to act as reference sensors measuring unwanted noise. Six reference accelerometers were placed on the test-section supports (solid aluminum blocks) to measure the test-section boundary vibrations at both ends in all three coordinate directions. Figure 13.2 shows the location and number of reference sensors used in the signal removal process. Acceleration is shown on the left and wall pressure is shown on the right. In addition, two reference pressure sensors were installed in the same axial plane as the most upstream TBL sensor, and were installed 120° apart (similar to those of Lauchle and Daniels [2]). The separation distance between the reference pressure sensors and the all TBL pressure sensors results in uncorrelated TBL pressures for all frequencies of interest. Two reference accelerometers were placed in the midst of the TBL pressure sensor arrays to measure local pipe vibrations.

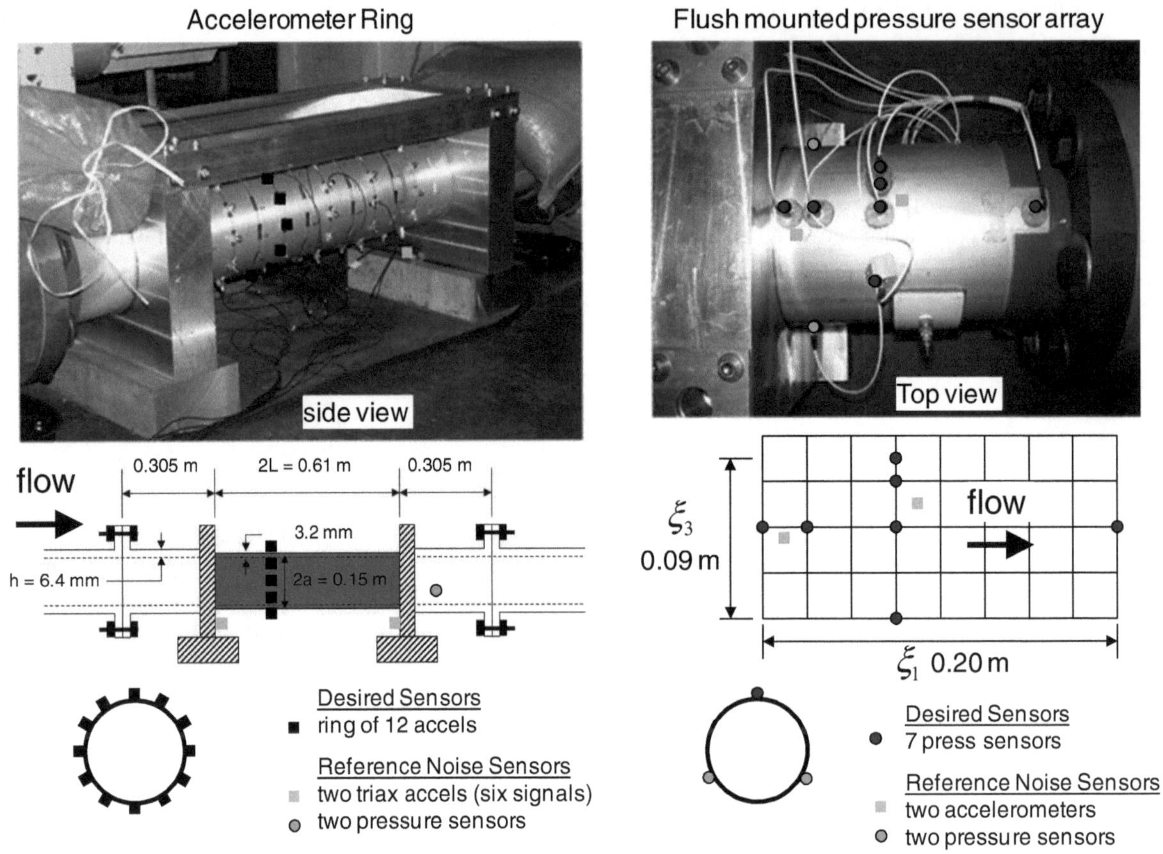

Fig. 13.2 Experimental setup, flow measurement arrays, and reference noise sensor locations

Transfer function data required for the modal analysis can be created from the noise-free cross spectra using an appropriate sensor location (or multiple locations) as the reference location(s) from

$$H_{xy}(f) = \frac{G_{xy}(f)}{\left|G_{xy}(f)\right|^{1/2}}. \tag{13.10}$$

13.3 Results

Figure 13.3 shows a spectrogram from a single accelerometer within the circumferential array shown in Fig. 13.2. This plot shows evidence of cylinder vibration before and after flow start and stop and identifies evidence of facility pumps and fans turning on and off during a 4 min data capture period. The 41 s time intervals labeled "less noise" and "more noise" are used in subsequent analyses to demonstrate the noise removal process and results.

It was desirable to remove energy from the measured vibration levels of the 0.61 m long test-section associated with the vibration of adjacent structures. Therefore, accelerometers were placed on the test-section supports in all three coordinate directions and used as reference noise sensors for the circumferential ring of accelerometers. Two pressure sensors were also used as reference noise sensors to remove any acoustic induced vibration. The results of removing the vibration energy correlated with six reference noise accelerometers and two reference pressure sensors are shown in Fig. 13.4. Results for the auto-spectrum of vibration for a single accelerometer are shown on the left side of the figure. The different curves show the progression of results as each reference noise signal was removed in succession. The top curve represents the vibration levels as measured. The bottom curve represents the remaining vibration energy not correlated with any of the eight reference noise sensors. Very little energy in the as-measured vibration spectrum was correlated with the two pressure sensors (p1 and p2), so the resulting spectra after those signals were removed changed very little. Various changes to the vibration spectra occurred

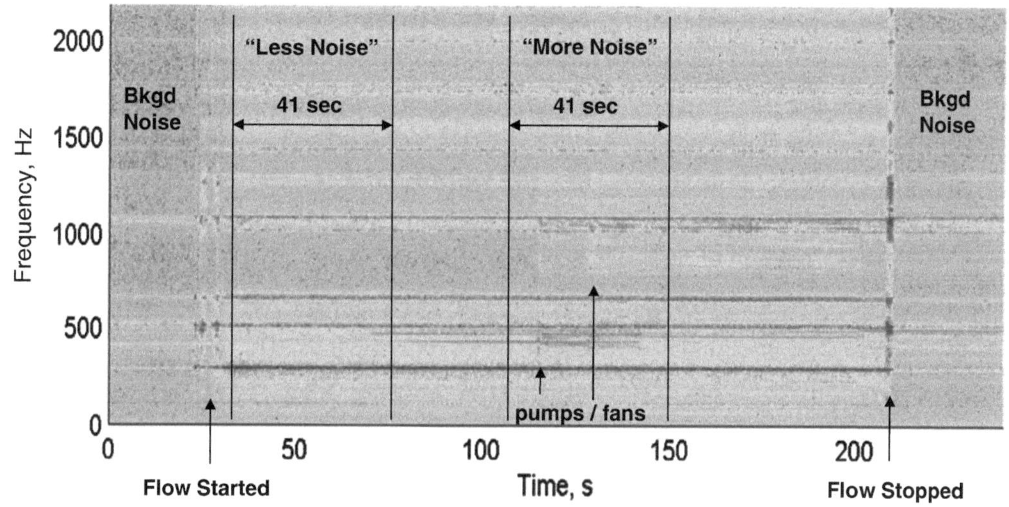

Fig. 13.3 Accelerometer spectrogram showing different levels of extraneous noise in time domain data

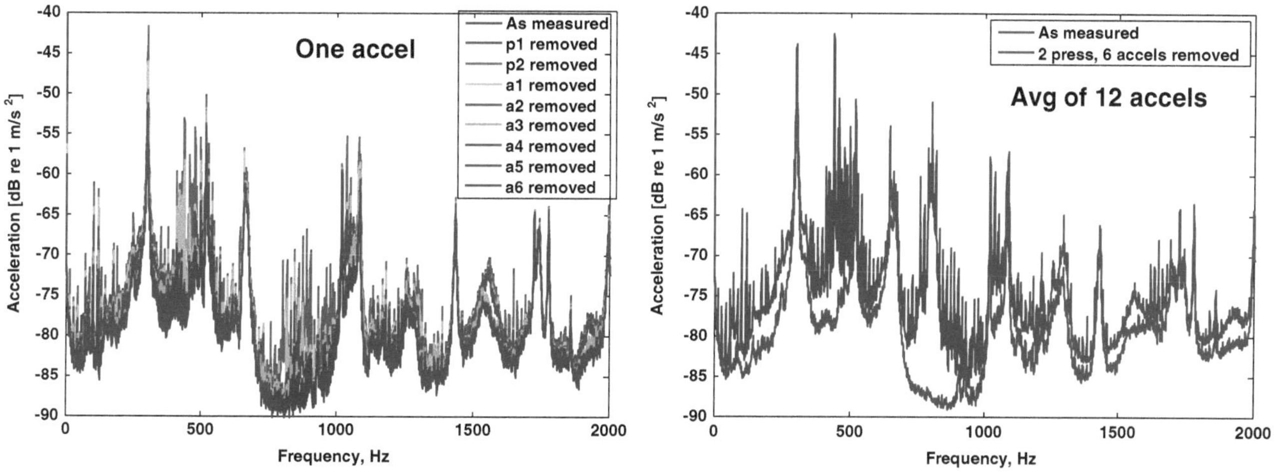

Fig. 13.4 Accelerometer auto-spectra showing progression of noise removal process (*left*) and final before and after noise removal results (*right*)

as each of the six reference accelerometer signals (a1–a6) were removed. Different vibration peaks correlated differently with reference accelerometers oriented in different directions. Some cylinder vibration peaks were reduced by less than 1 dB and some were reduced by as much as 10 dB. The final before and after plot of the noise removal process is shown on the right side of Fig. 13.4 for the average of all 12 accelerometers in the circumferential array.

The different colored curves in the plots shown in Fig. 13.5 represent the circumferential Fourier components of the total measured cylinder vibration spectrum. One must have the cross-spectra between all 12 accelerometers in the circumferential array to determine the individual spatial Fourier components. Figure 13.5 demonstrates the Fourier analysis for two different time segments identified in Fig. 13.3 before and after the conditioned spectral density noise removal technique is applied. For the "more noise" case on the top, a significant difference is evident before and after noise removal. For the "less noise" case on the bottom, very little difference is evident before and after noise removal. The clearly identifiable Fourier components following noise removal indicates that valid cross-spectra was retained. Despite the presence of more or less noise in the originally measured data, the final result is very similar. This provides strong evidence that the noise removal technique works as intended.

In addition to unwanted noise in the accelerometer data, unwanted noise was present in the wall pressure data. Because acoustic pressures and vibration induced pressures produce coherent signals over long distances and the TBL induced wall pressures do not, any portion of a wall pressure sensor signal coherent with a reference noise pressure signal (at a sufficient distance) can be removed leaving only the TBL induced signals. Figure 13.6 shows a progression of results similar to Fig. 13.4 for a wall pressure sensor. Two accelerometers and two pressure sensors were used as reference noise sensors. After applying the noise removal to the as-measured pressure data, the wall pressure levels above 100 Hz were slightly

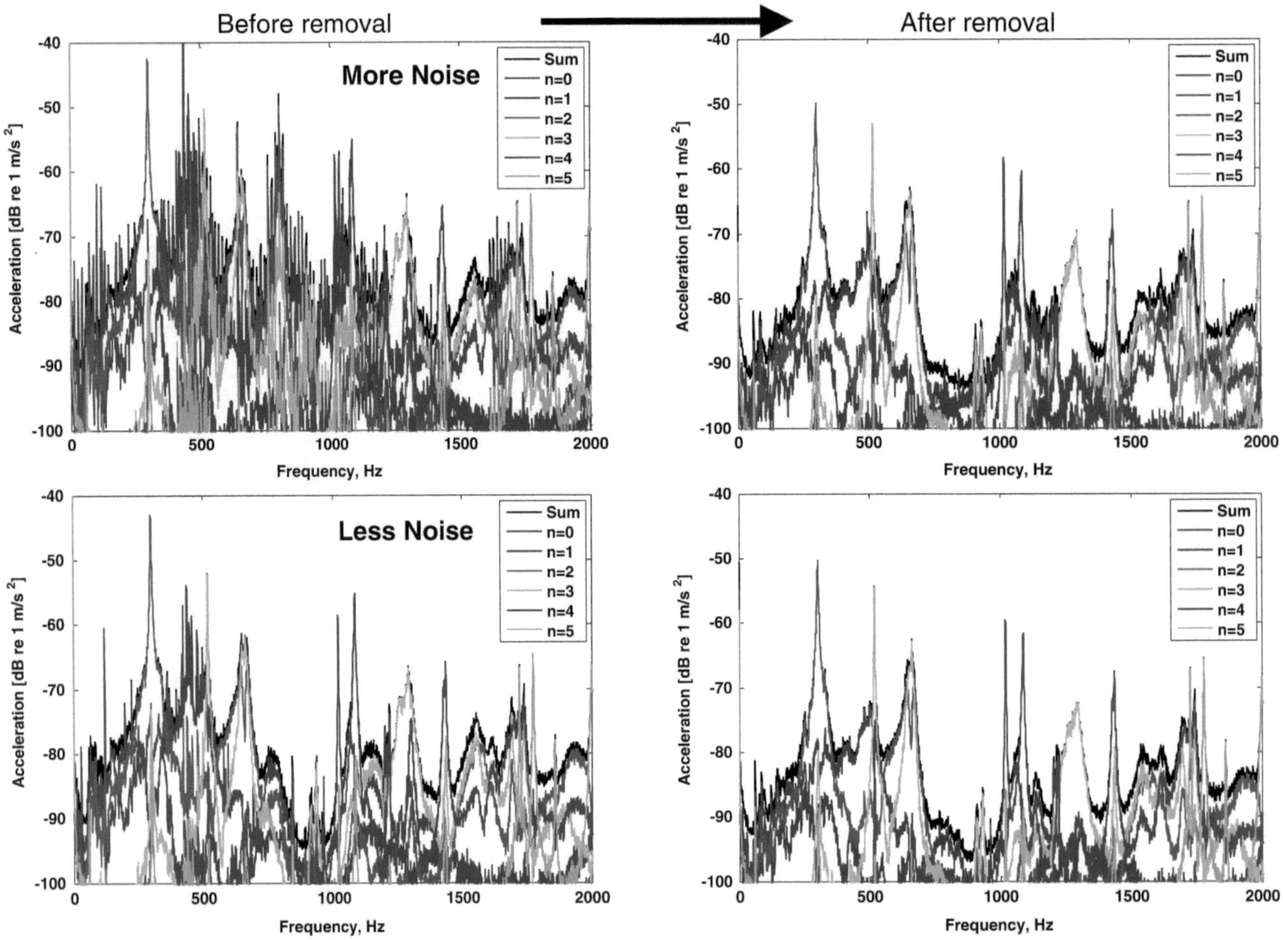

Fig. 13.5 Accelerometer cross-spectra decomposed into circumferential Fourier components. Before and after noise removal results applied to two different time segments identified in Fig. 13.3: noisy data (*top two plots*), less noisy data (*bottom two plots*)

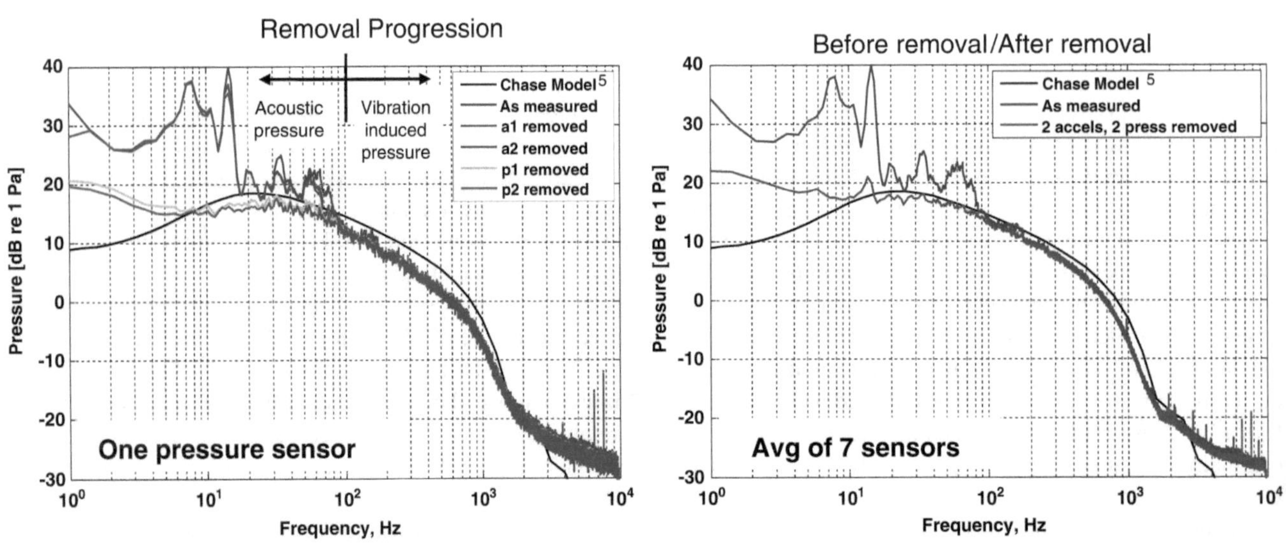

Fig. 13.6 Pressure auto-spectra showing progression of noise removal process (*left*) and final before and after noise removal results (*right*)

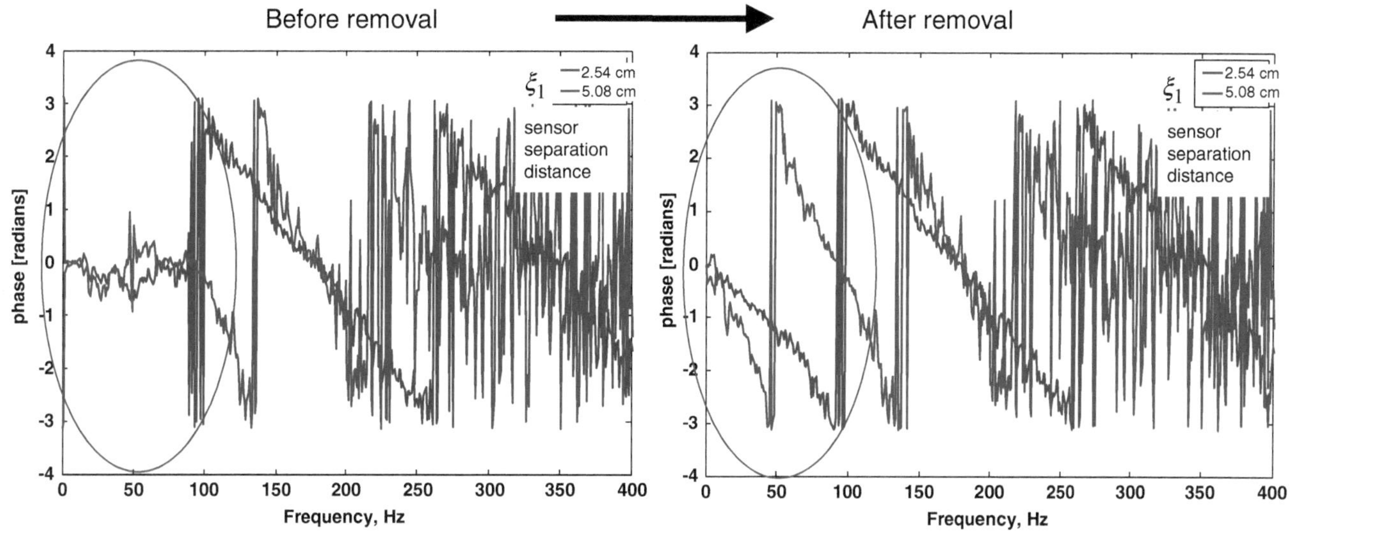

Fig. 13.7 Cross-spectra phase between wall pressure sensors separated in the streamwise direction

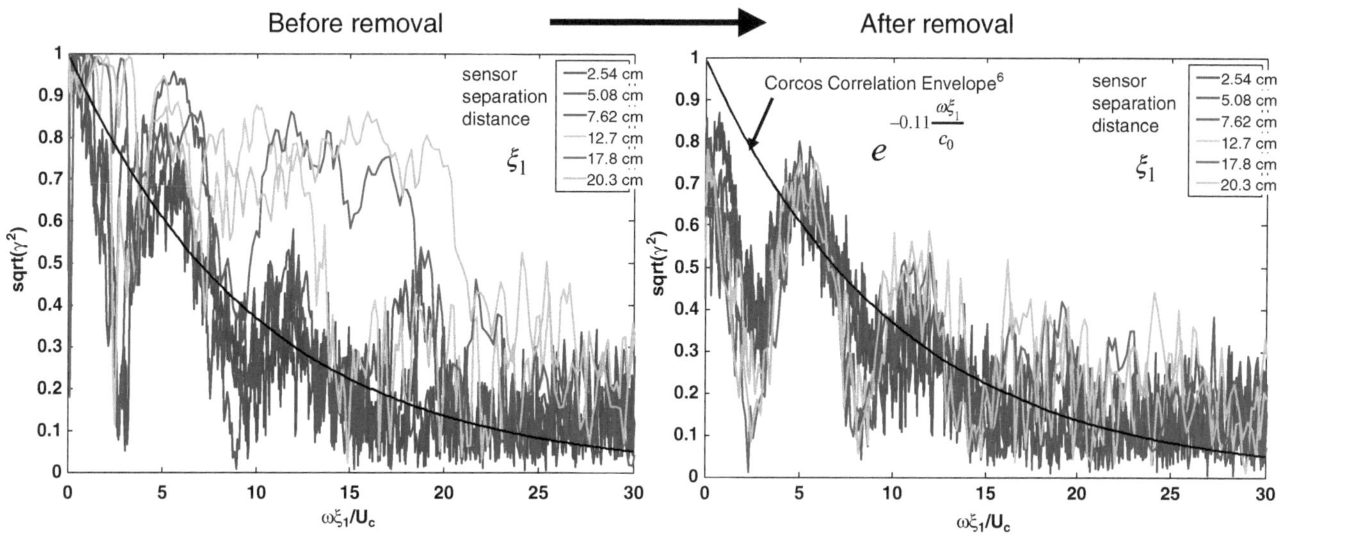

Fig. 13.8 Coherence between wall pressure sensors separated in the streamwise direction

reduced due to correlation between the pressure sensor and the reference accelerometers indicating the presence of vibration induced pressures. Similarly, the wall pressure levels below 100 Hz were significantly reduced due to correlation between the pressure sensor and the reference pressure sensors indicating the presence of acoustic pressures. A theoretical model of the expected TBL wall pressure spectrum from Chase [5] is given as a reference.

Figures 13.7 and 13.8 demonstrate example results using the signal removal technique on measured wall pressure cross-spectra. Figure 13.7 shows the improved phase recovered below 100 Hz between flush mounted TBL pressure sensors in the streamwise direction. Figure 13.8 shows the Corcos [6] correlation function compared with improved coherence recovered between TBL pressure sensors in the streamwise array. Both of these examples provide strong evidence that the noise removal technique works as intended. The full cross-spectra between signals of interest was retained providing improved magnitude and phase.

13.4 Conclusions

Measurements of cross-spectral data between accelerometers can be used in place of traditional transfer function data for an operational modal analysis when the input force cannot be easily measured. Unwanted noise often accompanies these measurements in an operational setting. A noise removal technique based on conditioned spectral densities is presented which allows an unlimited number of unwanted correlated signals to be removed from a set of measured cross-spectra. In its simplest form, this technique is related to the coherent output power and optimal filtering discussed by Naguib et al. [3]. An advantage of this technique over others is that multiple reference noise signals can be removed and the effect of removing each signal can be evaluated. This technique requires measuring reference noise sensors which contain signals that correlate with unwanted noise in the sensors of interest. Reference noise sensors must not contain any signals which correlate with signals of interest in the measured sensors. Measured cylinder vibration and wall pressure cross-spectral data are markedly improved using this noise removal technique. This technique can be applied to any dynamic measurements which meet the necessary criteria.

References

1. Bendat JS, Persol AG (1986) Random data – analysis and measurement procedures. Wiley, New York
2. Lauchle GC, Daniels MA (1987) Wall-pressure fluctuations in turbulent pipe flow. Phys Fluids 30(10):3019–3024
3. Naguib AM, Gravante SP, Wark CE (1996) Extraction of turbulent wall-pressure time-series using an optimal filtering scheme. Exp Fluids 22(1):14–22
4. Bonness WK, Capone DE, Hambric SA (2010) Low-wavenumber turbulent boundary layer wall-pressure measurements from vibration data on a cylinder in pipe flow. J Sound Vib 329(20):4166–4180
5. Chase DM (1987) The character of the turbulent wall pressure spectrum at subconvective wavenumbers and a suggested comprehensive model. J Sound Vib 112:125–147
6. Corcos GM (1964) The structure of the turbulent pressure field in boundary-layer flows. J Fluid Mech 18(3):353–378

Chapter 14
Adaptive-Like Vibration Control in a Three-Story Building-Like Structure with a PZT Stack Actuator

G. Silva-Navarro, F. Beltrán-Carbajal, and L.G. Trujillo-Franco

Abstract This work deals with the application of an adaptive-like vibration control in a three story building-like structure submitted to ground vibrations, by using on-line algebraic identification and robust modal control techniques based on Multi-Positive Position Feedback Control schemes. The building-like structure with a PZT stack actuator is modeled and analyzed via Euler-Lagrange method and the experimental validation is performed through experimental modal analysis techniques. The ground excitation is estimated by using on-line algebraic identification methods and an adaptive-like vibration control is synthesized to reduce as possible the overall motion on the building-like structure. The proposed estimation scheme and adaptive vibration controller are evaluated and validated on an experimental setup to illustrate the open-loop and closed-loop system performance.

Keywords Active vibration control • Adaptive control • Algebraic identification • Modal parameters • Positive position feedback

14.1 Introduction

First appeared in the early 1950s, adaptive control has been a constant subject of research because of its important role in aeronautics and control of time parameter changing systems and processes [1]. A major fact involved on adaptive control is that the on-line system parameter identification, used for tuning proposes on the adaptive controllers, most of them have requirements and limitations such as the need of persistent excitation and slow convergence of the estimation. In other words, the adaptive control schemes have the natural handicap of the on-line identification methods proper characteristics and limitations [2–5].

On the other hand, active vibration control in structures involves a systematic set of issues such as specialized sensors, actuators, and, at the very first stages of its implementation, the need of an accurate understanding of the inherent dynamic characteristics of the structure by means of a mathematical model and, for these purposes, the modal model of the structure is very descriptive and useful when the final goal of the modeling process is to perform a response forecasting in the context of harmonic excitations [6, 12–16]. In [7] an on-line and time domain algebraic identification approach for the estimation of modal parameters on multiple degrees-of-freedom mechanical structures is proposed, this identification scheme implies the possibility of a fast and adaptive-like tuning of a specific controller.

In this paper, an algebraic on line time domain identification in conjunction with a Positive Position Feedback control scheme is evaluated on an experimental building-like structure for reducing the overall system response resulted from ground excitation. The resulting adaptive-like control scheme consists of identification of the characteristic polynomial of the overall structure, considering that one of the outputs is available for the implementation of the parameter estimation scheme. The identification of the ground excitation forces is then possible by using the identified parameters.

G. Silva-Navarro (✉) • L.G. Trujillo-Franco
Departamento de Ingeniería Eléctrica, Sección de Mecatrónica, Centro de Investigación y de Estudios Avanzados del I.P.N., Av. IPN No. 2508, Col. S.P. Zacatenco, CP 07360 México, D.F., México
e-mail: gsilva@cinvestav.mx; forest_6@hotmail.com

F. Beltrán-Carbajal
Departamento de Energía, Universidad Autónoma Metropolitana, Unidad Azcapotzalco, Av. San Pablo No. 180, Col. Reynosa Tamaulipas, CP 02200 México, D.F., México
e-mail: fbeltran@correo.azc.uam.mx

M. Mains (ed.), *Topics in Modal Analysis, Volume 10*, Conference Proceedings of the Society for Experimental Mechanics Series, DOI 10.1007/978-3-319-15251-6_14

14.2 Three Story Building-Like Structure

The experimental setup consists on a three story building like structure as shown in Fig. 14.1, which consists of three aluminum alloy plates coupled by two flexible aluminum columns. Here x_i, $i = 1, 2, 3$, are the displacements of three masses representing the floors or Degrees-Of-Freedom (DOF) of the structure, respectively. The columns are modeled as flexural springs with equivalent stiffness k_i and the structural damping ratios c_i are considered as proportional or Rayleigh damping [8, 9]. The structure is mounted over a frictionless bearing rail such that the ground excitation motion on the x axis is performed by an electromagnetic shaker *Labworks*®, model ET-139, which moves the overall structure for both, experimental modal analysis purposes and ground excitation forces for the performance evaluation of vibration control schemes.

The simplified mathematical model of this flexible mechanical system of 3 DOF under harmonic excitation f is given by the ordinary differential equation

$$M\ddot{x} + C\dot{x} + Kx = f(t) + B_f u_{pzt}(t), \quad B_f, x, f \in R^3, \ u_{pzt} \in R \tag{14.1}$$

where $x \in R^3$ is the vector of generalized coordinates (displacements) of each floor respect to the main frame reference, u_{pzt} is the voltage control input applied to the PZT stack actuator. Moreover, $f_{pzt}(t) = B_f u_{pzt}(t)$ represents the control force to be used for active vibration control of the overall building-like structure. In addition, B_f is the input matrix, and M, C and K are symmetric inertia, damping and stiffness $n \times n$ matrices, respectively, given by

$$M = \begin{bmatrix} m_1 & 0 & 0 \\ 0 & m_2 & 0 \\ 0 & 0 & m_3 \end{bmatrix}, \quad C = \begin{bmatrix} c_1 + c_2 & -c_2 & 0 \\ -c_2 & c_2 + c_3 & -c_3 \\ 0 & -c_3 & c_3 \end{bmatrix}, \quad K = \begin{bmatrix} k_1 + k_2 & -k_2 & 0 \\ -k_2 & k_2 + k_3 & -k_3 \\ 0 & -k_3 & k_3 \end{bmatrix} \tag{14.2}$$

It is easy to verify that system (14.1) is completely controllable and observable, as long as K is positive definite and $C \equiv 0$, and asymptotically stable when C is positive definite (see, e.g., Inman [10]). It is important to take into consideration that the harmonic excitation by the ground (base) motion $z(t) = Z sin(\omega_0 t)$, provides the force f as defined in expression (14.1) (see [20]), hence f is given by:

$$f(t) = -Me\ddot{z}(t) \tag{14.3}$$

Here $e = [1, 1, 1]^T \in R^3$ is used to describe the displacements of each mass due to the ground acceleration $\ddot{z}(t)$ thus, each component of the excitation input vector f_i is given by:

$$f_i = -m_i\ddot{z}(t) = -m_i \frac{d^2 (Z \sin(\omega_0 t))}{dt^2} = m_i \omega_0^2 Z \sin(\omega_0 t), \qquad i = 1, 2, 3 \tag{14.4}$$

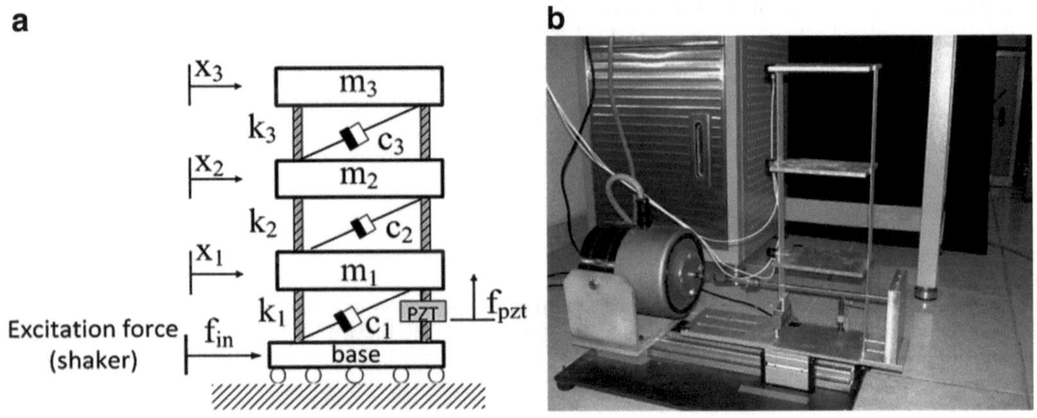

Fig. 14.1 Three story building-like structure with PZT stack actuator. (**a**) Schematic diagram. (**b**) Experimental setup

where ω_o and Z are the excitation frequency and amplitude of the ground motion, respectively. It is well-known that, the mathematical model (14.1) (considering at first that $u_{pzt}(t) \equiv 0$) can be transformed to modal (principal) coordinates q_i, $i = 1, 2, \cdots, n$, as follows (see, e.g., [6])

$$\ddot{q}_i + 2\zeta_i \omega_i \dot{q}_i + \omega_i^2 q_i = \Psi^T f \tag{14.5}$$

with

$$x(t) = \Psi q(t) \tag{14.6}$$

where ω_i and ξ_i denote the natural frequencies and damping ratios associated to the i-th vibration mode, respectively, and Ψ is the so-called 3×3 modal matrix given by

$$\Psi = \begin{bmatrix} \psi_{11} & \psi_{12} & \psi_{13} \\ \psi_{21} & \psi_{22} & \psi_{23} \\ \psi_{31} & \psi_{32} & \psi_{33} \end{bmatrix} \tag{14.7}$$

Therefore, the modal analysis representation of the vibrating mechanical system (14.1) is expressed by

$$\ddot{q}_i + 2\xi_i \omega_i \dot{q}_i + \omega_i^2 q_i = \psi_{1i} f_1 + \psi_{2i} f_2 + \psi_{3i} f_3 \tag{14.8}$$

In notation of Mikusiński operational calculus [17, 18], this modal model is then described as

$$\left(s^2 + 2\xi_i \omega_i s + \omega_i^2\right) q_i(s) = \psi_{1i} f_1(s) + \psi_{2i} f_2(s) + \psi_{3i} f_3(s) + p_{o,i} + p_{1,i} s \tag{14.9}$$

$$q_i(s) = \frac{\psi_{1i} f_1(s) + \psi_{2i} f_2(s) + \psi_{3i} f_3(s) + p_{o,i} + p_{1,i} s}{\left(s^2 + 2\xi_i \omega_i s + \omega_i^2\right)} \tag{14.10}$$

where $p_{o,i}$ are constants depending on the system initial conditions at the time $t_0 \geq 0$. From (14.4) and (14.7), one then obtains that

$$x_i(s) = \sum_{j=1}^{3} \frac{\psi_{ij} \psi_{ji} f_j(s) + \psi_{ij} \left(p_{o,i} + p_{1,i} s\right)}{\left(s^2 + 2\xi_j \omega_j s + \omega_j^2\right)} \tag{14.11}$$

Therefore, the physical displacements x_i are given by

$$p_c(s) x_i(s) = \sum_{k=1}^{3} \left[\sum_{j=0}^{4} \left[\left(\beta_{ji} s^j f_k(s)\right] \right] + r_{0i} + r_{1i} s + r_{2i} s^2 + r_{3i} s^3 + r_{4i} s^4 + r_{5i} s^5 \tag{14.12}$$

with

$$p_c(s) = s^6 + a_5 s^5 + a_4 s^4 + a_3 s^3 + a_2 s^2 + a_1 s + a_0 \tag{14.13}$$

where $p_c(s)$ is the characteristic polynomial of the mechanical system, r_{ij} are constants which depend on the initial conditions of the system as well as the modal matrix components ψ_{ij} and the constants β_{ji} depend on the modal matrix components ψ_{ij} and the roots of the $p_c(s)$. The roots of the characteristic polynomial (14.10) provide the damping factors and damped natural frequencies and, hence, the natural frequencies and damping ratios of the flexible structure.

14.3 On-Line Algebraic Identification of the Harmonic Excitation

Consider the mathematical model (14.8), where only measurements of some position variable x_i is available to be used in the synthesis of the on-line algebraic identification scheme for the estimation of the coefficients a_k as reported in [1]. We consider, at first, that $f_k(t) \equiv 0$ and the system has a change in its initial conditions, for identifying the characteristic polynomial of the system. Now, consider the mathematical model (14.9), where the coefficients a_k have been estimated as described before, if the ground acceleration produces the forces f_k and we consider them as harmonic and sinusoidal, we have, in terms of operational calculus

$$f_k(s) = m_k \omega_0^2 Z \frac{\omega_0^2}{s^2 + \omega_0^2} \tag{14.14}$$

Note that the frequency ω_0 and the amplitude Z are the same for each component of the input force vector, therefore substituting (14.11) in (14.9) one obtains:

$$p_c(s) x_i(s) = \frac{B_t(s) Z \omega_0^2}{s^2 + \omega_0^2} + r_{0i} + r_{1i}s + r_{2i}s^2 + r_{3i}s^3 + r_{4i}s^4 + r_{5i}s^5 \tag{14.15}$$

where $B_t(s) = \beta_4 s^4 + \beta_3 s^3 + \beta_2 s^2 + \beta_1 s + \beta_1$ is a polynomial which coefficients depend on the mass of each floor and the modal matrix components. By multiplying (14.15) by $\left(s^2 + \omega_0^2\right)$ is obtained:

$$\left(s^2 + \omega_0^2\right)\ p_c(s) x_i(s) = B_t(s) Z \omega_0^2 + \left(s^2 + \omega_0^2\right)\left(r_{0i} + r_{1i}s + r_{2i}s^2 + r_{3i}s^3 + r_{4i}s^4 + r_{5i}s^5\right) \tag{14.16}$$

$$\left(s^8 + \widehat{a}_5 s^7 + \widehat{a}_4 s^6 + \widehat{a}_3 s^5 + \widehat{a}_2 s^4 + \widehat{a}_1 s^3 + \widehat{a}_0 s^2\right) x_i(s) = -\omega_0^2 p_c(s) + \Gamma(s) \tag{14.17}$$

with $\Gamma(s) = \lambda_0 + \lambda_1 s^2 + \lambda_2 s^3 + \lambda_3 s^4 + \lambda_4 s^5 + \lambda_5 s^6 + \lambda_6 s^7$. To eliminate the influence of the constants λ_k, the Eq. (14.17) is derived eight times with respect to the complex variable s, the result is multiplied by s^{-8} and transformed back to the time domain resulting in

$$N(t) = \omega_0^2 D(t) \tag{14.18}$$

where

$$\begin{aligned} N(t) = {} & \sum_{k=0}^{8} (-1)^{8-k} \frac{(8)!(8)!}{k!(8-k)!(8-k)!} \int_{t_0}^{(k)} (\Delta t)^{8-k} x_i(t) + \widehat{a}_5 \sum_{k=0}^{7} (-1)^{8-k} \frac{(8)!(7)!}{k!(8-k)!(8-1-k)!} \int_{t_0}^{(k+1)} (\Delta t)^{8-k} x_i(t) + \\ & \widehat{a}_4 \sum_{k=0}^{6} (-1)^{8-k} \frac{(8)!(6)!}{k!(8-k)!(8-2-k)!} \int_{t_0}^{(k+2)} (\Delta t)^{8-k} x_i(t) + \cdots + \widehat{a}_1 \sum_{k=0}^{1} (-1)^{8-k} \frac{(8)!}{k!(8-k)!(1-k)!} \int_{t_0}^{(k)} (\Delta t)^{8-k} x_i(t) + \widehat{a}_0 \int_{t_0}^{(8)} (\Delta t)^{8} x_i(t) \end{aligned} \tag{14.19}$$

$$\begin{aligned} D(t) = {} & \sum_{k=0}^{6} (-1)^{6-k} \frac{(8)!(6)!}{k!(8-k)!(8-2-k)!} \int_{t_0}^{(k)} (\Delta t)^{6-k} x_i(t) + \widehat{a}_5 \sum_{k=0}^{5} (-1)^{6-k} \frac{(8)!(5)!}{k!(8-k)!(8-1-k)!} \int_{t_0}^{(k+1)} (\Delta t)^{6-k} x_i(t) \\ & + \widehat{a}_4 \sum_{k=0}^{6} (-1)^{8-k} \frac{(8)!(4)!}{k!(8-k)!(8-2-k)!} \int_{t_0}^{(k+2)} (\Delta t)^{6-k} x_i(t) + \cdots + \widehat{a}_1 \sum_{k=0}^{1} (-1)^{6-k} \frac{(8)!}{k!(8-k)!(8-k)!} \int_{t_0}^{(k)} (\Delta t)^{6-k} x_i(t) + \widehat{a}_0 \int_{t_0}^{(8)} (\Delta t)^{6} x_i(t) \end{aligned} \tag{14.20}$$

where $\Delta t = t - t_0$ and $\int_{t_0}^{(N)} \phi(t)$ are iterated integrals of the form $\int_{t_0}^{t}\int_{t_0}^{\sigma_1} \cdots \int_{t_0}^{\sigma_{N-1}} \phi(\sigma_N)\, d\sigma_N \cdots d\sigma_1$ with $\int_{t_0}^{(1)} \phi(t) = \int_{t_0}^{(t)} \phi(\sigma)\, d\sigma$, $\int_{t_0}^{(0)} \phi(t) = \phi(t)$, and N a positive integer. Due to the fact that $\omega_0^2 > 0$ we have that

$$|D(t)|\, \omega_0^2 = |N(t)| \tag{14.21}$$

Thus, we can get an expression for estimating the excitation frequency ω_0. To avoid singularities and, at the same time getting smoother behavior for the estimation the Eq. (14.20) is integrated twice with respect to the time

$$\widehat{\omega}_0 = \sqrt{\frac{n_i}{d_i}}, \qquad t > t_0 \tag{14.22}$$

where

$$n_i = \int_{t_0}^{(2)} |N_1|, \quad d_i = \int_{t_0}^{(2)} |D_1|, \qquad t > t_0 \tag{14.23}$$

Note that, it is possible to identify the excitation frequency using any of the outputs x_i obtaining a similar result. Moreover, the estimation of the excitation frequency $\widehat{\omega}_0$ is independent on the amplitude Z and the system initial conditions.

14.4 An Adaptive-Like Positive Position Feedback Control Scheme

The final goal of implementing a vibration control scheme is to minimize the undesired overall movement in presence of harmonic excitation of some degree of freedom of the structure or, in the best case, all of them. The positive position feedback control scheme is a well-known dynamic output feedback controller as a modal control method for vibration attenuation [9, 20]. This control scheme adds an additional degree-of-freedom to the original system considered as a virtual passive absorber, that is, a second order low pass filter. In order to make clear how the PPF controller works, one considers the simplified model of the three story building-like structure (14.1) perturbed by ground motion. Here the PZT stack actuator is placed between the base and the first level of the structure, so that, this actuates directly over the first floor dynamics, however, the output variable to be used for active vibration control is the displacement of the third floor as shown in Fig. 14.2.

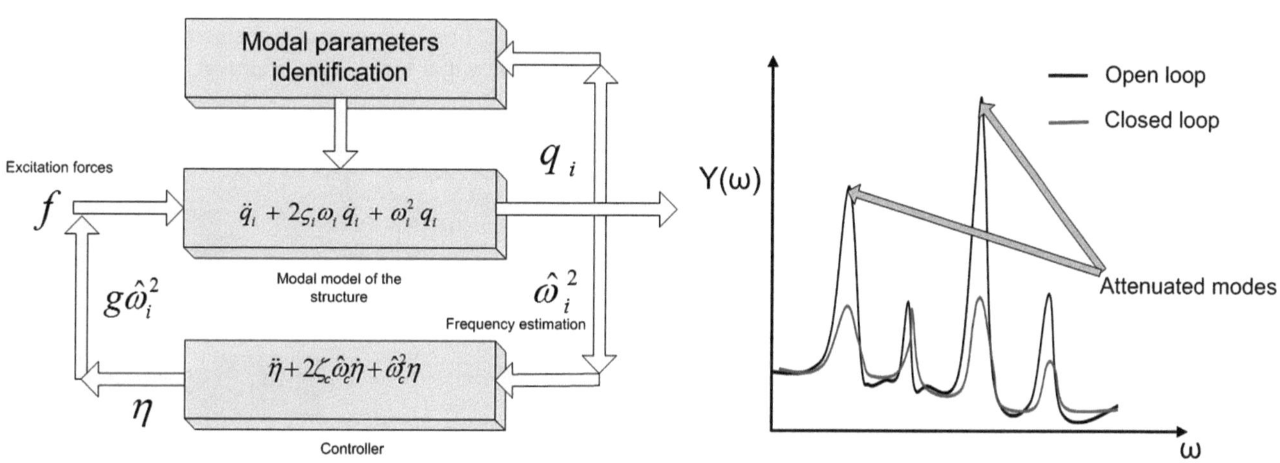

Fig. 14.2 Block diagram of an adaptive-like PPF vibration absorption scheme

The position or displacement of the third floor is positively feed to the virtual passive absorber, and the position of the compensator in this case, the PZT actuator is positively feedback to the building-like structure. In our particular case the output variable to be used for identification of the system parameters and active vibration control purposes is the displacement of the third floor x_3, whereas the control effort is produced by the PZT stack actuator attached to the first lower lateral column, therefore, this particular vibration control problem corresponds to a non-collocated sensor and actuator case [9, 20]. Taking into consideration that the excitation frequency ω_0 is identified by using the algebraic identifier (14.24) and, afterwards, this estimation is used to tune a PPF controller to minimize/attenuate this undesired perturbation, then we get that

$$\begin{aligned} M\ddot{x} + C\dot{x} + Kx &= f(t) + B_f u_{pzt}(t), \qquad && x, f \in R^3, u_{pzt} \in R \\ \ddot{\eta} + 2\xi_f \widehat{\omega}_0 \dot{\eta} + \omega_f^2 &= g\widehat{\omega}_0^2 B_0^T x, && \eta \in R \\ u_{pzt} &= g\omega_0^2 \eta \end{aligned} \tag{14.24}$$

Because $B_0 = [1\ 0\ 0]$ the output variable for control is x_3, which couples the virtual passive absorber by the terms $g\widehat{\omega}^2{}_0 B_0^T$ and $g\widehat{\omega}^2{}_0\eta$, where g is a design control gain. The virtual passive absorber has relative damping ξ_f and natural frequency ω_f. The overall closed-loop system is described, in compact form, as follows

$$\begin{bmatrix} M & 0 \\ 0 & 1 \end{bmatrix} \begin{bmatrix} \ddot{x} \\ \ddot{\eta} \end{bmatrix} + \begin{bmatrix} C & 0 \\ 0 & 2\zeta_f \widehat{\omega}_0 \end{bmatrix} \begin{bmatrix} \dot{x} \\ \dot{\eta} \end{bmatrix} + \begin{bmatrix} K & -B_f g\widehat{\omega}^2{}_0 \\ -gB_0^T \widehat{\omega}^2{}_0 & \widehat{\omega}^2{}_0 \end{bmatrix} \begin{bmatrix} x \\ \eta \end{bmatrix} = \begin{bmatrix} f \\ 0 \end{bmatrix} \tag{14.25}$$

Due to the fact that the mass matrix M is symmetric and positive definite, the overall mass matrix in (14.25) is also symmetric and positive definite, besides, the proportional damping matrix C is also symmetric and positive definite and, as a consequence, the overall damping matrix C in (14.25) has similar properties. Even though the stiffness matrix K is symmetric and positive definite in the above closed loop system, the sensor and actuator are not collocated (i.e., $B_f \neq B_0$). Therefore, it is necessary to select an appropriate gain g value and a fast and stable estimation of the frequency $\widehat{\omega}_0$ to guarantee the closed-loop asymptotic stability, that is, it is necessary that the matrix

$$\widehat{K}_{CL} = \begin{bmatrix} K & -B_f g\widehat{\omega}^2{}_0 \\ -gB_0^T \widehat{\omega}^2{}_0 & \widehat{\omega}^2{}_0 \end{bmatrix} \tag{14.26}$$

be positive definite [9]. Thus, in presence of (bounded) ground motion, the dynamic system response will asymptotically converge to a bounded steady-state response.

Now, for experimental validation, consider that the building like structure shown in Fig. 14.1 is excited by harmonic ground acceleration with frequency ω_0 resulting in undesired vibrations, specially in the third floor. The experimental results for validating the closed loop performance of the PPF controller and the algebraic identifiers for the on line estimation of the characteristic polynomial and the excitation frequency are shown in Fig. 14.2. The excitation was a ground sinusoidal acceleration of approximately $Z_0 = 1.8\ g$ excitation with frequency $\omega_0 = 9\ Hz$, whereas the design parameters for the PPF controller are $\xi_f = 0.25$ and $g = 1.8$ the natural frequency of the passive absorber is selected as $\omega_f = \widehat{\omega}_0$, using the algebraic identifier (14.24).

Figure 14.3 shows the identification and vibration absorption sequence, first, the system's characteristic polynomial coefficients are fast and effectively estimated in a period of $T_0 = 0.2\ s$ as shown in Fig. 14.3a. After that, these coefficients

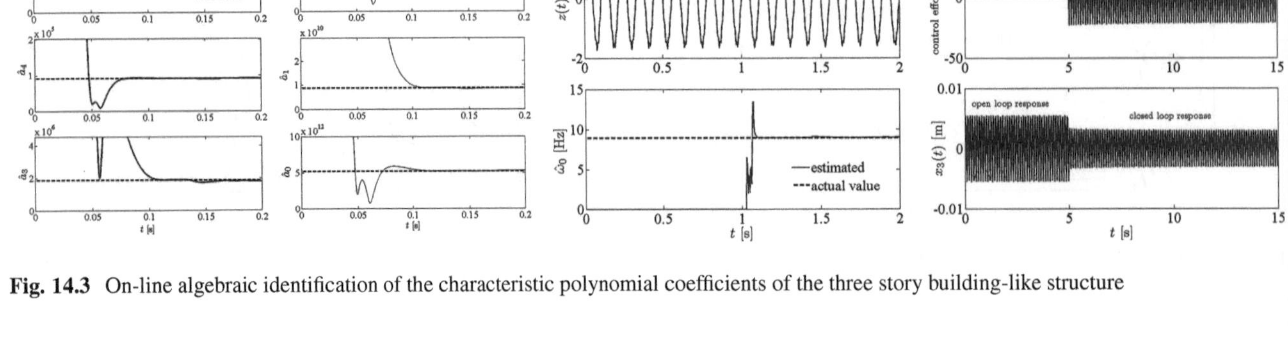

Fig. 14.3 On-line algebraic identification of the characteristic polynomial coefficients of the three story building-like structure

are used in the identifier (14.24) for estimating the excitation frequency, although the excitation frequency is ready after a short period of time as small as 1.3 s, the controller is activated after 5 s to guarantee the convergence and stabilization of the estimation with the possibility for this delay to be reduced. Here we achieve a strong attenuation of 40 % with small control efforts considering that we use a piezostack actuator also shown in Fig. 14.3b. Note that, it is possible to use the presented adaptive-like PPF vibration absorption scheme in pre-compensated control approaches, where the control strategy is strongly related to cancel undesired harmonic excitations and the on line estimation of amplitude and phase of the perturbation is required (see [11]).

14.5 A Multi Positive Position Feedback Control Approach

If we consider that the building-like structure has dominant vibration modes, even though this flexible structure is a distributed parameters system, we can make a natural extension of the PPF control scheme. For experimental modal analysis purposes, we perform a sinusoidal sweep on the building-like structure in an interest bandwidth. The experimental FRF is shown in Fig. 14.4. Three dominant modes are shown by means of acceleration measurements on the third floor.

The multi-PPF control scheme consists of the simultaneous attenuation of vibration modes [9], in our particular case, and as we can observe in Fig. 14.4, we are interested in attenuate three main or dominant vibration modes, corresponding to $\omega_{n1} = 8.89\ Hz$, $\omega_{n1} = 28.19\ Hz$ and $\omega_{n3} = 38.34\ Hz$. Here, we estimate the three natural frequencies of the building-like structure as exposed in [1] and then, the Multi-PPF controller is tuned to perform the multi mode attenuation. In matrix form we have

$$\begin{bmatrix} M & 0 \\ 0 & 1 \end{bmatrix} \begin{bmatrix} \ddot{x} \\ \ddot{\eta} \end{bmatrix} + \begin{bmatrix} C & 0 \\ 0 & 2Z_f \widehat{W} \end{bmatrix} \begin{bmatrix} \dot{x} \\ \dot{\eta} \end{bmatrix} + \begin{bmatrix} K & -B_f G \widehat{W}^2 \\ -GB^T \widehat{W}^2 & \widehat{W}^2 \end{bmatrix} \begin{bmatrix} x \\ \eta \end{bmatrix} = \begin{bmatrix} f \\ 0 \end{bmatrix} \tag{14.27}$$

where $Z_f \in R^{3\times 3}$ is the design diagonal damping matrix. In this case, the three virtual absorbers are coupled to the primary system (14.1) by the positive injection term $G\hat{W}^2 B^T x$ where G is the diagonal design gain matrix. Finally, $\hat{W}$ is a diagonal matrix containing the three estimated natural frequencies to tune the Multi-PPF control scheme. Once again, to guarantee closed-loop stability the matrix

$$K_{MPPF} = \begin{bmatrix} K & -B_f G \widehat{W}^2 \\ -GB^T \widehat{W}^2 & \widehat{W}^2 \end{bmatrix} \tag{14.28}$$

must be positive definite, which is achieved by selecting the appropriate values of the gains in matrix G, and the correct and fast convergence of the estimations in the estimated frequency matrix $\widehat{W}$.

Some experiments were performed on the experimental setup to validate the performance of the natural frequency identification considering that the system is firstly excited only by an initial conditions change. The fast and effective

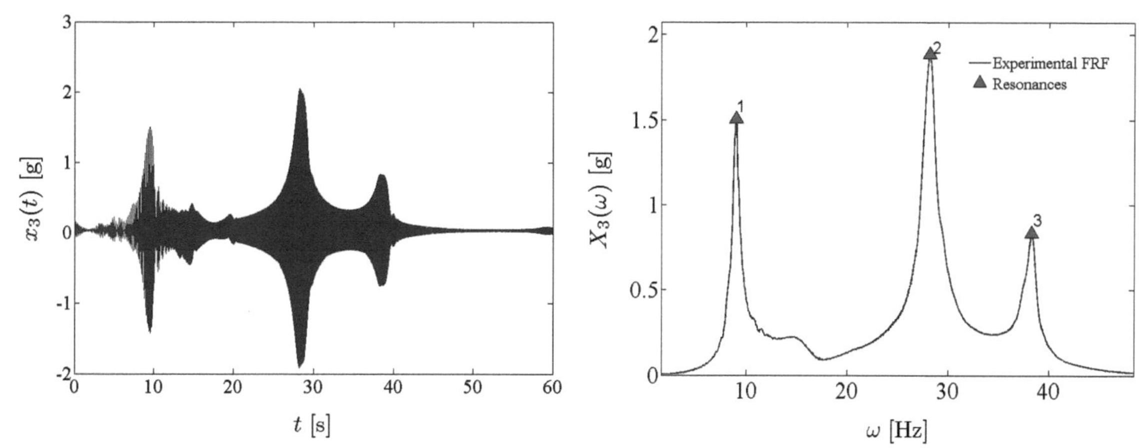

Fig. 14.4 Experimental FRF computed from a sine sweep from 0.1 to 60 Hz

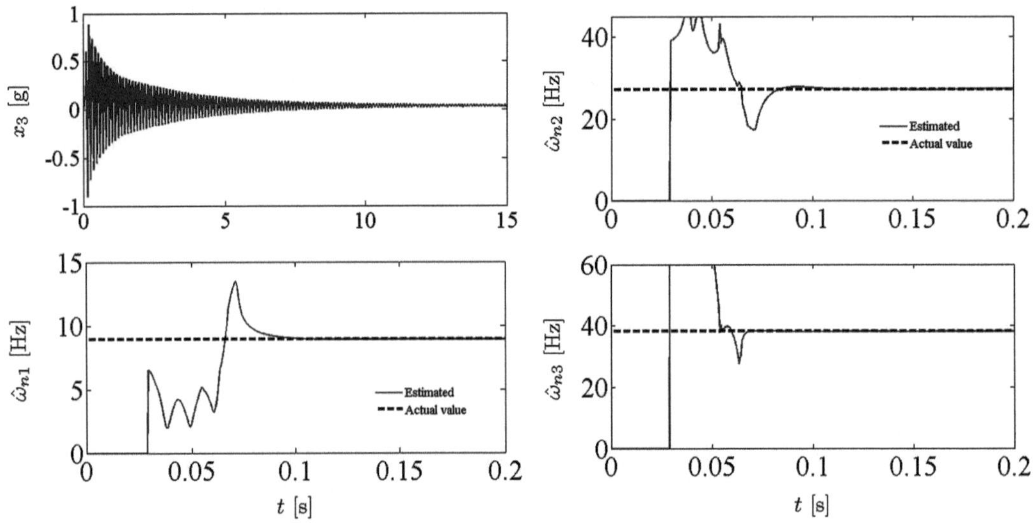

Fig. 14.5 Free vibration response of the third floor x_3 and on-line algebraic estimation of natural frequencies

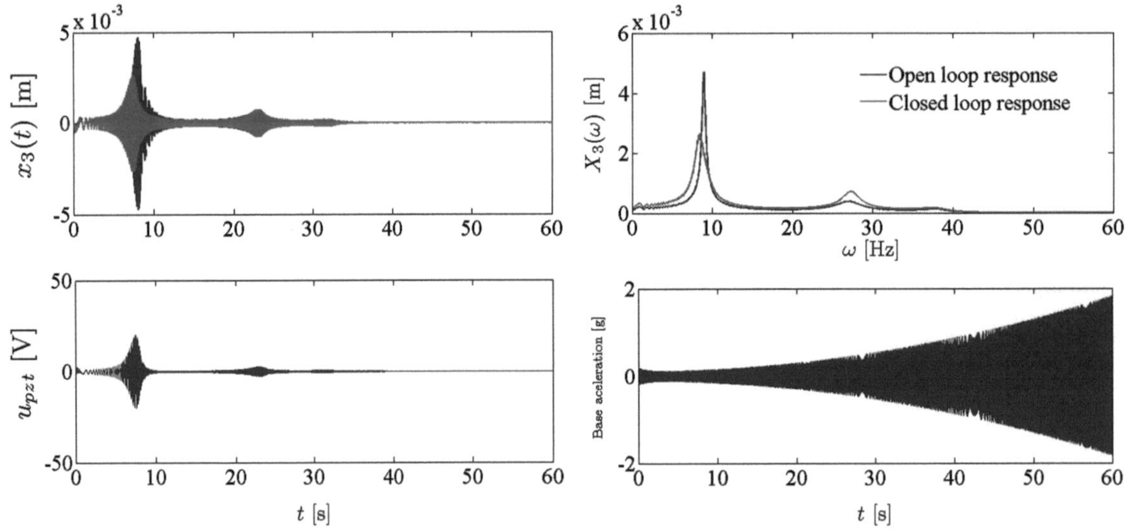

Fig. 14.6 Closed-loop system performance with adaptive like multi-PPF control scheme

estimations are shown in Fig. 14.5. With these parameters we implement the Multi-PPF vibration absorption scheme and validate the closed-loop system performance by applying a sinusoidal sweep in order to make visually clear the attenuation of the undesired vibrations on the third floor, as shown in Fig. 14.6. Here, the MPPF control parameters were selected as $g_1 = 1.8,\ g_1 = 0.12,\ g_3 = 0.5, \xi_{f1} = 0.15,\ \xi_{f2} = 0.12,\ \xi_{f3} = 0.15$ and the three natural frequencies equal to the on-line estimations as $\omega_1 = \widehat{\omega}_{n1}, \omega_2 = \widehat{\omega}_{n2}$ and $\omega_3 = \widehat{\omega}_{n3}$. Note how the first lateral mode is attenuated about 50 % with respect to the open-loop response, employing small control efforts.

14.6 Conclusions

It is proposed an adaptive-like vibration absorption scheme by using a time domain algebraic identification approach for on-line estimation of the harmonic ground excitation and natural frequencies for a multiple degrees-of-freedom mechanical system. The values of the coefficients of the characteristic polynomial of the mechanical system are firstly estimated in real-time, and then it is possible to estimate the excitation frequency. In the design process, for both, identification and control implementations, it is considered that only measurements of some position output variable are available for the identification scheme. However, the identification and control can be implemented using acceleration measurements. In general, the simulation and experimental results show a satisfactory performance of the proposed identification approach with fast and effective estimations and good attenuation properties using the adaptive-like vibration controller on the first vibration mode.

References

1. Astrom KJ, Wittenmark B (1995) Adaptive control, 2nd edn. Addison-Wesley, Reading, MA
2. Isermann R, Munchhof M (2011) Identification of dynamic systems. Springer, Berlin
3. Ljung L (1987) Systems identification: theory for the user. Prentice-Hall, Upper Saddle River
4. Soderstrom T, Stoica P (1989) System identification. Prentice-Hall, New York
5. Fliess M, Sira-Ramírez H (2003) An algebraic framework for linear identification. ESAIM Contr Optim Calc Var 9:151–168
6. Heylen W, Lammens S, Sas P (2003) Modal analysis, theory and testing. Katholieke Universiteit Leuven, Leuven, Belgium
7. Beltran-Carbajal F, Silva-Navarro G, Trujillo-Franco LG (2014) Evaluation of on-line algebraic modal parameter identification methods. In: Proceedings of the 32nd international modal analysis conference (IMAC XXXII), Orlando, FL, vol 8, pp 145–152
8. Chopra AK (1995) Dynamics of structures theory and applications to earthquake engineering. Prentice-Hall, Englewood Cliffs, NJ
9. Friswell MI, Inman DJ (1999) The relationship between positive position feedback and output feedback controllers. Smart Mater Struct 8: 285–291
10. Inman DJ (2006) Vibration with control. Wiley, Chichester, England
11. Beltran-Carbajal F, Silva-Navarro G (2013) Adaptive-like vibration control in mechanical systems with unknown parameters and signals. Asian J Control 15(6):1613–1626
12. Vu VH, Thomas M, Lafleur F, Marcouiller L (2013) Towards an automatic spectral and modal identification from operational modal analysis. J Sound Vib 332:213–227
13. Yang Y, Nagarajaiah S (2013) Output-only modal identification with limited sensors using sparse component analysis. J Sound Vib 332: 4741–4765
14. Chakraborty A, Basu B, Mitra M (2006) Identification of modal parameters of a MDOF system by modified L-P wavelet packets. J Sound Vib 295:827–837
15. Chen-Far H, Wen-Jiunn K, Yen-Tun P (2004) Identification of modal parameters from measured input and output data using a vector backward auto-regressive with exogenous model. J Sound Vib 276:1043–1063
16. Fladung WA, Phillips AW, Allemang RJ (2003) Application of a generalized residual model to frequency domain modal parameter estimation. J Sound Vib 262:677–705
17. Mikusiński J (1983) Operational calculus, vol 1, 2nd edn. PWN & Pergamon, Warsaw
18. Glaetske HJ, Prudnikov AP, Skrnik KA (2006) Operational calculus and related topics. Chapman and Hall/CRC, Boca Raton
19. Beltran-Carbajal F, Silva-Navarro G (2013) Algebraic parameter identification of multi-degree-of-freedom vibrating mechanical systems. In: Proceedings of the 20th international congress on sound and vibration (ICSV20), Bangkok, Thailand, pp 1–8
20. Silva-Navarro G, Enriquez-Zarate J, Belandria-Carvajal ME (2014) Application of positive position feedback control schemes in a building-like structure. In: Proceedings of the 32nd international modal analysis conference (IMAC XXXII), Orlando, FL, vol 4, pp 469–477

Chapter 15
A Fast Maximum Likelihood-Based Estimation of a Modal Model

Mahmoud El-kafafy, Giampiero Accardo, Bart Peeters, Karl Janssens, Tim De Troyer, and Patrick Guillaume

Abstract In this paper, the ML-MM estimator, a multivariable frequency-domain maximum likelihood estimator based on a modal model formulation, will be represented and improved in terms of the computational speed and the memory requirements. Basically, the design requirements to be met in the ML-MM estimator were to have accurate estimate for both of the modal parameters and their confidence limits and, meanwhile, having a clear stabilization chart which enables the user to easily select the physical modes within the selected frequency band. The ML-MM method estimates the modal parameters by directly identifying the modal model instead of identifying a rational fraction polynomial model. In the ML-MM estimator, the confidence bounds on the estimated modal parameters (i.e., frequency, damping ratios, mode shapes, etc.) are derived directly by inverting the so-called Fisher information matrix and without using many linearization formulas that are normally used when identifying rational fraction polynomial-based models. Another advantage of the ML-MM estimator lies in its potential to overcome the difficulties that the classical modal parameter estimation methods face when fitting an FRF matrix that consists of many (i.e., 4 or more) columns, i.e., in cases where many input excitation locations have to be used in the modal testing. For instance, the high damping level in acoustic modal analysis requires many excitation locations to get sufficient excitation of the modes. In this contribution, the improved ML-MM estimator will be validated and compared with some other classical modal parameter estimation methods using simulated datasets and real industrial applications.

Keywords Modal model • Maximum likelihood • Modal parameters • Frequency-domain • Estimator

15.1 Introduction

Over the last decades, a number of algorithms have been developed to estimate modal parameters from measured frequency or impulse response function data. A very popular implementation of the frequency-domain linear least squares estimator optimized for the modal parameter estimation is called Least Squares Complex Frequency-domain (LSCF) estimator [1]. The LSCF estimator uses a discrete-time common denominator transfer function parameterization. In [2], the LSCF estimator is extended to a poly-reference case (pLSCF). The pLSCF estimator uses a right matrix fraction description (RMFD) model. LSCF and pLSCF estimators were optimized both for the memory requirements and for the computation speed. The main advantages of those estimators are their speed and the very clear stabilization charts they yield even in the case of highly noise-contaminated frequency response functions (FRFs). Regardless of their very clear stabilization charts, these estimators have two main drawbacks. The first drawback is that the damping estimate is not always reliable where it decreases with increasing the noise level, and this situation becomes worse for the highly damped and weakly excited modes [3]. A second drawback is that the confidence bounds on the estimates are unavailable. Although these intervals can be constructed after the estimation as shown in [4, 5], these estimates are always higher than the maximum likelihood estimator (MLE). LSCF and pLSCF estimators are essentially deterministic curve fitting algorithms in which the estimation process is achieved

M. El-kafafy (✉)
Vrije Universiteit Brussel (VUB), Pleinlaan 2, B-1050 Brussel, Belgium

Siemens Industry Software, Interleuvenlaan 68, B-3001 Leuven, Belgium

Faculty of engineering – Mattaria, Helwan University, Helwan, Egypt
e-mail: melkafaf@vub.ac.be

G. Accardo • B. Peeters • K. Janssens
Siemens Industry Software, Interleuvenlaan 68, B-3001 Leuven, Belgium

T. De Troyer • P. Guillaume
Vrije Universiteit Brussel (VUB), Pleinlaan 2, B-1050 Brussel, Belgium

M. Mains (ed.), *Topics in Modal Analysis, Volume 10*, Conference Proceedings of the Society for Experimental Mechanics Series, DOI 10.1007/978-3-319-15251-6_15

without using information on the statistical distribution of the data. By taking knowledge about the noise on the measured data into account, the modal parameters can be derived using the so-called frequency-domain maximum likelihood estimator (MLE) with significant higher accuracy compared to the ones developed in the deterministic framework. MLE for linear time invariant systems was introduced in [6]. A multivariable frequency-domain maximum likelihood estimator was proposed in [1] to identify the modal parameters together with their confidence intervals where it was used to improve the estimates that are initially estimated by LSCF estimator. In [7], the poly-reference implementation for MLE was introduced to improve the starting values provided by pLSCF estimator. Both of the ML estimators introduced in [1] and [7] are based on a rational fraction polynomial model, in which the coefficients are identified. The modal parameters are then estimated from the identified coefficients in a second step. In these estimators, the uncertainties on the modal parameters are calculated from the uncertainties on the estimated polynomial coefficients by using some linearization formulas. These linearization formulas are straightforward when the relation between the modal parameter and the estimated coefficients is explicitly known but can be quite involved for the implicit case. Moreover, they may fail when the signal-to-noise ratio is not sufficiently large [8]. In [9, 10], a non-linear least squares (NLS) estimation method based on the modal model formulation is introduced. The Gauss-Newton method is used to solve the NLS problem and the Multivariate Mode Indicator Function (MMIF) is used to generate initial values for the modal parameters of the modal model.

Recently, a multivariable frequency-domain modal parameter estimator called ML-MM [11–13] has been introduced. The key challenges behind introducing the ML-MM estimator were to keep the benefit of the well-known Polymax estimator while giving other additional features. The benefit of the Polymax estimator that the ML-MM estimator keeps is the construction of a very clear stabilization diagram in a very fast way. The other additional features that ML-MM estimator adds are the highly accurate modal parameters it estimates together with their confidence bounds. This method belongs to the class of maximum likelihood-based estimation techniques. Taking into account the uncertainty on the measurements, the ML-MM estimator estimates the modal parameters together with their confidence bounds by directly identifying the modal model in a maximum likelihood sense. In the ML-MM method, the Gauss-Newton method together with Levenberg-Marquardt loop is used to solve a NLS problem. Using Levenberg-Marquardt loop during the iterations forces the ML-MM cost function to converge. Identifying directly the modal model instead of a rational fractional polynomial model gives the advantage of having the confidence bounds on the estimated modal parameters directly without using linearization formulas which have to be used to extend the uncertainty from the estimated polynomial coefficients to the estimated modal parameters in case of identifying a rational fractional polynomial model. Another advantage found of the ML-MM estimator lies in its potential to overcome the difficulties that the classical modal parameter estimation methods face when fitting an FRF matrix that consists of many (i.e., 4 or more) columns, i.e., in cases where many input excitation locations have to be used in the modal testing. For instance, the high damping level in acoustic modal analysis requires many excitation locations to get sufficient excitation of the modes [14]. In this contribution, the ML-MM method will be represented, including an algorithm variant that significantly speeds up the execution of the method. The fast implementation will be validated and compared to the basic implementation of the method using simulated examples and industrial applications cases.

15.2 Maximum Likelihood Estimation Based on the Modal Model (ML-MM)

15.2.1 The Basic Implementation of the ML-MM Estimator

In this section, the theoretical formulation of the ML-MM estimator will be introduced. Assuming the different frequency response functions (FRFs) to be uncorrelated the (negative) log-likelihood function reduces to [15, 16]

$$\mathrm{K}_{ML-MM}(\theta) = \sum_{o=1}^{N_o} \sum_{k=1}^{N_f} E_o(\theta, \omega_k)\, E_o^H(\theta, \omega_k) \tag{15.1}$$

with N_f the number of the frequency lines, N_o the number of the measured responses (outputs), $()^H$ the complex conjugate transpose of a matrix, ω_k the circular frequency, θ the model parameters vector, and $E_o(\theta, \omega_k)$ the error equation corresponds to the oth output degree of freedom (DOF) as a row vector given as follows:

$$E_o(\theta, \omega_k) = \left[\frac{H_{o1}(\omega_k) - \hat{H}_{o1}(\theta, \omega_k)}{\sqrt{var(H_{o1}(\omega_k))}} \cdots \frac{H_{oN_i}(\omega_k) - \hat{H}_{oN_i}(\theta, \omega_k)}{\sqrt{var\left(H_{oN_i}(\omega_k)\right)}} \right] \in \mathbb{C}^{1\times N_i} \tag{15.2}$$

with $H_o(\omega_k) \in \mathbb{C}^{1\times N_i}$ the measured FRFs, $var(H_o(\omega_k)) \in \mathbb{R}^{1\times N_i}$ the variance of the measured FRFs, and $\hat{H}_o(\theta,\omega_k) \in \mathbb{C}^{1\times N_i}$ the FRFs represented by the displacement-over-force modal model formulation written as follows:

$$\hat{H}(\theta,\omega_k) = \sum_{r=1}^{N_m}\left(\frac{\psi_r L_r}{s_k-\lambda_r} + \frac{\psi_r^* L_r^*}{s_k-\lambda_r^*}\right) + \frac{LR}{s_k^2} + UR \tag{15.3}$$

with $\hat{\mathrm{H}}(\theta,\omega_k) \in \mathbb{C}^{N_o\times N_i}$ is the FRFs matrix with N_o outputs and N_i inputs, N_m is the number of the identified modes, $\psi_r \in \mathbb{C}^{N_o\times 1}$ the rth mode shape, λ_r the rth pole, $s_k = j\omega_k$, ()* stands for the complex conjugate of a complex number, $L_r \in \mathbb{C}^{1\times N_i}$ the rth participation factor, $LR \in \mathbb{C}^{N_o\times N_i}$ and $UR \in \mathbb{C}^{N_o\times N_i}$ the lower and upper residual terms. The model parameters vector $\theta = \left[\theta_{\psi_o}\ \theta_{LUR_o}\ \theta_L\ \theta_\lambda\right]^T$ is a column vector containing all the parameters of the modal model represented by Eq. 15.3. θ_{ψ_o} and θ_{LUR_o} are all the mode shapes elements, lower and upper residuals term elements correspond to output o. θ_{ψ_o}, θ_{LUR_o}, θ_L, and θ_λ are written as follows:

$$\begin{aligned}&\theta_{\psi_o} = \left[\psi_{o1}\ \psi_{o2}\ldots.\psi_{oN_m}\right],\ \ \theta_{LUR_o} = \left[LR_{o1}\ldots LR_{oN_i}\ UR_{o1}\ldots.UR_{oN_i}\right],\\ &\theta_L = \left[L_1\ L_2\ldots.L_{N_m}\right],\ \ \theta_\lambda = \left[\lambda_1\ \lambda_2\ldots.\lambda_{N_m}\right]\end{aligned} \tag{15.4}$$

The maximum likelihood estimates of θ are given by minimizing the ML-MM cost function presented by Eq. 15.1 using Gauss-Newton optimization algorithm together with Levenberg-Marquardt approach to ensure the convergence. The Gauss-Newton iterations are given by

(a) solve $\left(J_m^T J_m\right)\ vec(\delta_m) = -J_m^T\ E_m$ for vec(δ_m)
(b) set $\theta_{m+1} = \theta_m + \delta_m$

with

$$E_m = \begin{bmatrix} vec(E_1)\\ vec(E_2)\\ \vdots\\ vec(E_{N_o})\end{bmatrix} \in \mathbb{R}^{2N_iN_fN_o\times 1} \tag{15.5}$$

with $vec(.)$ the column-stacking operator, $J_m = \frac{\partial E(\theta_m)}{\partial \theta_m} \in \mathbb{R}^{2N_iN_fN_o\times 2N_m(N_o+N_i)+4N_iN_o}$ is the Jacobian matrix at iteration m and $vec(\delta_m) \in \mathbb{R}^{2N_m(N_o+N_i)+4N_iN_o\times 1}$ are the perturbations on the parameters. To have a relatively fast and numerically stable implementation, the Jacobian matrix J_m is written in the following (well-structured sparse form):

$$J = \begin{bmatrix} \Gamma_1 & 0 & \ldots & 0 & \Phi_1^L & \Phi_1^\lambda\\ 0 & \Gamma_2 & \ldots & 0 & \Phi_2^L & \Phi_2^\lambda\\ \vdots & \vdots & \ddots & \vdots & \vdots & \vdots\\ 0 & 0 & \ldots & \Gamma_{N_o} & \Phi_{N_o}^L & \Phi_{N_o}^\lambda\end{bmatrix} \tag{15.6}$$

with $\Gamma_o \in \mathbb{R}^{2N_iN_f\times 2(N_m+2N_i)}$ the error derivatives with respect to the mode shapes and the residual terms for output o, $\Phi_o^L \in \mathbb{R}^{2N_iN_f\times 2N_m(N_i-1)}$ is the error derivative with respect to the participation factors for output o and $\Phi_o^\lambda \in \mathbb{R}^{2N_iN_f\times 2N_m}$ is the error derivative with respect to the poles for output o. Taking into account the structure of that Jacobian matrix, the normal equation, $\left(J_m^T J_m\right)\ vec(\delta_m) = -J_m^T\ E_m$, can be written as follows:

$$\begin{bmatrix} R_1 & 0 & \ldots & 0 & S_1^L & S_1^\lambda\\ 0 & R_2 & \ldots & \vdots & S_2^L & S_2^\lambda\\ \vdots & 0 & \ddots & 0 & \vdots & \vdots\\ 0 & \ldots & 0 & R_{N_o} & S_{N_o}^L & S_{N_o}^\lambda\\ S_1^{L^H} & S_2^{L^H} & \ldots & S_{N_o}^{L^H} & \sum_{o=1}^{N_o} T_o^L & \sum_{o=1}^{N_o} T_o^{L\lambda}\\ S_1^{\lambda^H} & S_2^{\lambda^H} & \ldots & S_{N_o}^{\lambda^H} & \sum_{o=1}^{N_o} T_o^{\lambda L} & \sum_{o=1}^{N_o} T_o^{\lambda}\end{bmatrix} \begin{bmatrix} \Delta vec(\Psi_1)\\ \Delta vec(\Psi_2)\\ \vdots\\ \Delta vec(\Psi_{N_o})\\ \Delta vec(L)\\ \Delta\lambda\end{bmatrix} = -\begin{bmatrix} \Gamma_1^H vec(E_1)\\ \Gamma_2^H vec(E_2)\\ \vdots\\ \Gamma_{N_o}^H vec(E_{N_o})\\ -\\ \sum_{o=1}^{N_o} \Phi_o^{L^H} vec(E_o)\\ \sum_{o=1}^{N_o} \Phi_o^{\lambda^H} vec(E_o)\end{bmatrix} \tag{15.7}$$

The sub-matrices in Eq. 15.7 are given as following:

$$R_o = \Gamma_o^H \Gamma_o \in \mathbb{R}^{2(N_m+2N_i)\times 2(N_m+2N_i)} \quad S_o^L = \Gamma_o^H \Phi_o^L \in \mathbb{R}^{2(N_m+2N_i)\times 2N_m(N_i-1)} \quad S_o^\lambda = \Gamma_o^H \Phi_o^\lambda \in \mathbb{R}^{2(N_m+2N_i)\times 2N_m}$$
$$T_o^L = \Phi_o^{L^H} \Phi_o^L \in \mathbb{R}^{2N_m(N_i-1)\times 2N_m(N_i-1)} \quad T_o^\lambda = \Phi_o^{\lambda^H} \Phi_o^\lambda \in \mathbb{R}^{2N_m\times 2N_m} \quad T_o^{L\lambda} = \Phi_o^{L^H} \Phi_o^\lambda \in \mathbb{R}^{2N_m(N_i-1)\times 2N_m}$$
$$T_o^{\lambda L} = \Phi_o^{\lambda^H} \Phi_o^L \in \mathbb{R}^{2N_m\times 2N_m(N_i-1)} \tag{15.8}$$

The major gain in the calculation time comes from the use of the structure of the normal equations. Using some elimination and substitution procedures yields to

$$\Delta vec\,(\Psi_o) = -R_o^{-1}\ \left(\Gamma_o^H vec\,(E_o) + S_o^L\ \Delta vec(L) + S_o^\lambda\ \Delta\lambda\right) \tag{15.9}$$

$$\begin{bmatrix} \Delta \text{vec}(L) \\ \Delta\lambda \end{bmatrix} = M^{-1}\ D \tag{15.10}$$

$$M = \begin{bmatrix} M_1 & M_2 \\ M_3 & M_4 \end{bmatrix} = \begin{bmatrix} \sum_{o=1}^{N_o} T_o^L - S_o^{L^H} R_o^{-1} S_o^L & \sum_{o=1}^{N_o} T_o^{L\lambda} - S_o^{L^H} R_o^{-1} S_o^\lambda \\ \sum_{o=1}^{N_o} T_o^{\lambda L} - S_o^{\lambda^H} R_o^{-1} S_o^L & \sum_{o=1}^{N_o} T_o^\lambda - S_o^{\lambda^H} R_o^{-1} S_o^\lambda \end{bmatrix} \in \mathbb{R}^{2N_mN_i\times 2N_mN_i} \tag{15.11}$$

$$D = \begin{bmatrix} \sum_{o=1}^{N_o} \left(S_o^{L^H} R_o^{-1}\Gamma_o^H - \Phi_o^{L^H}\right) E_o \\ \sum_{o=1}^{N_o} \left(S_o^{\lambda^H} R_o^{-1}\Gamma_o^H - \Phi_o^{\lambda^H}\right) E_o \end{bmatrix} \in \mathbb{R}^{2N_mN_i\times 1} \tag{15.12}$$

Solving Eq. 15.10 leads to the update of the participation factors elements and the poles, while the update of the mode shapes and residual terms elements are obtained by substitution of Eq. 15.10 into Eq. 15.9. In this way, the direct inversion of the full matrix $(J_m^T J_m)$ can be avoided leading to significant reduction in the calculation time. To derive the uncertainty (confidence) bounds of the estimated modal parameters, the approximation given by [6] can be used as follows:

$$COV\,(\psi, LR, UR, L, \lambda) \approx \left[J^T J\right]^{-1} \tag{15.13}$$

J is the Jacobian matrix evaluated in the last iteration step of the Gauss-Newton algorithm. Taking into account the structure of the Jacobian matrix and by using the matrix inversion lemma [17], the uncertainties on the poles and the participation factors, can be derived as follows:

$$COV\,(L, \lambda) = M^{-1} \tag{15.14}$$

where M is given by Eq. 15.11. The uncertainty bounds of the estimated mode shapes and lower and upper residual terms can be derived as follows:

$$\begin{aligned} \text{COV}\,(\psi_o, LR_o, UR_o) &= R_o^{-1} + \beta_o \\ \beta_o &= A_o S_o^{L^H} R_o^{-1} + G_o S_o^{\lambda^H} R_o^{-1} \\ A_o &= R_o^{-1}\left(S_o^L\left[M_1 - M_2 M_4^{-1} M_3\right]^{-1} + S_o^\lambda\left(-\left[M_4 - M_3 M_1^{-1} M_2\right]^{-1} M_3 M_1^{-1}\right)\right) \\ G_o &= R_o^{-1}\left(S_o^L\left(-M_1^{-1} M_2\left[M_4 - M_3 M_1^{-1} M_2\right]^{-1}\right) + S_o^\lambda\left[M_4 - M_3 M_1^{-1} M_2\right]^{-1}\right) \end{aligned} \tag{15.15}$$

Since the presented ML-MM estimator is basically a non-linear optimization algorithm and thus iterative, it requires good starting values for all the entire modal parameters of the modal model (see Eq. 15.3). These starting values of all the modal parameters are obtained by applying the well-known Polymax estimator to the measured FRFs. The Polymax estimator is

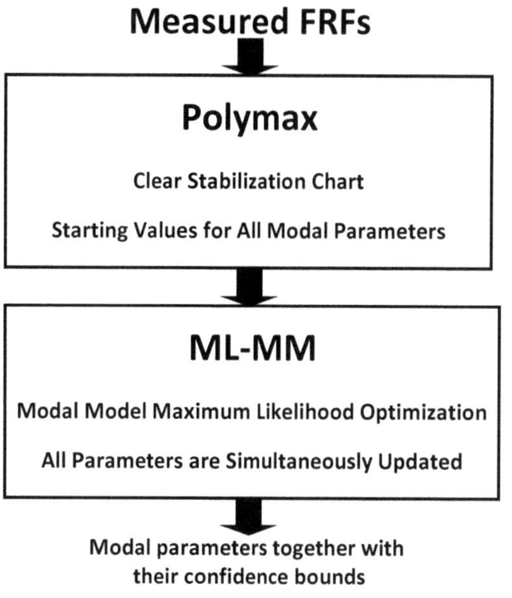

Fig. 15.1 Schematic representation of the basic ML-MM estimator

selected since it provides a very clear stabilization chart in a fast way. This will help the user to select the really physical vibration modes in the selected frequency band which in turns leads to obtain good staring values for all the modal parameters to start the iterative ML-MM estimator. Figure 15.1 shows a schematic description of the basic ML-MM estimator.

15.2.2 A Fast Implementation of the ML-MM Estimator

In this section, an alternative implementation of the ML-MM estimator will be discussed. The aim of this alternative implementation is to speed up the implementation of the algorithm. In Sect. 15.2.1, the basic implementation, in which all the modal parameters (i.e., mode shapes, participation factors, poles, upper and lower residual terms) of the modal model are simultaneously updated at each iteration of the Gauss-Newton optimization, is introduced. In Eq. 15.3, when the poles λ_r and the participation factors L_r are given, it is clear that the mode shapes ψ_r and the lower and upper residuals (LR, UR) can be calculated easily in a linear least squares sense. This can be done by writing Eq. 15.3 for all the values of the frequency axis of the FRF data, and basically the unknown parameters (i.e., ψ, LR, UR) are then found as the least-squares solution of these equations. This approach is basically what so-called the LSFD estimator [18, 19]. The size of the mode shapes vector and the lower and upper residual matrices is mainly dependent on the number of outputs N_o, and in modal testing the number of outputs (responses) N_o is normally much higher than the number of inputs (excitation points) N_i. So, one idea to significantly decrease the computational time of the ML-MM estimator is to exclude the mode shapes and the lower and upper residuals from the updated parameters vector $\theta = \left[\theta_\psi \ \theta_{LUR} \ \theta_L \ \theta_\lambda\right]^T$. This can be done as follows (see Fig. 15.2):

- At each iteration of the Gauss-Newton algorithm
  1. Only the participation factors θ_L and the poles θ_λ are updated
  2. The mode shapes and the lower and the upper residuals are calculated in linear least-squares sense using the LSFD estimator.

This will lead to huge reduction in the number of columns of the Jacobian matrix, hence the size of the normal equations matrix. Now, the Jacobian matrix $J_m = \partial E\theta_{(L,\lambda)_m} / \partial\theta_{(L,\lambda)_m} \in \mathbb{R}^{2N_i N_f N_o \times 2N_m (N_i+1)}$ is the derivative of the equation error with respect to the participation factors and the poles. It can be seen that the number of the columns of the Jacobian matrix has been reduced to $2N_m\,(N_i + 1)$ columns compared to the previous one which has $2N_m\,(N_o - 1) + 4N_i N_o$ columns. The new Jacobian matrix is given as:

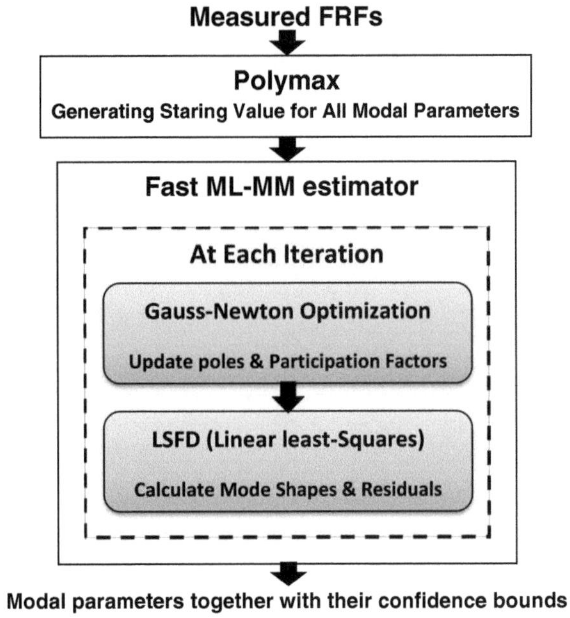

Fig. 15.2 Schematic representation of the fast ML-MM estimator

$$J = \begin{bmatrix} \Gamma_{L1} & 0 & \dots & 0 & \Phi^{\lambda}_{L1} \\ 0 & \Gamma_{L2} & \dots & 0 & \Phi^{\lambda}_{L2} \\ \vdots & \vdots & \ddots & \vdots & \vdots \\ 0 & 0 & \dots & \Gamma_{LN_i} & \Phi^{\lambda}_{LN_i} \end{bmatrix} \in \mathbb{R}^{2N_i N_f N_o \times 2N_m(N_i+1)} \tag{15.16}$$

with $\Gamma_{Li} \in \mathbb{R}^{2N_o N_f \times 2N_m}$ the equation error derivatives with respect to the participation factors of all the identified modes for the input location i where $i = 1, 2, ..N_i$ and $\Phi^{\lambda}_{Li} \in \mathbb{R}^{2N_o N_f \times 2N_m}$ is the equation error derivative with respect to the poles of all the identified modes for input i. Taking into account the structure of that Jacobian matrix, the reduced normal equations, $\left(J_m^T J_m\right) vec\left(\delta_m\right) = -J_m^T\ E_m$, can be written as follows:

$$\begin{bmatrix} R_{L1} & 0 & \dots & 0 & S_1^{L\lambda} \\ 0 & R_{L2} & \dots & 0 & S_2^{L\lambda} \\ 0 & 0 & \ddots & \vdots & \vdots \\ \vdots & \vdots & \ddots & R_{LN_i} & S_{N_i}^{L\lambda} \\ S_1^{L\lambda^H} & S_2^{L\lambda^H} & \dots & S_{N_i}^{L\lambda^H} & \sum_{i=1}^{N_i} T^{\lambda}_{Li} \end{bmatrix} \begin{bmatrix} \Delta L_1 \\ \Delta L_2 \\ \vdots \\ \Delta L_{N_i} \\ \Delta\lambda \end{bmatrix} = - \begin{bmatrix} \Gamma^H_{L1} vec\left(E_1\right) \\ \Gamma^H_{L2} vec\left(E_2\right) \\ \vdots \\ \Gamma^H_{LN_i} vec\left(E_{N_i}\right) \\ --- \\ \sum_{i=1}^{N_i} \Phi^{\lambda^H}_{Li} vec\left(E_i\right) \end{bmatrix} \tag{15.17}$$

with $R_{Li} = \Gamma_{Li}{}^H \Gamma_{Li} \in \mathbb{R}^{2N_m \times 2N_m}$, $S_i^{L\lambda} = \Gamma_{Li}{}^H \Phi^{\lambda}_{Li} \in \mathbb{R}^{2N_m \times 2N_m}$, $T^{\lambda}_{Li} = \Phi^{\lambda}_{Li}{}^H \Phi^{\lambda}_{Li} \in \mathbb{R}^{2N_m \times 2N_m}$, $vec\left(E_i\right) \in \mathbb{R}^{2N_f N_o \times 1}$ the error vector corresponds to input i and all the outputs at all the frequency lines, $\Delta L_i \in \mathbb{R}^{2N_m \times 1}$ a vector contains all the perturbations on the real and imaginary parts of the participation factors correspond to input i and all the identified modes, and $\Delta\lambda \in \mathbb{R}^{2N_m \times 1}$ a vector contains the perturbations on the real and imaginary parts of the identified poles. Using some elimination and substitution procedures yields the perturbations on the poles and the participation factors respectively as follows:

$$\Delta\lambda = \left[\sum_{i=1}^{N_i} T^{\lambda}_{Li} - S_i^{L\lambda^H} \left(R_{Li}\right)^{-1} S_i^{L\lambda}\right]^{-1} \left[\sum_{i=1}^{N_i} S_i^{L\lambda^H} \left(R_{Li}\right)^{-1} \Gamma^H_{Li} vec\left(E_i\right) - \Phi^{\lambda^H}_{Li} vec\left(E_i\right)\right] \tag{15.18}$$

$$\Delta L_i = -\left(R_{Li}\right)^{-1} \left[\Gamma^H_{Li} vec\left(E_i\right) + S_i^{L\lambda} \Delta\lambda\right] \tag{15.19}$$

Once these perturbations are calculated, the poles and the participation factors are updated to be used for a new iteration. Before moving to the new iteration, the updated poles and participation factors will be used to calculate the mode shapes and the lower and upper residuals in a linear least-squares sense using the LSFD estimator. Since in this fast implementation of the ML-MM estimator the mode shapes and the residual terms are assumed to be certain parameters (i.e., they are assumed to have no errors hence they are not updated during the Gauss-Newton iterations), the uncertainty bounds on those parameters cannot be retrieved. So, only the uncertainty bounds on the poles and the participation factors will be retrieved. These uncertainty bounds on the poles and the participation factors can be calculated using Eq. 15.13 substituting the Jacobian matrix with the one represented in Eq. 15.16. Taking into account the structure of the Jacobian matrix and by using the matrix inversion lemma, the uncertainties (variances) on the poles and participation factors can be derived respectively as follows:

$$cov\left(\lambda\right) = \left[\sum_{i=1}^{N_i} T_{Li}^{\lambda} - S_i^{L\lambda^H}\left(R_{Li}\right)^{-1} S_i^{L\lambda}\right]^{-1} \tag{15.20}$$

$$cov\left(L_i\right) = \left(R_{Li}\right)^{-1} S_i^{L\lambda} cov\left(\lambda\right) S_i^{L\lambda^H}\left(R_{Li}\right)^{-1} + \left(R_{Li}\right)^{-1} \tag{15.21}$$

15.3 Validations and Discussion

In this section, both implementations of the ML-MM estimator (i.e., basic and fast ones) will be validated and compared to some other estimators like the pLSCF (Polymax), polyreference Least-squares Complex Exponential (pLSCE) estimator, and the Maximum Likelihood common denominator polynomial model –based (ML-CDM) estimator. Polymax [2, 20] is a linear-least squares frequency domain estimator, which fits a right matrix fraction description (RMFD) polynomial model to the measured FRFs. Then, the poles and the participation factors are derived from the optimized denominator coefficients, while the mode shapes and the lower and upper residuals are estimated in a second step using the LSFD estimator [15.18]. The pLSCE estimator [21–23] is a linear least-squares time-domain estimator, which uses the impulse response function to retrieve the poles and the participation factors and then the mode shapes and the residual terms can be estimated in a second step by fitting the modal model equation (see Eq. 15.3) to the measured FRFs using the pre-estimated poles and the participation factors. The ML-CDM estimator [1] is a frequency-domain estimator fits a common-denominator polynomial model to the measured FRFs in a non-linear least squares sense using Gauss-Newton algorithm. The poles are retrieved as the roots of the denominator coefficients, and the mode shapes together with the participation factors are obtained by decomposing the residues matrices to rank-one matrices using singular value decomposition (SVD). These residues matrices are estimated by fitting the pole-residue model to the measured FRFs in a linear least-squares sense. Both ML-CDM and ML-MM estimator are maximum likelihood estimators. But the model which is being used to fit the measured FRFs is different. ML-MM, as it is mentioned before, fits directly the modal model to the measured FRFs, while ML-CDM fits a common-denominator polynomial model. Since both ML-MM and ML-CDM are maximum likelihood estimators, both of them can deliver the uncertainty bounds on the estimated modal parameters as long as the equation error is weighted by the variance of the measured FRFs. So, the comparison between the ML-MM and the ML-CDM will be extended to include also the predicted uncertainty while for the Polymax and the pLSCE estimators the comparison will be done in terms of the bias error and the root mean squares error (RMSE). So, in the following sections these estimators will be compared to each other using simulated FRFs and some real industrial applications.

15.3.1 Seven DOF System with Mixed Noise [White and Relative Noise]

In this section, simulated FRFs of a 7-DOF system will be used to validate and compare the proposed ML-MM estimator with the other estimators (i.e., Polymax, pLSCE, and ML-CDM). The first column of the FRF matrix will be used (i.e., $N_o = 7$, $N_i = 1$). The exact FRFs are contaminated with a combination of uncorrelated white and relative noise. At each frequency line, the noise variance is taken as a summation of constant variance, which corresponds to the white noise component, and a relative variance that corresponds to the relative noise component. Figure 15.3 shows the driving point FRF together with the noise standard deviation. To perform Monte-Carlo simulations, 500 disturbed data sets have been generated.

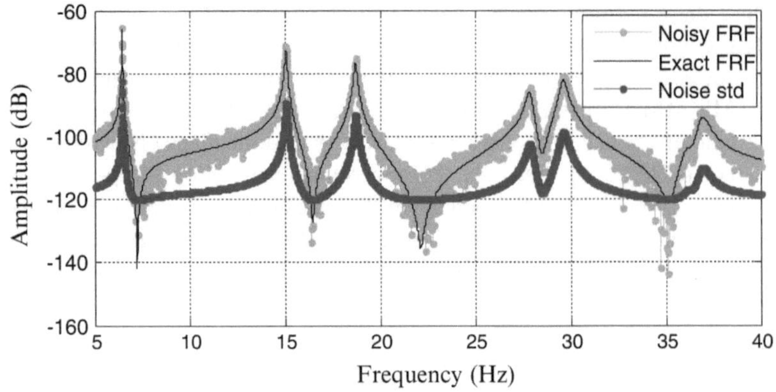

Fig. 15.3 The noisy FRFs (gray dots), the exact FRFs (thin-black line), and the FRFs standard deviation (thick-gray line) of the 7-DOF system

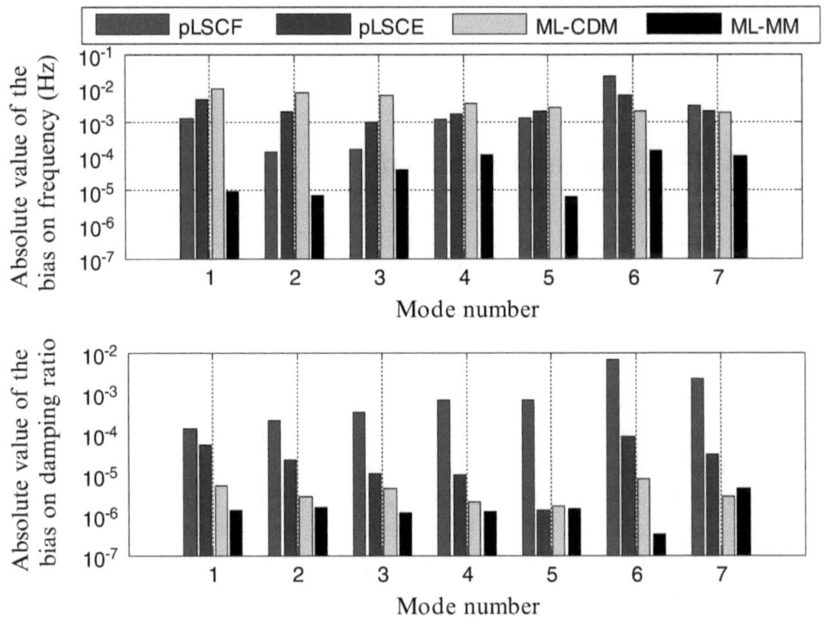

Fig. 15.4 The absolute value of the bias error on the estimated resonance frequencies (*top*) and the estimated damping ratios (*bottom*) (The order of the different methods in the figure legend corresponds to the order of the bars in the plot.)

Figure 15.4 shows the absolute value of the bias error on the estimated frequency and damping estimates. The order of the different methods in the figure legend corresponds to the order of the bars in the plot. From Fig. 15.4, it can be seen that the ML-MM estimator improves the frequency and damping estimates that are initially estimated by the pLSCF (PolyMAX) estimator. It can be seen that the ML-MM gives the lowest bias error for, almost, all the identified modes compared with the other estimators, and this is for the frequency and the damping estimates.

In terms of the RMSE of the frequency and damping estimates, Fig. 15.5 (The order of the different methods in the figure legend corresponds to the order of the bars in the plot.) shows that the performances of both ML-CDM and ML-MM estimators are very comparable. However, one can note that the ML-MM estimator gives a lower RMSE of the frequency and damping for the last two modes (i.e., 6th and 7th modes). These modes are not well excited compared to the other modes (examine Fig. 15.3, there are two peaks above 35 Hz correspond to the 6th and 7th modes). From Figs. 15.4 and 15.5, we can sum up that the ML-MM estimator gives estimates with lower bias error and lower variance in comparison with the other estimators under test.

Concerning the predicted uncertainty bounds of the estimated modal parameters, Fig. 15.6 compares the predicted standard deviation (std) (see Eq. 15.14 or 15.20 for the ML-MM estimator) to the sample standard deviation on the resonance frequency and the damping ratio of the 1st mode. In this figure, the predicted std is presented by the dots while the sample std is presented by the gray solid line. From this figure, one can see that the ML-CDM and the ML-MM estimators show a good agreement between the predicted and the sample (true) standard deviation on both the frequency and damping estimate. Although both ML-MM and ML-CDM show good agreement between the predicted and the sample std, the ML-MM estimator predicts the uncertainty bounds with a lower variability in comparison with the ML-CDM estimator (i.e., the black dots for the ML-MM are more clustered together in comparison with the ones for the ML-CDM).

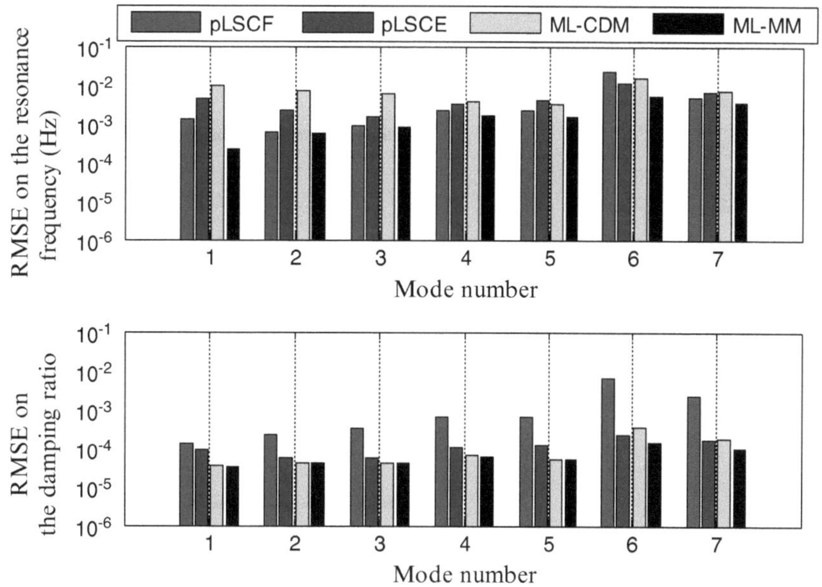

Fig. 15.5 The RMSE on the estimated resonance frequencies (*top*) and the estimated damping ratios (*bottom*) (The order of the different methods in the figure legend corresponds to the order of the bars in the plot.)

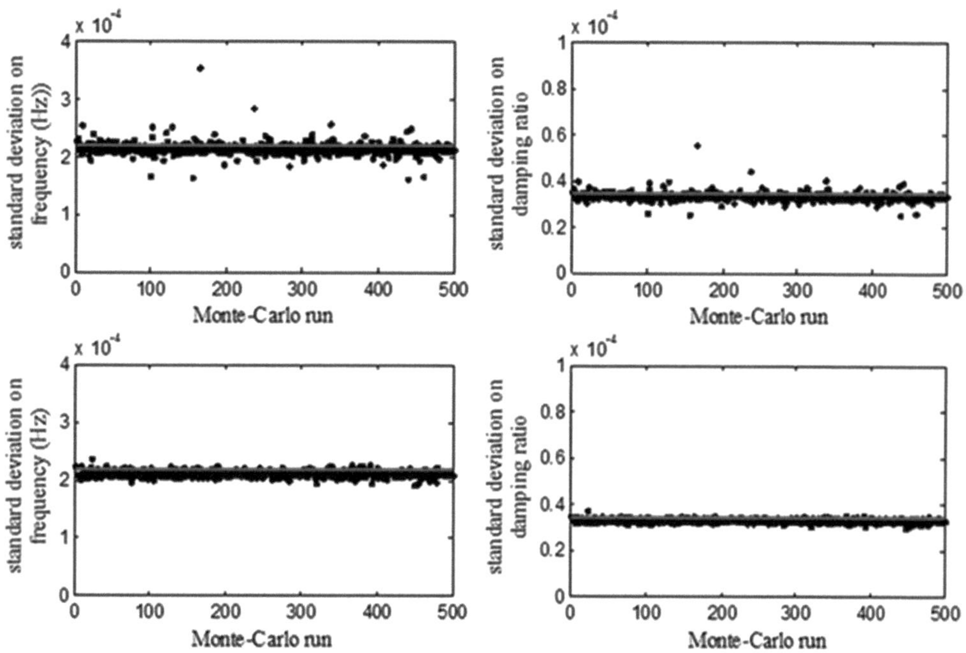

Fig. 15.6 Monte-Carlo simulations results of the 7-DOF system: comparison between the predicted (black dots) and sample (gray solid line) standard deviation on the frequency (*left*) and damping ratio (*right*) of the first mode: ML-CDM (*top*), and ML-MM (*bottom*)

In fact, this higher variability on the uncertainty prediction which the ML-CDM shows comes from the fact that it uses some complex linearization formulas to propagate the uncertainty from the estimated polynomial coefficients to the modal parameters (i.e., from the denominator coefficients to the frequency and damping estimates), which is not the case for the ML-MM estimator. The ML-MM estimator estimates the uncertainty on the modal parameters directly without the need to use these linearization formulas since it fits directly the modal model (not polynomial model) to the measured data. This is an advantage of the ML-MM estimator where the uncertainty on all the model parameters can be estimated in a direct and simple way compared with the polynomials-based estimators like the ML-CDM estimator. For the ML-MM estimator, the results shown in Fig. 15.6-bottom confirm the correctness of Eq. 15.14 to predict the uncertainty bounds on the poles, hence the frequency and damping estimates. Equation 15.15 shows that the ML-MM estimator can deliver the uncertainty bounds on the estimated mode shapes as well. To check the validity of this prediction, the predicted std and the sample std over the 500 Monte-Carlo runs are compared to each other for the estimated 7th mode shape elements, and these results are shown in Fig. 15.7. The relative predicted standard deviation presented with black dots is not in full agreement with

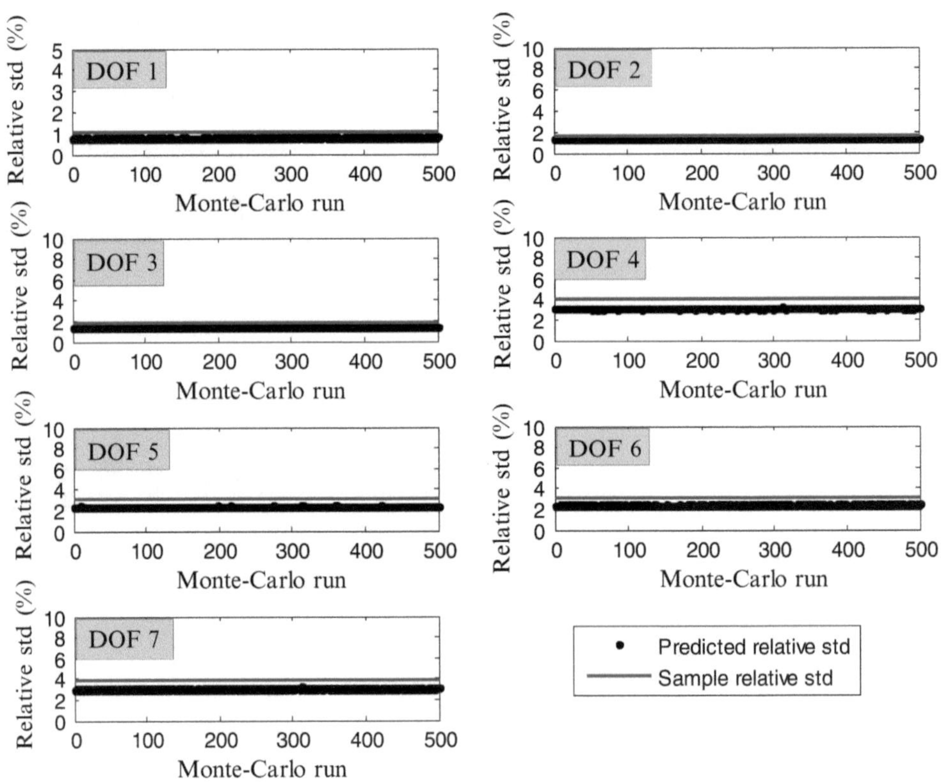

Fig. 15.7 Monte-Carlo simulations results of the 7-DOF system: comparison between the relative predicted (black dots) and relative sample (gray solid line) standard deviation of the seventh mode shape elements obtained by the ML-MM estimator

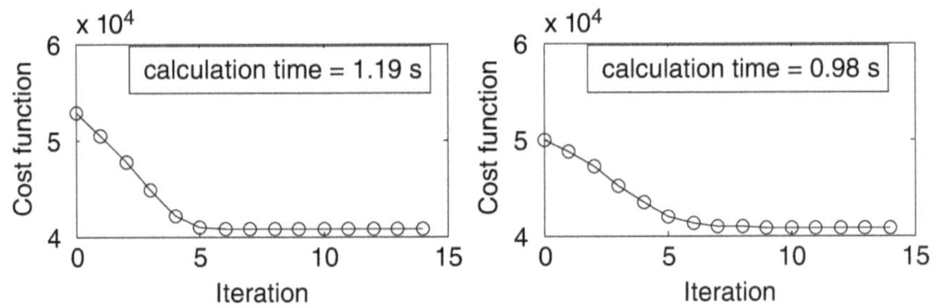

Fig. 15.8 Cost function convergence and the computational time: Basic ML-MM (*left*) and Fast ML-MM (*right*)

the relative sample standard deviation presented by a gray solid line, but they are still quit close to each other. Although it can be seen that the relative predicted standard deviation is slightly underestimated, by examining the relative bias error on the mode shapes it was found that the relative bias error of the mode shape elements is within the predicted 68 % ($\pm\sigma$) confidence bound, which can be considered as an indicator for the correctness of this predicted confidence bounds. All the ML-MM's results shown previously have been obtained using the basic (slow) implementation of the method. To validate the fast implementation introduced in section 2.2, the fast ML-MM will be compared to the basic (slow) implementation in terms of the computational speed, the bias error, and the predicted confidence bounds on the estimated resonance frequencies and damping ratios.

Figure 15.8 shows the cost function convergence and the computational time taken to reach convergence for both of the basic ML-MM and the fast ML-MM estimators. It can be seen that both implementations converge to the same value. Although both of them converge to almost the same value, the fast ML-MM is indeed faster since it takes only 0.98 s while the basic ML-MM takes 1.19 s. The small difference in the computational time is due to the fact that the system under test is very simple ($N_o = 7,\ N_i = 1, N_m = 7$). Concerning the accuracy of the estimated parameters, Table 15.1 shows that the fast implementation of the ML-MM estimator leads to almost the same estimates' accuracy as the ones obtained by the basic implementation where it can be seen that the bias error on the resonance frequency and damping ratios of the basic ML-MM and the fast ML-MM are very comparable. Figure 15.9 shows that the fast implementation gives a good prediction for the uncertainty on the resonance frequencies and the damping ratios. From this figure, one can see that there is a good agreement between the predicted standard deviation (black dots) and the sample standard deviation (solid line) and this is for the frequency and the damping estimate. Table 15.2 shows the mean value (over 500 Monte-Carlo runs) of the predicted standard deviation on the frequency and the damping estimates for both implementations of the ML-MM estimators (i.e.,

Table 15.1 Fast ML-MM versus basic ML-MM in terms of bias error

Mode	Bias error on resonance frequency $\times 1^{-3}$, Hz		Bias error on damping ratio $\times 1^{-3}$, %	
	Basic ML-MM	Fast ML-MM	Basic ML-MM	Fast ML-MM
1	−0.0095	−0.0095	−0.1278	−0.1277
2	0.0072	0.0073	−0.1455	−0.1455
3	0.0393	0.0393	0.1144	0.1141
4	0.1118	0.1131	0.1207	0.1206
5	0.0062	0.0066	0.1456	0.1478
6	−0.1389	−0.0624	−0.0344	0.5235
7	0.1028	0.0715	0.4576	0.7750

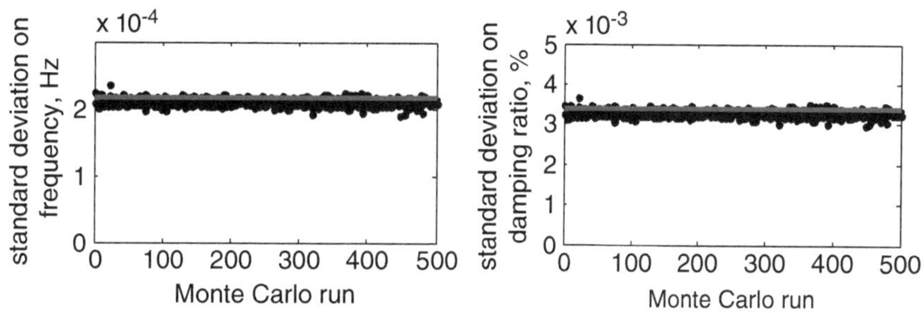

Fig. 15.9 Monte-Carlo simulations results of the 7-DOF system: comparison between the predicted (black dots) and sample (gray solid line) standard deviation on the frequency (*left*) and damping ratio (*right*) of the first mode obtained from the fast implementation of the ML-MM estimator

Table 15.2 Fast ML-MM versus basic ML-MM in terms of predicted uncertainty bounds

	Mean value of the predicted std on resonance frequency, Hz		Mean value of the predicted std on damping ratio, %	
	Basic ML-MM	Fast ML-MM	Basic ML-MM	Fast ML-MM
1	0.000211	0.000210	0.003298	0.003275
2	0.000541	0.000524	0.003604	0.003496
3	0.000683	0.000642	0.003666	0.003441
4	0.001433	0.001213	0.005112	0.004355
5	0.001237	0.001065	0.004194	0.003602
6	0.003924	0.002387	0.010730	0.006606
7	0.002647	0.001712	0.007255	0.004639

the basic and the fast implementations). In general, one can see that the fast ML-MM delivers estimates with a bit lower uncertainty bounds compared to the basic ML-MM. This agrees with the fact that adding additional parameters to the model increases the attainable uncertainty on the estimated model parameters [24]. In the basic ML-MM implementation, adding the mode shapes and the lower and upper residuals as uncertain parameters increases the uncertainty bounds on all the modal model's parameters. This explains why the basic ML-MM leads to estimates with higher uncertainty bounds compared to the fast ML-MM.

15.3.2 Flight Flutter Testing

In this section, the proposed ML-MM estimator will be validated and compared with some other estimators (i.e., the pLSCF (Polymax), the pLSCE and the ML-CDM estimators) using experimentally measured FRFs that were measured during a business jet in-flight testing. These types of FRFs are known to be highly contaminated by noise. During this test, both the wing tips of the aircraft are excited during the flight with a sine sweep excitation through the frequency range of interest by using rotating fans. The forces are measured by strain gauges. Next to these measurable forces, turbulences are also exciting the plane resulting in rather noisy FRFs. Figure 15.10 shows some typical coherence functions and the corresponding FRFs, which clearly show the noisy character of the data. During the flight, the accelerations were measured at nine locations while both the wing tips were excited (two inputs).

The different estimators (i.e., Polymax, pLSCE, ML-CDM, basic ML-MM and fast ML-MM) are applied to the data to extract the modal parameters of the modes lie within the selected band. Figure 15.11 shows the stabilization charts constructed by the Polymax and the time-domain pLSCE estimators. It can be seen that Polymax gives extremely clear stabilization chart in comparison to the pLSCE. The very clear stabilization char that Polymax yields was the main reason

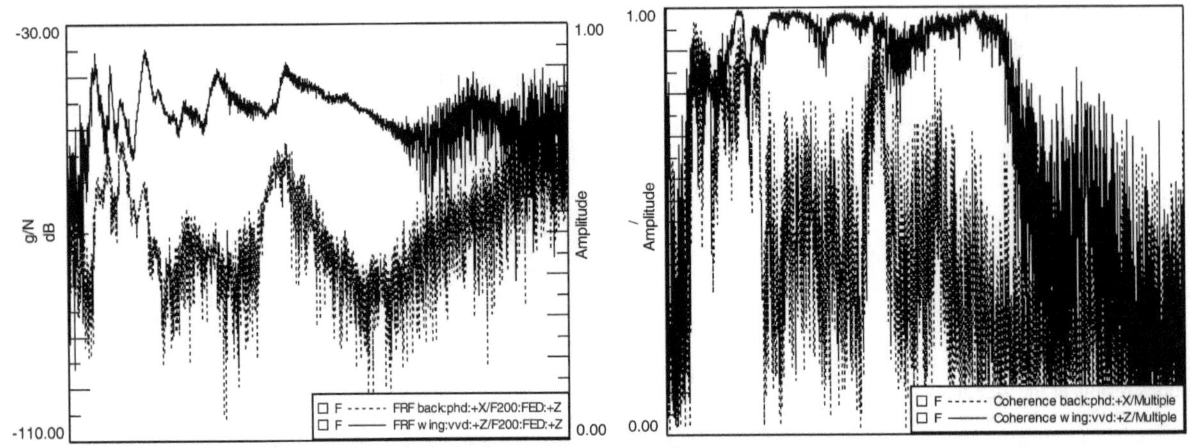

Fig. 15.10 Typical multiple coherence functions (*right*) and typical FRFs (*left*) for the tested business jet (the frequency axis is hidden for confidentiality)

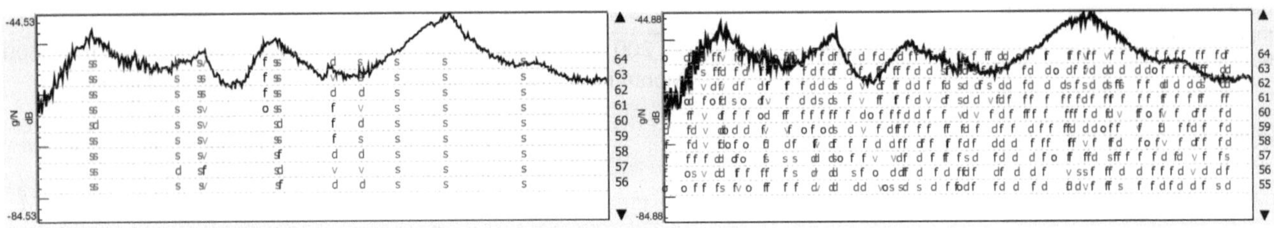

Fig. 15.11 The stabilization chart constructed for the business jet data set: Polymax (*left*) and pLSCE (*right*)

Fig. 15.12 The convergence of the ML-MM cost function during the iterations and the computational time taken for 30 iterations: basic ML-MM (*left*) and fast ML-MM (*right*)

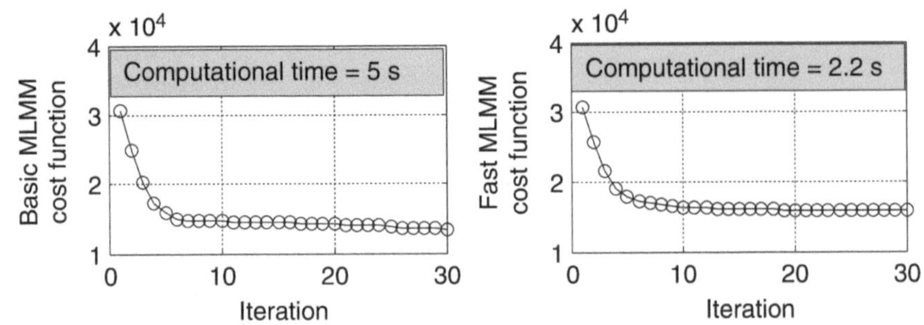

behind selecting Polymax as starting values generator for the ML-MM estimator (see Figs. 15.1 and 15.2). The cost function convergence and the computational time for both the basic ML-MM and the fast ML-MM estimators are shown in Fig. 15.12. One can see that both of them converge to almost the same value, while the computational time of the fast ML-MM is less than the one of the basic ML-MM. A comparison between the different estimators was made on basis of the estimated modal model. An easy but very popular way to check the quality of the estimated modal model is to look to the quality of the fit between the measured FRFs and the modal model synthesized FRFs. Figure 15.13 shows a comparison between all the estimators in terms of the quality of the fit between one of the measured FRFs and the corresponding synthesized one. It can be seen that the ML-CDM estimator resulted in perfect fit of the pole-residues model (i.e., ML-CDM (Residue matrices)). Nevertheless this common denominator based algorithm loses the fit quality by converting the pole-residue model to the modal model to obtain the mode shapes and the participation factors by reducing the residues to a rank-one matrix using singular value decomposition (SVD) (i.e., ML-CDM (SVD on residue matrices)). This degradation on the fit quality is clear in Fig. 15.13. It can be seen from Fig. 15.13 that both ML-MM estimators (i.e., basic and fast) outperform the linear least-squares estimators (pLSCF and pLSCE), and the ML-CDM (SVD on residue matrices) while keeping the residues as rank-one matrices (i.e., multiplication of the mode shape column vector and the participation factors row vector, see Eq. 15.3).

The MAC between the mode shapes of the basic and the fast ML-MM estimators is shown in Fig. 15.14. It shows that there is a good agreement between the mode shapes estimated by each technique. Some of the mode shapes estimated by

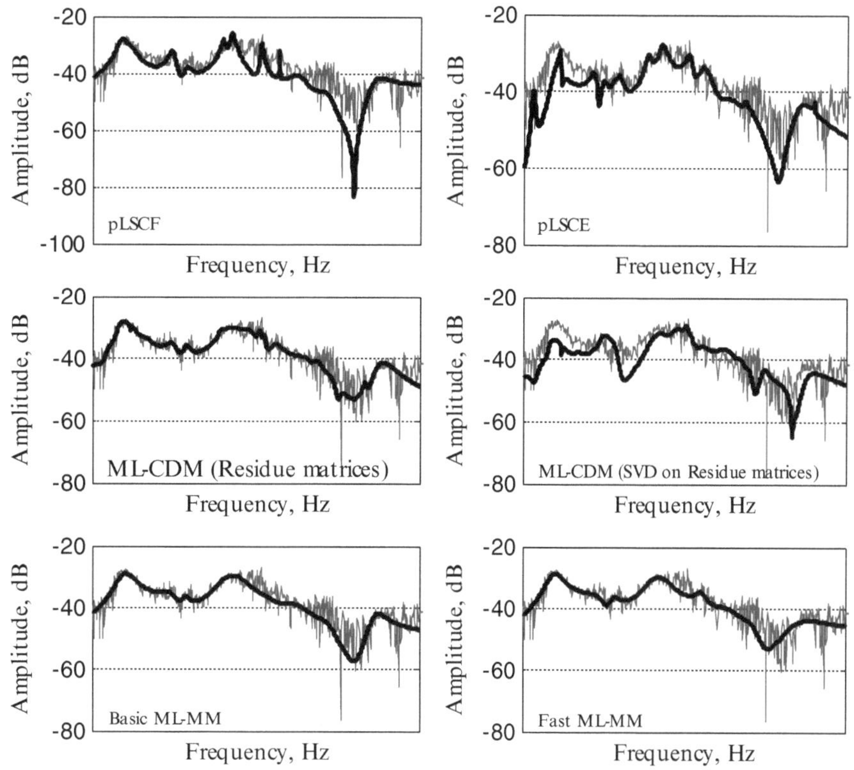

Fig. 15.13 A comparison between the measured (thin-gray line) and the synthesized (thick-black line) FRF (the shown FRF measured from a sensor at the tail of the aircraft)

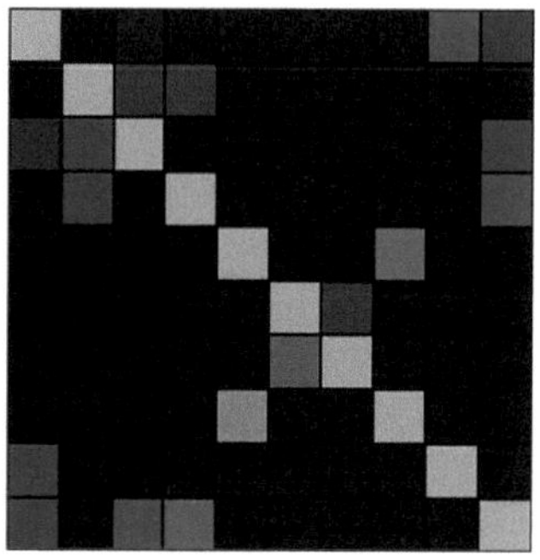

Fig. 15.14 MAC between the basic ML-MM and the fast ML-MM

the fast ML-MM estimater are shown in Fig. 15.15. Since the variances of the measured FRFs were available in this data set, this enabled us to get, from the ML-MM estimator, the uncertainty bounds on the estimated resonance frequencies and damping ratios. The uncertainty bounds on the resonance frequencies and the damping ratios together with their uncertainty bounds estimated by both the basic and the fast ML-MM are shown in Table 15.3. Figure 15.16 shows the sum of the measured FRFs together with the location of the selected modes, and the sum of the synthesized FRFs. The first remark we can draw from Table 15.3 is that the estimates of the fast ML-MM are estimated with lower uncertainty bounds compared with the ones estimated by the basic ML-MM. As it was noticed and discussed in the simulation part (Sect. 15.3.1), having lower uncertainty bounds using the fast ML-MM agrees with the fact that decreasing the number of the uncertain parameters in the model decreases the uncertainty bounds on those parameters [24]. Secondly, it can be seen that modes 1,3 and 9, which are very dominant and have large contribution to the FRFs sum compared to the other modes, are estimated with lower uncertainty bounds either by the basic ML-MM or the fast ML-MM, and this is in particular for the frequency estimates. However, the basic and the fast ML-MM give the damping estimate of the first mode with higher uncertainty bounds compared to some other modes. The reason behind that could be at the first peak there are two very close modes and normally the estimation of the very close modes is accompanied with high level of uncertainty. Moreover, this mode is estimated with a relatively high level of damping. Also, it can be noticed that modes 2, 7, and 10 are estimated with a relatively higher uncertainty bounds which agrees with the fact that those modes do not have a large contribution to the FRF sum indicating that they are weakly excited/observed.

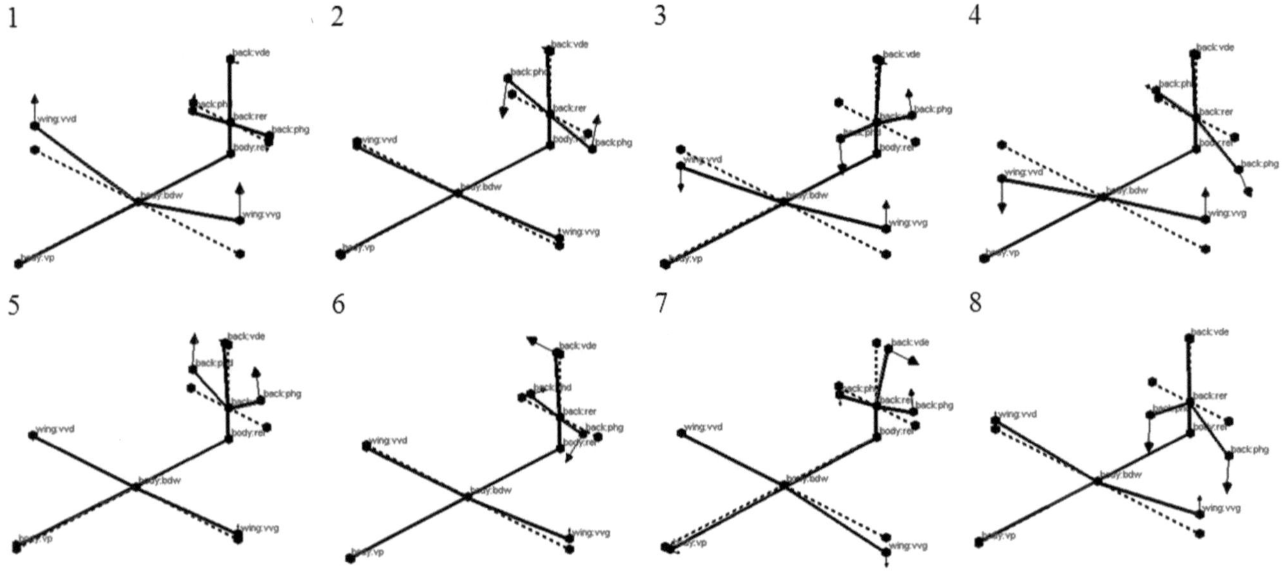

Fig. 15.15 Some mode shapes estimated by the Fast ML-MM estimator (from 1 to 8: increasing the natural frequency)

Table 15.3 Estimated resonance frequencies and damping ratios together with their predicted standard deviation

	Predicted standard deviation on resonance frequencies ± σ Hz (frequency values are hidden for confidentiality reason)		Estimated damping ratios together with their predicted standard deviation ± σ %			
	Basic ML-MM	Fast ML-MM	Basic ML-MM		Fast ML-MM	
1	±0.0061	±0.0038	8.0988	±0.1147	8.3960	±0.0662
2	±0.0207	±0.0049	8.0942	±0.2443	7.7829	±0.0615
3	±0.0085	±0.0033	3.2759	±0.0989	3.5581	±0.0386
4	±0.0169	±0.0044	3.7907	±0.1739	3.4158	±0.0423
5	±0.0182	±0.0059	6.1909	±0.1267	5.1761	±0.0554
6	±0.0192	±0.0070	4.7593	±0.1971	5.6421	±0.0650
7	±0.0180	±0.0078	2.6941	±0.1410	2.2237	±0.0609
8	±0.0142	±0.0047	6.1545	±0.0920	5.7802	±0.0331
9	±0.0050	±0.0025	3.4552	±0.0332	3.3909	±0.0159
10	±0.0171	±0.0076	5.7227	±0.0976	6.3655	±0.0423

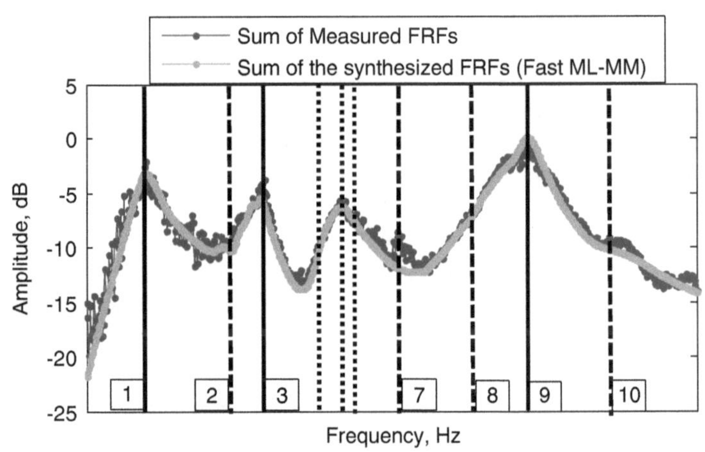

Fig. 15.16 Sum of the FRFs with the Fast ML-MM fit

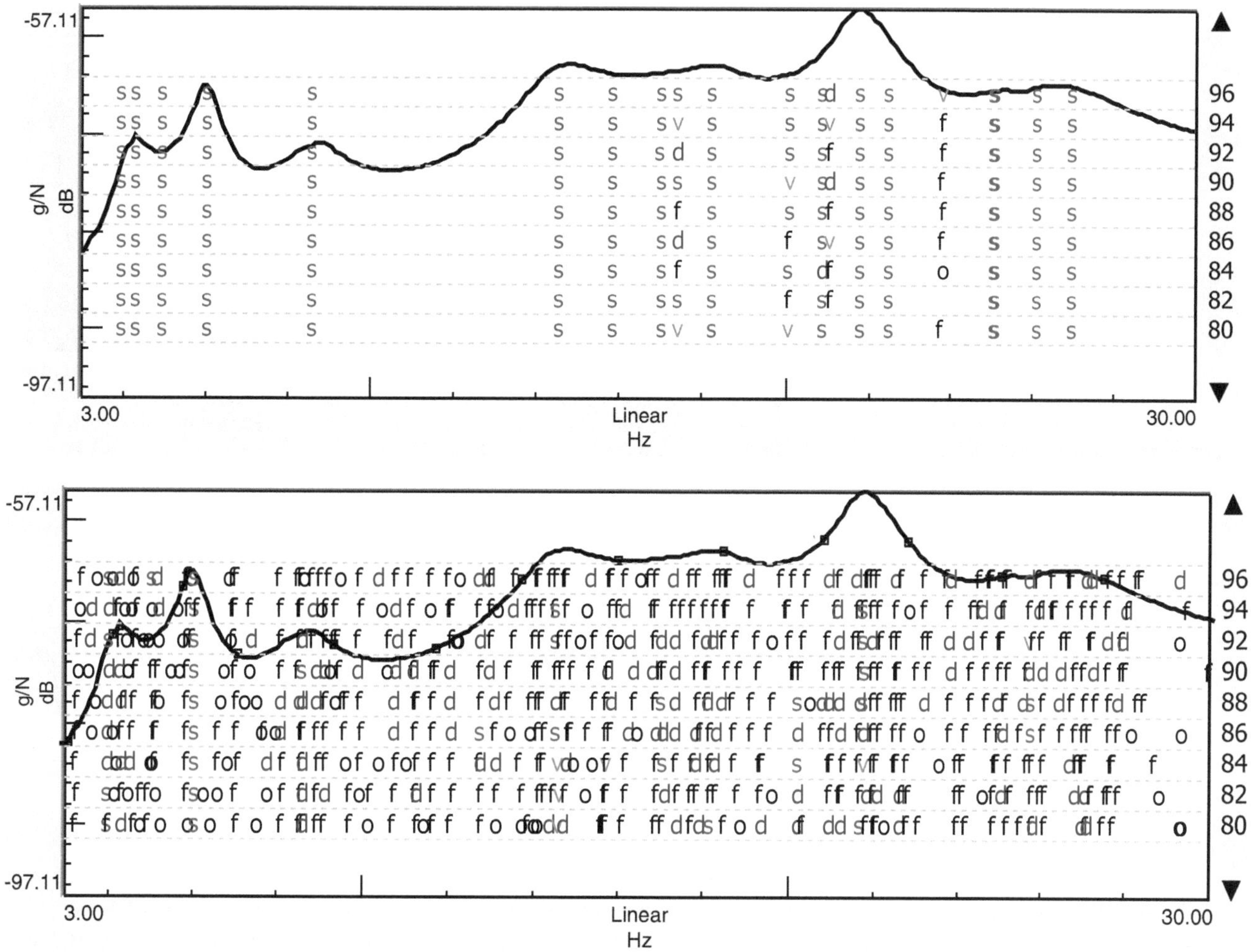

Fig. 15.17 Stabilization chart constructed by Polymax (*top*) and pLSCE (*bottom*) for measurements on a fully trimmed car

Table 15.4 Mean fitting error and correlation between the synthesized and measured FRFs for the fully trimmed car

		Mean fitting error %	Mean fitting correlation %
pLSCF (PolyMAX)		10.01	93.24
pLSCE		9.43	93.51
ML-CDM1	Using directly the residue matrices	1.72	99.11
ML-CDM2	Modal model (Applying SVD on residue matrices to have the mode shapes and participation factors)	34.40	78.90
Basic ML-MM		7.61	95.01
Fast ML-MM		7.49	95.09

15.3.3 Fully Trimmed Car

In automotive engineering, experimental modal analysis (EMA) is considered as a "commodity" tool and accurate models are needed for modelling and finite element updating. A typical example of a challenging modal analysis application is the structural analysis of a trimmed car body. In this example, data from a fully trimmed car was used. The accelerations of the fully equipped car were measured at 154 locations, while four shakers were simultaneously exciting the structure. This gives a total of 616 FRFs. More details about this test setup can be found in [25]. The different estimators under test are applied to the measured FRFs to identify the modes within the frequency band [3.30 Hz]. In Fig. 15.17, the stabilization charts constructed by the Polymax and the time-domain pLSCE estimators are shown. Indeed, as we saw before in the in-flight data set Polymax outperforms the time-domain pLSCE in terms of the clarity of the stabilization chart. In Table 15.4,

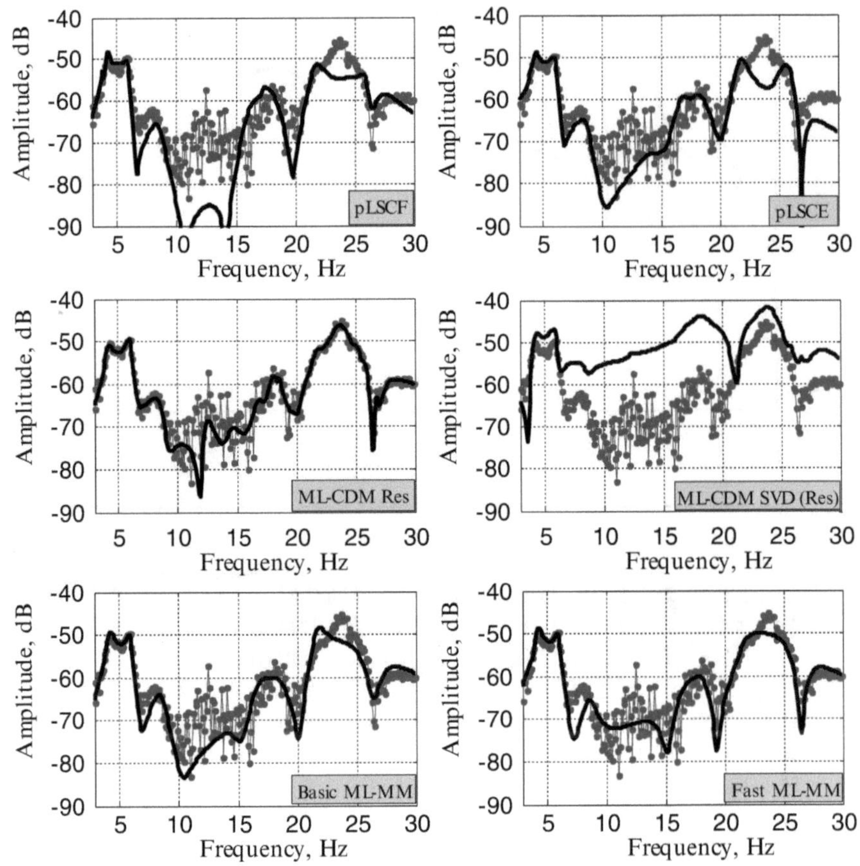

Fig. 15.18 Comparison between the modal models obtained by different estimators under test for one FRF of the fully trimmed car data set (Measured FRF: dotted gray line Synthesized FRF: solid black line)

Figs. 15.18 and 15.19 a comparison between the different estimators is presented in terms of the quality of the fit between the measured (dotted-gray line) and the synthesized (solid-black line) FRFs. The linear least-squares estimators (i.e., pLSCF and pLSCE) give a comparable mean fitting error and correlation. For the common denominator polynomial –based maximum likelihood estimator (ML-CDM), the quality of the fitting is dramatically degraded by reducing the residue matrices to a rank-one matrix to get the mode shapes and the participation factors. Comparing the degradation of the fitting quality of the ML-CDM in this case to the case of the previous example (i.e., inflight data set in Sect. 15.3.2), it seems that the degradation of the fitting quality by transferring common denominator model into modal model tends to be more problematic for the highly-damped case, e.g., fully trimmed car.

The basic ML-MM and the fast ML-MM give very comparable results, and they outperform the other estimators in terms of the quality of the fit between the measured and the synthesized FRFs. Figure 15.20 shows the cost function convergence and the computational time for both the basic and the fast ML-MM estimators. With this relatively huge data set ($N_i = 4$, $N_o = 154$), one can see clearly the difference in the time taken by each implementation. The time taken by the fast ML-MM is almost 1/4 the time taken by the basic ML-MM. Figure 15.21 shows that there is a good agreement between the mode shapes estimated by both the basic ML-MM and the fast ML-MM. The resonance frequencies and damping ratios of the estimated modes are shown in Table 15.5 while some typical mode shapes estimated by the fast ML-MM estimator are shown in Fig. 15.22.

15.3.4 Acoustic Modal Analysis of a Car Cavity

This experimental case concerns the cabin characterization of fully trimmed sedan car (see Fig. 15.23) using acoustic modal analysis. The acoustic modal analysis of a car cavity (cabin) is, in general, accompanied by using many excitation sources (up to 12 sources in some cases) and the presence of highly-damped complex modes [10, 14]. Due to the high level of the modal damping in such application, the many excitation locations are required to get sufficient excitation of the modes across the entire cavity. It has been observed that the classical modal parameter estimation methods have some difficulties in fitting an FRF matrix that consists of many (i.e., 4 or more) columns, i.e., in cases where many input excitation locations have been used in the experiment [10, 14].

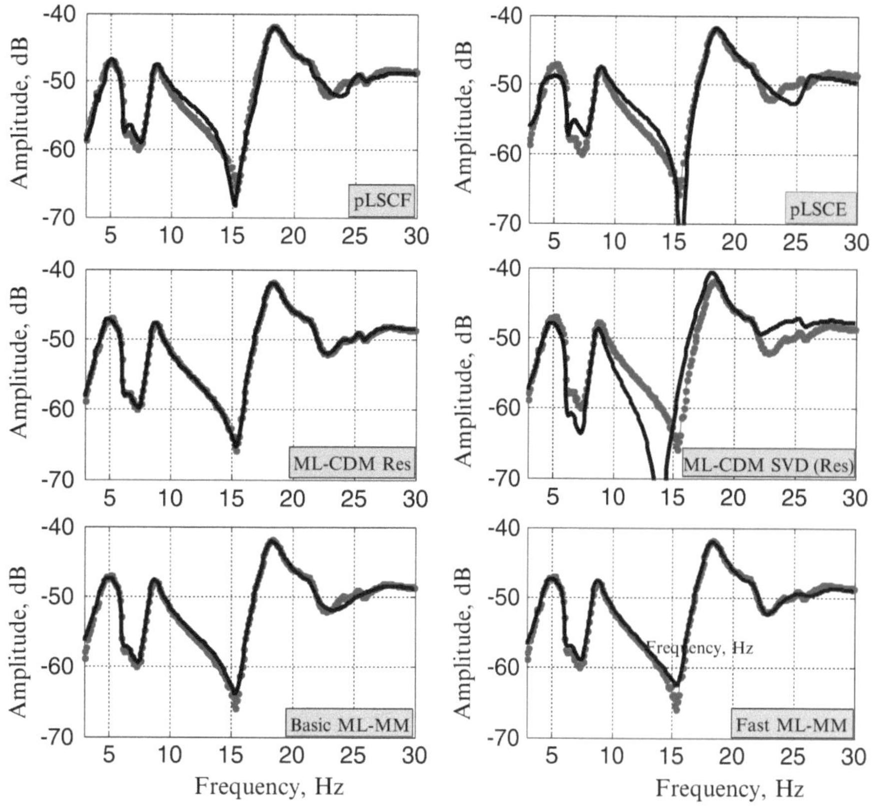

Fig. 15.19 Comparison between the modal models obtained by different estimators under test for one FRF of the fully trimmed car data set (Measured FRF: dotted gray line Synthesized FRF: solid black line)

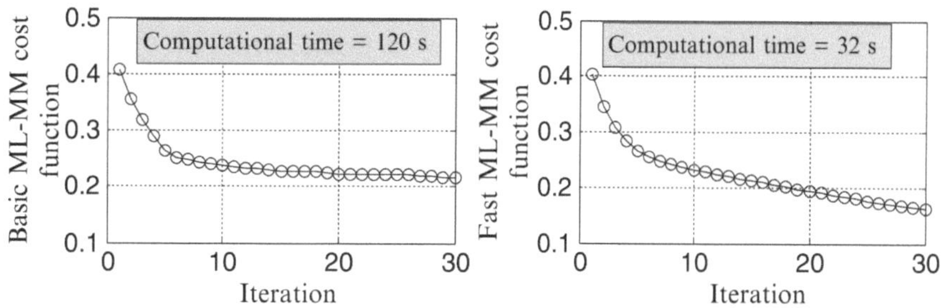

Fig. 15.20 Cost function convergence and the computational time for the basic ML-MM (*left*) and the fast ML-MM (*right*)

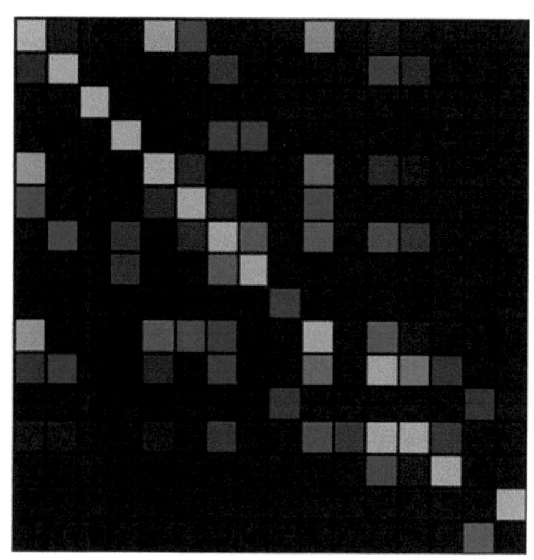

Fig. 15.21 MAC between the mode shapes estimated by the basic ML-MM and the fast ML-MM

Table 15.5 The basic ML-MM and the fast ML-MM modal results for the fully trimmed car

	Frequency (Hz)		Damping (%)		
	Basic ML-MM	Fast ML-MM	Basic ML-MM	Fast ML-MM	Mode description
1	3.9438	3.9384	5.5217	6.0005	Rigid body – lateral mode
2	4.2084	4.1986	8.5429	8.0733	Rigid body – longitudinal mode
3	4.8883	4.8597	17.9462	17.8124	Rigid body – vertical mode
4	5.9841	5.9862	4.3450	4.3345	Rigid body – pitch mode
5	8.5271	8.5240	6.6041	6.5893	Rigid body – roll (yaw) mode
6	14.4818	14.4779	5.9311	5.9037	Vertical front differential mode
7	17.0069	16.9638	7.2244	7.6584	Powertrain rolling coupled to body torsional
8	18.0902	18.1343	4.8237	5.1341	Powertrain vertical coupled to body bending
9	21.6577	21.6730	2.6857	2.5681	Body bending coupled to powertrain bending (in phase)
10	22.2000	22.3321	4.3209	4.5910	Body torsional coupled to powertrain bending
11	25.0733	25.2210	6.9513	6.4886	Body torsional coupled to powertrain torsional (out of phase)
12	26.8808	26.9057	5.7851	5.6318	Body bending coupled to powertrain bending (out of phase)

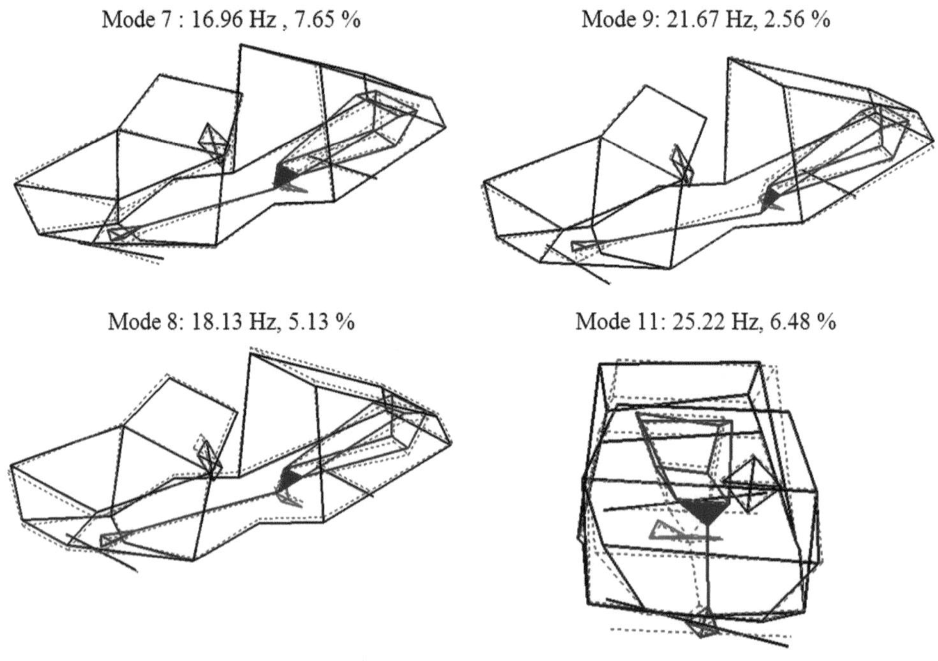

Fig. 15.22 Some typical mode shapes estimated by the fast ML-MM estimator applied to the fully trimmed car data set

Fig. 15.23 Picture of Sedan car under test (*left*) and its CAE cavity model (*right*)

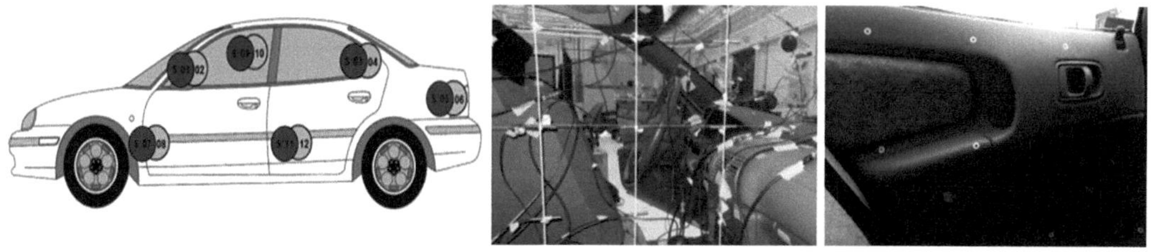

Fig. 15.24 Excitation sources (loudspeakers) locations inside the car cavity (*left*), microphones roving array inside the cavity (*middle*), and measurement points on boundary surface (*right*)

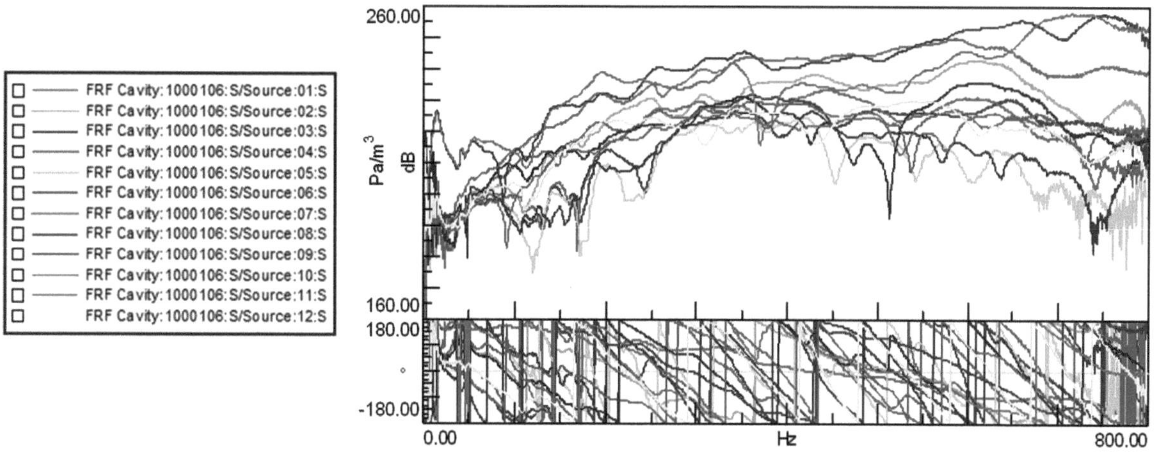

Fig. 15.25 Typical FRF of one point measured in the cavity due to all the 12 excitation sources

Multiple inputs multiple output (MIMO) test were carried out inside the cavity of the Sedan car where 34 microphones located both on a roving array with spacing equal to around 20 cm and near to boundary surfaces captured the responses simultaneously. A total of 18 runs were performed to measure the pressure distribution over the entire cavity (both cabin and trunk) resulting in 612 response locations ($N_o = 612$). For each run, up to 12 loudspeakers ($N_i = 12$) switched on sequentially were used for acoustic excitation. The excitation sources locations and some of the measurement points are show in Fig. 15.24. Continuous random white noise was chosen as excitation signal and the FRFs were measured up to 800 Hz using H_1 estimator with 150 averages. More details about the measurements procedure can be found in [26]. Some typical measured FRFs are shown in Fig. 15.25.

For this case study, because of the time limitation the pLSCE and the ML-CDM estimators will be excluded from the comparison. Since we have anyway to use the pLSCF (Polymax) to generate starting values for the ML-MM estimator, the comparison will be done between the pLSCF (Polymax) and the ML-MM estimators. A frequency band between 30 and 200 Hz has been selected to perform the acoustic modal estimation. Due to a memory requirement issue, it was not possible to apply the basic ML-MM to the full data set. Therefore, for the comparison between the basic ML-MM and the fast ML-MM in terms of their cost functions convergence and their computational time only 10 columns of the FRF matrix, i.e., all the measured responses due to 10 excitation sources are used. The Polymax has been applied to the FRFs to generate starting values of all the modal parameters (i.e., poles, mode shapes, participation factors and lower and upper residuals) to start both the basic ML-MM and the fast ML-MM estimators. In Fig. 15.26 – top, the cost function convergence and the computational time for both estimators are shown. It can be seen that although both estimators converge to a comparable value of the cost function the fast ML-MM is very fast compared to the basic ML-MM where it takes only 14 min whereas the basic implementation takes 81 min. Figure 15.26–bottom shows that both the basic ML-MM and the fast ML-MM give the same fitting quality where it can be seen that the sum of the synthesized FRFs either from the basic implementation or from the fast implementation of the ML-MM perfectly fits the measured one. Figure 15.26 clearly shows that the proposed fast implementation of the ML-MM method yields the same improved modal model as the one obtained from the basic implementation while outperforming the basic implementation in terms of the computational time and the memory requirements.

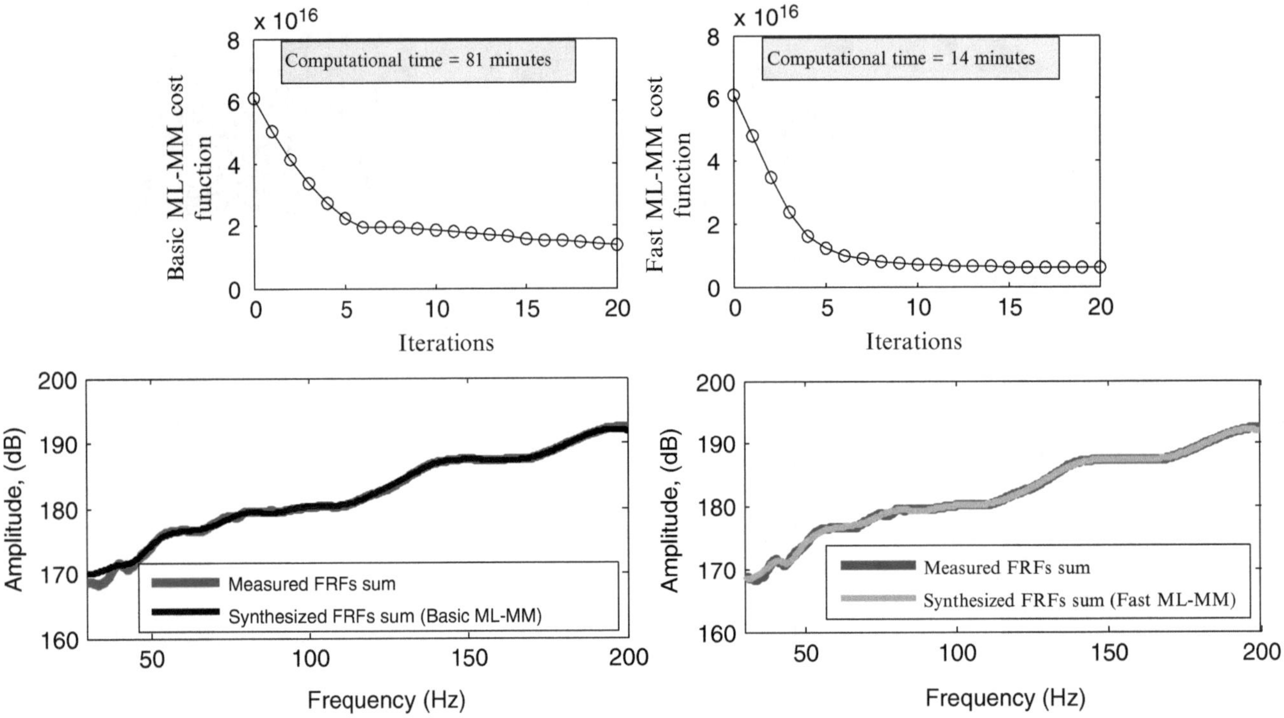

Fig. 15.26 *Top*: decreasing of the cost function of the basic ML-MM (*left*) and the fast ML-MM (*right*) with the computational time taken. *Bottom*: the fit between the measured FRFs sum and the synthesized FRFs sum for the basic ML-MM (*left*) and the fast ML-MM (*right*)

Now, both the Polymax and the fast ML-MM will be applied to the full data set (i.e., taking into account all the excitation sources). The stabilization chart has been constructed by Polymax as it is shown in Fig. 15.27 to generate starting values for the modal parameters of the corresponding modes within the selected band. Then, the initial modal model estimated by Polymax is iteratively optimized by the fast ML-MM method.

Figure 15.28 shows the decrease of the ML-MM cost function at each iteration. The analysis was stopped after 20 iterations. Figure 15.30 shows that the initially modal model estimated by Polymax is drastically improved during the iterations of the ML-MM estimator. It can be seen from that figure that the ML-MM synthesis results are superior to the Polymax synthesis results. Some typical mode shapes identified with Polymax and ML-MM estimators are shown and compared with the numerical ones in Fig. 15.30. One can see that the estimated mode shapes by Polymax and the ML-MM are comparable. However, the mode shapes estimated by the ML-MM method is more similar to the numerical mode shapes. It can be seen that there are some differences between Polymax and the ML-MM method in terms of the frequencies and damping estimates. Actually, these differences in the values of the frequency and the damping estimates are reflected as an improvement in the goodness of the fit between the measured and the synthesized FRFs (see Fig. 15.29). Moreover, it was found that the mode shapes estimated by the ML-MM method shows a good independency as compared to the ones of Polymax. This can be evidenced by looking to the AutoMAC matrix for both Polymax and ML-MM that are shown graphically in Fig. 15.31. It can be seen from this figure that the off diagonal elements of the ML-MM AutoMAC matrix have lower values as compared to the ones of Polymax.

15.4 Conclusions

The basic implementation of the ML-MM (Maximum Likelihood estimator based on modal model) estimator is represented and improved in terms of the memory requirements and the computational time. The proposed fast implementation of the estimator, the fast ML-MM, has been validated and compared with the basic implementation (basic ML-MM) and with some

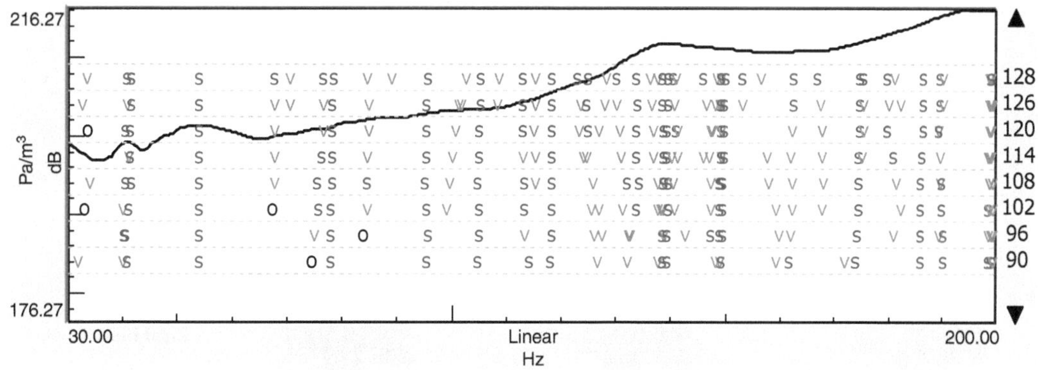

Fig. 15.27 Polymax stabilization chart when applied to the full data set of the FRFs measured inside the Sedan car cavity

Fig. 15.28 Decrease of the fast ML-MM cost function at each iteration cavity

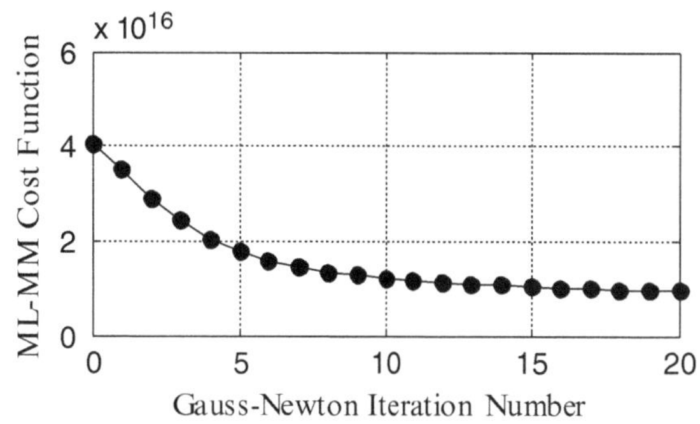

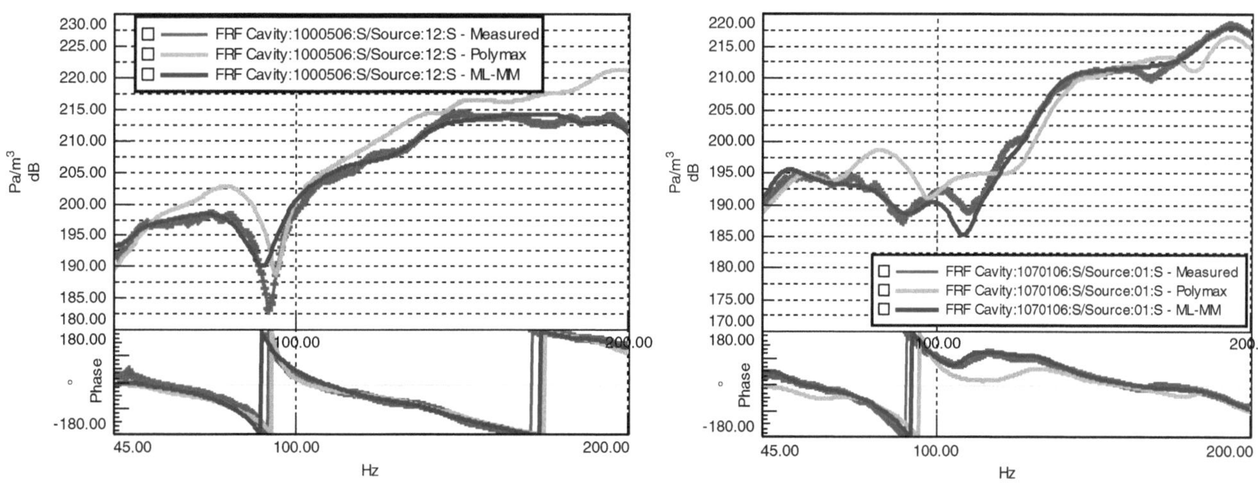

Fig. 15.29 Improved FRF curve fit when using fast ML-MM as compared to Polymax

other estimators (e.g., Polymax, pLSCE and ML-CDM) suing simulated example and three real industrial applications. From the presented results, it is found that the fast implementation of the ML-MM estimator gives very comparable performance as compared with the basic implementation in terms of the accuracy of the estimated modal parameters and the predicted confidence bounds while outperforming the basic ML-MM in terms of the computational time and the memory requirements. As compared with Polymax, pLSCE and ML-CDM, the fast ML-MM estimator results in better estimates for the modal parameters, and it is capable to properly deal with high modal densities, highly damped systems and FRF matrices with many references (excitation sources) and provides superior FRF synthesis results.

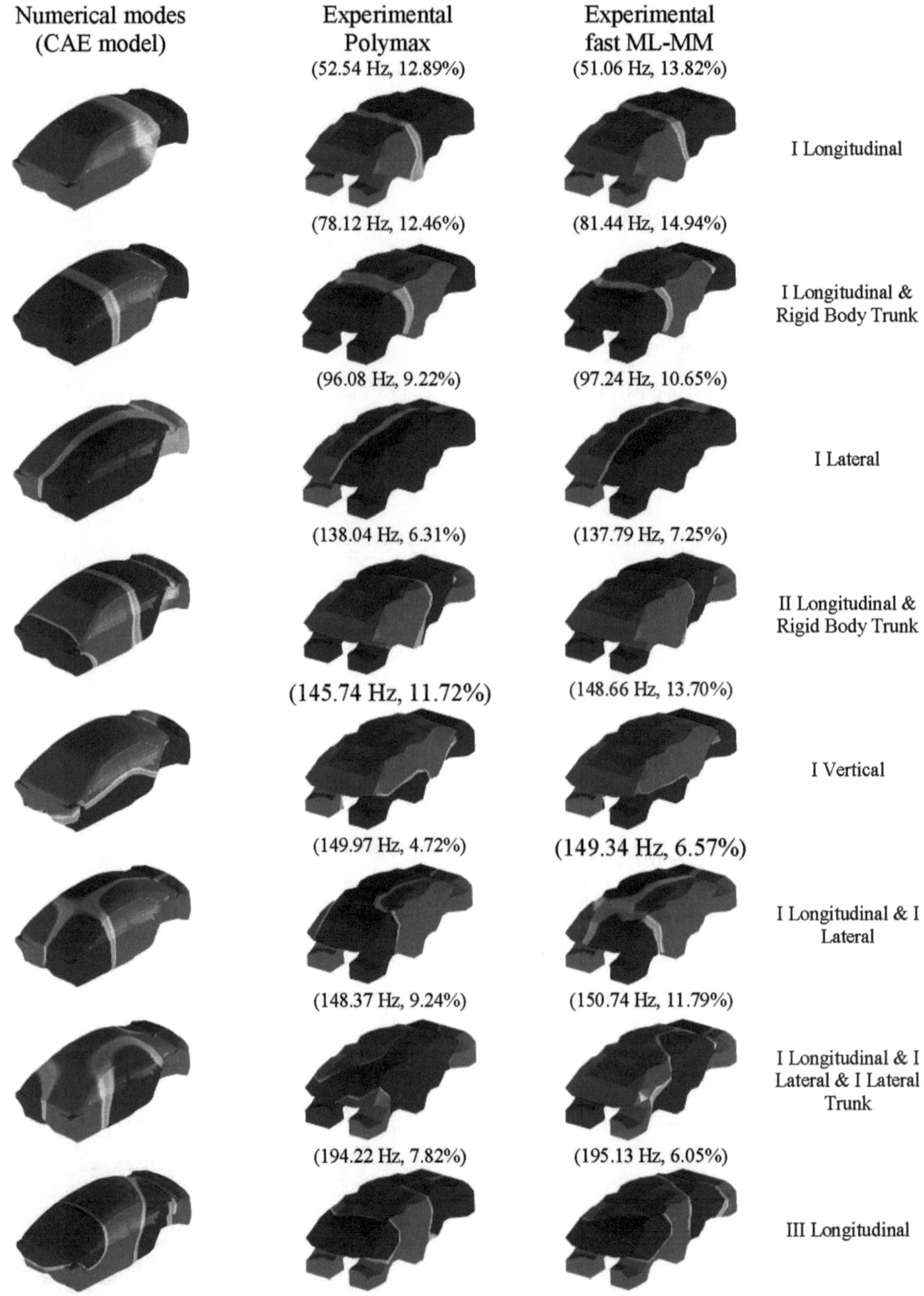

Fig. 15.30 Some typical acoustic mode shapes of the Sedan interior cabin identified by Polymax and ML-MM as compared to the ones predicted by the CAE model

Acknowledgment The financial support of the IWT (Flemish Agency for Innovation by science and Technology), through its Innovation mandate IWT project 130872, is gratefully acknowledged. The financial support of the FP7 Marie Curie ITN EID project "ENHANCED" (Grant Agreement No. FP7-606800) is gratefully acknowledged.

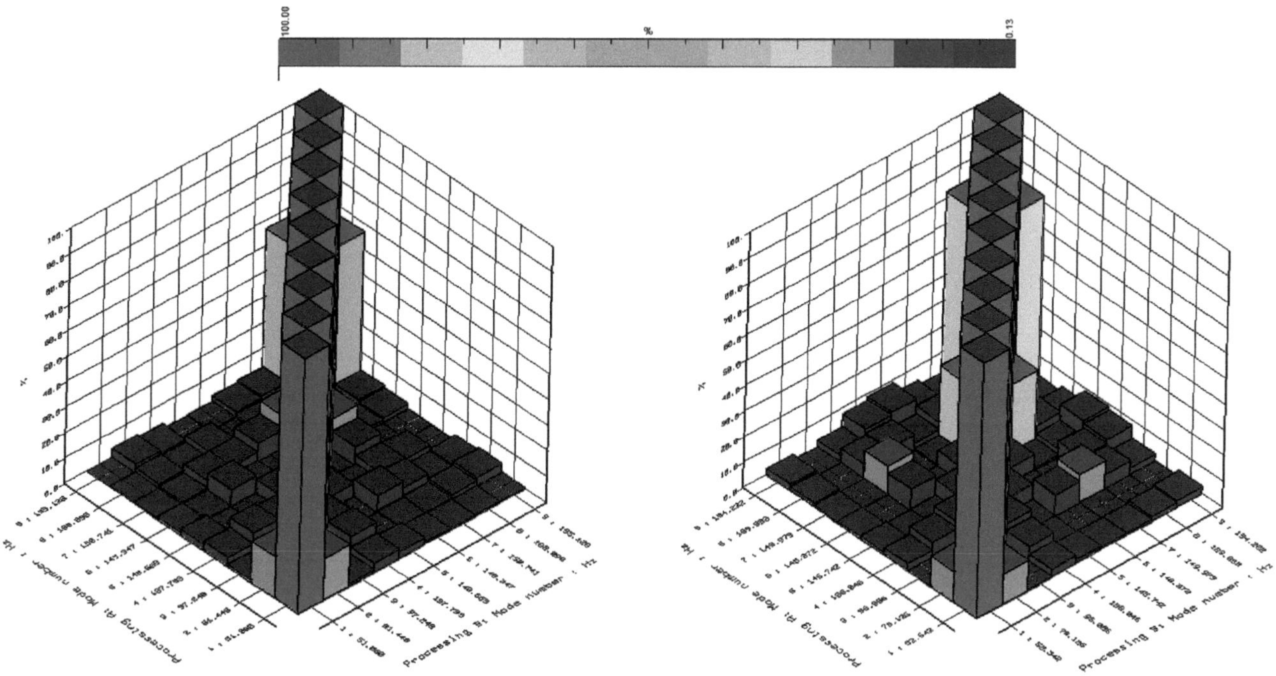

Fig. 15.31 Auto MAC: Polymax (*right*) and ML-MM (*left*)

References

1. Guillaume P, Verboven P, Vanlanduit S (1998) Frequency-domain maximum likelihood identification of modal parameters with confidence intervals. In: The proceeding of the 23rd international seminar on modal analysis, Leuven
2. Guillaume P, Verboven P, Vanlanduit S, Van der Auweraer H, Peeters B (2003) A poly-reference implementation of the least-squares complex frequency domain-estimator. In: The proceeding of the 21st international modal analysis conference (IMAC), Kissimmee
3. Böswald M, Göge D, Füllekrug U, Govers Y (2006) A review of experimental modal analysis methods with respect to their applicability to test data of large aircraft structures. In: The proceeding of the international conference on noise and vibration engineering (ISMA), Leuven
4. De Troyer T, Guillaume P, Pintelon R, Vanlanduit S (2009) Fast calculation of confidence intervals on parameter estimates of least-squares frequency-domain estimators. Mech Syst Signal Process 23(2):261–273
5. De Troyer T, Guillaume P, Steenackers G (2009) Fast variance calculation of polyreference least-squares frequency-domain estimates. Mech Syst Signal Process 23(5):1423–1433
6. Schoukens J, Pintelon R (1991) Identification of linear systems: a practical guide to accurate modeling. Pergamon Press, Oxford
7. Cauberghe B, Guillaume P, Verboven P (2004) A frequency domain poly-reference maximum likelihood implementation for modal analysis. In: The proceeding of 22nd international modal analysis conference, Dearborn/Detroit
8. Pintelon R, Guillaume P, Schoukens J (2007) Uncertainty calculation in (operational) modal analysis. Mech Syst Signal Process 21(6): 2359–2373
9. Tsuji H, Enomoto T, Maruyama S, Yoshimura T (2012) A study of experimental acoustic modal analysis of automotive interior acoustic field coupled with the body structure. SAE technical paper 2012-01-1187. doi:10.4271/2012-01-1187
10. Tsuji H, Maruyama S, Yoshimura T, Takahashi E (2013) Experimental method extracting dominant acoustic mode shapes for automotive interior acoustic field coupled with the body structure. J Passeng Cars Mech Syst 6(2):1139–1146. doi:10.4271/2013-01-1905
11. El-Kafafy M, De Troyer T, Guillaume P (2014) Fast maximum-likelihood identification of modal parameters with uncertainty intervals: a modal model formulation with enhanced residual term. Mech Syst Signal Process 48(1–2):49–66
12. El-Kafafy M, De Troyer T, Peeters B, Guillaume P (2013) Fast maximum-likelihood identification of modal parameters with uncertainty intervals: a modal model-based formulation. Mech Syst Signal Process 37:422–439
13. El-Kafafy M, Guillaume P, De Troyer T, Peeters B (2012) A frequency-domain maximum likelihood implementation using the modal model formulation. In: The proceeding of 16th IFAC symposium on system identification, SYSID, Brussels
14. Peeters B, El-Kafafy M, Accardo G, Knechten T, Janssens K, Lau J, Gielen L (2014) Automotive cabin characterization by acoustic modal analysis. The JSAE Annual Congress, At Japan, Volume: 114-2014543
15. Guillaume P (1992) Identification of multi-input multi-output systems using frequency-domain models. Ph.D. thesis in department of Electrical, Vrije Universiteit Brussel (VUB), Brussels
16. Guillaume P, Pintelon R, Schoukens J (1996) Parametric identification of multivariable systems in the frequency-domain – a survey. In: The proceeding of the international conference on noise and vibration engineering, Leuven
17. Kailath T (1980) Linear systems. Prentice-Hall, Upper Saddle River
18. Heylen W, Lammens S, Sas P (1997) Modal analysis theory and testing. Katholieke Universiteit Leuven, Department Werktuigkunde, Heverlee

19. Verboven P (2002) Frequency-domain system identification for modal analysis. Ph.D. thesis in mechanical engineering department, Vrije Universiteit Brussel (VUB), Brussels
20. Peeters B, Van der Auweraer H, Guillaume P, Leuridan J (2004) The PolyMAX frequency-domain method: a new standard for modal parameter estimation? Shock Vib 11(3–4):395–409
21. Brown DL, Allemang RJ, Zimmerman R, Mergeay M (1979) Parameter estimation techniques for modal analysis. SAE transactions, Paper no 790221, pp 828–846
22. Vold H (1990) Numerically robust frequency domain modal parameter estimation. Sound Vib 24(1):38
23. Vold H, Kundrat J, Rocklin G, Russel R (1982) A multi-input modal estimation algorithm for mini-computers. SAE Trans 91(1):815–821
24. Pintelon R, Schoukens J (2001) System identification: a frequency domain approach. IEEE Press, Piscataway
25. Van der Auweraer H, Liefooghe C, Wyckaert K, Debille J (1993) Comparative study of excitation and parameter estimation techniques on a fully equipped car. In: The proceeding of the International modal analysis conference (IMAC), Kissimmee
26. Accardo G, El-kafafy M, Peeters B, Bianciardi F, Brandolisio D, Janssens K, Martarelli M (2015) Experimental acoustic modal analysis of an automotive cabin. In: The proceeding of international modal analysis conference (IMAC XXXIII). Springer, Orlando

Chapter 16
An Improved Implementation of the Orthogonal Polynomial Modal Parameter Estimation Algorithm Using the Orthogonal Complement

William Fladung and Håvard Vold

Abstract The rational fraction polynomial algorithm is the entry-level model of the high-order, frequency-domain modal parameter estimation methods. However, it has some well-known issues with numerical ill-conditioning for a high model order and a wide frequency range. Among the alternatives that have been proposed over the years to address this shortcoming is a change of basis functions from power polynomials to orthogonal polynomials. While this approach does cure the numerical ill-conditioning issues, this algorithm has not yet achieved mainstream acceptance, with the reasons for this reluctance typically cited being additional complication or increased computation time. This paper introduces an improved implementation of the orthogonal polynomial algorithm that uses the orthogonal complement, coupled with QR decomposition, to greatly reduce the time of the accumulation phase. The neat trick performed by the orthogonal complement is to get all of the overdetermination possible without having to do all of the work.

Keywords Modal parameter estimation • Orthogonal polynomial • Orthogonal complement

16.1 The Notoriously Ill-Conditioned Rational Fraction Polynomial Algorithm

The transpose of the multiple-input, multiple-output (MIMO), multiple degree of freedom (MDOF), rational fraction polynomial (RFP) frequency response function (FRF) model

$$\underset{N_i \times N_o}{[H(j\omega_i)]^T} = \left[\sum_{k=0}^{m} (j\omega_i)^k \underset{N_i \times N_i}{[\alpha_k]}\right]^{-1} \left[\sum_{k=0}^{n} (j\omega_i)^k \underset{N_i \times N_o}{[\beta_k]}\right], \tag{16.1}$$

is the basis of the RFP frequency-domain modal parameter estimation algorithm [1], where H is the FRF, N_i is the number of inputs (or references), and N_o is the number of outputs (or responses). The model order is m, and $n = m + r - 2$, where $r \geq 0$ and for $r > 0$ residual terms are included in the numerator. The α-polynomial in the denominator is the characteristic equation, and its roots are the poles of the system. The number of poles of this high-order model is mN_i, which is the model order times the size of the coefficient matrices. Equation (16.1) is valid for all frequencies ω_i, and a least-squares solution can be formulated for the unknown alpha and beta coefficients. Simple enough – except that the RFP algorithm is notoriously ill-conditioned for a wide frequency range and a high model order.

This numerical ill-conditioning comes from the $(j\omega_i)^k$ power polynomial terms that raise the frequencies to increasingly higher powers, as shown in Fig. 16.1 (left) for a frequency range of 1–100 Hz and a model order of 20. The condition number of the Vandermonde matrix formed from the power polynomials is listed in the plot legend, where a good condition number is relatively close to unity. The frequency can be normalized, e.g., from -1 to $+1$, but that only partially alleviates the problem, because now the small, normalized frequencies trend toward zero as they are raised to increasingly higher powers, as shown in Fig. 16.1 (right) for the same frequency range and model order. In either case, a matrix with a high condition number is a sign of pending numerical ill-conditioning troubles because it adversely affects the invertibility (or pseudo-invertibility) of the least-squares solution. This gives poor accuracy for solution of the alphas, which in turn gives poor estimates of the poles. Now, there's nothing fundamentally wrong with the theory behind Eq. (16.1); it's just that the computer has a finite word size. This is why RFP is typically limited to a narrow frequency range with a few modes. This is also why so many people have sought alternatives over the past 30+ years.

W. Fladung (✉) • H. Vold
ATA Engineering, Inc., 13290 Evening Creek Drive South, Suite 250, San Diego, CA 92128, USA
e-mail: bfladung@ata-e.com

M. Mains (ed.), *Topics in Modal Analysis, Volume 10*, Conference Proceedings of the Society for Experimental Mechanics Series, DOI 10.1007/978-3-319-15251-6_16

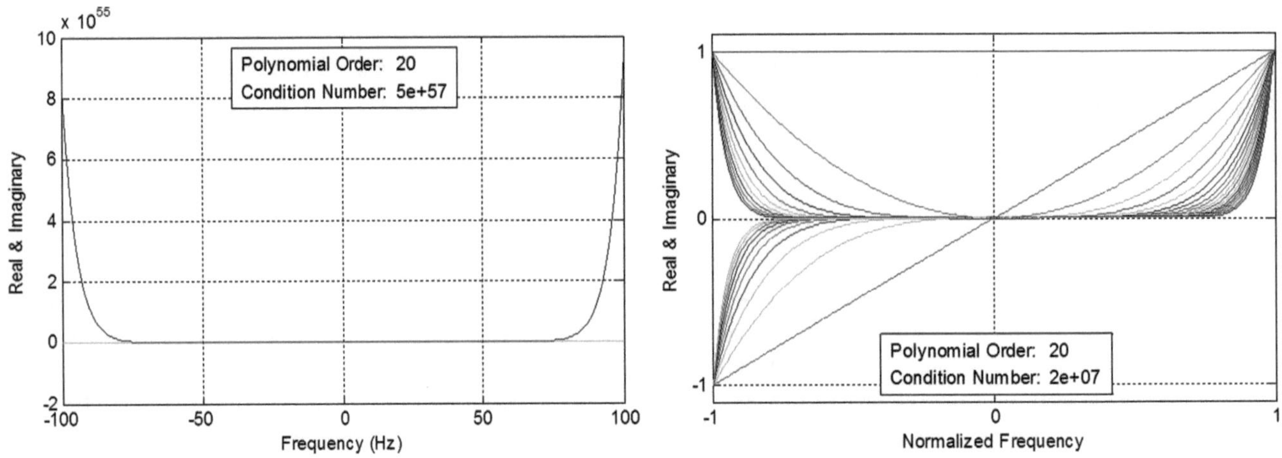

Fig. 16.1 Power polynomials for frequencies in rad/s (*left*) and frequencies normalized -1 to $+1$ (*right*)

16.2 A Change of Basis to Orthogonal Polynomials

The folklore surrounding the RFP method tells us that we have to process the frequency range in narrow sections for a few modes at a time. However, another option would be to avoid the numerical issues inherent with the high-order, frequency-domain power polynomial model with a change of basis functions. Ideally, these new basis functions should be bounded over a chosen interval and have favorable numerical conditioning behavior. A set of orthogonal polynomials has these traits as well as another property that allows the size of the least-squares problem for the alphas to be greatly reduced [2–8].

16.3 Some Basics on Orthogonal Polynomials

Let $p_k(s_i)$ be a function of the discrete complex variable s_i. Now let $p_k(s_i)$ also be a polynomial of order k.

$$p_k\left(s_i\right) = a_k s_i^k + a_{k-1} s_i^{k-1} + \cdots + a_1 s_1 + a_o, \tag{16.2}$$

which is a standard power polynomial. The inner (or dot) product of polynomials is defined as

$$\left\langle p\left(s_i\right), q\left(s_i\right)\right\rangle = \sum_{i=1}^{N} \overline{p}\left(s_i\right) q\left(s_i\right) \text{ and } \left\langle p, p\right\rangle = \|p\|^2 \geq 0 \in \Re, \tag{16.3}$$

where $\overline{p}\left(s_i\right)$ is the complex conjugate of $p(s_i)$ and N is the number of data points (i.e., spectral lines). For *orthogonal polynomials* $\varphi_k(s_i)$ normalized such that $\|\varphi_k\|^2 = 1$,

$$\left\langle \varphi_k\left(s_i\right), \varphi_m\left(s_i\right)\right\rangle = \begin{cases} 1, & \text{for } k = m \\ 0, & \text{for } k \neq m \end{cases}. \tag{16.4}$$

For a set of discrete-valued polynomials (orthogonal or otherwise) defined over N data points, there are N basis functions that define the N-dimensional polynomial space. The standard power polynomials are linearly independent basis functions, but they are not orthogonal. Furthermore, the highest-order orthogonal polynomial that can be defined for an interval containing N data points is $N-1$, since the lowest order is zero.

A set of orthogonal polynomials $\varphi_k, \varphi_{k-1}, \ldots, \varphi_1, \varphi_0$ is constructed such that

$$s_i^k = b_k \varphi_k\left(s_i\right) + b_{k-1}\varphi_{k-1}\left(s_i\right) + \cdots + b_1\varphi_1\left(s_i\right) + b_0\varphi_0\left(s_i\right) = \sum_{j=0}^{k} b_j \varphi_j\left(s_i\right), \tag{16.5}$$

which is a weighted summation of the orthogonal polynomials. This property is used to express the rational fraction polynomial FRF model in an orthogonal polynomial basis instead of the power polynomial basis.

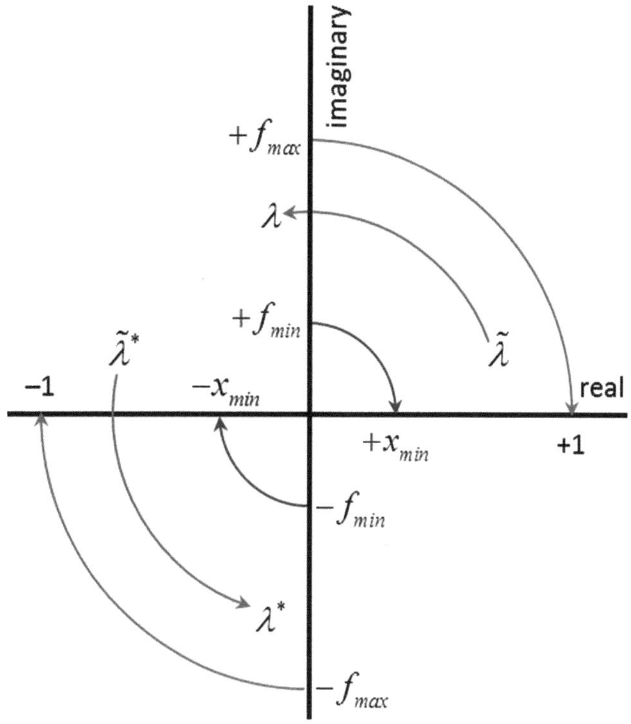

Fig. 16.2 Frequency mapping for the orthogonal polynomial interval

16.4 Frequency Mapping for the Orthogonal Polynomial Interval

When changing the basis functions to orthogonal polynomials, the solution space is also mapped from the frequency range on the imaginary axis to -1 to $+1$ on the real axis, which is just a scaling and a rotation by 90° on the complex plane. There is no restriction on the evenness of the frequency spacing, and the interval of -1 to $+1$ on the real axis is just one particular choice. The orthogonal polynomials could be defined over any interval in the complex plane, either on the real axis or on the imaginary axis, or somewhere in between – and it doesn't necessarily have to be a straight line. However, by choosing an interval on the real axis, the orthogonal polynomials will be real-valued.

The orthogonal polynomials will be constructed on the interval of $x \in [-1, +1]$. The double-sided frequency range, with the mirrored negative frequencies, is scaled by dividing by the maximum frequency f_{max}. The poles $\tilde{\lambda}$ computed in the $x \in [-1, +1]$ interval are mapped back to the poles λ in the actual frequency range of the FRF as $\lambda = j\tilde{\lambda} f_{max}$. The j is necessary because by mapping the frequency range to the orthogonal polynomial interval, the solution space was rotated on the complex plane as depicted in Fig. 16.2. Multiplying by j rotates the poles computed with respect to the interval of the orthogonal polynomial characteristic equation on the real axis back to the actual frequency range on the imaginary axis.

16.5 Constructing the Orthogonal Polynomials

The orthogonal polynomials used in the algorithm do not have a well-known name but instead are constructed at the discrete frequency values using a Gram-Schmidt orthogonalization procedure. From the preceding section, the frequency range has been mapped to the orthogonal polynomial interval of $x \in [-1, +1]$, with no restrictions on the frequency spacing. The orthogonal polynomials $\varphi_n(x_i)$ are constructed recursively, with each successive polynomial being one order higher than the preceding.

The zeroth-order basis function φ_0 is a constant normalized to have unity length (i.e., a 2-norm of one). Multiplying φ_0 by x creates a first-order polynomial φ_1. The projection of φ_1 onto φ_0 is subtracted from φ_1, and then it is normalized to have unity length. The higher-order basis functions are constructed in the same manner except that subtracting the projections onto the preceding two polynomial orders is necessary for orthogonalization [2]. The orthogonal polynomials are built in steps so as to collect the scaling factors and scaled projections that will be needed to form the orthogonal polynomial companion matrix using the recurrence relationship for defining the orthogonal polynomials, which is

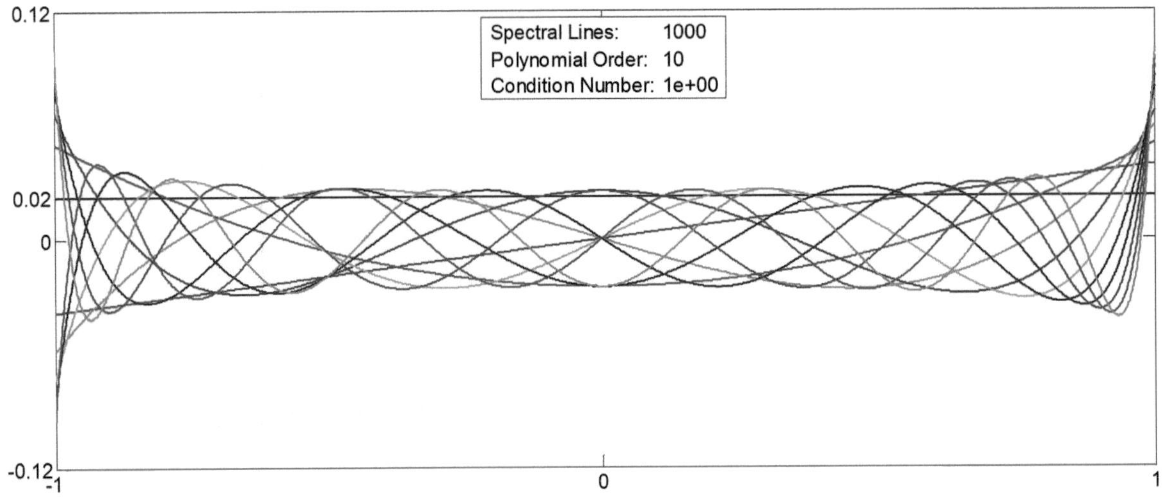

Fig. 16.3 Orthogonal polynomials for a baseband frequency range

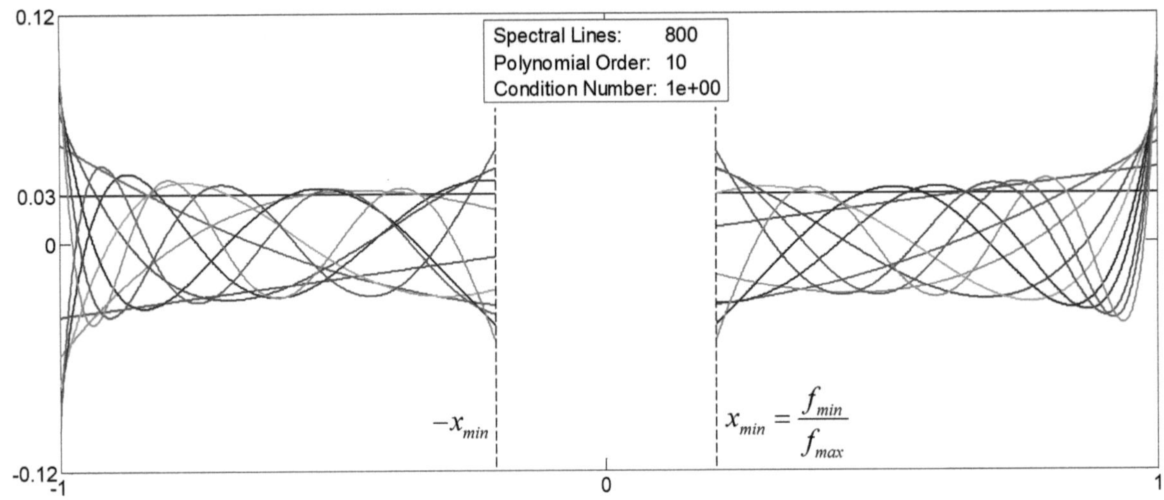

Fig. 16.4 Orthogonal polynomials for a zoom-band frequency range

$$\varphi_n(x_i) = w_n x_i \varphi_{n-1}(x_i) - q_{n,n-1}\varphi_{n-1}(x_i) - q_{n,n-2}\varphi_{n-2}(x_i), \tag{16.6}$$

where w_n is the scaling factor for unity-length normalization and $q_{n,n-k}$ is a scaled projection (i.e., the scalar projection of φ_n onto φ_{n-k} multiplied by w_n).

So what do orthogonal polynomials constructed with the procedure outlined above look like and do they have the traits we wanted? In Fig. 16.3 are orders zero through ten of orthogonal polynomials for an evenly spaced frequency range with a minimum frequency of zero and one thousand spectral lines. The zeroth-order is a constant; the first-order is a straight, sloped line; the second-order is bowl-shaped; the higher orders have an oscillatory behavior; and alternating orders are even and odd functions. A common characteristic of these functions is that they tend to ramp up at the edges, with the ratio of the amplitudes at the edges to the middle typically being less than ten. While their amplitudes are not constant over the interval, they are bounded and the condition number is a perfect one, which is why they eliminate the ill-conditioning problem plaguing the rational fraction polynomial.

In Fig. 16.4 are orders zero through ten of orthogonal polynomials for an evenly spaced frequency range with a minimum frequency not at zero. Here the basis functions are only defined on the interval of -1 to $-x_{min}$ and x_{min} to $+1$ corresponding to the zoom-band frequency range, so there is a gap in the middle. The functions ramp up at the inner and outer edges, but they are still bounded and have a condition number of one.

In Fig. 16.5 are orthogonal polynomials for uneven frequency spacing, plotted as dots to emphasize that the orthogonal polynomials are not just defined over the interval from -1 to $+1$ but at specific points in this interval. These unevenly spaced

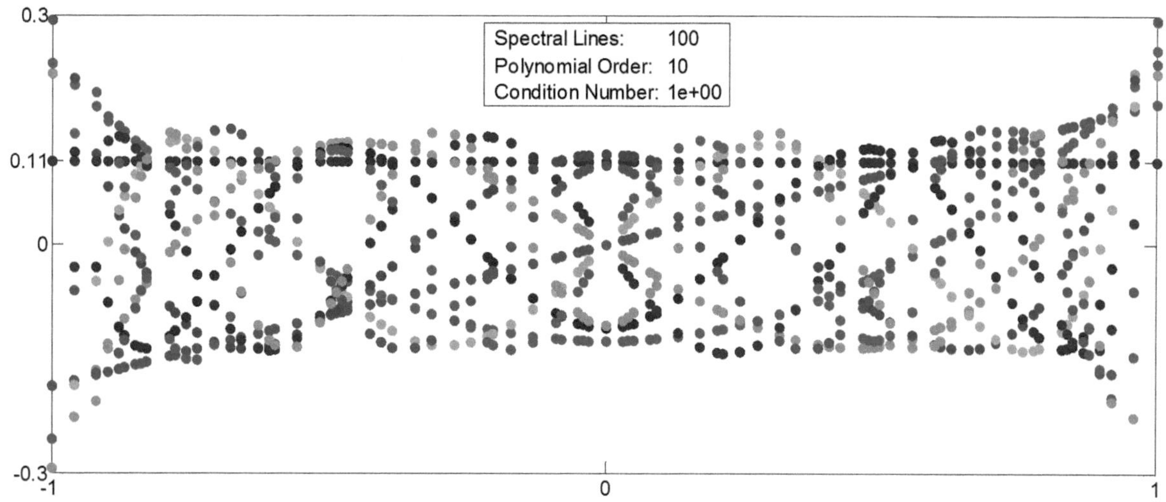

Fig. 16.5 Orthogonal polynomials for unevenly spaced frequency

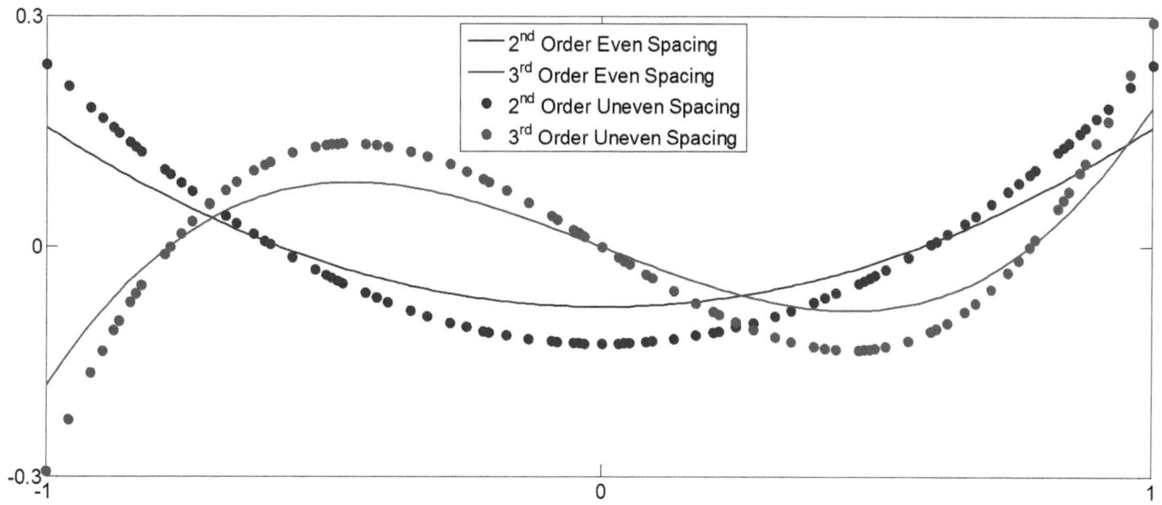

Fig. 16.6 Orthogonal polynomials for evenly and unevenly spaced frequency

functions have the same basic shape as those for even frequency spacing, as shown in Fig. 16.6 for the second and third orders. However, they are not the same as generating evenly spaced functions and then down-selecting to a subset of points, as that would nullify the orthogonality.

In Fig. 16.7 are orders zero through ten of orthogonal polynomials for an evenly spaced baseband frequency range with an excluded frequency band. Again, the basis functions are only defined at the points in the interval corresponding to spectral lines in the frequency range, so there is a gap at the excluded frequency band. The functions ramp up at the edges of the interval and also at the edges of the gap, but they are still bounded and have a condition number of one.

These plots of orthogonal polynomials were meant to show that a set of basis functions with favorable numerical conditioning traits can be constructed, without any loss of generality, for a baseband or zoom-band frequency range, with even or uneven frequency spacing and excluded frequency bands as well.

16.6 The Orthogonal Polynomial Companion Matrix

In the earliest implementations of the orthogonal polynomial modal parameter estimation algorithm [2, 3], the alpha coefficients found in the orthogonal polynomial basis were transformed back to the power polynomial basis to compute the poles from a standard companion matrix – only to find that by doing so, the numerical ill-conditioning reappeared [4]. Shortly

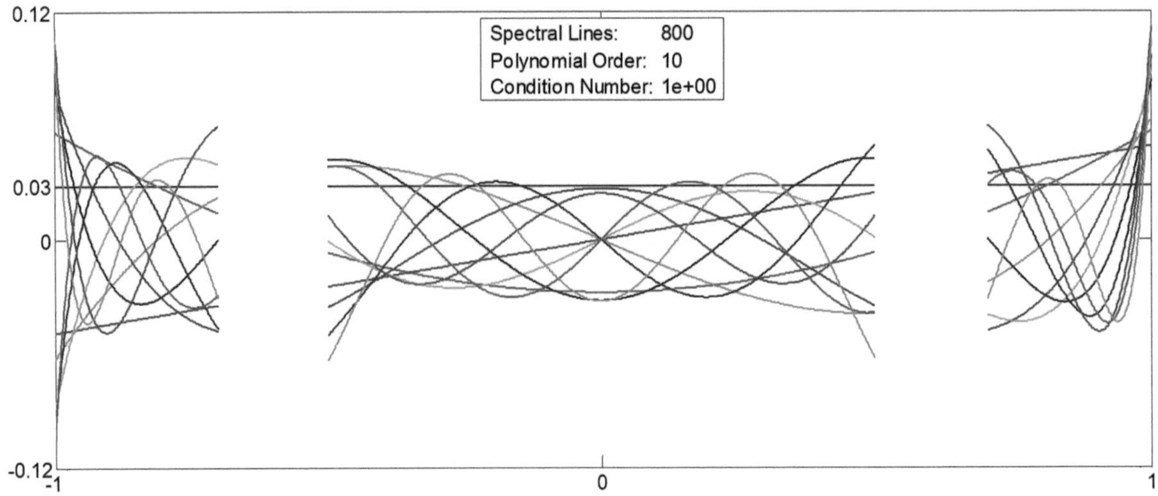

Fig. 16.7 Orthogonal polynomials with an excluded frequency band

thereafter, it was discovered that only if the poles were estimated from a companion matrix formed from the orthogonal polynomial alpha coefficients would the numerical ill-conditioning be truly vanquished [5]. The key to formulating a companion matrix is to use the characteristic equation and recurrence relationships to get a matrix times a vector equal to a scalar times the same vector, i.e., $Ax = \lambda x$. The slight twist on this theme for the orthogonal polynomial companion matrix is that we get a generalized eigenvalue problem in the form of $Ax = \lambda Bx$. Starting with general definition of the eigenvalue and eigenvector for the matrix orthogonal polynomial

$$\left(\sum_{k=0}^{m} [\alpha_k]\, \varphi_k\left(\lambda_r\right)\right)\{L_r\} = 0, \tag{16.7}$$

where L_r is an $N_i \times 1$ modal participation vector, expand the characteristic equation as

$$\left(\alpha_0\varphi_0 + \alpha_1\varphi_1 + \cdots + \alpha_{m-1}\varphi_{m-1} + \alpha_m\varphi_m\right) L_r = 0. \tag{16.8}$$

Since the characteristic equation is homogeneous, it can be normalized such that any of the alpha coefficients is an identity and still have the same roots. The historic choice is to let $\alpha_m = I$, and in matrix form this becomes

$$\begin{bmatrix} 0 & I & & 0 \\ 0 & 0 & \ddots & 0 \\ 0 & 0 & & I \\ -\alpha_0 & -\alpha_1 & \cdots & -\alpha_{m-1} \end{bmatrix} \begin{Bmatrix} \varphi_0 L_r \\ \varphi_1 L_r \\ \vdots \\ \varphi_{m-1} L_r \end{Bmatrix} = \begin{Bmatrix} \varphi_1 L_r \\ \varphi_2 L_r \\ \vdots \\ \varphi_m L_r \end{Bmatrix}, \tag{16.9}$$

which is a matrix times a vector equal to a slightly different vector. We also need the recurrence relationship for defining the orthogonal polynomials in Eq. (16.6), written here in terms of a pole $\tilde{\lambda}_r$ instead of the interval x_i.

$$\varphi_k\left(\tilde{\lambda}_r\right) = w_k \tilde{\lambda}_r \varphi_{k-1}\left(\tilde{\lambda}_r\right) - q_{k,k-1}\varphi_{k-1}\left(\tilde{\lambda}_r\right) - q_{k,k-2}\varphi_{k-2}\left(\tilde{\lambda}_r\right) \tag{16.10}$$

Combining Eqs. (16.9) and (16.10) yields a generalized eigenvalue formulation for the orthogonal polynomial companion matrix as

$$(C + Q)\, x = \lambda_r W x, \tag{16.11}$$

$$\text{where } C = \begin{bmatrix} 0 & I & & 0 \\ 0 & 0 & \ddots & 0 \\ 0 & 0 & & I \\ -\alpha_0 & -\alpha_1 & \cdots & -\alpha_{m-1} \end{bmatrix}, x = \begin{Bmatrix} \varphi_0(\lambda_r) L_r \\ \varphi_1(\lambda_r) L_r \\ \vdots \\ \varphi_{m-1}(\lambda_r) L_r \end{Bmatrix},$$

$$W = \begin{bmatrix} w_1 I & 0 & 0 & 0 \\ 0 & \ddots & 0 & 0 \\ 0 & 0 & \ddots & 0 \\ 0 & 0 & 0 & w_m I \end{bmatrix}, \text{ and } Q = \begin{bmatrix} q_{1,0} I & 0 & 0 & 0 & 0 \\ q_{2,0} I & q_{2,1} I & 0 & 0 & 0 \\ 0 & q_{3,1} I & \ddots & 0 & 0 \\ 0 & 0 & \ddots & q_{m-1,m-2} I & 0 \\ 0 & 0 & 0 & q_{m,m-2} I & q_{m,m-1} I \end{bmatrix}$$

W is a diagonal matrix of scaling factors, Q is bi-diagonal matrix of scaled projections, and the eigenvectors have the customary form of stacked partitions of the modal participation vector multiplied by scalars.

If α_0 is taken as the identity coefficient, the generalized eigenvalue formulation is

$$(I + CQ)\ x = \lambda_r C W x \text{ where } C = \begin{bmatrix} -\alpha_1 & \cdots & -\alpha_{m-1} & -\alpha_m \\ I & & 0 & 0 \\ 0 & \ddots & 0 & 0 \\ 0 & & I & 0 \end{bmatrix} \tag{16.12}$$

and W, Q, and the eigenvector are the same as in Eq. (16.11).

16.7 Expressing the Rational Fraction Polynomial FRF Model in Orthogonal Polynomials

Now that we have a set of orthogonal polynomials and companion matrix formulations to match, the rational fraction polynomial FRF model can be expressed in these basis functions. For one output (response p) and N_i inputs, the multiple-input, single-output (MISO) rational fraction polynomial FRF model can be written in terms of the Laplace variable s instead of the frequency variable $j\omega$ as

$$\underset{N_1\times 1}{\{H_p(s_i)\}^T} = \underset{N_1\times N_1}{[A(s_i)]^{-1}} \underset{N_1\times 1}{[B(s_i)]} \text{ or } [A(s_i)]\{H_p(s_i)\}^T = [B(s_i)], \tag{16.13}$$

$$\text{where } [A(s_i)] = \sum_{j=0}^{m} \underset{N_i\times N_i}{[\alpha_j]} s_i^j,\ [B(s_i)] = \sum_{j=0}^{n} \underset{N_i\times 1}{[\beta_j]} s_i^j,$$

and the denominator polynomial $[A(s_i)]$ is the characteristic equation of the system from which the poles are to be estimated. Using the property for orthogonal polynomials from Eq. (16.5), it can be shown that

$$[A(s_i)] = \sum_{j=0}^{m} s_i^j [\alpha_j] = \sum_{j=0}^{m}\sum_{k=0}^{j} b_k \varphi_k(s_i) [\alpha_j] = \sum_{j=0}^{m} [\widehat{\alpha}_j] \varphi_j(s_i) \text{ and} \tag{16.14}$$

$$[B(s_i)] = \sum_{j=0}^{n} s_i^j [\beta_j] = \sum_{j=0}^{n}\sum_{k=0}^{j} b_k \varphi_k(s_i) [\beta_j] = \sum_{j=0}^{n} [\widehat{\beta}_j] \varphi_j(s_i). \tag{16.15}$$

Thus, the rational fraction polynomial form of the FRF model can be expressed with orthogonal polynomials, just with a different set of alpha and beta coefficient matrices. Just as the α power polynomial is the characteristic equation of the system, so too is the $\widehat{\alpha}$ orthogonal polynomial, and the roots of either characteristic equation are the poles of the system.

16.8 Removing the Betas from the Problem

The MISO rational fraction polynomial FRF model can be now be written in the orthogonal polynomial basis as

$$\sum_{j=0}^{m} \underset{N_i \times N_i}{\left[\widehat{\alpha}_j\right]} \underset{1\times 1}{\varphi_j\left(s_i\right)} \underset{N_i \times 1}{\left\{H_p\left(s_i\right)\right\}^T} = \sum_{j=0}^{n} \underset{N_i \times 1}{\left[\widehat{\beta}_j\right]} \underset{1\times 1}{\varphi_j\left(s_i\right)}. \tag{16.16}$$

A least-square solution for the alphas and betas could be formulated in the orthogonal polynomial basis using Eq. (16.16). However, the orthogonality property of the basis functions can be used to remove the betas from the problem by taking the dot product (over the indices of s_i) of both sides with $\varphi_k(s_i)$, which is an orthogonal polynomial of order k.

$$\left\langle \varphi_k\left(s_i\right), \sum_{j=0}^{m}\left[\widehat{\alpha}_j\right]\varphi_j\left(s_i\right)\left\{H_p\left(s_i\right)\right\}^T\right\rangle = \left\langle \varphi_k\left(s_i\right), \sum_{j=0}^{n}\left[\widehat{\beta}_j\right]\varphi_j\left(s_i\right)\right\rangle \tag{16.17}$$

$$\text{For } k = n+1, \left\langle \varphi_{n+1}\left(s_i\right), \sum_{j=0}^{n}\left[\widehat{\beta}_j\right]\varphi_j\left(s_i\right)\right\rangle = \sum_{j=0}^{n}\left[\widehat{\beta}_j\right]\left\langle\varphi_{n+1}\left(s_i\right), \varphi_j\left(s_i\right)\right\rangle = 0 \tag{16.18}$$

since $\left\langle\varphi_{n+1}\left(s_i\right), \varphi_j\left(s_i\right)\right\rangle = 0$ because φ_{n+1} is orthogonal to the polynomials of lower order, i.e., φ_j for $j = 0,1,\ldots,n$ in the summation. This is true for all $k > n$, and it greatly reduces the size of the least-squares problem for the alphas.

This is a second benefit of orthogonal polynomials, with the first being that they are a remedy for numerical ill-conditioning associated with the power polynomials. Take, for example, a typical, medium-size modal test with 4 references, 200 responses, 1,600 spectral lines, and a maximum model order of 50, with $n = m-2$ for no residual terms in the numerator. If the betas are not removed, the unknowns array of the least-squares solution would have over 40,000 elements, of which only 816 are needed to solve for the poles – this is a 98 % size reduction.

16.9 The Least-Squares Problem for the Alphas

We now have for all $k > n$

$$\left\langle \varphi_k\left(s_i\right), \sum_{j=0}^{m}\left[\widehat{\alpha}_j\right]\varphi_j\left(s_i\right)\left\{H_p\left(s_i\right)\right\}^T\right\rangle = \sum_{j=0}^{m}\left[\widehat{\alpha}_j\right]\left\langle \varphi_k\left(s_i\right), \varphi_j\left(s_i\right)\left\{H_p\left(s_i\right)\right\}^T\right\rangle = 0. \tag{16.19}$$

Expanding the second summation, while dropping the matrix and function of s_i notation for brevity, rearranging into a matrix form, and transposing to the traditional $Ax = b$ form, with A taller than it is wide, we have

$$\underset{1\times(m+1)N_i}{\left[\left\langle\varphi_k, \varphi_0 H_p\right\rangle \quad \left\langle\varphi_k, \varphi_1 H_p\right\rangle \cdots \left\langle\varphi_k, \varphi_{m-1} H_p\right| \left.\varphi_k, \varphi_m H_p\right\rangle\right]} \underset{(m+1)N_i \times N_i}{\begin{bmatrix} \widehat{\alpha}_0^T \\ \widehat{\alpha}_1^T \\ \vdots \\ \widehat{\alpha}_{m-1}^T \\ \hline \widehat{\alpha}_m^T \end{bmatrix}} = 0 \quad . \tag{16.20}$$

Note that Eq. (16.20) is in the form of

$$[A \mid -b]\left[\frac{x}{1}\right] = Ax - b = 0 \rightarrow Ax = b. \tag{16.21}$$

As Eq. (16.20) is arranged, we will let $\widehat{\alpha}_m$ be an identity and take the last column of the horizontal array over to the right-hand side with a minus sign to be the b in $Ax = b$. To have $\widehat{\alpha}_0$ be the identity coefficient, all that has to be done is to move the first column of the horizontal array to the end and $\widehat{\alpha}_0$ to the bottom of the unknowns. Because the correlation matrix for the least-squares problem will be accumulated with QR decomposition, the derivation will continue with the right-hand side included at the end of the horizontal array and all of the alphas among the unknowns.

Equation (16.20) is valid for all φ_k with $k > n$, and additional equations are generated for $k = n+1, n+2, \ldots, n+N_k$, as

$$\left[\begin{array}{cccc|c} \langle\varphi_{n+1},\varphi_0 H_p\rangle & \langle\varphi_{n+1},\varphi_1 H_p\rangle & \cdots & \langle\varphi_{n+1},\varphi_{m-1} H_p\rangle & \langle\varphi_{n+1},\varphi_m H_p\rangle \\ \langle\varphi_{n+2},\varphi_0 H_p\rangle & \langle\varphi_{n+2},\varphi_1 H_p\rangle & \cdots & \langle\varphi_{n+2},\varphi_{m-1} H_p\rangle & \langle\varphi_{n+2},\varphi_m H_p\rangle \\ \vdots & \vdots & & \vdots & \vdots \\ \vdots & \vdots & & \vdots & \vdots \\ \langle\varphi_{n+N_k},\varphi_0 H_p\rangle & \langle\varphi_{n+N_k},\varphi_1 H_p\rangle & \langle\varphi_{n+N_k},\varphi_{m-1} H_p\rangle & & \langle\varphi_{n+N_k},\varphi_m H_p\rangle \end{array}\right] \begin{bmatrix} \widehat{\alpha}_0^T \\ \widehat{\alpha}_1^T \\ \vdots \\ \widehat{\alpha}_{m-1}^T \\ \widehat{\alpha}_m^T \end{bmatrix} = 0. \tag{16.22}$$

Equation (16.22) is for only one response p, and more equations are accumulated into the correlation matrix by including the other responses.

16.10 Assembling the Correlation Matrix for the Least-Squares Problem

Assembling the correlation matrix is a matter of evaluating the $\langle\varphi_k, \varphi_j H_p\rangle$ dot product terms in Eq. (16.22) for each response, which could be done easily enough row by row and column by column. But that would be rather slow, so it has been reformulated for a more efficient implementation.

The dot product of two polynomials as defined in Eq. (16.3) can be calculated as the inner product of two vectors as

$$\langle p(s_i), q(s_i)\rangle = \sum_{i=1}^{N} \overline{p}(s_i)\, q(s_i) = \left[\overline{p}(s_1) \cdots \overline{p}(s_N)\right] \begin{bmatrix} q(s_1) \\ \vdots \\ q(s_N) \end{bmatrix} = p^H q \tag{16.23}$$

where p^H is the conjugate transpose of p. The $\phi_j H_p$ terms in Eq. (16.22) are the element-by-element multiplication of ϕ_j with each column of $H_p = \left[\, H_{p1} \cdots H_{pN_i} \,\right]$, or

$$\varphi_j H_p = \left[\varphi_j \circ H_{p1} \cdots \varphi_j \circ H_{pN_i}\right]_{N_s \times N_i}, \tag{16.24}$$

where N_s is the number of spectral lines. Using the $p^H q$ form of the inner product from Eq. (16.23), the partition for each response in the correlation matrix can be rearranged as

$$\underbrace{\begin{bmatrix} \varphi_{n+1}^H \\ \vdots \\ \varphi_{n+N_k}^H \end{bmatrix}}_{N_k \times N_s} \underbrace{\left[\left[\varphi_0 \circ H_{p1} \cdots \varphi_0 \circ H_{pN_i}\right] \cdots \left[\varphi_m \circ H_{p1} \cdots \varphi_m \circ H_{pN_i}\right]\right]}_{N_s \times (m+1) N_i}, \tag{16.25}$$

which is a standard matrix multiplication. Now we can recognize that the columns of the matrix on the right in Eq. (16.25) are all combinations of $\left[\, \varphi_0 \cdots \varphi_m \,\right]$ and the columns of H_p.

16.11 Accumulating the Correlation Matrix with QR Decomposition

Instead of assembling a potentially very large correlation matrix in its entirety, it is accumulated one response at a time using QR decomposition. For an $m \times n$ complex matrix A, with $m \geq n$, the "economy-size" QR decomposition is $A = QR$, where R is an $n \times n$ upper triangular matrix and Q is $m \times n$ with orthonormal columns. The normal equations for the least-squares solution of $Ax = b$, where A is complex, are

$$Ax = b \rightarrow A^H Ax = A^H b \rightarrow x = \left(A^H A\right)^{-1} A^H b. \tag{16.26}$$

If $A = QR$, the normal equations become

$$(QR)^H QRx = (QR)^H b \rightarrow R^H Q^H QRx = R^H Q^H b \rightarrow Rx = Q^H b \rightarrow x = R^{-1} Q^H b. \tag{16.27}$$

However, Eq. (16.22) is in the form of

$$Ax - b = 0, \text{ or } \underset{m \times n+1}{\begin{bmatrix} A & -b \end{bmatrix}} \underset{n+1 \times 1}{\begin{bmatrix} x \\ 1 \end{bmatrix}} = 0\,, \tag{16.28}$$

where A is $m \times n$, x is $n \times 1$, and b is $m \times 1$. The economy-size QR decomposition of $\begin{bmatrix} A & -b \end{bmatrix}$ is

$$\underset{m \times n+1}{\begin{bmatrix} A & -b \end{bmatrix}} = \underset{m \times n+1}{Q} \underset{n+1 \times n+1}{R} \rightarrow R = Q^H \begin{bmatrix} A & -b \end{bmatrix}. \tag{16.29}$$

Expanding the columns of both sides of Eq. (16.29) as

$$\begin{bmatrix} R_1 & \cdots & R_n & R_{n+1} \end{bmatrix} = Q^H \begin{bmatrix} A_1 & \cdots & A_n & -b \end{bmatrix} \rightarrow R_{n+1} = -Q^H b \tag{16.30}$$

shows that the last column contains the right-hand side of the normal equations. The solution for x then becomes

$$x = R^{-1} Q^H b = -\begin{bmatrix} R_1 & \cdots & R_n \end{bmatrix}^{-1} R_{n+1}. \tag{16.31}$$

As can be seen above, since the correlation matrix is arranged as an augmented $\begin{bmatrix} A & -b \end{bmatrix}$ matrix, the Q matrix is not explicitly needed because it is implicitly contained in the last column of the upper triangular factorization of the R matrix.

For each response, a partition of the correlation matrix as shown in Eq. (16.22) and computed by Eq. (16.25) is accumulated into the upper triangular QR factorization. It can be shown that this block-by-block approach is equivalent to the QR decomposition of the entire correlation matrix with all the blocks stacked together.

16.12 How Many Orthogonal Polynomials Should Be Used?

The size of the A partition of the correlation matrix formulated from Eq. (16.22) for N_o responses will be $N_o N_k \times m N_i$, regardless of how it is assembled or accumulated. For the least-squares problem to be overdetermined, there must be more equations than unknowns, or $N_k \geq m N_i / N_o$, but so far we have not considered what the value of N_k should be.

If the algorithm had been formulated from Eq. (16.16), the size of the correlation matrix would have been $N_o N_s \times m N_i + (n+1)\, N_o$, which is uniquely determined by the number of references, responses, and spectral lines, the model order, and residuals. However, the number of spectral lines is not explicitly a factor in the $N_o N_k \times m N_i$ size of the correlation matrix after removing the betas because the $\langle \varphi_k, \varphi_j H_p \rangle$ dot products are a summation over all of the spectral lines, which collapses the frequency dimension to a size of one. Thus, the formulation in Eq. (16.22) allows for a degree of freedom in choosing the overdetermination, but this still doesn't answer the question about N_k.

It could be defined to give a chosen overdetermination factor F_{OD} such that $N_k = F_{OD} \times m N_i / N_o$, which is rounded up to the next integer. But recall that N_{s-1} is the highest-order orthogonal polynomial that can be defined over an interval containing N_s spectral lines, and the orthogonal polynomials going down the rows of the correlation matrix partitions are $\varphi_{n+1}, \ldots, \varphi_{n+N_k}$. This means that $n + N_k \leq N_s - 1$ and $N_k \leq N_s - (n+1)$. Ideally, we would want to use the maximum overdetermination possible with $N_k = N_s - (n+1)$, but this could be several 100 to a few 1,000 orthogonal polynomials, and it could take some time to generate all of them and apply them to the partitions. However, there is a way to get all the overdetermination without having to do all the work, which finally brings us to the *new* part of the orthogonal polynomial algorithm.

16.13 The Orthogonal Complement: The New Part of the Orthogonal Polynomial Algorithm

Let $P = [P_1 \ P_2]$ be an $N_s \times N_s$ unitary matrix such that $P^H P = PP^H = I$ and $I - P_1 P_1^H = P_2 P_2^H$. Next, partition the full set of N_s orthogonal polynomials as

$$P_1 = \underset{N_s \times n+1}{\left[\varphi_0 \cdots \varphi_n\right]}, \text{ the lower-order orthogonal polynomials, and} \tag{16.32}$$

$$P_2 = \underset{N_s \times N_s-(n+1)}{\left[\varphi_{n+1} \cdots \varphi_{N_s-1}\right]}, \text{ the higher-order orthogonal polynomials,} \tag{16.33}$$

where $n+1$ is the minimum order of the orthogonal polynomials that removes the betas and the φ's are scaled to unity length so that $P^H P \doteq PP^H = I$.

According to Wikipedia (and to any number of linear algebra textbooks), "The orthogonal complement of a subspace W of a vector space V ... is the set $W^\perp$ of all vectors in V that are orthogonal to every vector in W."[9] This means that P_2 is the orthogonal complement of P_1, i.e., $P_2 = P_1^\perp$, since every φ is orthogonal to all φ's of a lower order. That is, all the φ's in P_2 are orthogonal to every φ in P_1.

Now form an orthogonal projection T such that $T = I - P_1 P_1^H = P_2 P_2^H$ and $\tilde{M} = TM$ for *any* matrix M. We then have

$$\tilde{M} = M - P_1 P_1^H M = P_2 P_2^H M. \tag{16.34}$$

Now let's relate this to the implementation of the orthogonal polynomial algorithm. The partition of the correlation matrix for one response p is given in Eq. (16.25), and let $N_k = N_s - (n+1)$ so that all available orthogonal polynomials are used for maximum overdetermination; we then have

$$\begin{bmatrix} \varphi_{n+1}^H \\ \vdots \\ \varphi_{N_s-1}^H \end{bmatrix} \left[\left[\varphi_0 \circ H_{p1} \cdots \varphi_0 \circ H_{pN_i}\right] \cdots \left[\varphi_m \circ H_{p1} \cdots \varphi_m \circ H_{pN_i}\right] \right] = P_2^H M. \tag{16.35}$$

As we can see, the matrix on the left in Eq. (16.35) with the higher-order orthogonal polynomials from Eq. (16.33) going down the rows is P_2^H, and the matrix on the right can be that *any* matrix M.

For the orthogonal polynomial algorithm, we want to accumulate $P_2^H M$ for each response with QR decomposition. But to do this, all N_s orthogonal polynomials would need to be generated. The orders 0 to m are needed to fill the columns of M, and the orders $n+1$ to $N_s - 1$ are needed for the rows of P_2^H. However, by using the orthogonal complement relationship in Eq. (16.34), we can instead compute $\tilde{M} = M - P_1 P_1^H M$, which requires far fewer φ's to effectively use all of them for maximum overdetermination – although there is one catch, since $\tilde{M} = P_2 P_2^H M$, but we want $P_2^H M$, not $P_2(P_2^H M)$.

However, the correlation matrix is being accumulated with QR decomposition, which is not unique – but it is unique up to pre-multiplication by a unitary matrix; that is, the QR decomposition of UA, where U is a unitary matrix, has *almost* the same upper triangular factorization as the QR decomposition A. When used in the normal equations, $A^H U^H UA = A^H A$ and all evidence of the unitary matrix vanishes. Luckily, $P = [P_1 \ P_2]$ is a unitary matrix and

$$P^H \tilde{M} = P^H \left(P_2 P_2^H M\right) = \begin{bmatrix} P_1^H \\ P_2^H \end{bmatrix} P_2 \left(P_2^H M\right) = \begin{bmatrix} P_1^H P_2 \\ P_2^H P_2 \end{bmatrix} P_2^H M = \begin{bmatrix} 0 \\ I \end{bmatrix} P_2^H M = \begin{bmatrix} 0 \\ P_2^H M \end{bmatrix}. \tag{16.36}$$

The matrix at the end of Eq. (16.36) is almost what we want, but the zeros on top do not affect the economy-size upper triangular QR factorization because

$$\begin{bmatrix} 0 & I \\ I & 0 \end{bmatrix}\begin{bmatrix} 0 \\ A \end{bmatrix} = \begin{bmatrix} A \\ 0 \end{bmatrix} \text{ and } \begin{bmatrix} 0 & I \\ I & 0 \end{bmatrix} \text{ is an orthogonal matrix, so triuqr}\left(\begin{bmatrix} 0 \\ A \end{bmatrix}\right) \to \text{triuqr}\left(\begin{bmatrix} A \\ 0 \end{bmatrix}\right) \to \text{triuqr}(A) \quad (16.37)$$

where *triuqr* is shorthand notation for the upper triangular QR factorization. This means that the upper triangular QR factorization of $\tilde{M}$ is *almost* the same as upper triangular QR factorization of $P_2^H M$ because

$$\text{triuqr}\left(\tilde{M}\right) \to \text{triuqr}\left(P^H \tilde{M}\right) \to \text{triuqr}\left(\begin{bmatrix} 0 \\ P_2^H M \end{bmatrix}\right) \to \text{triuqr}\left(\begin{bmatrix} P_2^H M \\ 0 \end{bmatrix}\right) \to \text{triuqr}\left(P_2^H M\right) \quad (16.38)$$

Using this orthogonal complement formulation, the highest-order orthogonal polynomial that will needed to be generated is the maximum of m and $n = m + r - 2$. Orders 0 to m are needed for the columns of M, and orders 0 to n are needed for P_1.

16.14 Some Correlation Matrix Accumulation Time Trials

Some time trials were run to evaluate the different options for implementing the accumulation of the least-squares correlation matrix for the orthogonal polynomial algorithm. The trials were conducted in Matlab R2014a on a laptop PC running the Windows 7 Professional 64-bit operating system, with a hyperthreaded quad-core Intel® Core™ i7 3.2 GHz CPU and 16 GB of RAM. The FRF datasets used for the time trials were small, medium, large, and extra-large, with the number of inputs (N_i), outputs (N_o), and spectral lines (N_s) shown below the x-axis in the bar graphs below. All time trials were run for a model order of 50, with no residuals in the numerator, and using all $N_s - 1$ orthogonal polynomials for maximum overdetermination.

Figure 16.8 shows the results for the three smallest datasets using the $\langle \varphi_k, \phi_j H_p \rangle$ dot product in Eq. (16.22), the $P_2^H M$ matrix multiplication in Eq. (16.35), and the $M - P_1 P_1^H M$ orthogonal complement in Eq. (16.34). Note that the y-axis is log-scale and the dot product method is several orders of magnitude slower than the other two. The conclusion here is that the dot product implementation is definitely not a viable option.

Figure 16.9 shows the results for the several more datasets using only the $P_2^H M$ matrix multiplication and $M - P_1 P_1^H M$ orthogonal complement methods. Note that the accumulation times have been normalized to the orthogonal complement

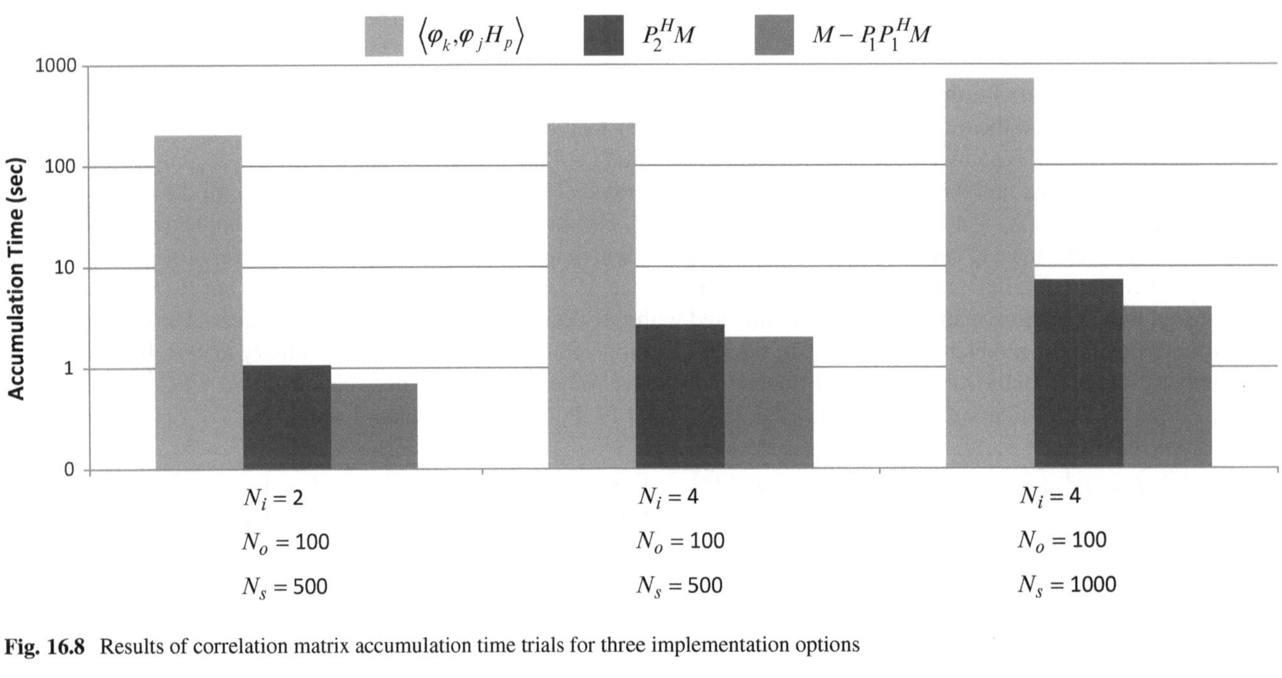

Fig. 16.8 Results of correlation matrix accumulation time trials for three implementation options

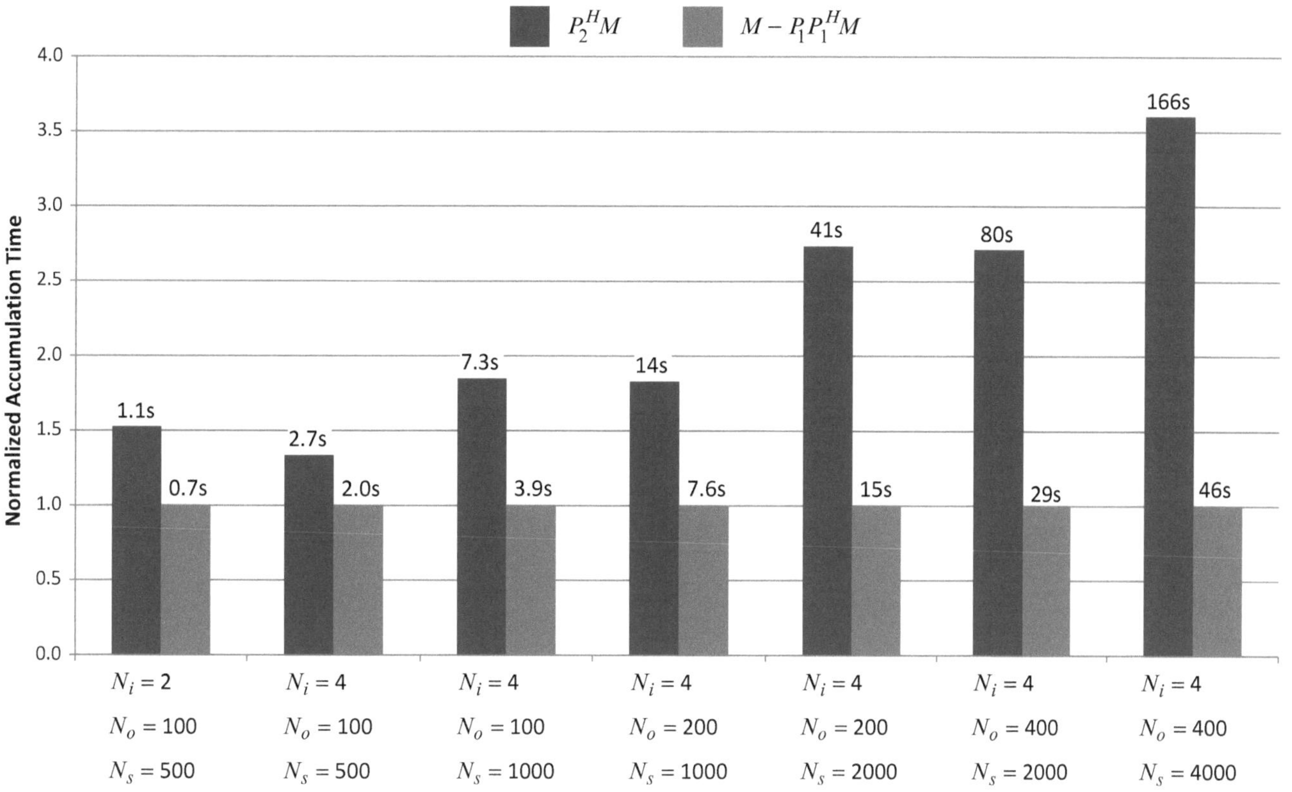

Fig. 16.9 Results of correlation matrix accumulation time trials for two implementation options

method so that all the datasets could be shown on the same plot. The accumulation times in seconds are shown above the corresponding bars. The orthogonal complement method is clearly faster, with the advantage becoming more relevant as the number of spectral lines increases, where the time savings can be measured in minutes for the largest dataset.

16.15 Concluding Remarks

Formulating the high-order, frequency-domain modal parameter estimation algorithm in an orthogonal polynomial basis has two important benefits. The first is that it cures the numerical ill-conditioning problem associated with the power polynomials for high model order and a wide frequency range. This was the primary reason why anybody decided to go to all this trouble in the first place back in the 1980s. The second is that it greatly reduces the size of the least-squares problem by removing the beta terms since we won't have to solve for a bunch of unknowns that aren't needed to find the poles.

While there is some choice in the matter, the frequency mapping proffered here scales and rotates the solution space from the frequency range of the FRF on the imaginary axis to the interval of -1 to $+1$ on the real axis. Then a set of real-valued orthogonal polynomial basis functions can be generated over this interval for any baseband or zoom-band frequency range without any restrictions on the frequency spacing. To estimate the poles in the chosen interval, an orthogonal polynomial companion matrix can be formulated for normalization of the characteristic equation by either the highest- or zeroth-order coefficient. A few prudent decisions allow for the efficient implementation of the algorithm, such as reformulating the dot products that remove the betas into a matrix multiplication, accumulating the correlation matrix with QR decomposition so that it is never formed in its entirety, and using the orthogonal complement to get all available overdetermination in less time.

References

1. Richardson PM, Formenti DL (1982) Parameter estimation from frequency response measurements using rational fraction polynomials. In: Proceedings, international modal analysis conference, pp 167–182
2. Van der Auweraer H, Leuridan J (1987) Multiple input orthogonal polynomial parameter estimation. Mech Syst Signal Process 1(3):259–272
3. Shih CY, Tsuei YG, Allemang RJ, Brown DL (1988) A frequency domain global parameter estimation method for multiple reference frequency response measurements. Mech Syst Signal Process 2(4):349–365
4. Shih CY (1989) Investigation of numerical conditioning in the frequency domain modal parameter estimation methods. Doctoral dissertation, University of Cincinnati, p 127
5. Vold H (1990) Numerically robust frequency domain modal parameter estimation. Sound and Vibration, Jan 1990, p 3
6. Vold H (1990) Statistics of the characteristic polynomial in modal analysis. In: Proceedings, international seminar on modal analysis, pp 53–57
7. Rolain Y, Pintelon R, Xu KQ, Vold H (1995) Best conditioned parametric identification of transfer function models in the frequency domain. IEEE Trans Autom Control 40(11):1954–1960
8. Vold H, Richardson M, Napolitano K, Hensley D (2008) Aliasing in modal parameter estimation – a historical look and new innovations. In: Proceedings, international modal analysis conference, p 15
9. Orthogonal Complement. Wikipedia. 26 June 2014. http://en.wikipedia.org/wiki/. Accessed 2 Oct 2014. Orthogonal_complement

Chapter 17
An Orthogonal View of the Polyreference Least-Squares Complex Frequency Modal Parameter Estimation Algorithm

William Fladung and Håvard Vold

Abstract The polyreference least-squares complex frequency (PLSCF) modal parameter estimation algorithm has gained some popularity since the introduction of its single-reference predecessor shortly after the turn of this millennium. It is a z-domain (i.e., discrete time) method that uses a complex exponential frequency mapping from the imaginary frequency axis to the unit circle on the complex plane. While it operates directly on frequency response functions, this method has been interpreted to be essentially equivalent to the polyreference time-domain algorithm, with the application of the discrete Fourier transform implicit in its formulation. Another way to view this algorithm is that its basis functions are a set of orthogonal polynomials evaluated around the unit circle. This paper shows that the PLSCF method can be implemented as an orthogonal polynomial algorithm by a simple substitution of the basis functions. Furthermore, the PLSCF method is extended for applicability to uneven frequency spacing by generating the z-domain basis functions with the same procedure that is used for the traditional Laplace-domain orthogonal polynomials. The paper also illustrates how PLSCF, the orthogonal polynomial algorithm, and their ancestor the rational fraction polynomial method all start from the same place but move to different neighborhoods to do their work.

Keywords Modal parameter estimation • Orthogonal polynomial • Complex frequency

17.1 Introduction

The rational fraction polynomial (RPF) [1] algorithm is the entry-level model of the high-order, frequency-domain modal parameter estimation methods. However, it has some well-known issues with numerical ill-conditioning for a high model order and a wide frequency range. The remedies for this shortcoming have been frequency mapping and a change of basis functions. The traditional, Laplace-domain orthogonal polynomial (LDOP) [2][1] algorithm scales and rotates the solution space from the frequency range on the imaginary axis to the interval of -1 to $+1$ on the real axis and substitutes orthogonal polynomials for the power polynomials. The polyreference least-squares complex frequency (PLSCF) algorithm uses a different frequency mapping and basis functions to address the numerical ill-conditioning.

17.2 One View of PLSCF

PLSCF is based on a z-domain model, where the z-domain is to the discrete time domain as the Laplace domain is to the continuous time domain. In the published works on the algorithm [3–6], the reduced normal equations are derived for the linear least-squares solution for the coefficients of the numerator and denominator polynomials of the rational fraction polynomial model of the frequency response function (FRF). The polynomial coefficients are assumed to be real-valued, which is enforced by including the conjugate FRF at the negative frequencies. A discrete time-domain model for complex trigonometric basis functions of the form $e^{jk\omega\Delta t}$ is chosen for a fast implementation of the reduced normal equations that has better numerical conditioning than the continuous time power polynomials. This choice also imposes even frequency

[1] Fladung and Vold [2] is a companion piece to this paper and should be considered a prerequisite for much of the discussion herein.

W. Fladung (✉) • H. Vold
ATA Engineering, Inc., 13290 Evening Creek Drive South, Suite 250, San Diego, CA 92128, USA
e-mail: bfladung@ata-e.com

M. Mains (ed.), *Topics in Modal Analysis, Volume 10*, Conference Proceedings of the Society for Experimental Mechanics Series, DOI 10.1007/978-3-319-15251-6_17

spacing and produces terms in the reduced normal equations that are equivalent to the inverse discrete Fourier transform of the FRF. For a nontrivial solution of the reduced normal equations, it is proposed to normalize by the highest-order denominator polynomial coefficient. The solution is obtained by first eliminating the numerator variables and solving for the denominator coefficients from the condensed set of equations. In fact, only the denominator polynomial, which is the characteristic equation, is necessary to estimate the poles and modal participation factors from a standard companion matrix.

Because the application of the discrete Fourier transform essentially creates the impulse response functions (IRFs) in the reduced normal equations, PLSCF can be viewed as the frequency-domain equivalent of the polyreference time-domain (PTD) algorithm. Although PLSCF processes FRFs instead of IRFs, it is in essence a discrete time-domain algorithm and as such it is subject to aliasing because the frequencies of the z-domain poles, which must wrap around the unit circle on the complex plane, are then mapped back into the original frequency range. However, this choice of complex trigonometric basis functions does offer very favorable numerical conditioning behavior, and PLSCF can accommodate a wide frequency range containing many modes.

17.3 An Orthogonal View of PLSCF

Another way to view the PLSCF algorithm is in terms of frequency mapping and basis functions. The normalized frequency $\tilde{\omega}_i \in [-1, +1]$ on the imaginary axis is mapped to the unit circle on the complex plane with

$$z_i = e^{j\pi\tilde{\omega}_i}. \tag{17.1}$$

The positive frequency is mapped to the top half of the unit circle and the negative frequency is mapped to the bottom, as shown in Fig. 17.1.

The basis functions are z^k, which are polynomials of a complex variable given by the frequency mapping in Eq. 17.1. By Euler's formula, the complex exponentials $e^{j\pi k\tilde{\omega}}$ are real cosine functions and imaginary sine functions going around the unit circle, as shown in Fig. 17.2 for k from zero to five.

These sine and cosine functions are periodic in the interval of $-\pi$ to π and are also orthogonal over this interval. That is, the set of z^k for $k = 0, 1, 2, \ldots$ is a set of orthogonal polynomials. In addition, this set of complex trigonometric basis functions has favorable attributes to address the numerical conditioning issues.

From this point of view, PLSCF can be thought of as a z-domain orthogonal polynomial (ZDOP) algorithm and can be implemented in the same manner as the LDOP algorithm – just with a different frequency mapping and set of basis functions. The numerator terms can be removed to greatly reduce the size of the least-squares problem, the correlation matrix can be accumulated response by response with QR decomposition, and the orthogonal complement can be used to get all possible

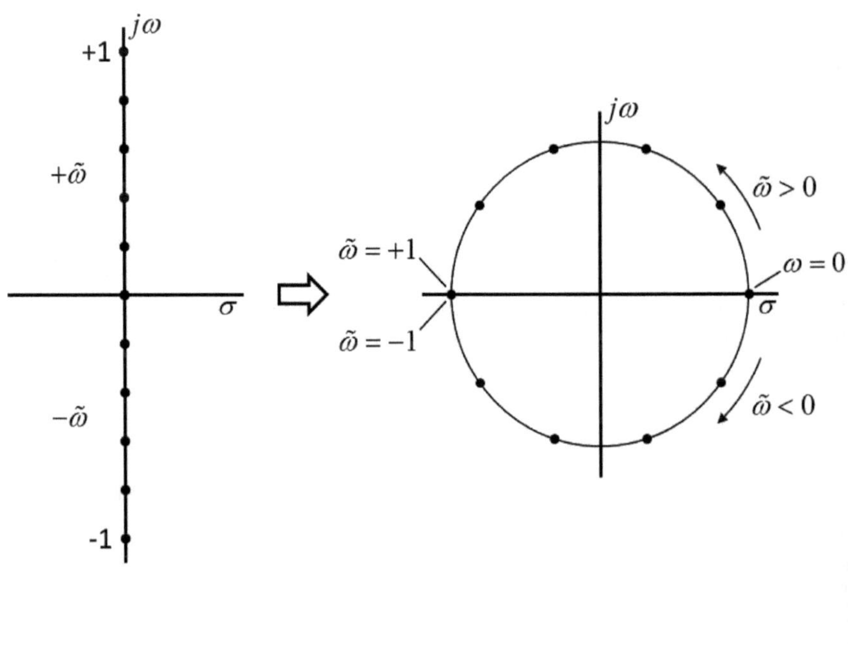

Fig. 17.1 The complex frequency mapping for PLSCF

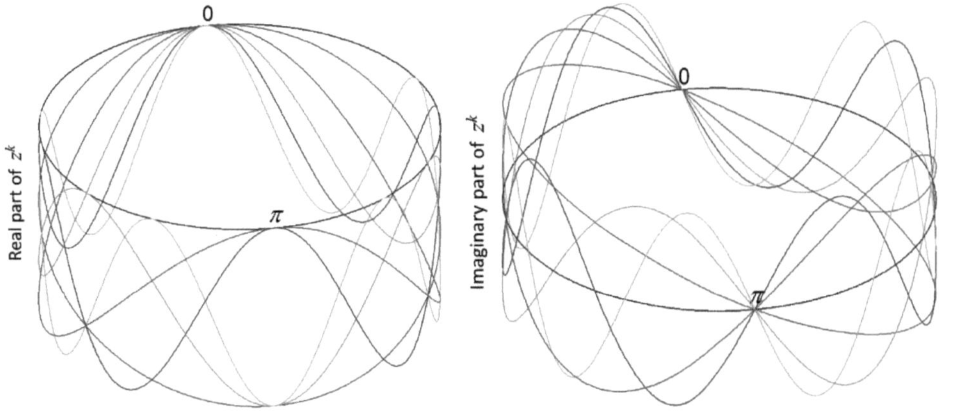

Fig. 17.2 Z-domain basis functions, cosines as the real part (*left*) and sine as the imaginary part (*right*)

overdetermination with less effort. It's just a simple substitution of the z-domain orthogonal polynomials for the Laplace-domain orthogonal polynomials.

The z-domain orthogonal polynomials $\xi_k(z_i) = z_i^k$ for $k = 0, 1, \ldots, N_s$ are constructed recursively, with each successive polynomial being one order higher than the preceding. Here N_s is the number of spectral lines, and the highest-order polynomial that can be defined for an interval containing N_s data points is $N_s - 1$, since the lowest order is zero – but not nearly that many are needed when using the orthogonal complement. The zeroth-order basis function $\xi_0 = z^0$ is a real constant normalized to have unity length (i.e., a 2-norm of one). Then the higher orders are generated recursively as

$$\xi_k(z_i) = z_i^k = z_i \times z_i^{k-1} = z_i \xi_{k-1}(z_i) \quad \text{for } k \geq 1, \text{ where } z_i = e^{j\pi\tilde{\omega}_i}. \tag{17.2}$$

17.4 Frequency Mapping

The key to implementing PLSCF as an orthogonal polynomial algorithm is getting the frequency mapping just right. If the frequency range $f \in [f_{min}, f_{max}]$ is normalized by just dividing by f_{max}, the resulting z^k basis function will not be orthogonal if {f}{min}>0 because the mapped interval must wrap all the way around the unit circle for the sines and cosines to be periodic. Rather, the frequency range must first be shifted to a baseband of $f' \in [0, f'_{max}]$, where $f'_{max} = f_{max} - f_{min}$. Then the shifted frequency range is scaled from -1 to 1, where the shifted and scaled interval of $\tilde{\omega} \in [-1, +1]$ can be mapped from the imaginary axis to the unit circle on the complex plane with Eq. 17.1, as shown in Fig. 17.3. There is one other small but important detail that must not be overlooked, which is that there cannot be coincident mapped points at 0 and $\pm\pi$, otherwise the sines and cosines will not be periodic. This means that the first and last spectral lines must be omitted when making the mirrored negative frequency.

The z-domain poles z_r are mapped back to Laplace-domain poles $\tilde{\lambda}_r$ in the shifted and scaled interval of $[-1, +1]$ as

$$z_r = e^{\pi\tilde{\lambda}_r} \quad \rightarrow \quad \tilde{\lambda}_r = \frac{\ln(z_r)}{\pi}, \tag{17.3}$$

then rescaled to the shifted frequency range with $\lambda'_r = f'_{max}\tilde{\lambda}_r$, and finally shifted back to the original frequency range with $\lambda_r = \lambda'_r \pm jf_{min}$ by moving the imaginary parts of the conjugate poles in the right direction.

17.5 Companion Matrices

The other differences for the ZDOP algorithm are a simplification of the LDOP generalized eigenvalue problem to the "standard" companion matrix and replacing the Laplace-domain poles λ_r with z-domain poles z_r. For the formulation that normalizes the characteristic equation with the highest-order coefficient α_m as an identity, the eigenvalue problem becomes

$$(C + Q)x = z_r Wx \quad \rightarrow \quad (C + 0)x = z_r Ix \quad \rightarrow \quad Cx = z_r x, \tag{17.4}$$

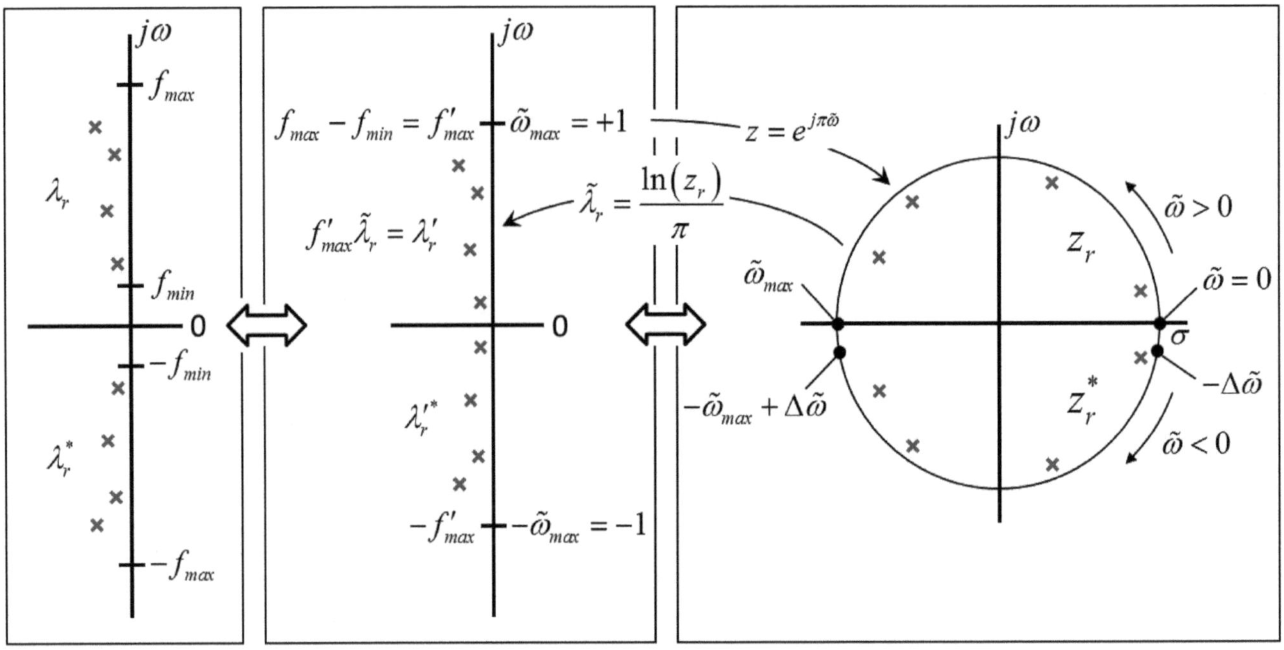

Fig. 17.3 Frequency mapping for PLSCF as an orthogonal polynomial algorithm

$$\text{where } C = \begin{bmatrix} 0 & I & & 0 \\ 0 & 0 & \ddots & 0 \\ 0 & 0 & & I \\ -\alpha_0 & -\alpha_1 & \cdots & -\alpha_{m-1} \end{bmatrix}, x = \begin{Bmatrix} L_r \\ z_r L_r \\ \vdots \\ z_r^{m-1} L_r \end{Bmatrix}$$

and L_r is the modal participation vector. For the LDOP generalized eigenvalue problem, W is a diagonal matrix of scaling factors and Q is a bi-diagonal matrix of scaled projections. However, since the z^k basis functions are scaled to have unity length, all of the scaling factors are unity and $W = I$, and since the sines and cosines are inherently orthogonal, all the projections are zero and $Q = 0$. The eigenvectors have the customary form of stacked partitions of the modal participation vector multiplied by scalars.

For the formulation that normalizes the characteristic equation with the zeroth-order coefficient α_0 as an identity, the eigenvalue problem becomes

$$(I + CQ)\,x = z_r C W x \quad \rightarrow \quad (I + 0)\,x = z_r C I x \quad \rightarrow \quad x = z_r C x \quad \rightarrow \quad z_r^{-1} x = C x, \tag{17.5}$$

$$\text{where } C = \begin{bmatrix} -\alpha_1 & \cdots & -\alpha_{m-1} & -\alpha_m \\ I & & 0 & 0 \\ 0 & \ddots & 0 & 0 \\ 0 & & I & 0 \end{bmatrix}$$

and the eigenvalues as computed are the reciprocal of the z-domain poles.

17.6 Characteristic Equation Normalization

Whichever way PLSCF is implemented, it bears the characteristics of the discrete time domain, such as the potential for aliasing. It also inherits another trait that makes a big difference in the "speed" and "clarity" of consistency diagrams: while the algorithm can be formulated for a normalization of the characteristic equation by either the highest- or zeroth-order

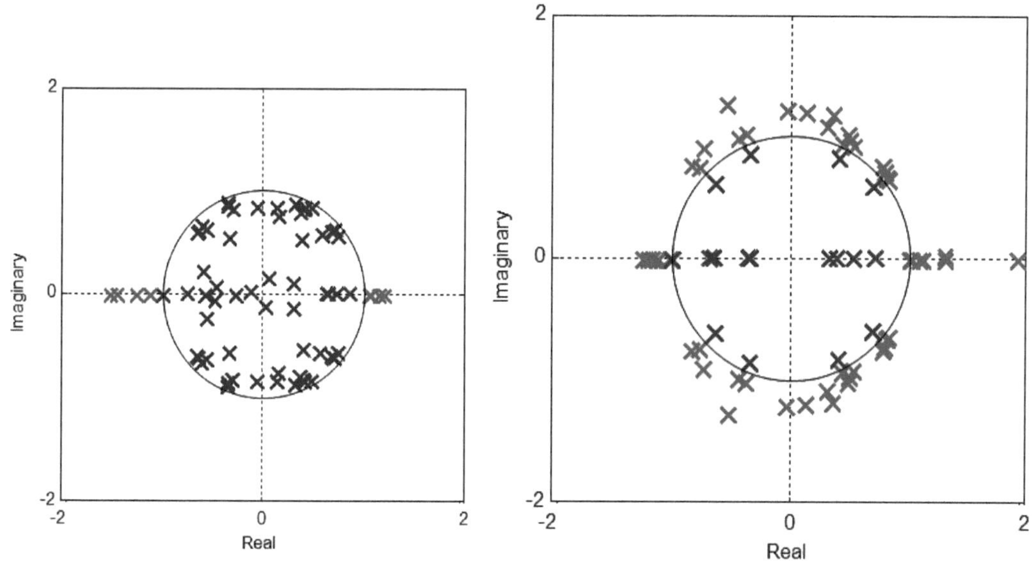

Fig. 17.4 Z-domain poles for highest-order coefficient normalization (*left*) and zeroth-order coefficient normalization (*right*)

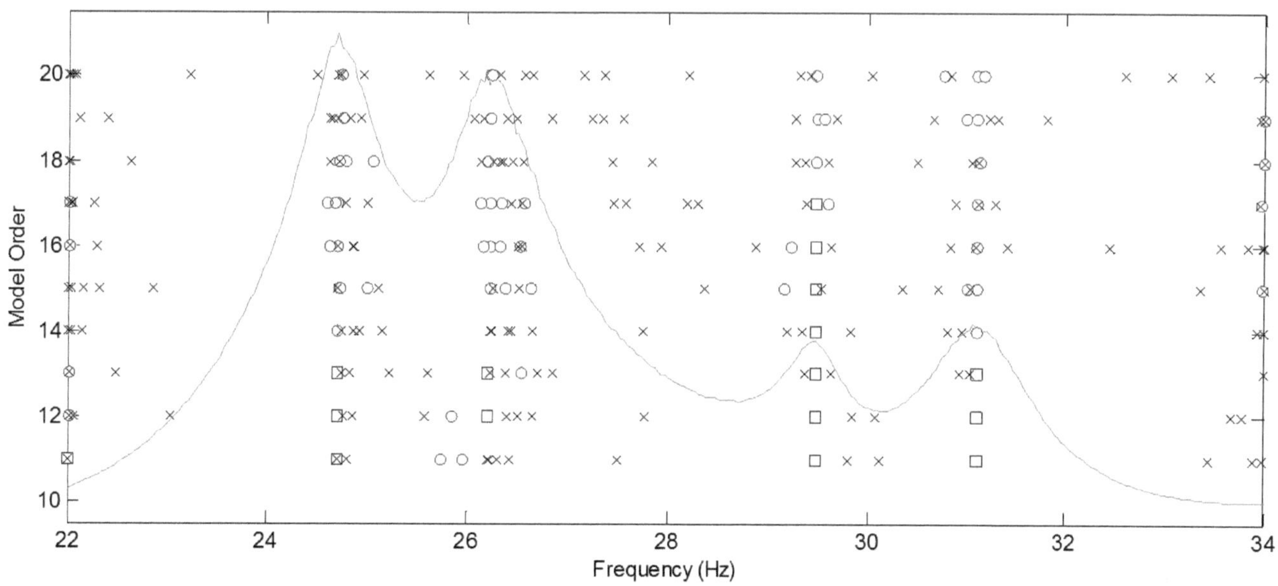

Fig. 17.5 Consistency diagram for highest-order coefficient normalization

coefficient, it has been shown that in the discrete time domain one choice is decidedly better than the other [7]. The α_m normalization estimates all (or at least most) of the poles as stable (i.e., with a magnitude less than one in the z-domain or with real parts in the left half-plane in the Laplace domain). However, the α_0 normalization estimates the consistent, system poles as stable but (almost all of) the computational poles as unstable, and these unstable poles can easily be discarded before assembling the consistency diagram.

In Fig. 17.4 are plotted the z-domain poles for the model order 20 row from the consistency diagrams in Figs. 17.5 and 17.6. The blue x's are the stable poles inside the unit circle, and the red x's are the unstable poles outside the unit circle. There are four modes in the frequency range, and for the α_0 normalization only these four poles are inside the unit circle (except for the ones near the real axis, which get pushed to the edges of the frequency range and can usually be ignored). However, for the α_m normalization, there are many more poles inside the unit circle – both the system poles and the computational poles – which leads to a rather messy consistency diagram as shown in Fig. 17.5.

Figure 17.5 is the consistency diagram for the α_m normalization, and Fig. 17.6 is the consistency diagram for the α_0 normalization. The green circles are consistent poles (frequency and damping), the blue squares are consistent poles and

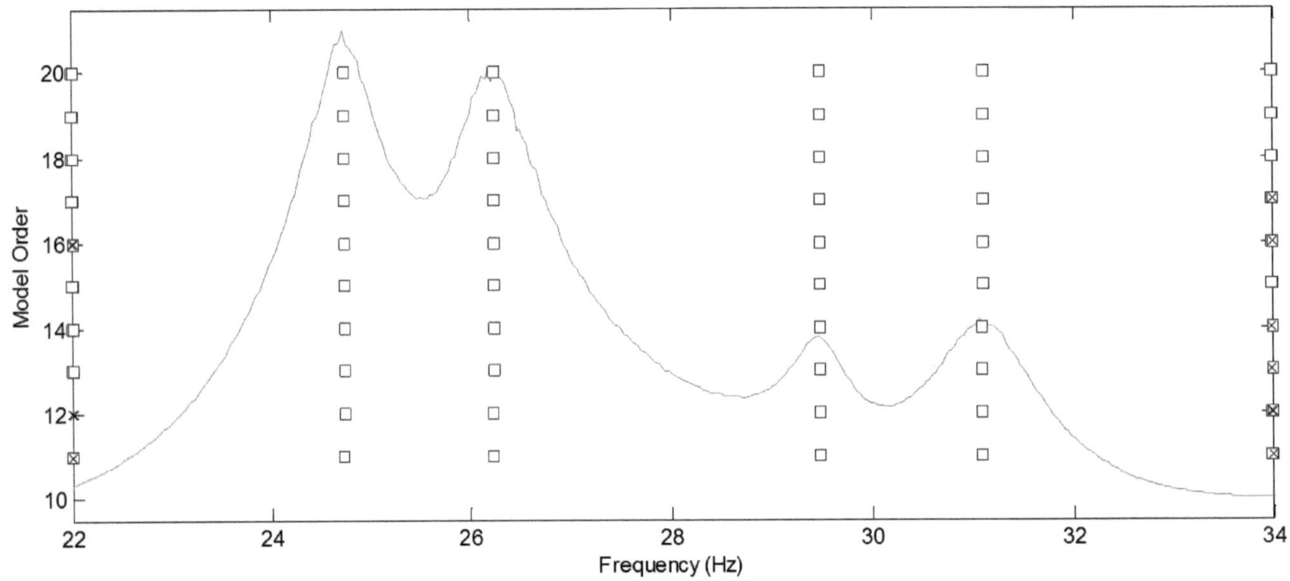

Fig. 17.6 Consistency diagram for zeroth-order coefficient normalization

vectors (residues), and the black x's are everything else inside the unit circle, as the unstable poles have already been discarded. The only thing that was changed in the processing for these two consistency diagrams was the choice of the coefficient for the characteristic equation normalization – but that made quite a difference.

17.7 Uneven Frequency Spacing

There's one small detail that we've overlooked thus far. First consider the consistency diagram in Fig. 17.7, which was generated using the PLSCF method implemented as an orthogonal polynomial algorithm with the recommended α_0 normalization. The blue squares indicate consistent poles and vectors, and black x's are everything else inside the unit circle but not too close to the real axis. Most any practitioner of the modal arts would consider this to be a very agreeable outcome. Now consider the consistency diagram in Fig. 17.8, which is the same dataset except that some of the spectral lines between the resonances have been removed. The x-ticks on the consistency diagram indicate the unevenly spaced spectral lines. Something certainly went wrong for the three modes around 500 Hz, and even though some consistent poles were found elsewhere, they are consistently wrong. As shown in the inset of Fig. 17.8, where the squares from the two consistency diagrams are plotted (but not the x's), the damping is too low. What happened here was that while the sines and cosines forming the basis functions are inherently orthogonal when the frequency is evenly spaced, the orthogonality is lost when evaluated at unevenly spaced discrete values.

The uneven frequency spacing in the example given above was meant to simulate a stepped sine test in which the frequency resolution is finer near the resonances and coarser away from the resonances. Although most FRFs are produced via the FFT and have evenly spaced spectral lines, there could be some other occasions in which a dataset intended for modal parameter estimation might have uneven frequency spacing. For example, you might want to exclude a frequency band that is overly noisy or remove harmonics of the electrical line frequency if these are troublesome. Also, if there is a mode that is contaminated by leakage, excluding the spectral lines near the resonance can give a better chance for a reasonable estimate of the pole. As will be shown next, the PLSCF algorithm can be generalized and extended to allow for uneven frequency spacing as well.

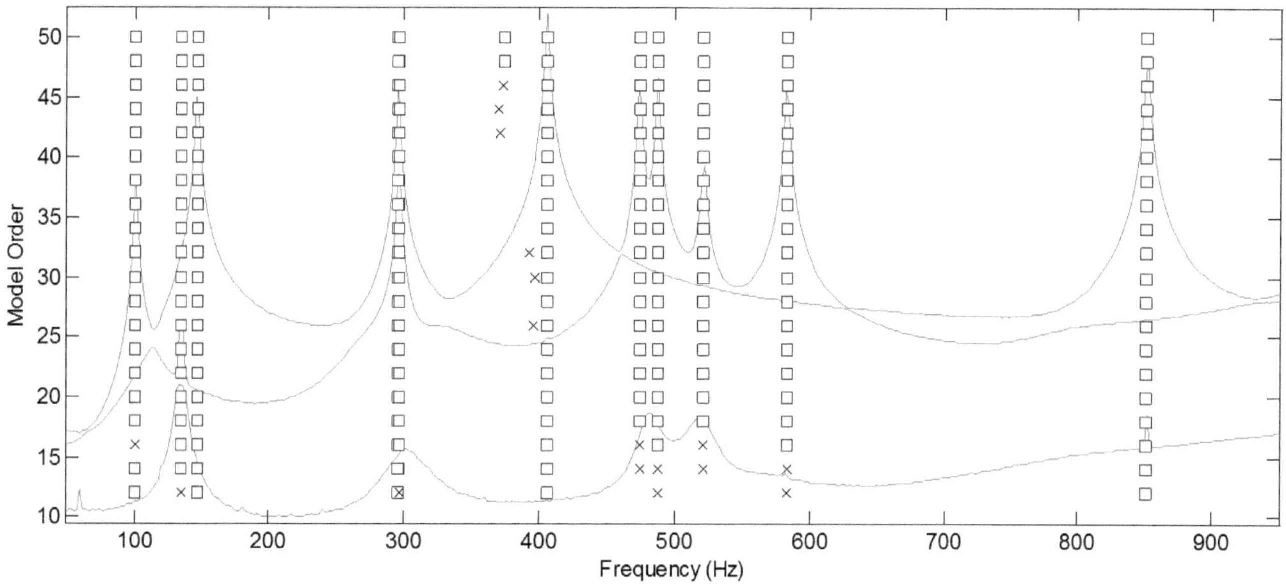

Fig. 17.7 Consistency diagram for even frequency spacing using trigonometric basis functions

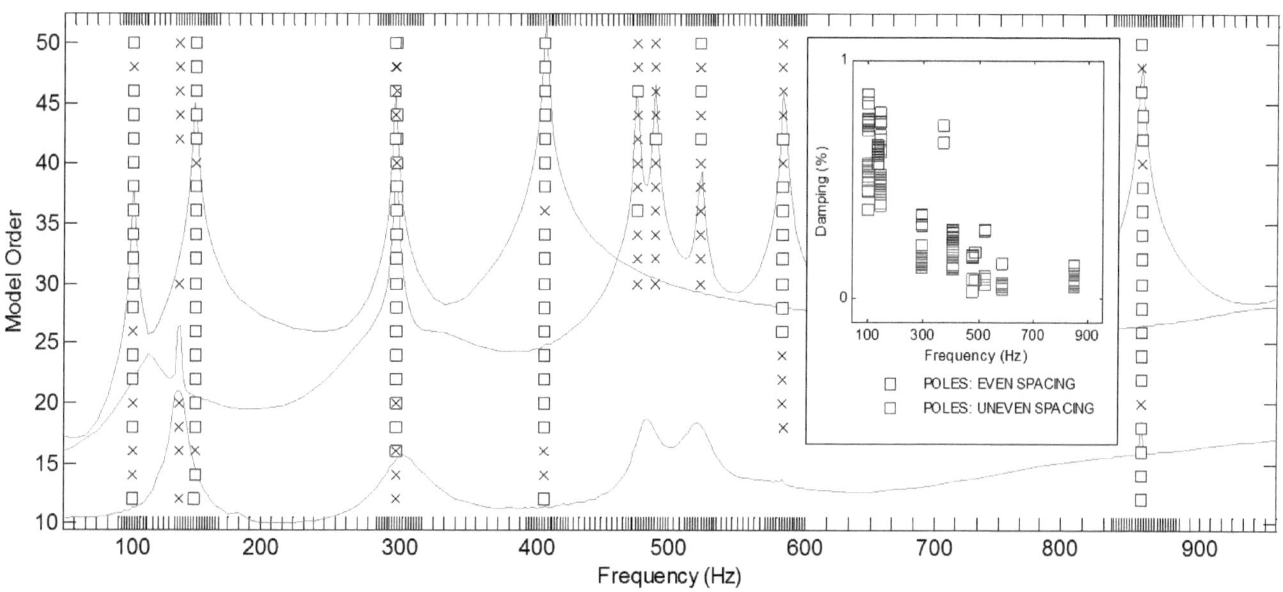

Fig. 17.8 Consistency diagram for uneven frequency spacing using trigonometric basis functions

17.8 The "Reorthogonalized" PLSCF

This limitation of even frequency spacing can be overcome by "reorthogonalizing" the complex polynomial basis functions in the z-domain using the same Gram-Schmidt approach that generates the Laplace-domain orthogonal polynomials. Constructing the orthogonal polynomials for the LDOP algorithm over the interval of -1 to $+1$ on the real axis was just one particular choice; they could have been defined over any interval in the complex plane, either on the real axis or on the imaginary axis, or somewhere in between – and it didn't have to be a straight line. So why not the unit circle?

For the Laplace-domain orthogonal polynomials, the frequency range was mapped to the interval of $x \in [-1, +1]$, with no restrictions on the frequency spacing. For the z-domain orthogonal polynomials, the frequency range is shifted and scaled to the interval of $\tilde{\omega} \in [-1, +1]$ and mapped to the unit circle as given in Eq. 17.1, now also with no restrictions on the frequency spacing. Again, the orthogonal polynomials $\xi_k(z_i)$ are constructed recursively, with each successive orthogonal polynomial being one order higher than the preceding one.

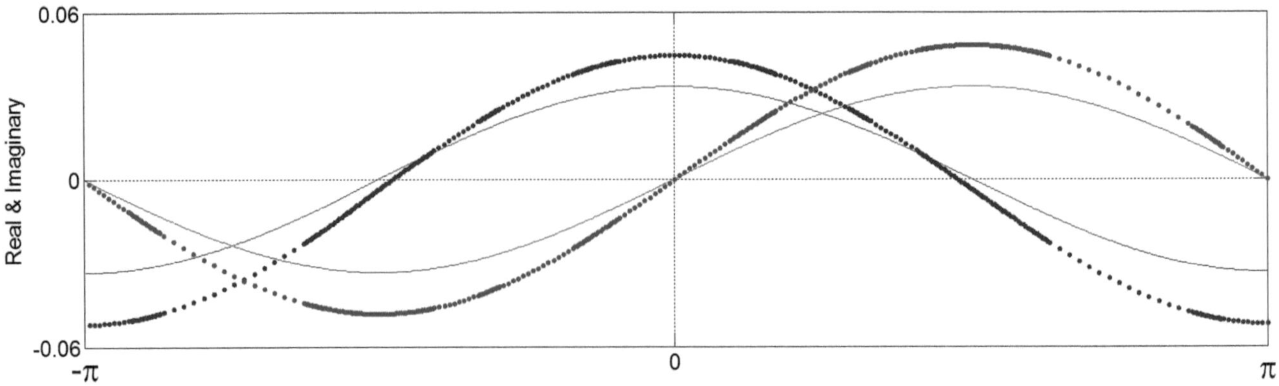

Fig. 17.9 First-order z-domain basis functions for even and uneven frequency spacing

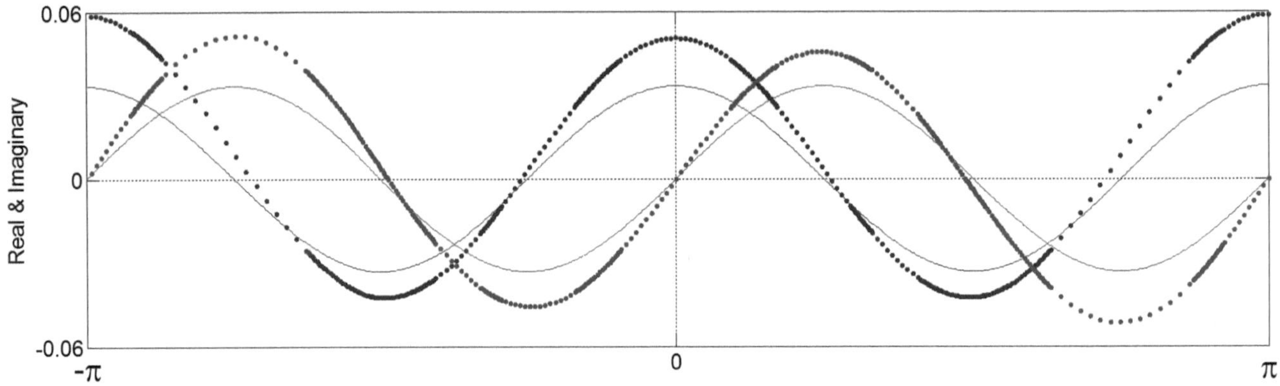

Fig. 17.10 Second-order z-domain basis functions for even and uneven frequency spacing

The zeroth-order basis function ξ_0 is still a real constant normalized to have unity length. Then the higher orders are generated recursively as

$$\xi_k(z_i) = w_k z_i \xi_{k-1}(z_i) - \sum_{n=0}^{k-1} q_{k,n} \xi_n(z_i) \quad \text{for } k \geq 1, \quad \text{where } z_i = e^{j\pi\tilde{\omega}_i}, \tag{17.6}$$

w_k is a scaling factor for unity-length normalization, and $q_{k,n}$ is a scaled projection (i.e., the scalar projection of φ_k onto φ_n multiplied by w_k). Note that there is one slight difference here: while the Laplace-domain orthogonalization required subtracting the projections onto only the preceding two orders, Eq. 17.6 contains the full recurrence from order $k-1$ to 0 [8].

So what do these reorthogonalized basis functions look like and, more importantly, do they have favorable numerical conditioning properties? The zeroth-order polynomial is still a real constant, but the higher orders are more interesting. In Figs. 17.9 and 17.10, the real (blue) and imaginary (green) parts of the first- and second-order polynomials are plotted in dots for the uneven frequency spacing from Fig. 17.8. The thin, solid lines are the trigonometric functions for the even frequency spacing from Fig. 17.7 (real part cosine, blue and imaginary part sine, green). The reorthogonalized functions have the same general pattern as the corresponding cosines and sines; i.e., the first order has one cycle over the interval of $-\pi$ to π and the second order has two cycles – they're just reshaped to impose orthogonality with the points over which they are defined.

Figures 17.11 and 17.12 show the real and imaginary parts of orders zero through ten of the reorthogonalized basis functions for the uneven frequency spacing from Fig. 17.8. While the trigonometric functions would have constant amplitude of the interval, these do not. However, they are still nicely bounded.

Figure 17.13 shows the real and imaginary parts of all 51 orders of the reorthogonalized basis functions used to estimate the poles in Fig. 17.14, plotted as dots to accentuate the unevenness of the frequency spacing and show the overall envelope. The condition number of the Vandermonde matrix formed from these polynomials is a perfect one, which means that they will have very favorable numerical conditioning properties. Incidentally, the condition number of the trigonometric basis functions is also unity.

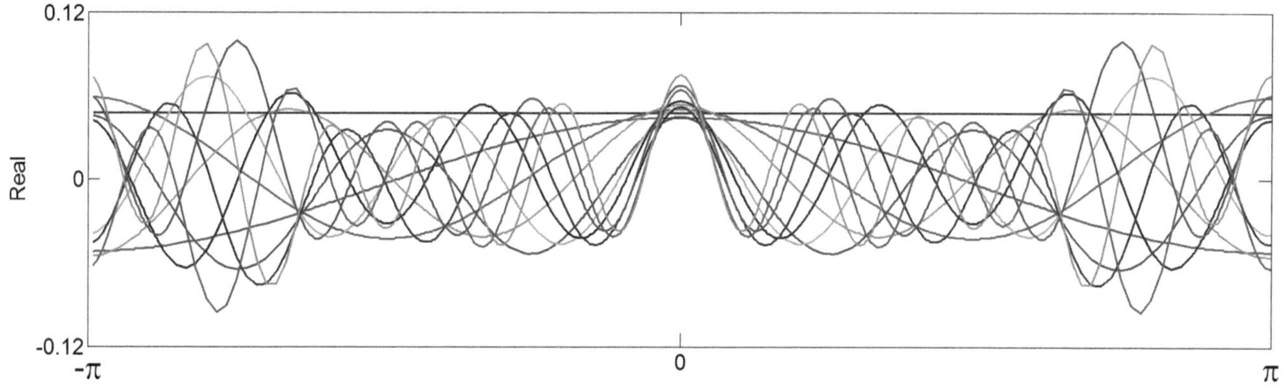

Fig. 17.11 Real part of orders zero through ten of z-domain basis functions for uneven frequency spacing

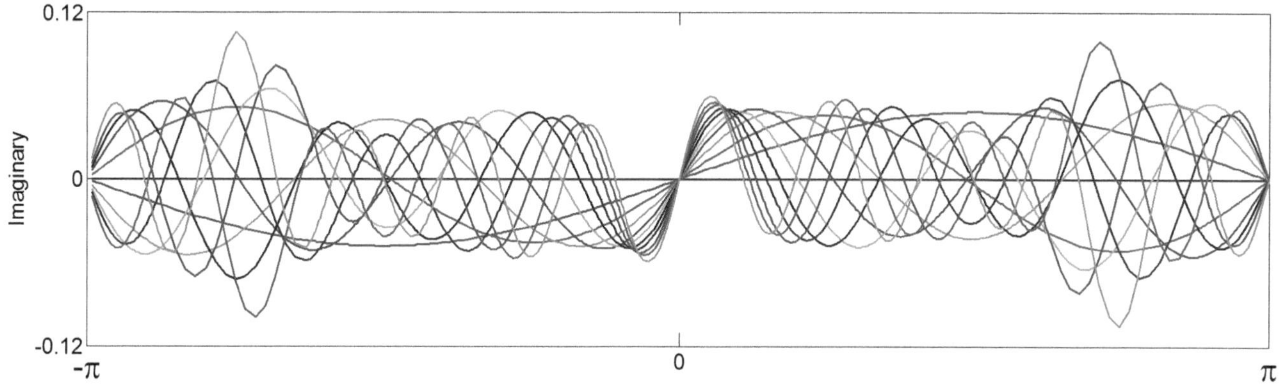

Fig. 17.12 Imaginary part of orders zero through ten of z-domain basis functions for uneven frequency spacing

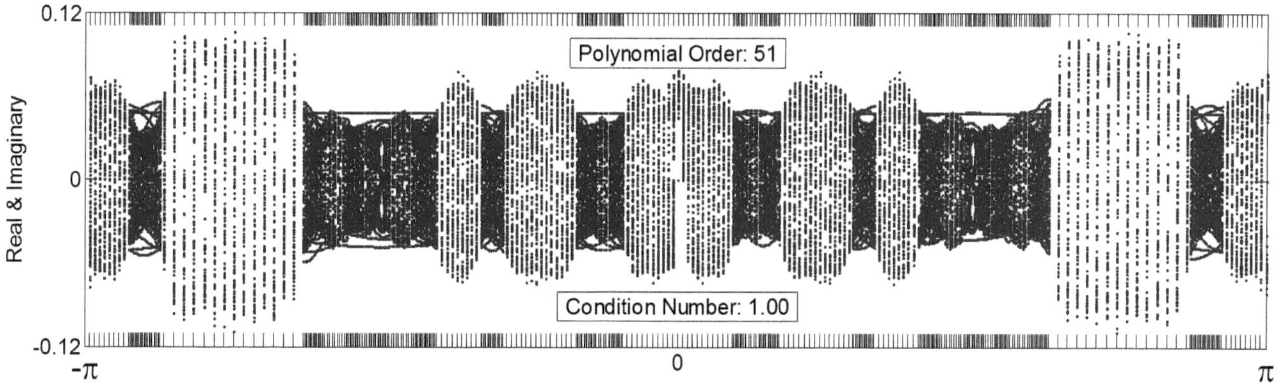

Fig. 17.13 All 51 orders of z-domain basis functions for uneven frequency spacing

To estimate the poles with these reorthogonalized basis functions, we revert back to the generalized eigenvalue problem formulations for the LDOP algorithm given in Eqs. 17.4 and 17.5. W is the same diagonal matrix of scaling factors w_k, but Q is a lower triangular matrix (instead of bi-diagonal) because of the full recurrence for subtracting the scaled projections $q_{k,n}$ in Eq. 17.6.

$$W = \begin{bmatrix} w_1 I & 0 & 0 & 0 \\ 0 & \ddots & 0 & 0 \\ 0 & 0 & \ddots & 0 \\ 0 & 0 & 0 & w_m I \end{bmatrix} \quad \text{and} \quad Q = \begin{bmatrix} q_{1,0} I & 0 & \cdots & 0 \\ q_{2,0} I & q_{2,1} I & \cdots & 0 \\ \vdots & \vdots & \ddots & \vdots \\ q_{m,0} I & q_{m,1} I & \cdots & q_{m,m-1} I \end{bmatrix} \tag{17.7}$$

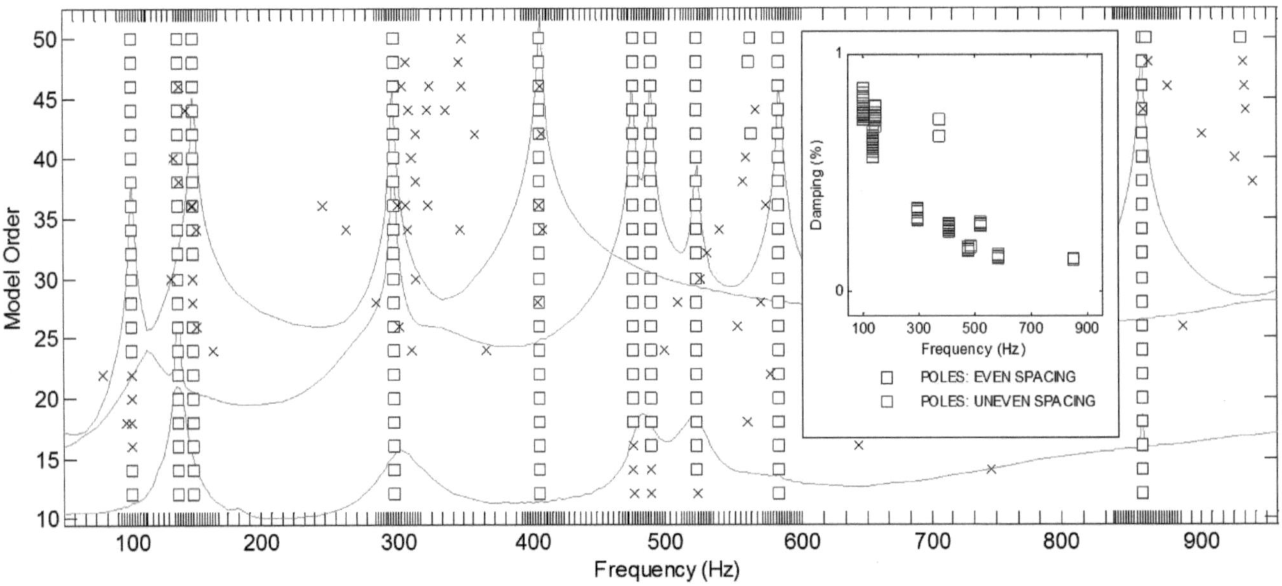

Fig. 17.14 Consistency diagram for uneven frequency spacing using reorthogonalized basis functions

The results from going to the trouble of creating a set of orthogonal basis functions for the uneven frequency spacing are shown in Fig. 17.14. This consistency diagram looks much better than the one in Fig. 17.8, and we're getting the right answers, too, as shown in the inset. Although it did seem to sprout some more extraneous poles (and there are a few more under the inset), those are not consistent and would not normally be plotted.

17.9 A Unified View

In the unified matrix polynomial approach (UMPA), modal parameter estimation algorithms are viewed from a common mathematical framework in order to study their similarities, differences, and numerical characteristics [9]. The underlying high-order model for the methods that process FRFs is based on the rational fraction polynomial model of an FRF.

$$H(j\omega_i) = \frac{\beta_n (s_i)^n + \beta_{n-1}(s_i)^{n-1} + \cdots + \beta_1 s_i + \beta_0}{\alpha_m (s_i)^m + \alpha_{m-1}(s_i)^{m-1} + \cdots + \alpha_1 s_i + \alpha_0} = \frac{\sum_{k=0}^{n} \beta_k (s_i)^k}{\sum_{k=0}^{m} \alpha_k (s_i)^k}, \tag{17.8}$$

where a distinction is made between the measured frequency ω_i of the FRF and the generalized frequency s_i of the model. To make another level of abstraction, the numerator can be expressed as some generic polynomial P_β of the generalized frequency variable, and the denominator can be another generic polynomial P_α, also of the generalized frequency variable.

$$H(j\omega_i) = \frac{P_\beta(s_i)}{P_\alpha(s_i)} = \underbrace{\frac{\sum_{k=0}^{n} \tilde{\beta}_k (j\tilde{\omega}_i)^k}{\sum_{k=0}^{m} \tilde{\alpha}_k (j\tilde{\omega}_i)^k}}_{\text{RFP}} = \underbrace{\frac{\sum_{k=0}^{n} \widehat{\beta}_k \varphi_k(x_i)}{\sum_{k=0}^{m} \widehat{\alpha}_k \varphi_k(x_i)}}_{\text{LDOP}} = \underbrace{\frac{\sum_{k=0}^{n} \overline{\beta}_k z_i^k}{\sum_{k=0}^{m} \overline{\alpha}_k z_i^k}}_{\text{PLSCF}} = \underbrace{\frac{\sum_{k=0}^{n} \overline{\beta}'_k \xi_k(z_i)}{\sum_{k=0}^{m} \overline{\alpha}'_k \xi_k(z_i)}}_{\text{ZDOP}} \tag{17.9}$$

The initials under each of the fractions in Eq. 17.9 indicate the modal parameter estimation algorithm represented by that polynomial model. The tildes, hats, bars, and primes on the alphas and betas imply that algorithms will not necessarily produce the same polynomial coefficients and hence will produce different estimates of the poles as well.

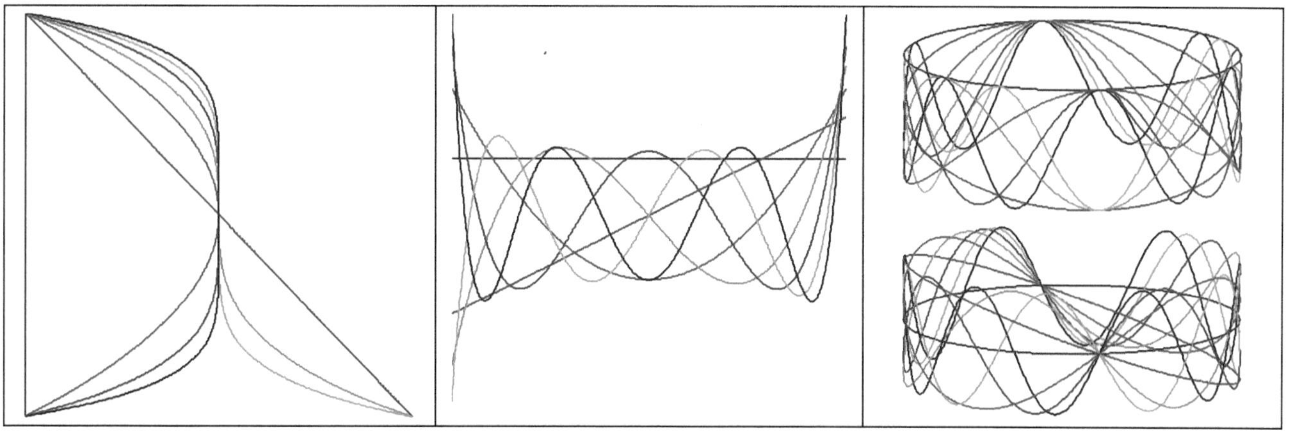

Fig. 17.15 Basis functions and frequency mapping for RFP (*left*), LDOP (*center*), and PLCSF and ZDOP (*right*)

For RFP, the frequency mapping is just a scaling of the frequency range such that $\tilde{\omega}_i \in [-1, +1]$, and the basis functions are power polynomials of $(j\tilde{\omega}_i)^k$ evaluated over the imaginary axis, as shown in Fig. 17.15 (left). For LDOP, the frequency range is scaled and rotated 90° clockwise to the real axis, and the basis functions are the Laplace-domain orthogonal polynomials $\varphi_k(x_i)$ evaluated over the interval $x_i \in [-1, +1]$, as shown in Fig. 17.15 (center). For PLSCF and ZDOP, the frequency range is shifted, scaled, and mapped from the imaginary axis to the unit circle on the complex plane. If the frequency is evenly spaced, PLSCF uses the trigonometric basis functions z_i^k; otherwise ZDOP uses the reorthogonalized basis functions $\xi_k(z_i)$, as shown in Fig. 17.15 (right).

From this point of view, all of these algorithms start from the same place but move to different neighborhoods to do their work. When they're done there, the poles that they produced are then mapped back to reality.

17.10 Conclusion

This paper has described how PLSCF can be viewed as the frequency-domain equivalent to PTD or as another version of an orthogonal polynomial algorithm operating a different solution space. From either point of view, it has the traits of a discrete time-domain algorithm and a prevalent choice for the characteristic equation normalization. The method was generalized and extended for uneven frequency spacing by constructing a new set of orthogonal basis functions using a procedure very similar to that used for the traditional Laplace-domain orthogonal polynomials. Although their provenance may differ and their implementations may diverge, all these methods – PLSCF, ZDOP, LDOP, and their ancestor RFP – all fit quite nicely into the UMPA framework – it's just a matter of frequency mapping and basis functions.

References

1. Richardson M, Formenti DL (1982) Parameter estimation from frequency response measurements using rational fraction polynomials. In: Proceedings of the international modal analysis conference, pp 167–182
2. Fladung W, Vold H (2015) An improved implementation of the orthogonal polynomial modal parameter estimation algorithm using the orthogonal complement. In: Proceedings of the international modal analysis conference, p 16
3. Van der Auweraer H, Guillaume P, Verboven P, Vanlanduit S (2001) Application of a fast-stabilizing frequency domain parameter estimation method. ASME J Dyn Syst Meas Control 123(4):651–658
4. Guillaume P, Verboven P, Vanlanduit S, Van der Auweraer H, Peeters B (2003) A polyreference implementation of the least-squares complex frequency domain estimator. In: Proceedings of the international modal analysis conference, p 12
5. Peeters B, Van der Auweraer H, Guillaume P, Leuridan J (2004) The PolyMAX frequency domain method: a new standard for modal parameter estimation? Shock Vib 11(3–4):395–409
6. Peeters B, Lowet G, Van der Auweraer H, Leuridan J (2004) A new procedure for modal parameter estimation. Sound Vib 5
7. Verboven P, Cauberghe B, Vanlanduit S, Parloo E, Guillaume P (2004) The secret behind clear stabilization diagrams: the influence of the parameter constraint on the stability of the poles. In: Proceedings of the society of experimental mechanics (SEM) annual conference, p 17

8. Rolain Y, Pintelon R, Xu KQ, Vold H (1995) Best conditioned parametric identification of transfer function models in the frequency domain. IEEE Trans Autom Control 40(11):1954–1960
9. Allemang RJ, Phillips AW (2004) The unified matrix polynomial approach to understanding modal parameter estimation: an update. In: Proceedings of the international conference on noise and vibration engineering, p 36

Chapter 18
Operational Modal Parameter Estimation from Short-Time Data Series

R. Arora, A. Phillips, and R. Allemang

Abstract The field of Operational Modal Analysis (OMA) has recently become an emerging research interest. OMA, also known as Response-Only Modal Analysis, extracts the modal parameters by processing only the system response data. In many experimental measurement situations, the system's output data series is short in length and buried under potentially correlated noise. In such cases, the separation of the noise from true data is challenging, generally resulting in inconsistent modal parameter estimates. For many of these measurement situations, it is believed that the correlated noise is a function of the system's operation and that a linear Auto-Regressive with eXogenous input (ARX) model exists that should describe the true system output. In this paper, a Nonlinear Auto-Regressive with eXogenous input (NARX) model based approach for estimating modal parameters from short output time series data is explored. In this approach, nonlinear terms are added to a linear ARX model to describe the noise and essentially filter out the system's true output data from the noisy time data series.

Keywords Parameter identification • Operational modal analysis • Non-linear system id • NARMAX • NARX

18.1 Introduction

OMA is a technique of estimating the modal parameters of a system only by processing its output data. Due to this reason, such a technique becomes especially useful in situations where measurement of input forces to the system is either impossible or extremely difficult. Therefore, it has been applied to large-scale structures where application of traditional Experimental Modal Analysis (EMA) methods poses difficulty.

In recent years, several signal processing techniques and modal parameter estimation (MPE) algorithms analogous to the EMA ones have been developed for the purpose of OMA. These methods utilize the ensemble averaging techniques like Root Mean Square (RMS) averaging to process the output data and the modal information is extracted from either the auto and cross correlation functions in the time domain or the output power spectrum in the frequency domain [1–3]. While the time domain MPE algorithms estimate the modal parameters by solving a set of least squares equations based on an Auto Regressive Moving Average with eXogenous input (ARMAX) model or an ARX model, a Rational Fraction Polynomial (RFP) model is the basis for the frequency domain algorithms [4–11]. The spatial domain algorithms, on the other hand, use the concept of linear superposition and the method of expansion theorem to decompose the spatial information at each point in the temporal axis and extract the mode shape vectors [2, 12–16]. The modal frequency and damping are estimated at a later stage after estimating the modal vectors. All these algorithms can be consistently formulated as a sequential least squares problem by the Unified Matrix Polynomial Approach (UMPA) as it provides a stage-wise and a step-wise mathematical explanation for the procedure of estimating the modal parameters [2, 17–19].

The above mentioned methods of MPE from the output-only data requires two major assumptions to be met, namely, a broadband, random and smooth nature of the input in the frequency range of interest and uniform spatial distribution of the forcing excitation. It has been shown that when the measured data does not absolutely comply with the assumptions of the forcing function regarding its spatial and temporal nature, these methods fail to provide a conclusive estimate of modal damping [2, 20]. Therefore, in situations when the system's output data series is short in length and is buried under noise, the ensemble averaging techniques used to filter the noise become difficult to implement [21]. Further, in the case of self-excitation problems like flight flutter or machining chatter, the noise is also correlated with the system's output thus making the averaging of the noise from true data rather difficult and resulting in inconsistent modal parameter estimates.

R. Arora • A. Phillips (✉) • R. Allemang
Structural Dynamics Research Laboratory, University of Cincinnati, PO Box 210072, Cincinnati, OH 45221-0072, USA
e-mail: arora.rahul.nitk@gmail.com; allyn.phillips@uc.edu; randall.allemang@uc.edu

M. Mains (ed.), *Topics in Modal Analysis, Volume 10*, Conference Proceedings of the Society for Experimental Mechanics Series, DOI 10.1007/978-3-319-15251-6_18

In the literature, the Nonlinear Auto Regressive Moving Average with eXogenous input (NARMAX) and NARX process based model identification methods have been shown to work well with short time data series and under the presence of noise, as in the case of the structural response of an aero-elastic system [22, 23]. Therefore, in this paper, a NARX model based approach, in-line with the concept of the time-domain UMPA, for estimating modal parameters using output time data series is discussed. In such an approach, nonlinear terms are added to a linear Auto-Regressive with eXogenous input (ARX) model to describe the noise and essentially filter out the system's true output data from the noisy data series. Since an ARX model describes the true output of the system, its modal parameters or the roots of its characteristic polynomial are computed by calculating the coefficients of the linear ARX terms.

18.2 Background

The NARMAX and NARX models [24] can be understood as the extensions of the ARMAX and ARX processes in order to account for the nonlinearity in the system. A number of physical phenomena have been successfully modeled and explained by the NARMAX and NARX model based approaches and these have been an active research topic in the field of system identification for more than two decades.

A general NARMAX process can be written as [25, 26]:

$$y(t) = f\left(y\left(t-1\right),\ldots,y\left(t-n_y\right),u\left(t-d\right),\ldots,u\left(t-n_u\right),e\left(t-1\right),\ldots,e\left(t-n_e\right)\right)+e(t) \tag{18.1}$$

Equation (18.1) relates the output at a given time instant to the past outputs, inputs, noise terms and the current measurement error $e(t)$. The function $f(\cdot)$ is an unknown nonlinear mapping function and n_y, n_u, n_e are the maximum output, input and noise lags respectively.

By excluding the noise lag terms, a NARX model can be written as [25, 27]:

$$y(t) = f\left(y\left(t-1\right),\ldots,y\left(t-n_y\right),u\left(t-d\right),\ldots,u\left(t-n_u\right)\right)+e(t) \tag{18.2}$$

Since the function $f(\cdot)$ is unknown, the NARMAX or NARX model identification involves estimating an appropriate structure of $f(\cdot)$ and computing the parameters of the model. In the literature, there are various implementations of the NARMAX models of which polynomial and rational representations, neural networks and wavelets are the most common ones. A polynomial type nonlinear map $f(\cdot)$ can be written as [26]:

$$f(x) = \sum_{k}^{n}\theta_k g_k\left(\mathbf{x}\right) \tag{18.3}$$

where θ_k are the coefficients of the polynomial, g_k is the polynomial of a chosen order and $\mathbf{x}$ is the vector denoted as

$$\mathbf{x} = \left[y\left(t-1\right),\ldots,y\left(t-n_y\right),u\left(t-d\right),\ldots,u\left(t-n_u\right),e\left(t-1\right),\ldots,e\left(t-n_e\right)\right] \tag{18.4}$$

The polynomial implementation of the NARMAX process shown in Eq. (18.3) can be reduced to a linear-in-parameter NARX model by excluding the past noise terms from the vector $\mathbf{x}$ shown in Eq. (18.4).

Several algorithms have been developed by researchers to identify the terms of a linear-in-parameter polynomial type NARX model. The key idea in all these algorithms is to evaluate each term from a pool of candidate terms and select the ones which best fit the data. Researchers have utilized both the least squares and the maximum likelihood approaches for identification of the model terms. The algorithms utilizing the least squares approach minimize the cost function or the Mean Square Error (MSE) of the model fit and select the terms which result in maximum reduction of the MSE [25, 28–34]. This process of selecting the terms by solving the least squares problem at each iteration can also be done recursively and in a numerically more efficient manner as shown in [35, 36].

Thus, the terms are selected successively from a candidate pool and added to the model until a balance between the goodness of the fit and the complexity of the resulting model is achieved. Information criteria like Akaike Information Criterion (AIC), Bayesian Information Criterion (BIC), which weigh the model fit against its complexity, are used to obtain a parsimonious model [37–41].

18.3 NARX Model Based OMA-MPE

18.3.1 Implementation of NARX Model

The main objective of taking a NARX process based approach is to describe the colored noise present in the measured data series. In this study, the system is assumed to be largely linear in the frequency range of interest and hence the dynamic response of the system can be modeled as an ARMAX process. The transfer function of a linear Single Input Single Output (SISO) system can be described in discrete form (z-domain) as an ARMAX model as follows:

$$H(z) = \frac{b_1 z^{-1} + b_2 z^{-2} + \ldots b_m z^{-m}}{a_0 z^0 + a_1 z^{-1} + a_2 z^{-2} + \ldots a_n z^{-n}} \tag{18.5}$$

Since it is only required to find the natural frequencies, which are the global properties of the system, the characteristic equation in the denominator (or the Auto Regressive part of the linear model shown in Eq. (18.5)) is to be solved for its roots. The characteristic polynomial equation in the z-domain for a SISO system can be written as:

$$a_0 z^0 + a_1 z^{-1} + a_2 z^{-2} + \ldots a_n z^{-n} = 0 \tag{18.6}$$

Let $y(k)$ be the response of the system at time k. Equation (18.6) therefore results in:

$$a_0 y(k) + a_1 y(k-1) + a_2 y(k-2) + \ldots a_n y(k-n) = 0 \tag{18.7}$$

Since the measured output is known at discrete time points, Eq. (18.7) can be solved by the method of least squares to compute coefficients a_i which can be used to compute the z-domain roots from Eq. (18.6). To account for the correlated measurement noise and the response due to unmeasured forces, the ARX process shown in Eq. (18.7) is extended to a linear-in-parameter NARX process by adding the non-linear functional terms. The resulting discrete time difference equation describing such a process can be written as:

$$a_0 y(k) + a_1 y(k-1) + a_2 y(k-2) + \ldots a_n y(k-n) + \sum_{i=1}^{l} \beta_i \theta_i = 0 \tag{18.8}$$

where,

$y(k)$ = measured response at time t = k
a_i = coefficient of characteristic polynomial equation
θ_i = non-linear functional terms of the form $\{[y(k-1)]^{c}[y(k-2)]^{c_2} \ldots [y(k-d)]^{c_d}\}$
β_i = coefficient of non-linear monomial term
d = total output lag to describe non-linear part of the process
n = model order
c_i = power of the monomial term where $i = 1, 2, \ldots d$

UMPA [17, 18] extends an ARX model describing a SISO system (LSCE type algorithm) to one describing a Multiple Input Multiple Output (MIMO) system (PTD or ERA type algorithm) by converting the scalar coefficient polynomial characteristic equation into a matrix coefficient polynomial characteristic equation. This approach of converting a scalar coefficient polynomial characteristic equation to a matrix coefficient polynomial characteristic equation is utilized to extend the single output NARX model shown in Eq. (18.8) to a multiple output NARX model. However, as opposed to the traditional EMA algorithms in which the size of the coefficient matrix is governed by the number of input and output degrees of freedom (DOFs), in OMA since the output of the system at the response locations is the only measurement being made, the size of matrix coefficients is governed by the number of output DOFs only. Therefore, a NARX model describing a MIMO system can be written as:

$$[a_0]_{N_o X N_o}\{y(k)\}_{N_o X 1} + [a_1]_{N_o X N_o}\{y(k-1)\}_{N_o X 1} + \ldots [a_n]_{N_o X N_o}\{y(k-n)\}_{N_o X 1} + \sum_{i=1}^{l}[\beta_i]_{N_o X N_o}\{\theta_i\}_{N_o X 1} = \{0\}_{N_o X 1} \tag{18.9}$$

where,
$\{y(k)\}_{N_oX1}$ = measured response vector at time t = k denoted as:

$$\{y(k)\}_{N_oX1} = [y_1(k),\ y_2(k)\ldots, y_{N_o}(k)]^T \tag{18.10}$$

$[a_i]_{N_o\ X\ N_o}$ = matrix coefficient of characteristic polynomial equation
$\{\theta_i\}_{N_o\ X\ 1}$ = column vector of non-linear functional terms denoted as:

$$\{\theta_i\}_{N_o\ X\ 1} = \begin{Bmatrix} (y_1(k-1))^{c_1}(y_1(k-2))^{c_2}\ldots(y_1(k-d))^{c_d} \\ (y_2(k-1))^{c_1}(y_2(k-2))^{c_2}\ldots(y_2(k-d))^{c_d} \\ \vdots \\ (y_{N_o}(k-1))^{c_1}(y_{N_o}(k-2))^{c_2}\ldots(y_{N_o}(k-d))^{c_d} \end{Bmatrix} \tag{18.11}$$

$[\beta_i]_{N_o\ X\ N_o}$ = matrix coefficient of vector of non-linear monomial terms
n = model order
d = total output lag to describe non-linear part of the process
c_i = power of the monomial term where $i = 1, 2, \ldots d$
N_o = number of output DOFs or response locations

Equation (18.9) is a linear-in-parameter NARX model which describes the response of the system at multiple response locations in terms of the past responses at the given output DOFs. Reformulating Eq. (18.9) as a set of linear least squares equations by normalizing the higher order coefficient $[a_0]$ results in:

$$\boldsymbol{Y} = \boldsymbol{AX} + \in \tag{18.12}$$

where,

$$\boldsymbol{Y} = -\begin{bmatrix} \{y(k)\}^T \\ \vdots \\ \{y(k+N)\}^T \end{bmatrix} \tag{18.13}$$

$$\boldsymbol{A} = \begin{bmatrix} \{y(k-1)\}^T & \ldots & \{y(k-n)\}^T & \{\theta_1(k)\}^T & \ldots & \{\theta_l(k)\}^T \\ \vdots & \ddots & \vdots & \vdots & \ddots & \vdots \\ \{y(k-1+N)\}^T & \ldots & \{y(k-n+N)\}^T & \{\theta_1(k+N)\}^T & \ldots & \{\theta_l(k+N)\}^T \end{bmatrix} \tag{18.14}$$

also denoted as:

$$\boldsymbol{A} = \begin{bmatrix} \begin{Bmatrix} | \\ \boldsymbol{p}_1 \\ | \end{Bmatrix} & \begin{Bmatrix} | \\ \boldsymbol{p}_2 \\ | \end{Bmatrix} \cdots & \begin{Bmatrix} | \\ \boldsymbol{p}_k \\ | \end{Bmatrix} & \begin{Bmatrix} | \\ \boldsymbol{p}_{k+1} \\ | \end{Bmatrix} & \cdots & \begin{Bmatrix} | \\ \boldsymbol{p}_s \\ | \end{Bmatrix} \end{bmatrix} \tag{18.15}$$

$$\boldsymbol{X} = \begin{Bmatrix} [a_1]_{N_o\ X\ N_o} \\ \vdots \\ [a_n]_{N_o\ X\ N_o} \\ [\beta_1]_{N_o\ X\ N_o} \\ \vdots \\ [\beta_l]_{N_o\ X\ N_o} \end{Bmatrix} \tag{18.16}$$

$$\in = \begin{bmatrix} \{\epsilon(k)\}^T \\ \vdots \\ \{\epsilon\,(k+N)\}^T \end{bmatrix} \tag{18.17}$$

$\{\epsilon(k)\}$ = residual error vector of size $N_o \times 1$ at time t = k denoted as:

$$\{\epsilon(k)\}_{N_o\ X\ 1} = [\epsilon_1(k),\ \epsilon_2(k)\ldots,\epsilon_{N_o}(k)]^T \tag{18.18}$$

In case of the MIMO NARX model, a complete column vector of non-linear monomial terms of the same structure of non-linearity is selected as shown in Eq. (18.11).

18.3.2 Selection of Model Terms

The linear terms associated with the matrix coefficients $[a_i]$ form the system's characteristic equation, which has the modal parameters buried in it. The selection of these terms is based on the model order which in turn is based on the number of poles to be computed. For a system with matrix coefficients, the number of poles computed is equal to the product of the size of the coefficients matrix N_o and model order n [17, 18]. Once a model order is chosen for the linear terms, the non-linear monomial term vectors of the NARX model from a candidate pool are selected such that the cost function of the model fit is minimized. The cost function of the model fit with m linear-in-parameter model term vectors is given as [35, 36, 42]:

$$J\,(\boldsymbol{A}_m) = trace\,\left(\boldsymbol{Y}^T\boldsymbol{Y} - \boldsymbol{Y}^T\boldsymbol{A}_m\left(\boldsymbol{A}_m{}^T\boldsymbol{A}_m\right)^{-1}\boldsymbol{A}_m{}^T\boldsymbol{Y}\right) \tag{18.19}$$

$\boldsymbol{A}_m$ in Eq. (18.19) is the intermediate information matrix with m vectors, including both the linear and non-linear monomials, having been selected and is denoted as:

$$\boldsymbol{A}_m = [\boldsymbol{p}_1,\ \boldsymbol{p}_2,\ldots,\ \boldsymbol{p}_m] \tag{18.20}$$

Since the selection of each new vector should minimize the cost function, a vector is selected such that the chosen vector, among all other monomial term vectors in the candidate pool, maximizes the contribution to the model fit. i.e., $\Delta\ J_{m+1}\ (\boldsymbol{A}_{m+1})$ is maximum. The constructional form of such a vector is shown in Eq. (18.11) and $\Delta\ J_{m+1}\ (\boldsymbol{A}_{m+1})$ is computed as:

$$\Delta\ J_{m+1}\ (\boldsymbol{A}_{m+1}) = J\,(\boldsymbol{A}_m) - J\,\left(\left[\boldsymbol{A}_m,\boldsymbol{p}_{m+1}\right]\right) \tag{18.21}$$

It is to be noted that although the model fit improves with each new selected term vector, the complexity of the model is also increased. Therefore, in order to obtain a parsimonious model, an information criterion is used which weighs the model fit against its complexity. In the literature, several information criteria for the selection of terms in modeling a data series are mentioned and their estimation characteristics are discussed. However, for the purpose of this research, only AIC and BIC are implemented as they have been more widely used for the identification of the NARX models.

AIC and BIC values for a least squares estimation is given as [41, 43, 44]:

$$AIC\ value = a\ log\left(\frac{RSS}{a}\right) + 2b \tag{18.22}$$

$$BIC\ value = a\ log\left(\frac{RSS}{a}\right) + a\log b \tag{18.23}$$

where,

a = number of observations or equations to form an over-determined set
b = number of parameters of the estimated model
RSS = residual sum of squared errors

For a NARX model with matrix coefficients as shown in Eq. (18.9), AIC and BIC values for an intermediate model $\boldsymbol{A}_m$ are computed by substituting RSS with $J(\boldsymbol{A}_m)$ (i.e., the cost function of the intermediate model computed in Eq. (18.19)), a with $m \times N_o$ (i.e., the total number of monomial term vectors in the intermediate model times the size of the coefficient matrix) and b with N (i.e., the total number of discrete time points used to formulate Eq. (18.12)). Since in both the information criteria the key idea is to penalize the complexity of the model, the selected model should have the minimum information criterion value. Therefore AIC/BIC values are calculated for each term $\boldsymbol{p}_{m+1}$ chosen from the candidate pool by maximizing $\Delta\ J_{m+1}\ (\boldsymbol{A}_{m+1})$ and the term is retained if the following relationship is satisfied:

$$(\text{AIC or BIC})_{\boldsymbol{A}_{m+1}} < (\text{AIC or BIC})_{\boldsymbol{A}_m} \tag{18.24}$$

18.3.3 Computation of Modal Parameters

Having selected the linear-in-parameter NARX model and identified the model terms as shown in the previous section, the next step towards computing modal parameters is to compute the coefficients or the parameters of the constructed model. For a system of equations as shown in Eq. (18.12), the model parameters are computed by the method of least squares as [42]:

$$\boldsymbol{X} = \left(\boldsymbol{A}^T \boldsymbol{A}\right)^{-1} \boldsymbol{A}^T \boldsymbol{Y} \tag{18.25}$$

It is important to note that although the parameter vector X contains both $[a_i]$ and $[\beta_i]$ coefficients, only the $[a_i]$ coefficients are associated with the characteristic polynomial equation of the system. Therefore, modal parameters are computed by forming a companion matrix using the $[a_i]$ coefficients and an Eigenvalue Decomposition (ED) is performed on the companion matrix to obtain the eigenvalues and eigenvectors from which the modal poles and the associated mode shapes are extracted [18].

In the case of traditional EMA UMPA methods, since the true model order is unknown due to the presence of noise on the measured data, the model order is iterated over a range and pole-consistency and pole-density diagrams are generated to select the system poles and the associated mode shapes [45]. The same approach of using the pole-consistency and pole-density diagrams is implemented in this method, which are generated by performing the following iterations:

1. The model order of the linear monomial terms is iterated over a range based on the estimate obtained from a Complex Mode Indicator Function (CMIF) or Singular Value Percentage Contribution (SVPC) [2, 13] plot of the output power spectrum matrix.
2. Since in a NARX model the accuracy of the model depends on the nonlinear terms selected from the candidate pool, the sufficiency of the candidate pool of the nonlinear monomial term vectors is an important concern [22]. For the linear-in-parameter polynomial type NARX model shown in Eq. (18.9), the sufficiency of the candidate pool is determined by the maximum output lag d and the power of the nonlinear monomial term, i.e., $\sum c_i$. Although an estimate of the above quantities can be obtained through *apriori* knowledge of the system to be analyzed, since the true values are unknown, an iterative approach is taken. Therefore, for a chosen model order, the candidate pool is varied by iterating the maximum output lag d and the power of the nonlinear monomial term $\sum c_i$.
3. Since the normalization of the high or low order coefficient of the characteristic polynomial has an effect on the location of computational poles due to the presence of noise [46], normalization of both the high and low order coefficients is performed for identification of the nonlinear model terms. In other words, the least squares equations are developed for both the normalizations to form $\boldsymbol{A}_m$ and compute the cost function $J(\boldsymbol{A}_m)$ and difference in the cost function $\Delta\ J_{m+1}\ (\boldsymbol{A}_{m+1})$. Further, both AIC and BIC are utilized for the selection of terms and poles are computed from both the selected models.
4. Finally, the least squares solution for computing the parameters of the model after the selection of terms is also performed for both the normalizations. Equation (18.26) shows the low order coefficient normalization for the MIMO system from which the set of linear equations can be formulated as shown in Eq. (18.12).

$$[a_0]_{N_o\ X\ N_o}\{y(k)\}_{N_o\ X\ 1} + \cdots + [a_{n-1}]_{N_o\ X\ N_o}\{y\ (k-n+1)\}_{N_o\ X\ 1} + \sum_{i=1}^{l}[\beta_i]_{N_o\ X\ N_o}\{\theta_i\}_{N_o\ X\ 1} = -\{y\ (k-n)\}_{N_o\ X\ 1} \tag{18.26}$$

Fig. 18.1 Flowchart showing the iterations performed to generate pole-stability diagrams

The above iterations can be represented algorithmically in a flowchart as shown in Fig. 18.1.

Having computed all the poles and the associated vectors for all the above iterations in the sequence shown in Fig. 18.1, the pole-density and pole-consistency diagrams are generated. Since the system is assumed to be largely linear in the frequency range of interest, the true structural poles of the system, excluding the computational poles due to noise, should remain consistent irrespective of any of the above mentioned iterations performed. Hence, the clusters of consistent system poles can easily be identified on a pole-density plot presented in the complex plane. Further, to select the poles with consistent modal vectors, the procedure of generating clear pole-consistency diagrams and autonomous MPE is followed as shown in [45].

18.4 Test Cases

To check the effectiveness of the NARX model based operational MPE method as developed in Sect. 18.3, the algorithm is implemented on three sets of analytically generated data and the results are presented in this section.

18.4.1 Test Case I

The first test data set is generated keeping in mind the conditions wherein the nonlinearity results from the closed loop interaction of the system with the ambience. To simulate such a condition, an impulse response of a system with nonlinearity is computed analytically for a chosen set of initial conditions as shown in Eq. (18.27).

$$\{y(k)\} = [a_1]_{6\times6}\{y\,(k-1)\}_{6\times1} + [a_2]_{6\times6}\{y\,(k-2)\}_{6\times1} + [\beta_1]_{6\times6}\left\{(y\,(k-2))^2\right\}_{6\times1} \tag{18.27}$$

where $\{y(k)\} = [0.1,\ 0.01,\ 0,\ 0,\ 0.01,\ 0.1]^T\ for\ k \le 0$

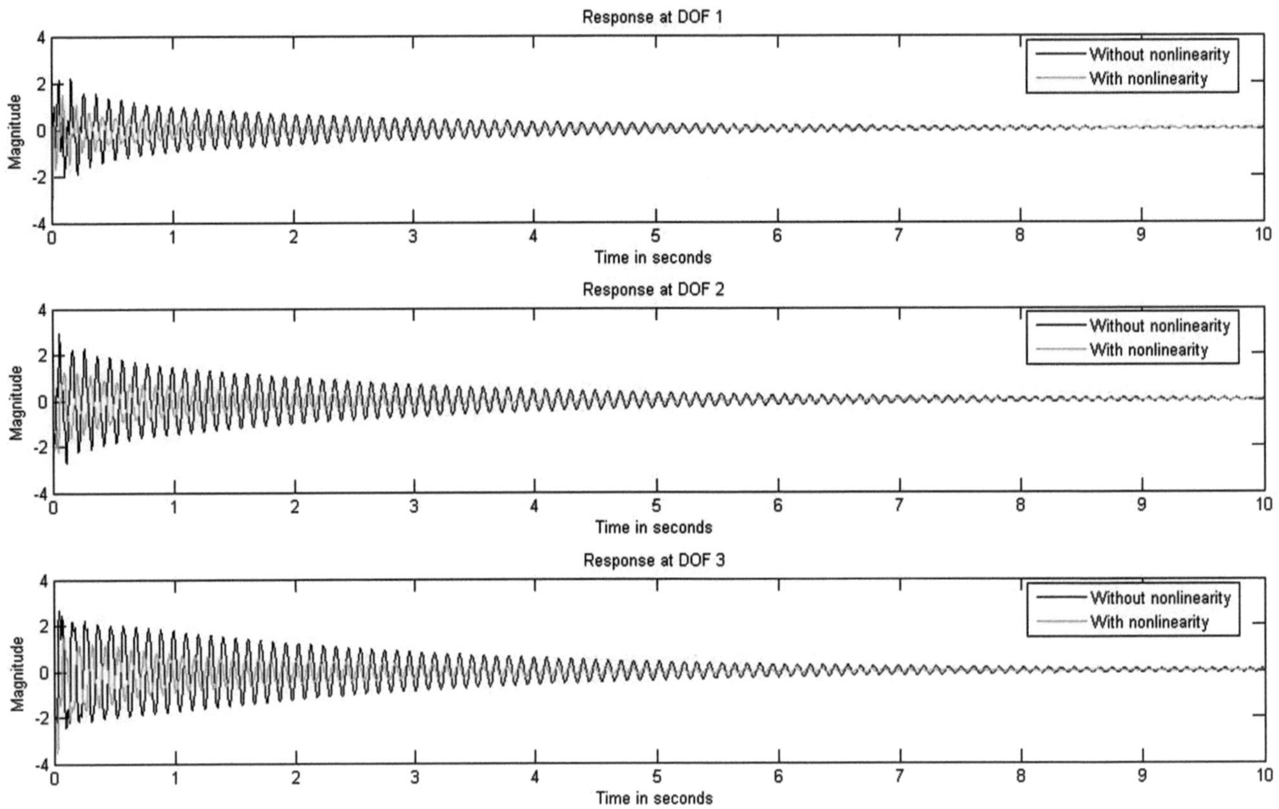

Fig. 18.2 Impulse response of system at three output DOFs for test case I

The coefficients $[a_1]$ and $[a_2]$ are computed from the M, C and K matrices of a light to moderately damped six DOF system. The coefficient $[\beta_1]$ is generated by trial and error such that the overall system is stable and the response of system does not grow exponentially. Also, to simulate real-world measurement, white random noise of the signal-to-noise Ratio (SNR) 10 dB is added to the resulting data series. Then an FFT is applied on the final time series and only the data in the frequency range of 0–128 Hz is inverse Fourier transformed to the time domain from which the modal parameters are estimated using a NARX model.

The resulting response data of the system at three output DOFs, which is used to estimate the modal parameters from the time data series, is shown in Fig. 18.2. After performing the iteration procedure mentioned in Sect. 18.3.3, a pole-density diagram is generated as shown in Fig. 18.3. The clusters of poles around the true pole values, which can be easily noticed from the figure, are zoomed in and plotted in Fig. 18.4. The consistent poles from these clusters are then found on the basis of higher levels of parameter consistency and a statistical evaluation of the resulting set is shown in Table 18.1. From the table it can be noticed that the estimates of modal frequency and damping are not only consistent but also reasonably accurate.

18.4.2 Test Case II

The second test data set is generated to simulate the conditions wherein the measurement noise present on the data is correlated with system output at previous time instants. Therefore, the data is generated by first computing the response of a very lightly damped six DOF system to a random broadband input excitation as shown in Eq. (18.28).

$$\{y(k)\} = \mathcal{F}^{-1}\ [\{H(\omega)\}_{6\times 6}\{F(\omega)\}_{6\times 1}\ ] \tag{18.28}$$

where,

$$H(\omega) = \left[Ms^2 + Cs + K\right]^{-1}\ for\ s = j\omega$$

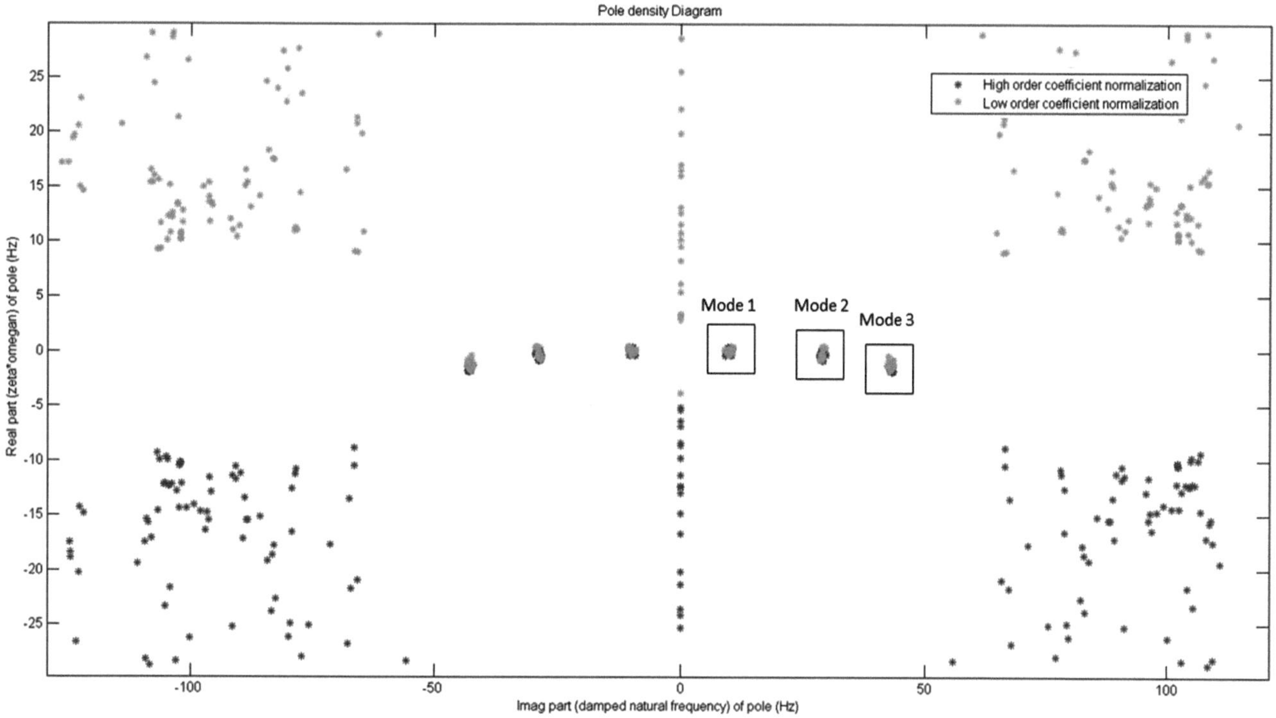

Fig. 18.3 Pole-density diagram generated utilizing NARX model for test case I

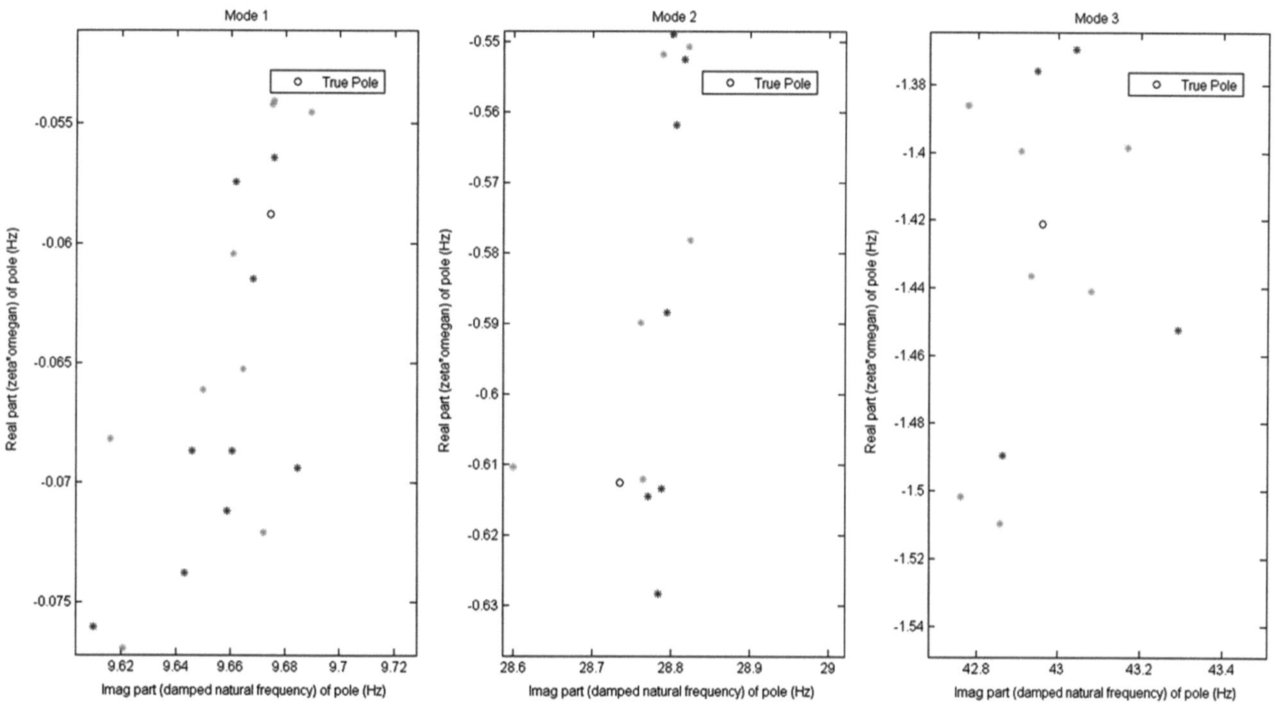

Fig. 18.4 Zoomed pole-density diagram for test case I

$$F(\omega) = \text{Random broadband excitation}$$

Second, correlated noise is added to the response data as shown in Eq. (18.29).

$$\left\{y'(k)\right\} = \{y(k)\} + [\beta_1]\left\{(y(k-1))^2\right\} \tag{18.29}$$

Table 18.1 Statistical evaluation of consistent modal poles estimated in test case I

	True pole (real), (Hz)	Mean of est. poles (real), (Hz)	Variance of est. poles (real), (Hz)	True pole (imag), (Hz)	Mean of est. poles (imag), (Hz)	Variance of est. poles (imag), (Hz)
Mode 1	−0.059	−0.066	0.0053	9.674	9.661	1.070
Mode 2	−0.613	−0.603	0.081	28.735	28.787	2.372
Mode 3	−1.421	−1.434	0.1971	42.958	42.935	1.619

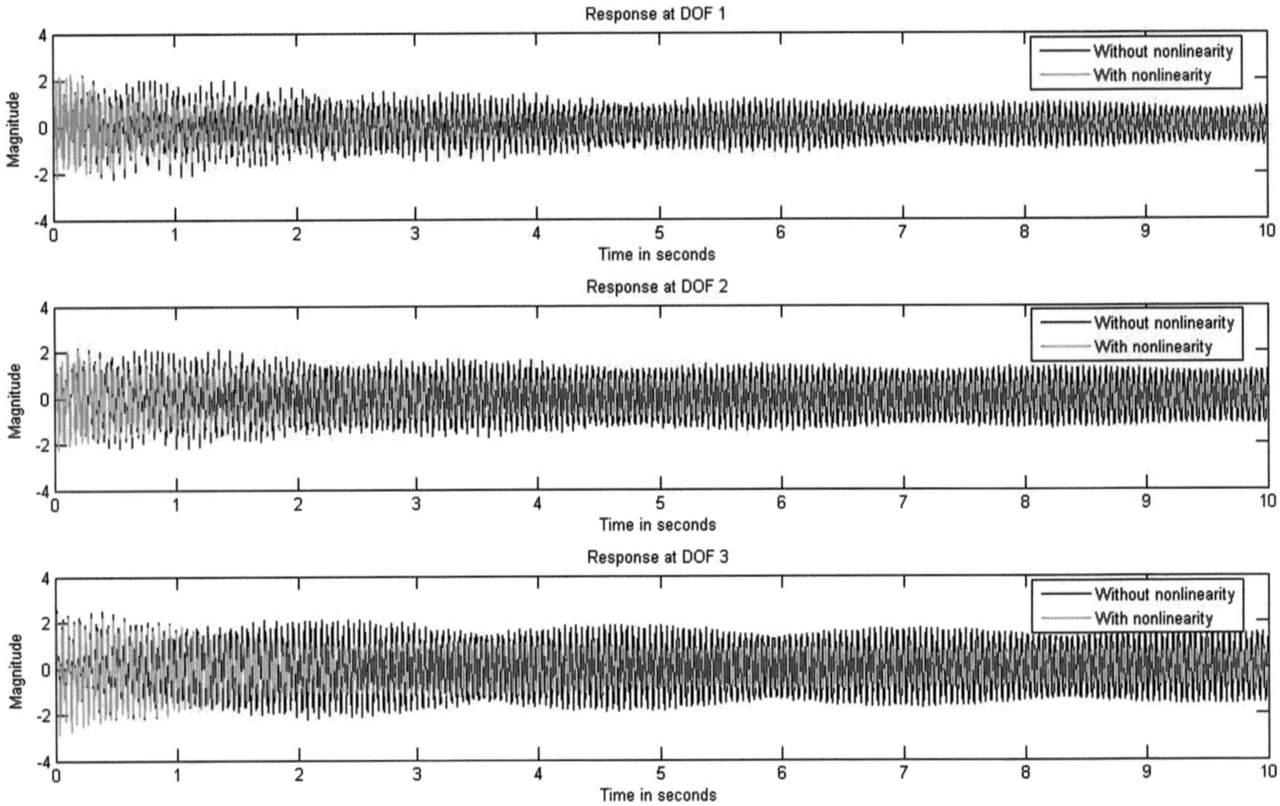

Fig. 18.5 Response of system at three output DOFs for test case II

The coefficient $[\beta_1]$ is chosen at random and scaled such that, along with additional random broadband noise, the resulting SNR of the time series $y'(k)$ is 10 dB. Finally an FFT is applied on the final time series $y'(k)$ and only the data in the frequency range of 0–128 Hz is inverse Fourier transformed for the purpose of MPE using the NARX model.

The response data of the system at the three output DOFs used for the NARX based MPE process is shown in Fig. 18.5. The resulting pole-density diagram and a zoomed-in plot showing the clusters of poles are shown in Figs. 18.6 and 18.7 respectively. As mentioned in the previous test case, a statistical evaluation of the set of consistent modal poles for the second case is presented in Table 18.2. For this case, the implementation of AIC and BIC in identification of the NARX model terms from the structural data is also compared by means of a consistency diagram of the modal poles computed using both the criteria as shown in Fig. 18.8. It can be noticed that all the poles spread over approximately the same space in the complex plane and AIC shows denser clusters for the first two modes.

18.4.3 *Test Case III*

The third test data set is generated to simulate conditions wherein the input, although broadband, random and smooth in the frequency range of interest, also has a certain color characteristic. Further, to check the effect of insufficient spatial excitation on MPE using the NARX model, the excitation is provided at only one input DOF. The system used in this case is the same light to moderately damped system used in the first test case and the response of the system is computed as shown in Eq. (18.30).

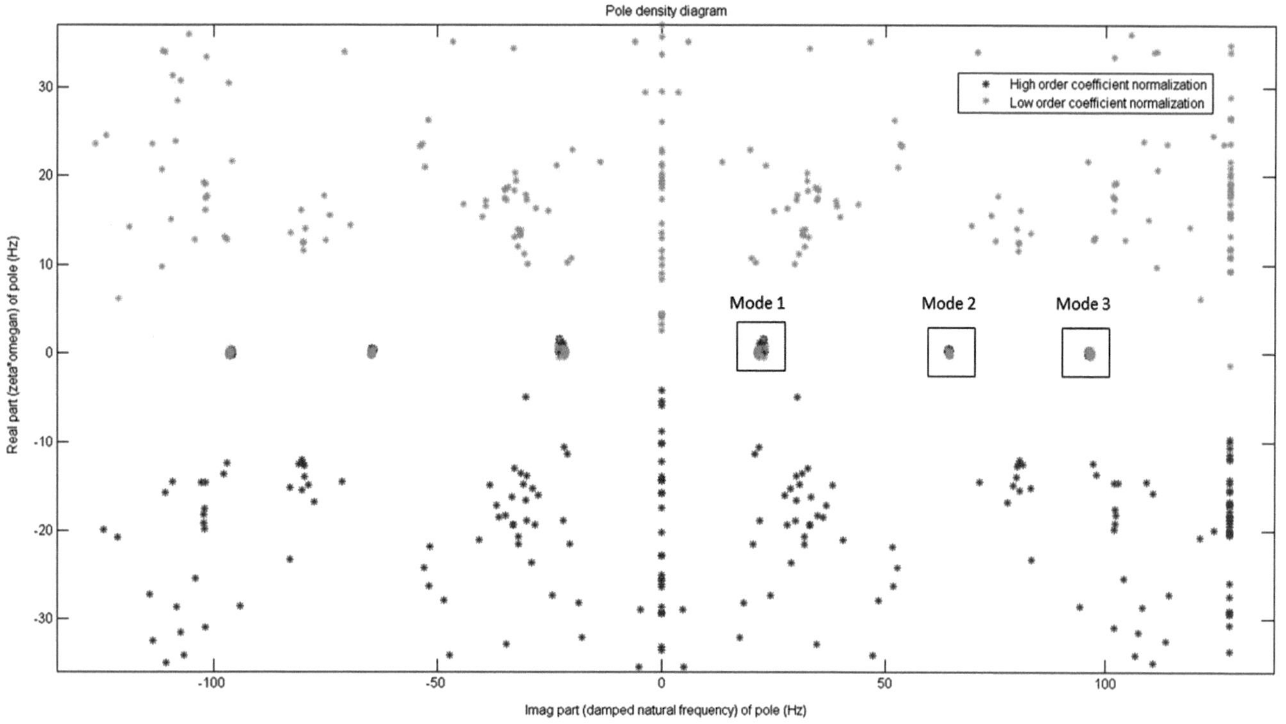

Fig. 18.6 Pole-density diagram generated utilizing NARX model for test case II

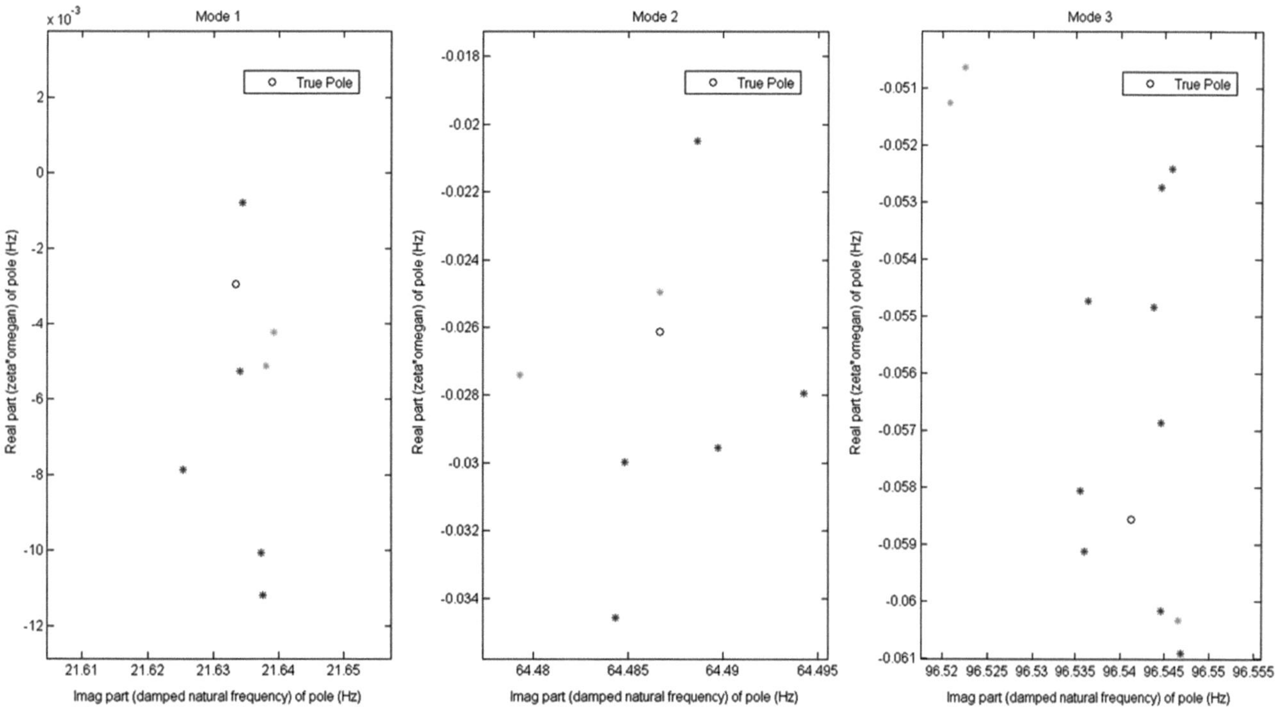

Fig. 18.7 Zoomed pole-density diagram for test case II

$$\{y(k)\} = \mathcal{F}^{-1}\ [\ \{H\ (\omega)\}_{6\times 6}\{F\ (\omega)\}_{6\times 1}\ ] \tag{18.30}$$

where, $F\ (\omega) = [f_1\ (\omega)\ ,\ 0,\ 0,\ 0,\ 0,\ 0]$ for all ω

Table 18.2 Statistical evaluation of consistent modal poles estimated in Test Case II

	True pole (real), (Hz)	Mean of est. poles (real), (Hz)	Variance of est. poles (real), (Hz)	True pole (imag), (Hz)	Mean of est. poles (imag), (Hz)	Variance of est. poles (imag), (Hz)
Mode 1	−0.0029	−0.0051	0.00041	21.633	21.636	2.952
Mode 2	−0.0261	−0.0248	0.0069	64.487	63.482	1.266
Mode 3	−0.0586	−0.0562	0.0033	96.541	96.539	0.989

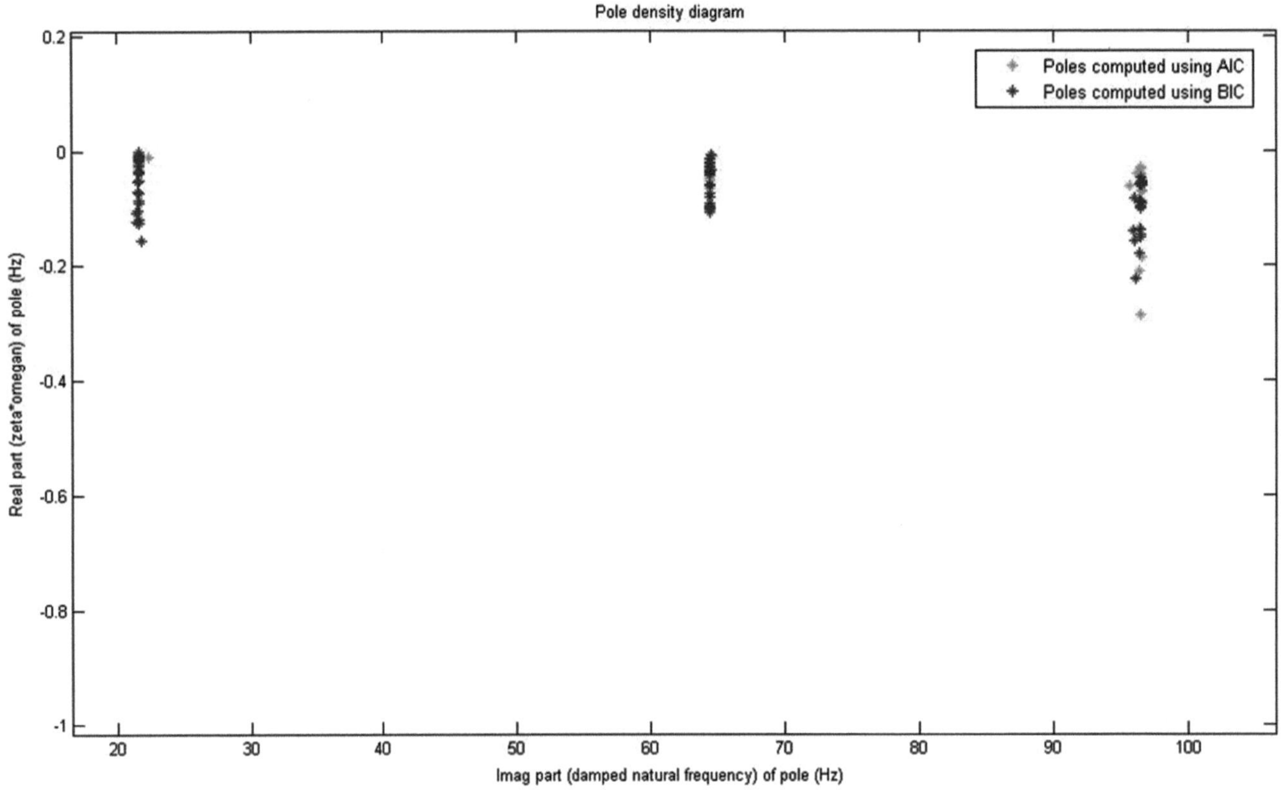

Fig. 18.8 Comparison of use of AIC and BIC in NARX model based MPE

The magnitude and phase of the excitation signal $f_1(\omega)$ in the frequency range of 0–128 Hz is shown in Fig. 18.9 and the resulting response of the system at three output DOFs with an addition of random broadband noise is shown in Fig. 18.10.

The poles are computed from the NARX model based MPE and plotted in the complex plane as shown in Fig. 18.11. The region near the pole clusters of each mode is zoomed in and plotted in Fig. 18.12. It can be noticed from the figure that the clusters around the true pole values are not as dense as in the previous test cases and very few modal poles are found in the left half plane when solved for by normalizing the low order coefficient. This drop in consistency of the modal parameters can also be readily noticed from the statistical evaluation of this set presented in Table 18.3.

18.5 Conclusions

The NARX model based approach, developed in line with the concept of UMPA, for estimating the modal parameters of a system from a short and noisy time series of the output-only data is presented in this paper. The MPE efficiency of the method is tested on three sets of analytically generated data and on the basis of the results shown in Sect. 18.4, the following conclusions can be made regarding this approach:

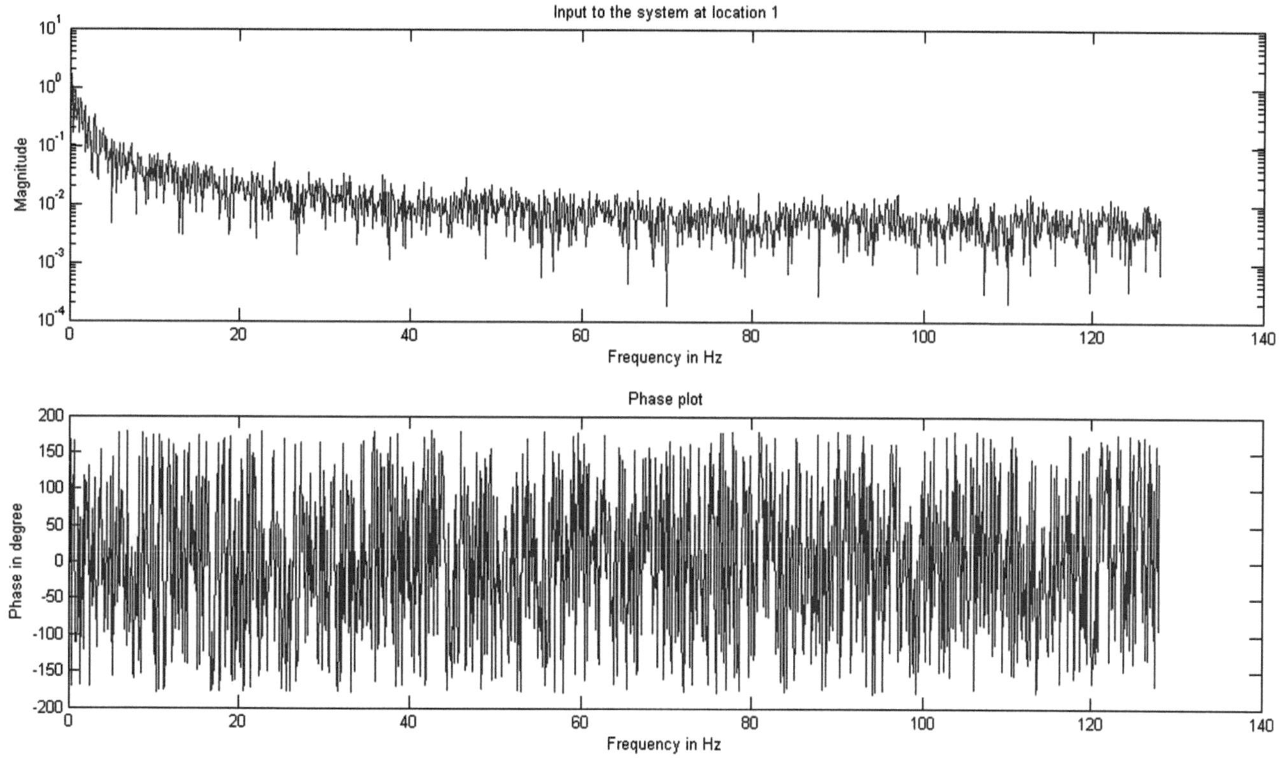

Fig. 18.9 Excitation signal provided to the system at one input DOF for test case III

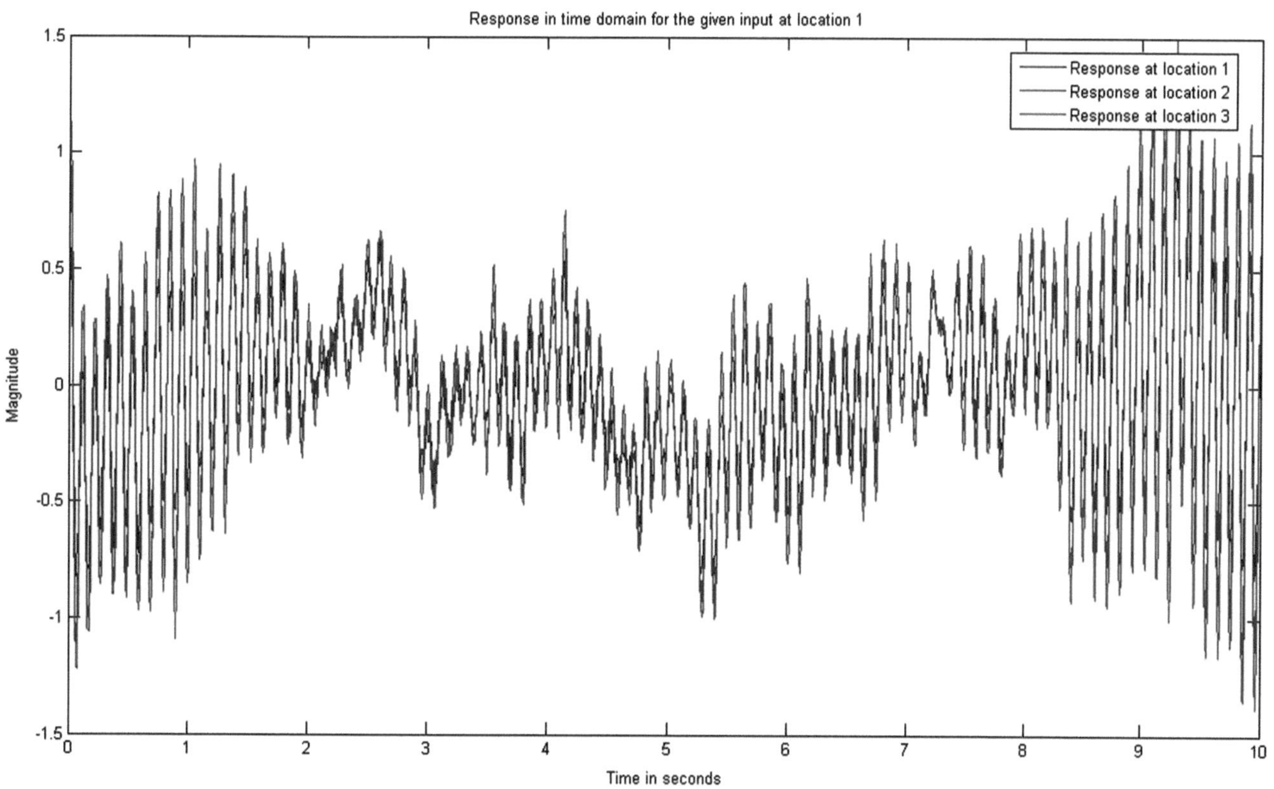

Fig. 18.10 System response at three output DOFs due to colored input for test case III

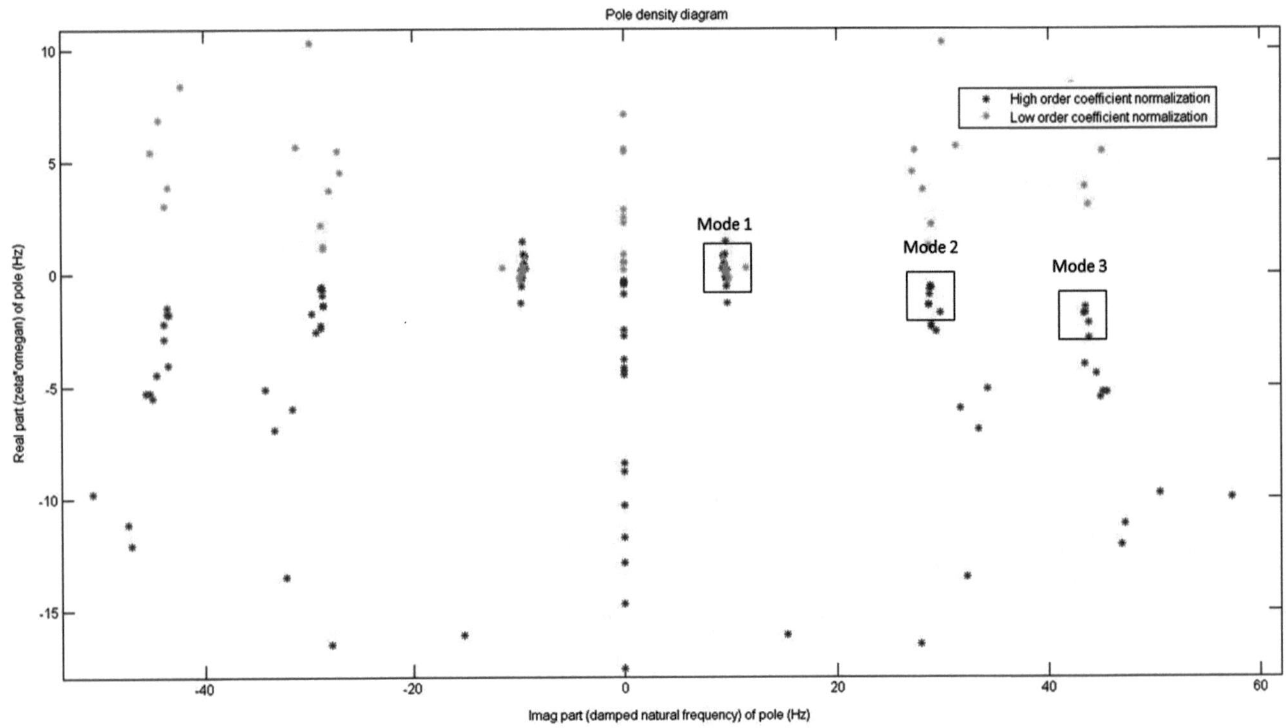

Fig. 18.11 Pole-density diagram generated utilizing NARX model for test case III

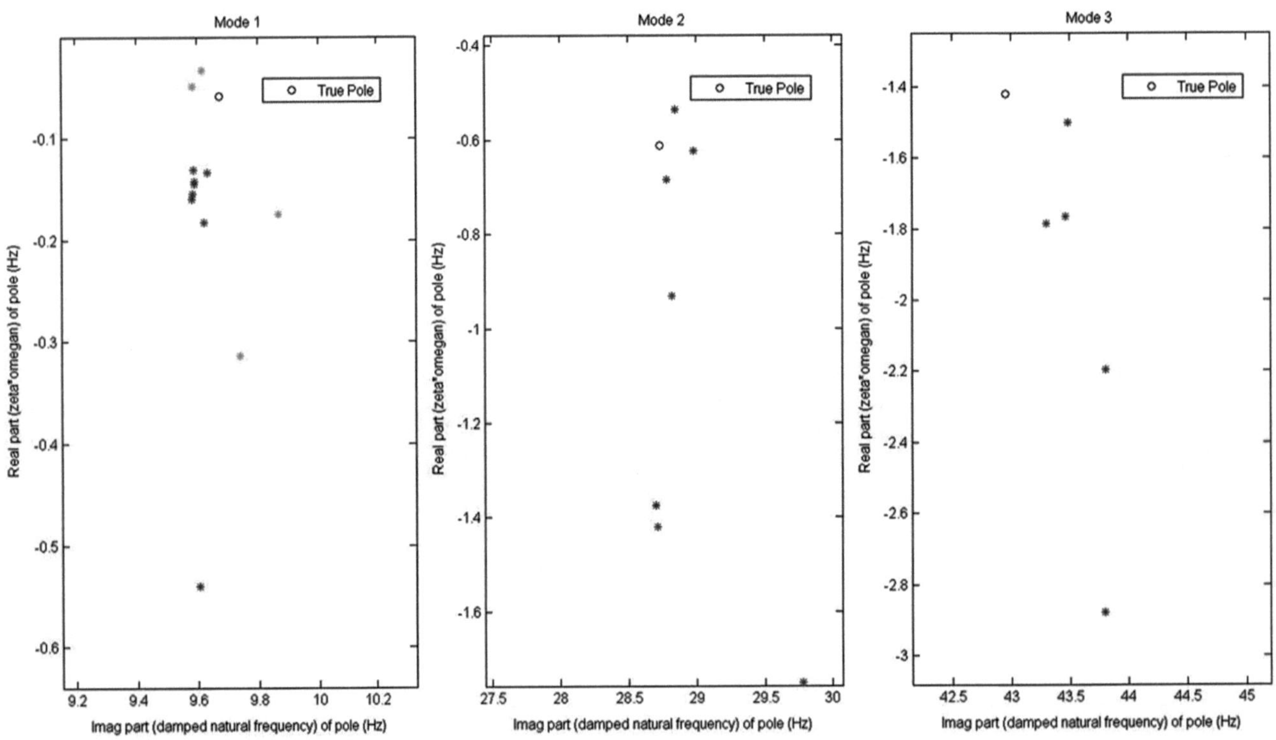

Fig. 18.12 Zoomed pole-density diagram for test case III

- The method is successfully able to describe the colored or correlated or biased noise present in the data series of system's response. Therefore the method yields good estimates of the modal parameters even from short data records on which the previously discussed averaging techniques cannot be performed effectively.

Table 18.3 Statistical evaluation of consistent modal poles estimated in test case III

	True pole (real), (Hz)	Mean of est. poles (real), (Hz)	Variance of est. poles (real), (Hz)	True pole (imag), (Hz)	Mean of est. poles (imag), (Hz)	Variance of est. poles (imag), (Hz)
Mode 1	−0.059	−0.068	0.0098	9.674	9.516	0.629
Mode 2	−0.613	−0.884	0.472	28.735	28.605	1.951
Mode 3	−1.421	−2.003	0.987	42.958	43.490	2.016

- When the system is insufficiently excited spatially by a colored input signal, the precision of estimation is adversely affected and there is a noticeable increase in the variance of the damping estimates. The same can also be noticed from the sparseness of the pole clusters in the pole-density diagram of the third test case.
- AIC and BIC are extended to the identification of the NARX model and in spite of the differences between the two criterion, both have been shown to work well with the structural data buried under noise. Also, since the use of a descriptive or a predictive information criterion does not affect the consistency of the modal parameters of a system as shown for the second test case, iterating the information criterion is justified for generating the pole-stability diagram.

Acknowledgments The authors would like to acknowledge the financial support of The Boeing Company for a portion of this work which has resulted in one Master's Thesis [47].

References

1. Phillips AW, Allemang RJ, Zucker AT (1998) An overview of MIMO-FRF excitation/averaging techniques. In: Proceedings of ISMA international conference on noise and vibration engineering, Katholieke Universiteit Leuven
2. Chauhan S (2008) Parameter estimation and signal processing techniques for operational modal analysis. Ph.D. dissertation, Department of Mechanical Engineering, University of Cincinnati
3. Ibrahim SR (1977) The use of random decrement technique for identification of structural modes of vibration. AIAA paper number 77-368
4. James GH, Carne TG, Lauffer JP (1995) The natural excitation technique (NExT) for modal parameter extraction from operating structures. Modal Anal Int J Anal Exp Modal Anal 10:260–277
5. Ljung L (1999) System identification: theory for the user, 2nd edn. Prentice-Hall, Upper Saddle River
6. Peeters B, De Roeck G (2001) Stochastic system identification for operational modal analysis: a review. ASME J Dyn Syst Meas Control 123
7. Brown DL, Allemang RJ, Zimmerman RJ, Mergeay M (1979) Parameter estimation techniques for modal analysis. SAE paper 790221
8. Ibrahim SR, Mikulcik EC (1977) A method for direct identification of vibration parameters from the free response. Shock Vib Bull 47(4):183–198
9. Vold H, Rocklin T (1982) The numerical implementation of a multi-input modal estimation algorithm for mini-computers. In: Proceedings of the 1st IMAC, Orlando
10. Juang JN, Pappa RS (1985) An eigensystem realization algorithm for modal parameter identification and model reduction. AIAA J Guid Control Dyn 8(4):620–627
11. Richardson M, Formenti D (1982) Parameter estimation from frequency response measurements using rational fraction polynomials. In: Proceedings of the 1st IMAC, Orlando
12. Allemang RJ, Brown DL (2006) A complete review of the complex mode indicator function (CMIF) with applications. In: Proceedings of ISMA international conference on noise and vibration engineering, Katholieke Universiteit Leuven
13. Phillips AW, Allemang RJ, Fladung WA (1998) The complex mode indicator function (CMIF) as a parameter estimation method. In: Proceedings of the 16th IMAC, Santa Barbara
14. Fladung WA (2001) A generalized residuals model for the unified matrix polynomial approach to frequency domain modal parameter estimation. Ph.D. dissertation, Department of Mechanical Engineering, University of Cincinnati
15. Brincker R, Zhang L, Andersen P (2000) Modal identification from ambient responses using frequency domain decomposition. In: Proceedings of the 18th IMAC, San Antonio
16. Brincker R, Ventura C, Andersen P (2000) Damping estimation by frequency domain decomposition. In: Proceedings of the 18th IMAC, San Antonio
17. Allemang RJ, Brown DL (1998) A unified matrix polynomial approach to modal identification. J Sound Vib 211(3):301–322
18. Allemang RJ, Phillips AW (2004) The unified matrix polynomial approach to understanding modal parameter estimation: an update. In: Proceedings of ISMA international conference on noise and vibration engineering, Katholieke Universiteit, Leuven
19. Chauhan S, Martell R, Allemang RJ, Brown DL (2006) Utilization of traditional modal analysis algorithms for ambient testing. In: Proceedings of the 23rd IMAC, Orlando
20. Martell R (2010) Investigation of operational modal analysis damping estimates. M.S. thesis, Department of Mechanical Engineering, University of Cincinnati
21. Vecchio A, Peeters B, Van der Auweraer H (2002) Application of advanced parameter estimators to the analysis of in-flight measured data. In: Proceedings of the 20th IMAC, Los Angeles

22. Dimitridis G (2001) Investigation of nonlinear aeroelastic systems. Ph.D. thesis, Department of Aerospace Engineering, University of Manchester
23. Kukreja SL (2008) Non-linear system identification for aeroelastic systems with application to experimental data. AIAA 7392
24. Leontaritis IJ, Billings SA (1985) Input output parameter models for nonlinear systems. Part 1: deterministic non-linear systems, Part 2: stochastic non-linear systems. Int J Control 41:303–344
25. Billings SA, Chen S (1989) Extended model set, global data and threshold model identification of severely nonlinear systems. Int J Control 50:1897–1923
26. Billings SA, Coca D (2002) Identification of NARMAX and related models. UNESCO encyclopedia of life support systems, Chapter 10.3, 1–11. http://www.eolss.net
27. Chen S, Billings SA (1989) Representations of nonlinear systems: the NARMAX models. Int J Control 49:1013–1032
28. Haber R, Keviczky L (1976) Identification of nonlinear dynamic systems. In: Proceedings of the 4th IFAC symposium on identification and system parameter estimation, Tbilisi, USSR, pp 79–126
29. Billings SA, Leontaritis IJ (1982) Parameter estimation techniques for nonlinear systems. In: Proceedings of the 6th IFAC symposium on identification and system parameter estimation, Washington, DC, pp 427–432
30. Billings SA (1989) Identification of nonlinear systems – a survey. IEEE Proc Part D 127:272–285
31. Korenberg MJ (1985) Orthogonal identification of nonlinear difference equation models. In: Proceedings of Midwest symposium on circuit and systems, Louisville, KY, pp 90–95
32. Korenberg MJ (1987) Fast orthogonal identification of nonlinear difference equation and functional expansion models. In: Proceedings of Midwest symposium on circuit and systems, Syracuse, NY, pp 270–276
33. Korenberg MJ (1989) A robust orthogonal algorithm for system identification and time-series analysis. J Biol Cybern 60:267–276
34. Billings SA, Chen S, Korenberg MJ (1988) Identification of MIMO non-linear systems using a forward-regression orthogonal estimator. Int J Control 50:2157–2189
35. Li K, Peng J, Irwin GW (2005) A fast nonlinear model identification method. IEEE Trans Autom Control 50(8):1211–1216
36. Li K, Peng J, Bai E (2006) A two-stage algorithm for identification of nonlinear dynamic systems. Automatica 42:1189–1197
37. Akaike H (1974) A new look at the statistical model identification. J R Stat Soc Ser B 36:117–147
38. Kass R, Raftery A (1995) Bayes factors. J Am Stat Assoc 90:773–795
39. Schwarz G (1978) Estimating the dimension of a model. Ann Stat 6:461–464
40. Burnham KP, Anderson DR (2004) Multimodel inference: understanding AIC and BIC in model selection. Sociol Methods Res 33:261–304
41. Cavanaugh JE (2009) 171:290 model selection, lecture VI: the Bayesian information criterion. Department of Biostatistics, Department of Statistics and Actuarial Science, University of Iowa
42. Haykin SS (2002) Adaptive filter theory (4th illustrated edn). Prentice Hall, Englewood Cliffs
43. Cavanaugh JE, Neath AA (2012) The Bayesian information criterion: background, derivation and applications. WIREs Comput Stat 4:199–203
44. Hu S (2007) Akaike information criterion. Retrieved Oct 2013, from North Carolina State University: http://bit.ly/OEM2B9
45. Phillips AW, Allemang RJ, Brown DL (2011) Autonomous modal parameter estimation: a. Methodology. b. Statistical considerations. c. Application examples. In: Modal analysis topics, vol 3. Springer, New York, pp 363–428
46. Allemang RJ (2008) Vibrations: experimental modal analysis. Lecture notes, UC-SDRL-CN-20-263-663/664, University of Cincinnati
47. Arora R (2014) Operational modal parameter estimation from short time-data series. M.S. thesis, Department of Mechanical and Material Engineering, University of Cincinnati

Chapter 19
Order Based Modal Analysis Versus Standard Techniques to Extract Modal Parameters of Operational Wind Turbine Gearboxes

S. Manzato, E. Di Lorenzo, A. Medici, F. Vanhollebeke, B. Peeters, and W. Desmet

Abstract Rotating machineries generally operate under very dynamic and complex conditions, during which structural nonlinearities, tonalities or other dynamic related phenomena may arise, affecting the assumptions made in the design phase. Techniques that allow accurately and confidently identifying the dynamic response are then of paramount importance in such complex scenarios. Wind turbine gearboxes are a typical example of such machines, as they operate under strong transient conditions caused by the turbulent and non-stationary wind speed as well as fluctuations in the electrical grid. If models can help predict these interactions, dedicated experiments need to be foreseen to characterize operational response and validate/improve the developed numerical models.

For these reasons, a comprehensive test campaign has been performed at the ZF Wind Power test rig on a multi megawatt gearbox in the framework of a research project. This paper describes the different experiments performed on the gearbox in as many operating conditions as possible (standstill, stationary or transient at different torque levels) as well as the data analysis techniques used to extract a dynamic model from the measured responses. In particular, accelerations acquired in the different conditions are processed using Experimental, Operational or Order Based Modal Analysis. The objective of the paper is to understand how the structural response varies in different operating conditions, and how different methods behave in extracting modal parameters in the analyzed operating conditions.

Keywords Rotating • Gearbox • Operational • Order analysis • Modal analysis

19.1 Introduction

In recent years, increasingly stricter regulations on noise and vibration (N&V) levels are imposing industries to accurately and thoroughly characterize the dynamic response of their machines, being it a washing machine, a car, a wind turbine or an aircraft. As an example, in the wind turbine industry, the installation of new farms quite close to residential areas is making the noise emission levels a matter of public interest and new standards have been developed to regulate and limit as much as possible these emissions. While airborne rotor noise is drastically reduced thanks to new control strategies and tip-speed limitation, the interaction between the drivetrain and the rest of the structure (tower, bedplate and blades) still poses major challenges as it can result in tonal components. In this situation, a root cause analysis requires both the excitation source (drivetrain) and the transmission paths (the rest of the turbine) to be accurately characterized from a dynamic perspective in their operating conditions [1, 2].

Noise and vibration measurements are nowadays mostly performed as quality estimation methods for gear mesh vibrations and or to verify that overall sound power levels comply with the standards. To properly characterize the dynamic behavior of the source and how forces are transmitted to the rest of the structure, however, dedicated and advanced measurements need to be performed. Current practice is to measure vibration levels on the structure, evaluate frequency or order spectra and analyze the so-called Operational Deflection Shape (ODS) at critical frequencies. If this approach generally allows identifying dominant vibration patterns, it fails to characterize whether the high response levels are due to excited resonances, high excitation or a combination of both, limiting the possibility to take corrective actions in the design phase.

S. Manzato (✉) • E. Di Lorenzo • A. Medici • B. Peeters
Siemens Industry Software NV, RTD Test Department, Interleuvenlaan 68, Leuven 3000, Belgium
e-mail: simone.manzato@siemens.com

F. Vanhollebeke
ZF Wind Power Antwerpen NV, G. Mercatorstraat 40, Lommel 3920, Belgium

W. Desmet
KULeuven, Division PMA, Celestijnenlaan 300B, Heverlee 3001, Belgium

M. Mains (ed.), *Topics in Modal Analysis, Volume 10*, Conference Proceedings of the Society for Experimental Mechanics Series, DOI 10.1007/978-3-319-15251-6_19

Also, as the accuracy of simulation models is continuously increasing, experimental data needs to be collected to both validate and improve these models and to extend their validity range. In particular, when dealing with rotating machineries, the resulting dynamics may depend on the applied torque and on the rotational shaft speed, thus a full validation can only be performed if also operational conditions are analyzed. Operational Deflection Shapes, providing only overall levels at specific frequencies, are seldom used as correlation metrics.

For these reasons, an extensive measurement campaign was performed on a prototype multi-MW gearbox at ZF Wind Power on their 13.2 MW test rig facility [3]. By using state-of-the-art measurement techniques, data at different operating conditions was collected with the objective to perform modal analysis not only using classical input–output techniques [4], but also more advanced output-only methods which allow estimating modal models in normal operational conditions where input forces cannot be measured [5]. Besides the full characterization of the gearbox, the performed campaign aimed at validating these advanced analysis techniques on a relevant industrial application case as well as providing guidelines on how these measurement can be optimized to be integrated in the current design process.

This paper is organized as follows: first, a theoretical background on the proposed measurement and estimation methods is presented. The soundness of the different methodology in identifying the modal parameters in the different conditions is demonstrated on a simulated example before applying it on the real measured data. Finally, the different methods will be reviewed and possible improvements proposed.

19.2 Theoretical Background

Classical experimental approaches commonly used to solve noise and vibrations problem can in the past fit into two groups:

- Techniques that attempt to characterize the dynamic properties of the system under study (i.e. Modal Analysis)
- Techniques that attempt to characterize the response signals measured while the machine is in operation (i.e. Order Tracking).

Obviously, techniques that attempt to characterize the dynamic properties of a machine are usually preferred as the processed results have true physical meaning with respect to the dynamic properties of the structure. However, the response of the structure in operating conditions might be significantly different than in the closed and controlled laboratory conditions [6]. For these reasons, significant effort have been put in the last two decades in the development of techniques to identify a modal model of the structure in operating conditions. In this paper, a technique combining Order Tracking and Modal Analysis to identify modal models from operating rotating machineries will be discussed and reviewed.

19.2.1 FRF Estimation from Sine Sweep Experiments

Modal identification of large and dynamically complex structures often requires a multipoint excitation to maximize the energy input. Consequently, sine sweep excitation is applied to concentrate more energy on each line of the frequency spectrum. However, as conventional estimation of FRFs generally requires the excitation signals to be uncorrelated, for multi-point sine sweep excitation several runs with altered excitation force patterns have to be performed [7]. In case of two-inputs excitation, two sweeps are required, respectively with the two sources exciting the structure first in phase and then out-of-phase. As the amplitudes and frequencies are the same, the two sweeps, if taken separately, will results in a correlated input matrix. A typical solution to estimate the MIMO FRFs is then to average over the sweeps as:

$$\left[\{\ddot{u}(\omega)\}_S\ \{\ddot{u}(\omega)\}_A\right] = [H(\omega)]\left[\{f(\omega)\}_S\ \{f(\omega)\}_A\right] \tag{19.1}$$

Where $\{\ddot{u}(\omega)\}_S$ and $\{\ddot{u}(\omega)\}_A$ are the dynamic responses due to the excitation runs with a symmetric force vector $\{f(\omega)\}_S$ and with an anti-symmetric force vector $\{f(\omega)\}_A$. By inverting equation (19.1), using the H1 estimator [8] or the on-line estimation feature available in most acquisition softwares, the FRF matrix H can be obtained. In this case, even though the excitation was chosen to excite either symmetric or antisymmetric modes, the FRFs contain contributions of both symmetric and antisymmetric modes. Consequently, the data analysis may be difficult and less clear.

In some cases and under very strong assumptions, so-called symmetric and anti-symmetric FRFs can also be derived. A simplified approach consists in assuming that the applied force and the contribution from the two inputs at all points are the same. Considering then the two-point excitation and assuming $|F_1| = |F_2|$, the following set of FRFs can be calculated:

$$\begin{aligned} \{\ddot{u}(\omega)\}_S &= [H(\omega)_{1S} + H(\omega)_{2S}]\{f(\omega)\}_S = [H(\omega)]_S\{f(\omega)\}_S \\ \{\ddot{u}(\omega)\}_A &= [H(\omega)_{1A} + H(\omega)_{2A}]\{f(\omega)\}_A = [H(\omega)]_A\{f(\omega)\}_A \end{aligned} \tag{19.2}$$

Where $[H(\omega)]_{i,S}$ and $[H(\omega)]_{i,A}$ are the FRFs between the input i and the considered output from the symmetric and the antisymmetric sweep cases. It can also be observed that, by combining the two equations in Eq. 19.2 and assuming that the absolute values of the input forces from the 2 shakers are the same, the MIMO FRF (Eq. 19.1) can be reconstructed as:

$$\begin{aligned} [H(\omega)]_1 &= 0.5\,[H(\omega)_S + H(\omega)_A] \\ [H(\omega)]_2 &= 0.5\,[H(\omega)_S + H(\omega)_A] \end{aligned} \tag{19.3}$$

Once the SIMO (Eq. 19.2) or MIMO (Eqs. 19.1, 19.2, and 19.3) FRFs are derived, standard modal identification algorithms can be applied to derive the modal model [4, 8]. In general, while the MIMO FRFs allow identifying in a single analysis all the modes, from the SIMO case only the symmetric or antisymmetric modes can be identified. The two models can then be of course put together to obtain the complete modal model.

19.2.2 *Operational Modal Analysis*

Experimental Modal Analysis techniques using standard input–output measurement and identification methods are widely used to estimate a modal model of the structure which is then used on one side to dynamically characterize the response of the structure and verify the agreement with the design assumptions and on the other side to correlate and update numerical models. However, during operation, the system properties may be different from those applied in the lab and the dynamic response may vary significantly. For these reasons, in the last 20 years Operational Modal Analysis methods have been developed to identify the structure in its operating conditions, without having to apply and measure a known excitation but relying on the environmental one. In [5] a reference description of the method, assumptions and processing steps are presented. Briefly, the weighted correlogram approach is used, to compute the power spectrum form the estimated correlation functions as:

$$S_{yy}(\omega) = \sum_{k=-L}^{L} w_k R_k \exp(-j\omega k\Delta t) \tag{19.4}$$

Where L is the total number of time lags used for the estimation of the correlations. As the correlation samples at negative time lags ($k < 0$) contain redundant information, it suffices to consider only the positive ones when computing the spectra. This leads to the so-called half spectra, of which even autospectra have a non-zero phase:

$$S_{yy}^{+}(\omega) = \frac{w_0 R_0}{2} + \sum_{k=1}^{L} w_k R_k \exp(-j\omega k\Delta t) \tag{19.5}$$

The power spectra can then be used for the estimation of the operational modal parameters using the following representation:

$$\left[S_{yy}^{+}(\omega)\right] = \sum_{i=1}^{N} \frac{\{\upsilon_i\}\langle g_i\rangle}{j\omega - \lambda_i} + \frac{\{\upsilon_i^*\}\langle g_i^*\rangle}{j\omega - \lambda_i^*} \tag{19.6}$$

where λ_i are the system poles (from which natural frequencies and damping are derived), $\{\upsilon_i\}$ are the mode shapes and $\langle g_i\rangle$ the operational reference factor, which replace the modal participation factors in case of output only data.

19.2.3 *Order Tracking*

The most commonly used methods for characterizing the response signal acquired on operating rotating machinery fall under the classification of order tracking. Order tracking is the estimation of the amplitude and phase of a response of a machine in function of its operational speed. Several techniques have been developed in the last decades to perform order tracking and two examples are discussed here.

The first method is a constant angle order tracking. In general, the analysis time interval to adequately characterize and extract orders should be large for low rotational speed and reasonably small for higher speeds. By resampling the acquired signal in the angle domain, a constant interval can be used as the angular increment per revolution is constant. The algorithm relies on a reference tacho signal to track the instantaneous rotational speed and perform the angular transformation. When DFT or FFT is used to process the resampled signals, the spectral analysis lines will represent constant-order components since the transform is based on angular domain rather than time domain. The order resolution can be found similarly to the frequency domain resolution of the normal Fourier transform. Defining o_s as the angular sampling rate and $\Delta\theta$ the angular increment, with of course $o_s = 1/\Delta\theta$, and N the total number of angular samples considered, the order resolution is given by:

$$\Delta o = \frac{1}{N \cdot \Delta\theta} = \frac{o_s}{N} \tag{19.7}$$

Also, similarly to the frequency domain, the angular Nyquist rate (or maximum order) that can be processed is defined as:

$$o_{Nyq} = \frac{o_s}{2} \tag{19.8}$$

An alternative methodology to extract orders is the Time Variant Discrete Fourier Transform (TVDFT), which is a combination of tacho-based resampling and DFT algorithms, with the advantage of significantly reducing the required computational time. The method is based upon a DFT with a kernel whose frequency varies as a function of time defined by the rpm of the machine. This kernel is a cosine or sine function of unity amplitude with an instantaneous frequency matching the one of the tracked order at each instant in time. It is generally formulated as:

$$a_n = \frac{1}{N}\sum_{n=1}^{N} x(n\Delta t)\cos\left(2\pi\int_0^{n\Delta t}(o_n \bullet \Delta t \bullet rpm/60)\,dt\right) \tag{19.9}$$

$$b_n = \frac{1}{N}\sum_{n=1}^{N} x(n\Delta t)\sin\left(2\pi\int_0^{n\Delta t}(o_n \bullet \Delta t \bullet rpm/60)\,dt\right) \tag{19.10}$$

where o_n is the analyzed order, a_n and b_n are the Fourier coefficients of the cosine and sine terms for o_n, $x(n\Delta t)$are the data samples, N is the number of data samples within an analysis block. The main difference with the previous method is that the order resolution varies as the rotational speed increases. However, the method is computationally more efficient and, in general, is able to more accurately track the phase of the order [9].

19.2.4 Order-Based Modal Analysis

The relationship between the structure modal model and an order can be investigated assuming the structure is excited by an unbalance-like force rotating at increasing frequency [10]. This can be modeled by two perpendicular forces of equal amplitude and in quadrature. The measured response $Y(\omega)$ in the frequency domain is:

$$Y(\omega) = H_{(:,fx)}(\omega)\,F_x(\omega) + H_{(:,fy)}(\omega)\,F_y(\omega) \tag{19.11}$$

where F represents the force and H the corresponding column of the transfer function matrix. Taking into account the relation between the two correlated forces, the response of the structure can be written as (considering only the positive frequency axis):

$$Y(\omega) \propto \omega_0^2\left(H_{(:,fx)}(\omega) - jH_{(:,fy)}(\omega)\right)\delta(\omega - \omega_0) \tag{19.12}$$

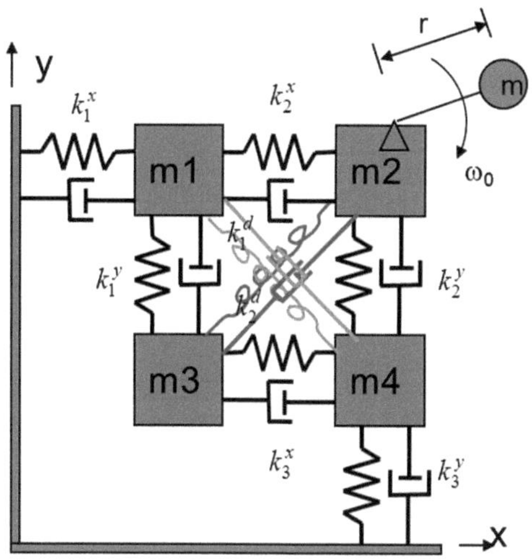

Fig. 19.1 Schematic representation of the anlytical model with rotating excitation

where ω_0 is the rotation speed. From this equation, it is clear that the measured output is proportional to the squared rotation speed and to a complex combination of two structural FRFs related to x and y excitation. A structural FRF can be decomposed in a modal sense as:

$$H_{(:,\bullet)}(\omega) = V(j\omega I - \Lambda)^{-1}L_\bullet + \frac{1}{\omega^2}LR_\bullet + UR_\bullet \tag{19.13}$$

where V, Λ, L are the complex-valued modal parameters (respectively, the mode shape matrix, the diagonal matrix containing the complex poles and the modal participation matrix). $LR_\bullet$ and $UR_\bullet$ are the real-valued lower and upper residuals, modeling the influences of the modes outside the considered frequency band. Finally, combining Eqs. 19.12 and 19.13 we obtain:

$$Y(\omega) \propto \omega^2\Big(V(j\omega I - \Lambda)^{-1}\left(L_x - jL_y\right) + \frac{1}{\omega^2}\left(LR_x - jLR_y\right) + \left(UR_x - jUR_y\right)\Big) \tag{19.14}$$

Equation 19.14 shows that modal analysis can be applied to displacement or acceleration orders taking into account the following statements:

- Displacement orders are proportional to the squared rotating speed and, therefore, acceleration orders are proportional to the 4th power of the rotating speed. In classical modal analysis, acceleration FRFs are proportional to the squared frequency axis.
- Complex upper and lower residuals, while in classical modal analysis they are real.
- Complex participation factor; also in classical modal analysis they can be complex.

It can be concluded that methods like PolyMAX [4] that estimates a right matrix-fraction model are robust against these statements, but methods working on the pole-residue formulation like the LSFD (Least Squares Frequency Domain) method need a new formulation in order to be applied.

19.3 Numerical Validation

To validate the proposed methodologies, an analytical example was built to demonstrate the ability of the different methods to extract the modal model of the structure during run up excitation. The model is shown in Fig. 19.1: it was an 8dof planar system with 4 masses. Forces were applied on mass 2. To simulate the effect of a rotating force, a rotating mass equivalent force was defined. The analytical matrices are converted to a discrete state space model and the transient acceleration responses to the rotating force were calculated. The unbalance-like rotating forces were defined as:

$$\begin{aligned} f_x &= r \cdot m \cdot \omega_0^2 \cos(\omega_0 t + \varphi) \\ f_y &= r \cdot m \cdot \omega_0^2 \sin(\omega_0 t + \varphi) \end{aligned} \tag{19.18}$$

Table 19.1 True modal parameters of the system in Fig. 19.1

Frequency [Hz]	Damping [%]
6.690	1.135
14.785	1.454
18.447	3.387
23.765	3.125
30.883	3.883
34.622	4.040
36.845	3.404
40.956	5.780

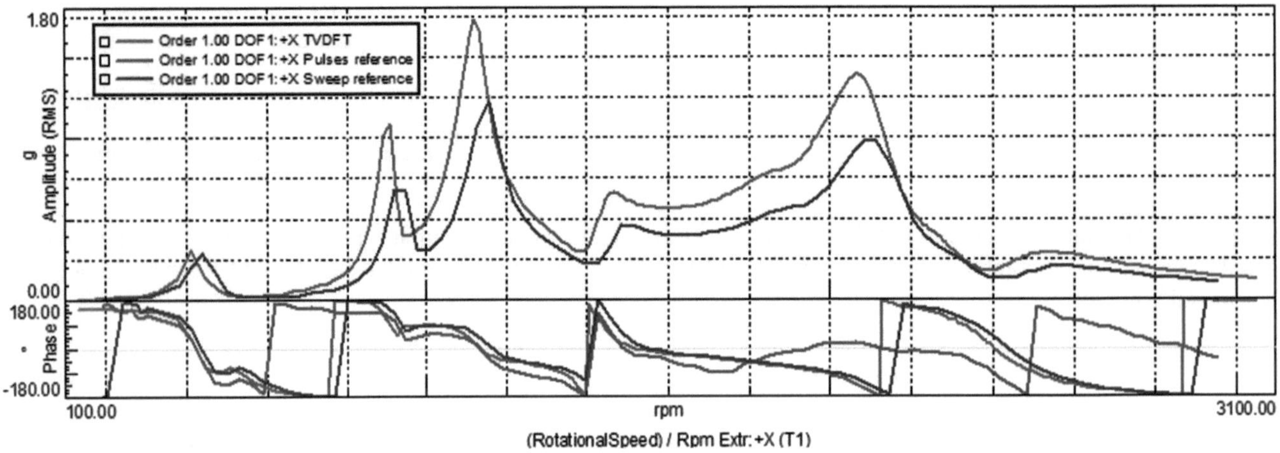

Fig. 19.2 Comparison of different order extraction methods

The response of the structure was simulated for a sweep over 120 s from 0 to 50 Hz. First, PolyMAX was applied on the FRFs calculated using as references the two forces in Eq. 19.18 and the frequency and damping values are listed in Table 19.1. These results are considered the reference for all other methods, as they are exactly the same as those obtained using the system matrices of the numerical model.

As the model was generated to replicate operational measurements, the knowledge of the input forces was then neglected and modal models are derived using the simulated acceleration signals only. Firstly, the data was processed into the power spectral matrix and Operational Modal Analysis was applied. The selected reference signals are the x and y accelerations on mass 2 in Fig. 19.1. In the case of a sine sweep (or a multi-sine sweep if different orders are excited), the excitation can be considered as a broad-band white noise in the frequency range of the sweep. However, as discussed in [10], the so-called "end of order" effect might arise: if any of the orders ends in the frequency range considered for the identification, a spurious peak will show in the spectra and it will be treated by the identification algorithm as a pole. In this case, however, as the excitation order ends at 50 Hz and all modes are below this value, the end-of-order pole falls out of the band of interest and does not pose any problem.

To apply Order Based Modal Analysis (OBMA), orders need to be extracted from the data. As all methods discussed in Sect. 19.2.3 requires the rotational speed as an input, an algorithm was used to extract it from a Time-Frequency diagram. For the constant angle order tracking method, also the tacho moments are required. As the so-called raw-tacho containing this information is generated by the software LMS Test.Lab during the acquisition, even though the data are simulated it needs to be estimated as well using the functions available in the software. For the TVDFT method, on the other end, only the rotational speed is needed. Compared to other applications of order tracking, OBMA requires that the phase is calculated with respect to a reference signal which is synchronous to the excitation (Eq. 19.12). In this case, both a sine sweep with frequency equal to the instantaneous rotational speed and the pulse train signal generated by a tacho sensor were used. An example of the extracted orders is given in Fig. 19.2. The amplitudes estimated with the two methods are quite different, mostly because of the size of the window used to extract the orders. However, it is the frequency shift which is more critical. The error is introduced after the rpm estimation, when additional steps are to be performed to derive also the tacho moments. Because of this, some kind of delay is introduced on the estimated rpm signal and not on the acceleration data, so that the result is a shifted frequency spectrum of approximately 0.6 Hz. TVDFT, relying only on the rpm estimation, is not affected; also, if the rotational speed was measured and the constant angle order tracking performed, this error would not appear.

Table 19.2 Overview of modal parameter identification results with the different techniques

OMA sweep			OBMA TVDFT			OBMA Angle pulse ref			OBMA angle sweep ref		
Freq.	Damp.	MAC	Freq.	Damp.	MAC	Freq.	Damp.	MAC	Freq.	Damp.	MAC
6.69	1.13	99.9	6.89	2.35	0.98	7.21	1.27	0.92	7.11	3.15	0.99
14.78	1.45	1.00	14.88	2.56	0.99	15.4	2.01	0.99	15.45	1.62	0.99
18.44	3.39	0.99	18.54	3.37	0.99	19.07	3.05	0.99	19.05	3.31	0.99
23.75	3.12	0.99	23.85	3.12	0.99	24.28	2.73	0.99	24.37	3.06	0.99
30.86	3.88	0.77	30.95	3.93	0.87	31.48	1.43	0.12	31.47	3.94	0.51
34.59	4.05	0.99	34.67	4.04	0.99	35.08	3.51	0.82	35.21	4.07	0.92
36.82	3.41	0.96	36.92	3.34	0.97	37.36	2.22	0.71	34.4	3.27	0.82
40.89	5.78	0.97	40.93	5.75	0.98	41.73	3.64	0.87	41.46	5.64	0.82

The results obtained for the different method are summarized in Table 19.2, where the MAC values are computed with respect to the true modes calculated assuming the input forces are known (Table 19.1). OMA and OBMA applied on TVDFT orders gives results which are very consistent with the reference ones. Constant angle based orders suffer from the frequency shifts related to the preprocessing steps applied to estimate the tacho moments. The effect of the reference signal used is however reflected in the results, with the sweep giving much better results than the pulse train signal. These results were then used used as a starting point for the analysis of the data acquired on the wind turbine gearbox that are discussed in the following section.

19.4 Operational Gearbox Analysis

With the main objectives of characterizing the gearbox dynamic response in different operating conditions and obtain experimental data for model validation and updating, a measurement campaign took place on the 13.2 MW dynamic test rig at ZF Wind Power in Lommel, Belgium (Fig. 19.3). Using this test rig, gearboxes can be tested under representative loading condition using parameterized load cases that can be programmed into the test rig controller. In this case, the following scenarios were tested:

- Shaker sine sweep during standstill and in stationary conditions at 1,200 rpms;
- Stationary conditions at 1,200 and 800 rpms;
- Run up from 200 to 1,500 rpms.

The operational conditions (stationary speed and run ups) were all performed under different torque loading (33 %, 66 % and 100 % of nominal torque).

As can be observed in Fig. 19.3 (right), an extensive grid of 250 points was measured using triaxial accelerometers. To measure all points, the whole test schedule was repeated seven times, roving the majority of the accelerometers while leaving a small group at fixed locations to allow data merging for a more efficient processing.

19.4.1 Gearbox Signature Analysis

Before analyzing the data and try to extract modal models by processing them into FRFs, Power Spectra or Orders, the typical signature of the gearbox and, specifically, of the gearbox installed on the test rig will be analyzed. This will help better understanding the results that will be obtained in the next section. Examples of Time-Frequency plots during a speed run up and stationary operation at rated speed are shown in Fig. 19.4.

For a gearbox in operating conditions, the response is typically dominated by frequency components that are multiples of the fundamental rotational speed and the main source of internal excitation. By knowing the gear ratios between the different stages and the number of teeth meshing at each stage, all these components can be characterized. Also, on top of these frequencies, components coming from pumps, cooling systems and other auxiliaries are also appearing in the response. These characteristics are visible in both diagrams in Fig. 19.4. During stationary operations, the rotational speed is constant as well as all its multiples, so that all orders appear as peaks at discrete frequencies. During run ups, as the rotational speed is increased, also all orders vary and they appear as diagonal lines.

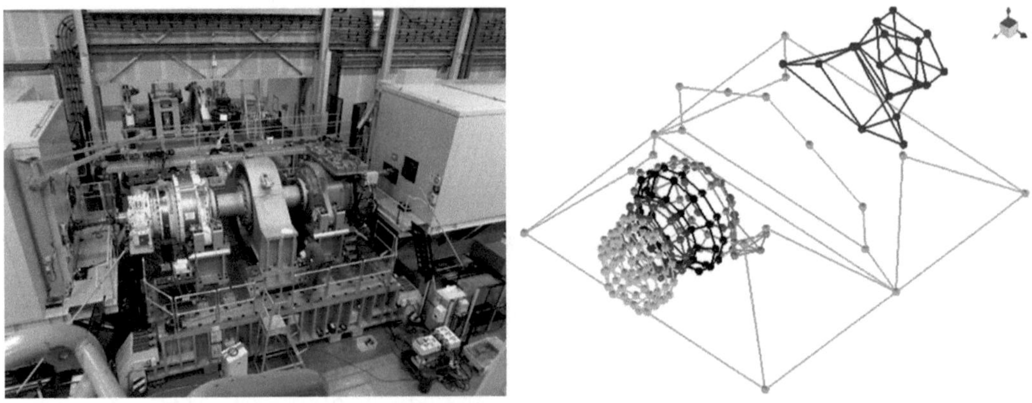

Fig. 19.3 ZF Wind Power dynamic test-rig with the two gearboxes in back to back configuration (*left*) and experimental measurement points (*right*)

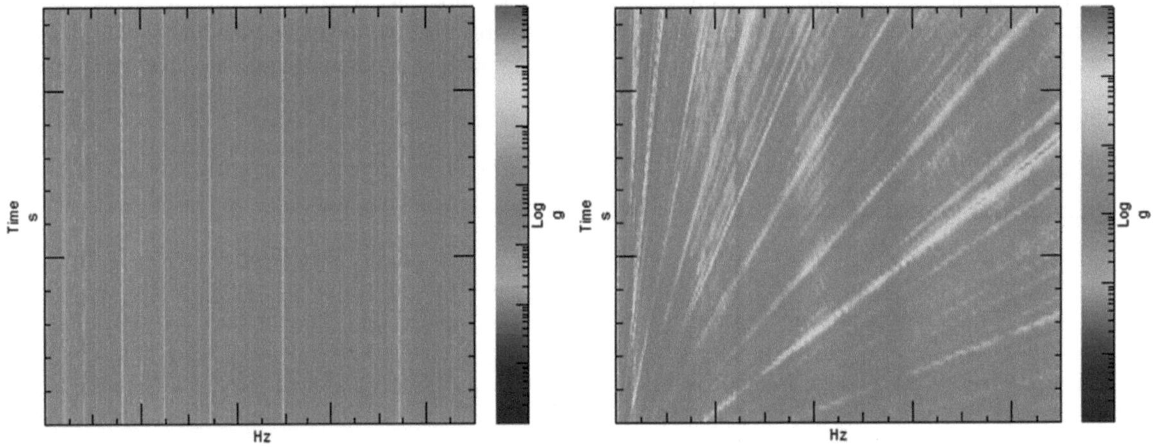

Fig. 19.4 Signature of the gearbox on the test rig in different operating condition. *Left*: stationary speed. *Right*: Run up

What complicates the analysis of the data for this particular case is that on the test rig two gearboxes are tested in a back to back configuration. As in this case the gearboxes were still prototypes, the two tested one had slightly different gear ratios, thus their rotational speeds were different. As vibrations transfer through the test rig and the main shaft from one gearbox to the other, all orders appear in couples at slightly different frequencies. This, as will be discussed, represents an additional challenge when processing the data to derive a modal model.

19.4.2 Shaker Data Analysis

First of all, the data acquired while exciting the gearbox with two inertial shakers are analyzed. The two used shakers were not identical and no force cell was placed in between the shakers and the structure to measure the applied forces. However, by knowing the mass of the shakers and measuring their acceleration, an estimate of the forces was calculated. The two drive signals were defined to obtain two frequency sweeps with the same force magnitude but with the shakers first in-phase and then out-of-phase, as described in Sect. 19.2.1. However, as no online estimation was performed, acquired signals had to be converted to FRFs a posteriori.

In Fig. 19.5 an example of the estimated forces during one of the measured sweeps is shown both in time and in frequency domain. Although the drives were calculated to obtain the same force amplitude on both exciters, measured accelerations (converted to forces) show some small differences. Also, peaks in the time and frequency plots may indicate that the measured acceleration is not only coming from the shaker but also from an interaction with the structure.

The measured accelerations and estimated forces were then processed into system FRFs. From Eq. 19.2 the symmetric and antisymmetric FRFs were derived for the SIMO case, while both Eqs. 19.1 and 19.3 were used for the MIMO FRFs,

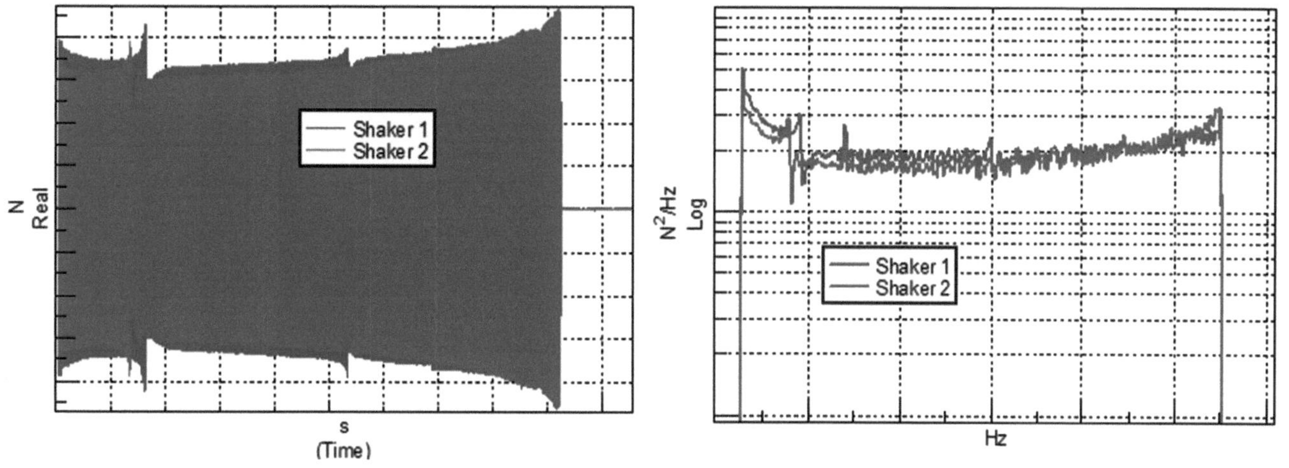

Fig. 19.5 Estimated force signals for the two shakers during a sine sweep in time and frequency domain

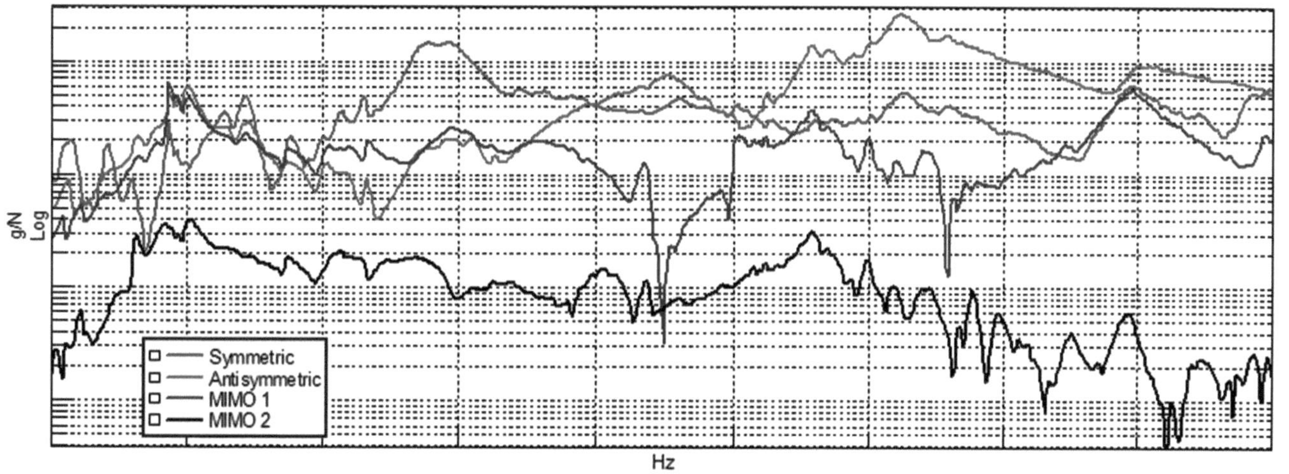

Fig. 19.6 Computed symmetric, antisymmetric and MIMO FRFs for the same point

giving, as expected, the same results. An example of the different FRFs for the same points is shown in Fig. 19.6. The low frequency dynamics is comparable for all FRFs, while less correspondence was generally found at higher frequencies, where however modes are not as well excited because of the low power of the two shakers. Also, the significant difference in level between the two MIMO FRFs is related to location of the analyzed sensor, much closer to shaker 1 than 2.

As they are supposed to be the "true" FRFs, the MIMO data was processed first. All FRFs were processed together but no clear stable poles were identified and the corresponding mode shapes were not clear. As a consequence, the identified modal model couldn't be considered reliable. On the other end, processing the symmetric and antisymmetric FRFs resulted in much clearer and stable poles. After analyzing the results, it was conclude that:

- in the MIMO case, the inability of the identification algorithm of identifying a reliable modal model can be related to a combination of small errors (non-equal forces, different forces in the different runs, see Fig. 19.5);
- in the SIMO case, all these errors are influencing less the results as the two analysis are performed independently;
- SIMO FRFs are derived under very strong approximations that need to be considered when interpreting the results.

Next to the standard FRF measurements at stand-still, a sine sweep was also applied at the gearbox when operating at constant rated speed. This method, generally called OMAX (Operational Modal Analysis with eXogenous excitation) [11] combines the ambient excitation with a known one, allowing to estimate FRFs and the modal parameters in operating conditions. An example of the signal measured in these testing conditions is shown in Fig. 19.7, while an example of the estimated FRFs is shown in Fig. 19.8. In the time-frequency representation, the sine sweep is barely visible on top of the background excitation. Despite the presence of the shakers, the response is still dominated by the rotation frequency and its

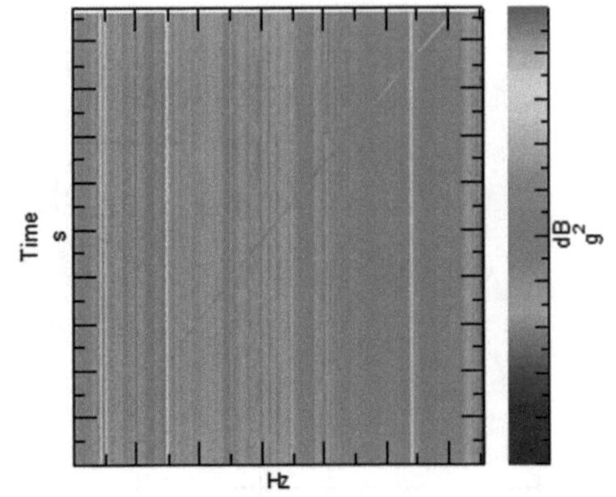

Fig. 19.7 OMAX test data with the applied sine sweep is barely visible on top of the operational excitation

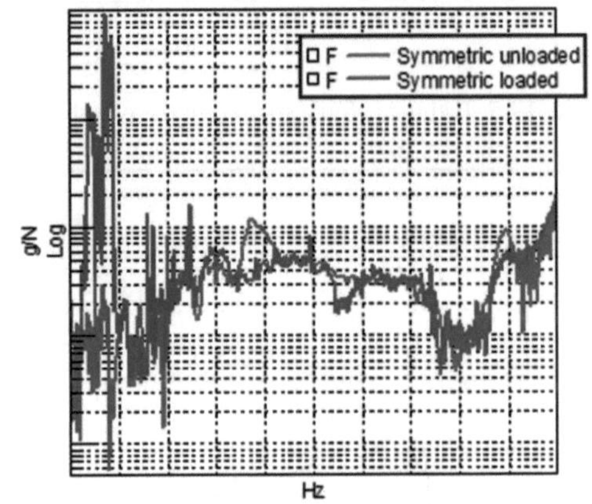

Fig. 19.8 Comparison of loaded and unloaded FRFs in the same conditions

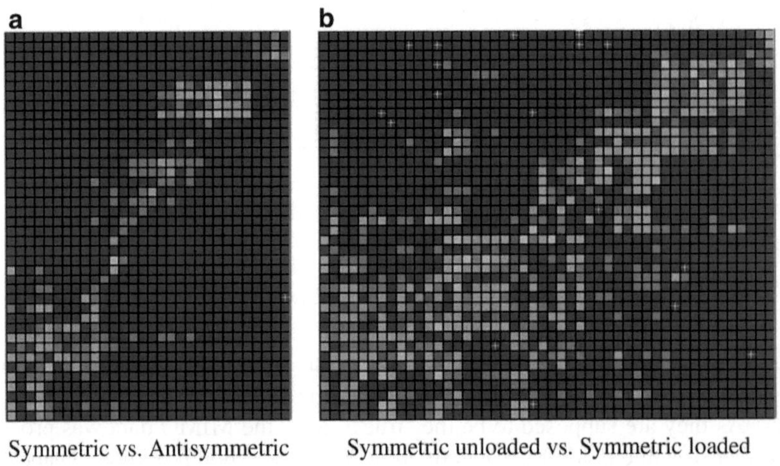

Fig. 19.9 Modal assurance criterion between modal model from different sets of FRFs

multiples, as discussed in Sect. 19.4.1. This reflects In a very poor coherence, in particular at lower frequencies, and in the persistence of the rotor harmonics also in the estimated FRFs.

After analyzing the calculated FRFs, the mode shapes obtained from the symmetric and antisymmetric FRFs at standstill and the symmetric OMAX FRFs are compared in Fig. 19.9 using the MAC. In both cases, modes are quite different as:

- Symmetric and antisymmetric sweeps excite, as expected, different modes, in particular if the structure has some geometrical symmetry;
- standstill conditions are calculated without applied torque, so the gear contact configuration is not repeatable. This problem becomes significant when comparing standstill and operational conditions;
- similarly to OMA, OMAX assumes the background excitation to be stochastic white noise in the frequency band of interest. However, as already discussed, the gearbox signature is dominated by harmonic components that also appear in the estimated FRFs, making the analysis much more complex.

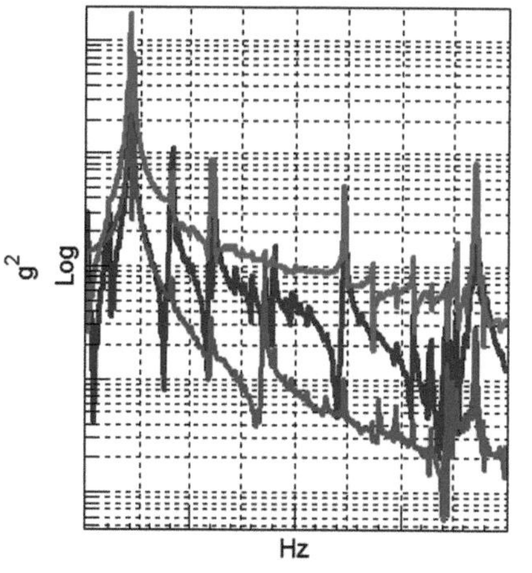

Fig. 19.10 Calculated cross powers during stationary operating conditions

- The MAC diagrams in Fig. 19.9 are shown for a maximum value of correlation of 0.4. Although this is typically considered a very small correlation level, all modes were also visually compared to verify the similarity also from a qualitative perspective. In particular, it was observed that also modes with a 0.2 MAC value where actually showing the same global deformation. This is probably due to the high number of points, the complexity of the test and the measurement noise which is propagated to the results.

19.4.3 Operational Data Analysis

Data in operational conditions were acquired at constant rotational speed as well as during run ups at different torque levels. In this case only data at 100 % torque is discussed. Fig. 19.10 shows a set of cross powers calculated from measured accelerations at constant rated rotational speed. All peaks in the spectrum correspond to harmonic components of the shafts rotational speeds and the gear mesh frequencies at the different stages. Because of the high density of discrete frequencies (and they all appear in pairs as the two gearboxes have different gear ratios) recognizing and thus identifying modes is almost impossible. Furthermore, due to the high number of these rotor harmonics and gear mesh frequencies and their dominance in the response spectrum, application of harmonic filters will not help in identifying more modes.

However, as demonstrated in Sect. 19.3, OMA can also be used on run up data, as soon as the sweeping orders excite the frequency band of interest. In this case, however, the response is dominated by orders ending in the frequency range of interest. This is a problem when applying OMA as the identification algorithm will identify peaks at these frequencies which are however not physical. In a typical gearbox, integer orders are related to the shafts rotation, while gear mesh frequency orders usually appear as non-integer multiples of the fundamental one. In this case, as already mentioned, the two gearboxes have a slightly different gear ratio and all orders are doubled. The number of end-of-orders is thus very high and, similarly to the harmonic component in stationary conditions, will completely dominate the measured response. In Fig. 19.11, an example of the computed cross powers is shown, with the vertical lines indicating some of the end-of-orders frequency lines for one of the two gearboxes. These deterministic components are seen as poles by the identification algorithms, but they actually represent, most of the time, only a forced response at that frequency. However, a modal model could still be obtained and was compared with the one derived from the Symmetric FRF during operation.

The MAC between the two models is shown in Fig. 19.12, where some correlation and similarity between the two modal set is observed. However, it still remains an issue whether correlating mode pairs are true modes or forced vibrations at rotor harmonics / end of order frequencies. Because of the complexity of the system, and the high modal/harmonic density, the general rule stating that non-physical harmonic-related poles show very low damping cannot be applied. Application of processing windows and interactions between modes and deterministic component make the distinction very difficult. Some recurring frequencies were however identified in almost all analyzed dataset giving confidence that these modes are the actual structural ones.

Finally, orders were extracted to apply Order-Based Modal Analysis. During the measurement campaign, 3 tacho signals were measured. One optical sensor was placed on each High Speed Shaft and one on the Low Speed Shaft. However, because of noise on the optical probes, only one of the signals was of sufficient good quality to extract the orders correctly.

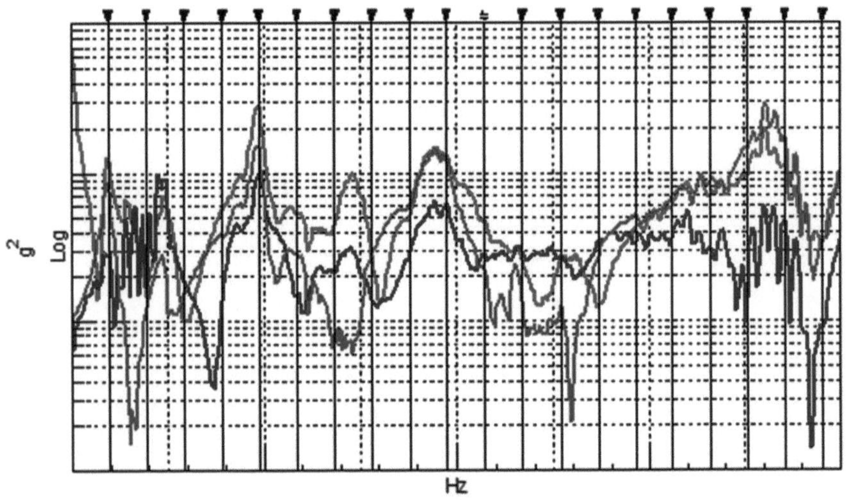

Fig. 19.11 Calculated cross powers during a run up. Vertical thick lines indicate some of the end-of-order frequencies

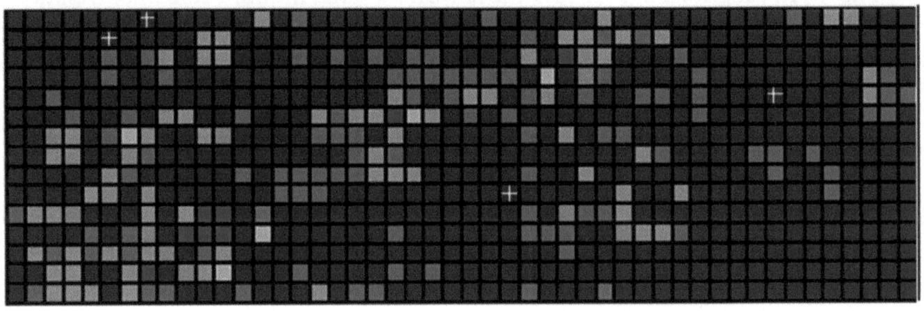

Fig. 19.12 MAC between OMA applied on run up data and EMA on OMAX FRF

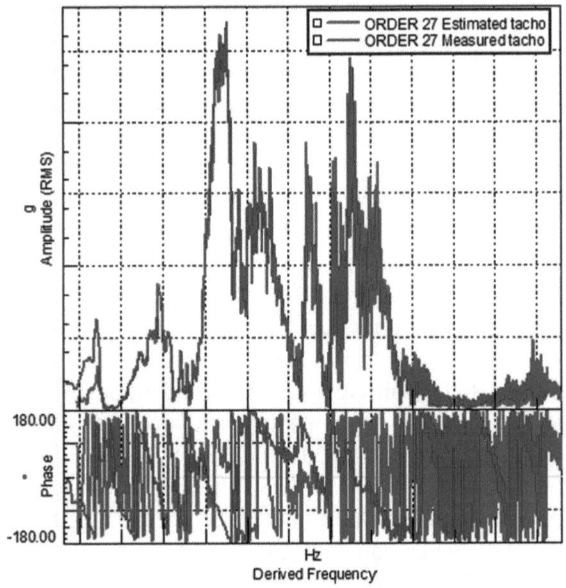

Fig. 19.13 Comparison of estimated orders with measured and estimated tacho signal used to generate the phase reference signal

Based on the results shown in Sect. 19.3, orders were calculated using the TVDFT method. To obtain a proper phase reference, the tacho signal was multiplied by the order to be extracted, the instantaneous angle calculated and finally the corresponding synchronous sine wave extracted. The importance of using the proper signal for the tracking and the phase reference is shown in Fig. 19.13. The phase of the order with the phase reference signal synchronous to the measured tacho is clearer than the one with the estimated tacho signal. Also, the importance of extracting different orders to properly identify the modes in a frequency band as broad as possible is shown in Fig. 19.14. The full modal model can then be assembled combining the models from the different orders together.

The results obtained applying Order-Based Modal Analysis to one of the dominant order are here presented. Fig. 19.15 shows the stabilization diagram obtained by applying PolyMAX on the approximately 700 measured orders. The stabilization

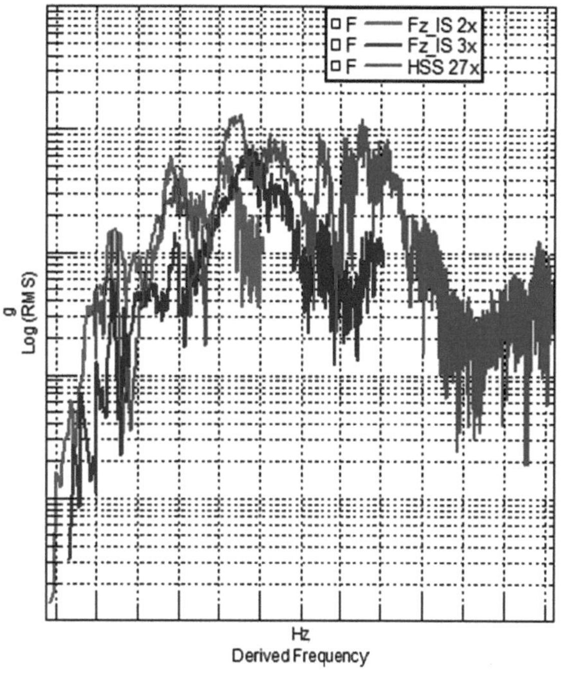

Fig. 19.14 Comparison of different offers extracted for the same point

diagram is quite clear and columns of stable poles can be easily identified. However, because of the noise in the orders, the synthesis is relatively poor (Fig. 19.16), resulting in noisy and non-consistent modes shapes. This can also be observed in Fig. 19.17. The mode shape involving the torque arms bending is compared. In both modes, the dominant fore-aft motion of the torque arms can be recognized, although the phase is slightly different. However, if MAC is computed, a value below 5 % is obtained, because of the general noise in the OBMA modal estimate. In general, however, a set of natural frequencies corresponding to those obtained in the EMA and OMA case were obtained, thus confirming that these frequencies are actual poles of the structure and not forced vibrations.

19.5 Conclusions

In this paper, the problem of identifying a modal model in operational conditions, and in particular in case of rotating machineries, was discussed. Typically, Operational Deflection Shapes at critical frequencies are analyzed to try to understand the cause of high vibrations or acoustic emissions. However, by just analyzing these features, it is impossible to understand whether the high motion is due to the input excitation, a resonance or a combination of both. For this reason, getting a modal model is of paramount importance, not only to identify the root causes of N&V problems, but also to update numerical models using operational information.

Different combinations of experiments and data analysis techniques were first validated on a numerical example and then applied on a gearbox controlled to operate at different conditions. FRFs were extracted using shakers sine sweep, both during stand-still and in operation, and standard Experimental Modal Analysis applied. Then, depending on the operating conditions (stationary or run up), Operational and/or Order-based Modal Analysis are applied. The theoretical examples have shown that all methods are able to properly identify the modes, at least those which are excited with sufficient energy. However, in real conditions, noise, measurement errors (even small errors on sensor orientation angles may cause huge inaccuracies when animating the modes) and the complexity of the test campaign, made identifying a reliable modal model quite complex, in particular in operational conditions. For these reasons, the measurements should be as accurate as possible as the methods are very sensitive to noise and other error sources (electrical disturbances, low excitation levels, inconsistency between the different runs). The importance of accurately measuring (or at least estimating) the rotational speed to perform order tracking was shown, together with the sensitivity of the different order tracking methods to the quality of the rpm signal. Finally, the importance of having a proper phase reference signal for the orders was shown. The most reliable signal was then identified as a sine sweep which instantaneous frequency is equal to the instantaneous rotational speed. In general, Order-based Modal Analysis was shown to be a reliable tool for modal parameter identification from data acquired on rotating machineries

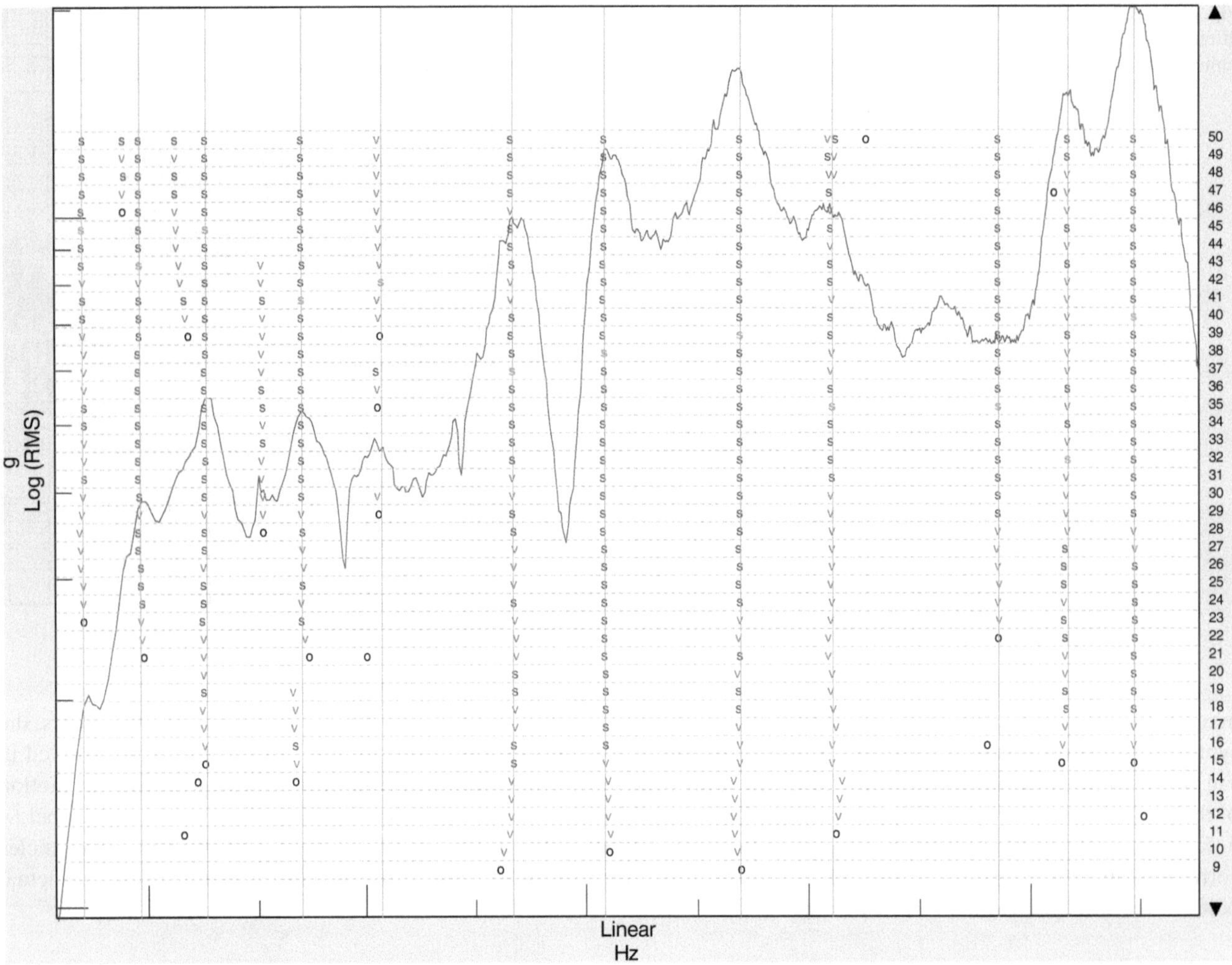

Fig. 19.15 Stabilization diagram obtained for the analyzed order

Fig. 19.16 Comparison of a measured vs. synthesized order

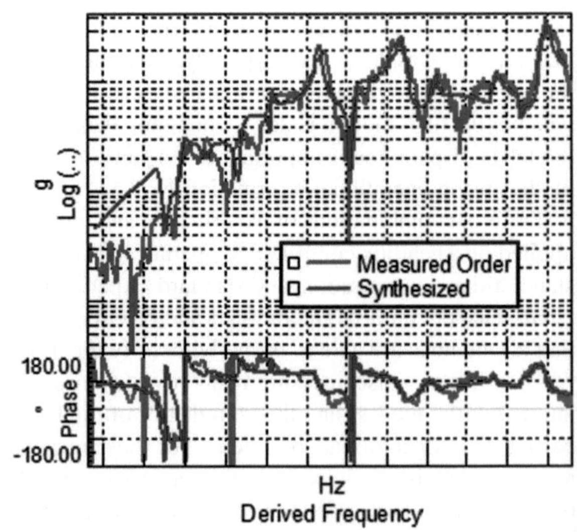

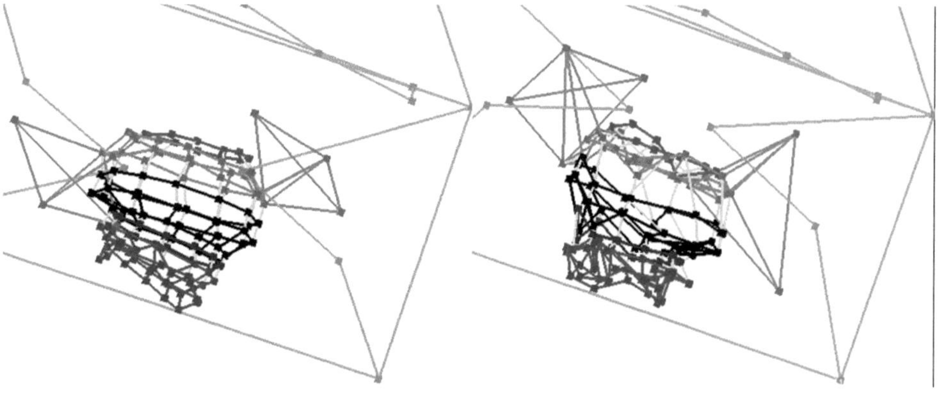

Fig. 19.17 Comparison between EMA mode shape (*left*) and OBMA (*right*)

during operation. Also, while the instrumentation is the same as the one used for Operational Deflection Shape, it allows to derive a modal model with little additional processing that can be used for root cause analysis and/or to update numerical models

Acknowledgments The research presented in this paper is carried out in the framework of the 7th FP Marie Curie ITN project 309395 "MAREWINT" (http://marewint.eu/). Furthermore, the research conducted in this paper is supported by following institutions:
Institute for Promotion of Innovation through Science and Technology in Flanders - Belgium (IWT Vlaanderen)
German Federal Ministry of Education and Research (BMBF) within the Framework Concept "Research for Tomorrow's Production" and managed by the Project Management Agency Karlsruhe (PTKA).
This project is supported by an Eureka label in the framework of international co-operation: www.eurekanetwork.org/project/-/id/7220.

References

1. Goris S, Vanhollebeke F, Ribbentrop A, Markiewicz M, Schneider L, Wartzack S, Hendrickx W, Desmet W (2013) A validated virtual prototyping approach for avoiding wind turbine tonality. In: Proceedings of 5th international conference on wind turbine noise, Denver
2. Helsen J (2012) The dynamics of high power density gear units with focus on the wind turbine application. Ph.D. dissertation, Katholieke Universiteit Leuven, Department of Mechanical Engineering, Leuven
3. Marrant B (2012) Validation of MBS multi-megawatt gearbox models on a 13.2 MW test rig. SIMPACK newsletter issue, May–June 2012
4. Peeters B, van der Auweraer H, Guillaume P, Leuridan J (2004) The PolyMAX frequency-domain method: a new standard for modal parameter estimation? Shock Vib 11:395–409
5. Peeters B, Van der Auweraer H, Vanhollebeke F, Guillaume P (2007) Operational modal analysis for estimating the dynamic properties of a stadium structure during a football game. Shock Vib 14(4):283–303
6. Blough JR (1998) Improving the analysis f operating data on rotating automotive components. Ph.D. dissertation, University of Cincinnati
7. Fuellekrug U, Boeswald M, Goege D, Govers Y (2008) Measurement of FRFs and modal identification in case of correlated multi-point excitation. Shock Vib 15(3):435–445
8. Heylen W, Lammens S, Sas P (1997) Modal analysis theory and testing. Katholieke Universiteit Leuven, Department Werktuigkunde, Leuven
9. Blough JR, Brown DL, Vold H (1997) The time variant discrete Fourier transform as an order tracking method. In: Proceedings of society of automotive engineers noise and vibration conference (SAE), paper no. 972006
10. Peeters B, Gajdatsy P, Aarnoutse P, Janssens K, Desmet W (2009) Vibro-acoustic operational modal analysis using engine run-up data. In: Proceedings of the 3rd international operational modal analysis conference (IOMAC), Ancona
11. Oliver-Escandell M-A, Leroy S, Peeters B, Roubertier J, Cordeau A, Lanslots J (2009) In-flight testing of a very large aircraft: a discussion on excitation and data analysis techniques. In: Proceedings on the international forum on aeroelasticity and structural dynamics IFASD, Seattle

Chapter 20
Modal Identification Results of Quasi-statically Tested RC Frames at Different Damage Levels

Ozgur Ozcelik, I. Serkan Misir, Carmen Amaddeo, Umut Yucel, and Erkan Durmazgezer

Abstract This paper presents modal parameter identification results of two half scale reinforced concrete (RC) portal frames experimentally tested under quasi-static cyclic loading. One of the RC frames is without infill and the other one is with hollow-fired clay brick infill. The frames are progressively damaged in in-plane direction by applying increasing drifts by a displacement controlled actuator at the floor level. At each pre-determined damage states (levels), the actuator is detached from the specimen, and dynamic tests are performed. In-plane dynamic excitation used for modal identification is applied by an electro-dynamic shaker placed on the slab level. Besides, impact hammer and ambient vibration tests are also used. NExT-ERA, EFDD, and SSI-Data are the modal identification techniques used for modal parameter estimations. The purpose of the paper is multi-fold: (i) identify modal parameters of code conforming RC frames with and without infill walls at different damage levels, (ii) follow the evolution of damage level vs. associated damping ratio estimations which can potentially be used for reference-free damage level detection, and (iii) quantify the effect of broadband excitation provided by an offline shaker tuning technique on modal identification results.

Keywords Reinforced concrete frames • Infill effect • Modal parameters • System identification • Shaker tests

20.1 Introduction

Civil engineering structures are exposed to different external effects that change their dynamic characteristics. Damage can be defined as changes affecting the structural performance of a system. The fact that damage can alter stiffness, mass, and/or energy dissipation capacity of a structure, which in turn results in detectable changes in its vibration signature, is the underlying principle of vibration-based structural health monitoring (SHM) [1]. Operational modal analysis, which does not require measurement of input, is used as a technology to extract modal parameters as well as identifying damage in structures. Modal parameters to be extracted are natural frequencies, mode shapes, damping ratios. In the literature numerous surveys exist on the recent developments of structural health monitoring techniques applied to civil engineering structures based on changes in their vibration characteristics [2].

An experimental study has been carried on two one-story one-bay half scale reinforced concrete (RC) frames with and without infill wall conditions under quasi-static lateral loads. At certain drift levels, the actuator was detached from the frames and a set of low level dynamic tests were conducted; the dynamic test program includes ambient vibration, impact and various white noise (WN) tests using a shaker. Using the dynamic data associated with certain damage levels, modal parameters of the RC frames were estimated. The aims of the study are to identify the modal parameters of code conforming frames with and without infill conditions as a function of different damage levels using three different output-only system identification methods, to investigate the effect of different excitations on modal parameter estimation results including the effect of offline tuning technique in reproducing truly low root mean square (RMS) level broadband excitation, and finally to follow the evolution of damage level vs. associated estimated damping ratios that may be potentially used for reference-free damage detection.

O. Ozcelik (✉) • I.S. Misir • U. Yucel
Dokuz Eylul University, Insaat Muh. Bol, Izmir 35160, Turkey
e-mail: oozcelik4all@gmail.com

C. Amaddeo • E. Durmazgezer
Gediz University, Insaat Muh. Bol, Izmir 35665, Turkey

M. Mains (ed.), *Topics in Modal Analysis, Volume 10*, Conference Proceedings of the Society for Experimental Mechanics Series, DOI 10.1007/978-3-319-15251-6_20

20.2 Test Setup and Program

20.2.1 Specimens and Test Program

Two one-story one-bay half scale reinforced concrete frames with and without infill wall conditions with a partial slab on the beam have been tested under quasi-static loads (Fig. 20.1). Columns and beam have dimensions of 150×250 mm^2 and 150×250 mm^2, respectively. The height and span width of the specimens are 151 cm and 250 cm, respectively. The specimens were designed by conforming the latest earthquake design code of Turkey enforcing strong column-weak beam design philosophy [3]. The concrete compressive strength used in specimens was around 50 MPa and the yield strength of the reinforcement was 420 MPa. The lateral force to the frames was applied by a displacement controlled actuator which was attached on one side to the specimen on the slab level as shown on the top right parts of Fig. 20.1a, b, and c, and on the other side a reaction wall (not shown in the figures). The actuator was connected to the slab level with an arbitrary gap between the slab and actuator head-plate so that the beam is not restraint axially by two end-plates and free to deform. 185 kN constant axial force (about 10 % of the axial load capacity of columns) was applied to each column by two hydraulic pistons to represent axial load due to upper stories.

In Fig. 20.2b, displacement pattern applied to the specimens is shown. Each drift level was repeated three times (i.e., three full cycles) in order to induce strength degradation within the same drift level. In the same figure with a circle symbol, sequence of dynamic tests is indicated. At 0 % and at the end of drift levels 0.20, 0.50, 1.0, 1.4, 2.2, and 3.5 % a set of dynamic tests were performed on the specimens in the order indicated in the figure (at the last dynamic test point when the actuator was detached there was a certain residual displacement – non-zero value). While performing the dynamic tests, the actuator was detached from the specimen to avoid any lateral stiffness contribution. Dynamic tests data is used to track the evolution of modal parameters (natural frequencies, mode shapes, and damping ratios) with respect to different damage levels for each specimen. The details of the dynamic tests are shown in Table 20.1.

Total of 142 dynamic tests have been performed on S1 and S2 including the repeatability tests. Here in Table 20.1, only the tests used for this study are shown. The data collected during all these tests were at least 8 min long. In ambient vibration tests no external excitation was applied to the specimens. In WN tests, an electro-dynamic shaker was used for external

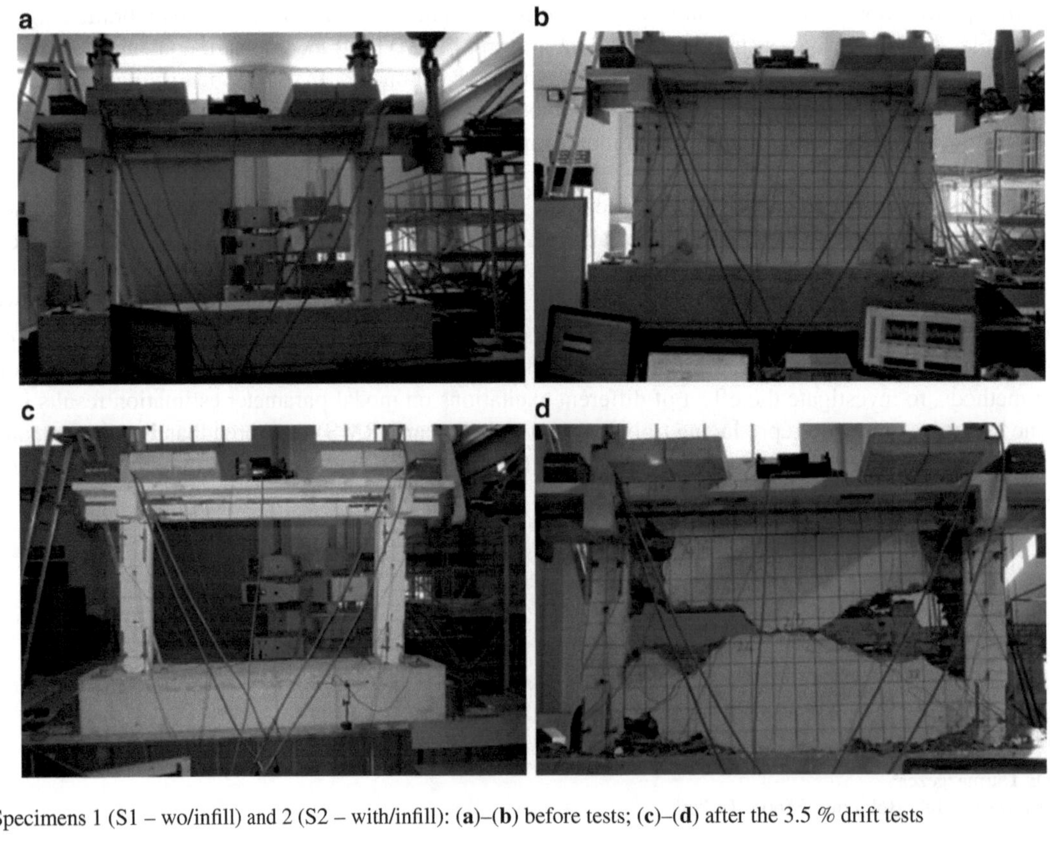

Fig. 20.1 Specimens 1 (S1 – wo/infill) and 2 (S2 – with/infill): (**a**)–(**b**) before tests; (**c**)–(**d**) after the 3.5 % drift tests

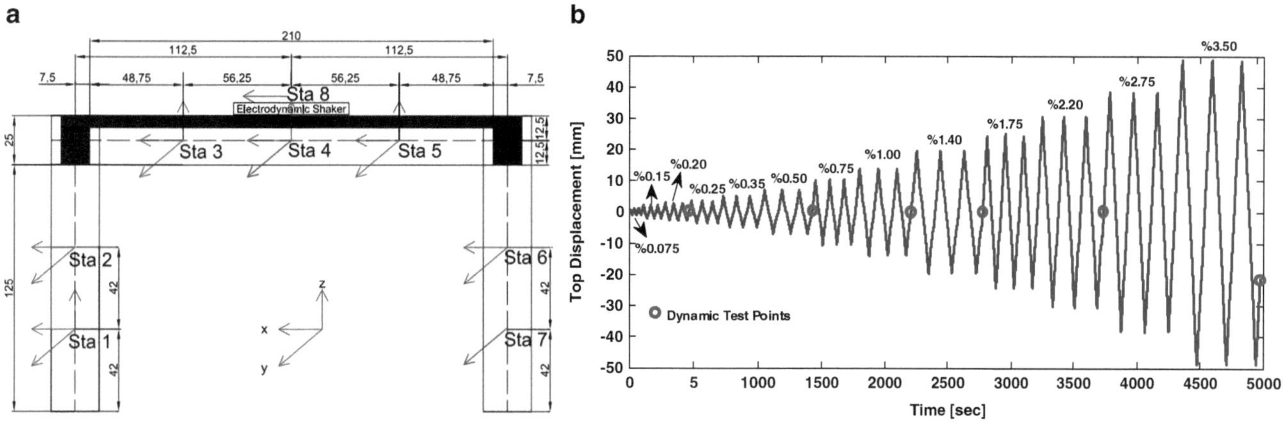

Fig. 20.2 (**a**) Accelerometer layout on both specimens and polarity plots of the sensors, (**b**) sequence of static and dynamic tests

Table 20.1 Tests performed on Specimen 1 (S1) and Specimen 2 (S2) at different damage levels

		Dynamic excitation type			
Specimen	Drift level (damage state)	Ambient	White noise (WN) wo/offline	WN w/offline-1[a]	WN w/offline-2[b]
S1 (S2)	No damage	NA (T3)	T31 (T1)	T32 (T6)	T32 (T6)
	0.2 %	T43 (T14)	T40 (T13)	T42 (T16)	T44 (T17)
	0.5 %	T58 (T26)	T54 (T25)	T57 (T28)	T59 (T29)
	1.0 %	T70 (T39)	T67 (T37)	T69 (T41)	T72 (T43)
	1.4 %	T85 (T56)	T82 (T55)	T84 (T59)	T86 (T60)
	2.2 %	T99 (T71)	T93 (T70)	T97 (T73)	T100 (T74)
	3.5 %	T111 (T81)	T108 (T82)	T110 (T84)	T112 (T85)

[a]Command signal to the shaker is compensated by offline tuning technique using transfer function estimate of the specimen-shaker system at the undamaged state
[b]Command signal to the shaker is compensated by offline tuning technique using transfer function estimate of the specimen-shaker system at different damaged states

excitation to the specimens, the input signal to the shaker was a clipped WN with a frequency band-width [0.5–100] Hz, the amplitude of the signal was manually adjusted using the gain knob on the signal amplifier unit, RMS value for these tests are in the order of 0.44 g, where g is the acceleration of gravity. In WN w/offline-1 and -2 tests, a control scheme called offline tuning was used to improve the signal reproduction fidelity of the shaker so that a truly broad-band excitation was achieved. Exciting the specimens with a truly broad-band signal is expected to increase the estimation accuracy. The details regarding the control scheme will be provided in the later sections, RMS values for these tests are in the order of 0.20 and 0.15 g, respectively. Note that in Table 20.1, dynamic tests performed at the end of the third cycles of each drift level were indicated.

20.2.2 Instrumentation

Both of the RC frames were instrumented with four tri-axial and six uni-axial piezoelectric type accelerometers; note that one uniaxial accelerometer was deployed on the uniaxial electro-dynamic force generator to measure the excitation (Fig. 20.2a), accelerometer stations are indicated with as "Sta". Electro-dynamic force generator is a shaker with broadband signal reproduction capability [4]. It is used to excite the specimen under different damage states with broadband excitations. The shaker is fitted with a reaction block to increase the force level transmitted to the specimen; the shaker was placed on top of the slab (Fig. 20.2a slab level). Also, 12 displacement transducers and 2 string pots were used to measure the static response of columns, beam, and wall elements during quasi-static tests, 2 more string pots were used to measure the top displacement along the in-plane direction (x-direction). Static measurements were done using a static data acquisition system whereas for dynamic measurements a separate 18 channel 24-bit dynamic data acquisition system was used. Here only dynamic test results are presented.

20.3 Damages Observed During Quasi-static Tests

Very thorough damage observations have been done for each specimen by means of visual observation and photos. Due to page limitations, here in Tables 20.2 and 20.3, only a brief description of the observed damages at certain drift ranges are given for S1 and S2, respectively. Very concise damage observations had to be given here, and while doing this, prominent features of the observed damages are kept so that modal parameter changes can be associated with these damages.

From damage observation given in these tables, it can be seen that both specimens gradually went through substantial amount of damage. One of the main purposes of this presented study is to associate these damages with the changes observed in the modal parameters estimated using output-only system identification methods. It can also be seen that infill wall has contributed substantially to the maximum lateral strength of the frames. It should be noted here that no out-of-plane deformation was observed throughout the tests.

20.4 Offline Tuning Technique to Increase Signal Fidelity

For modal parameter estimations using output-only methods, it is important that the system to be identified is excited by broadband excitation. In this study, an electro-dynamic force generator capable of broadband excitation was used. But a close look at the signal reproduction fidelity of the shaker shows that although the input signal is broadband, the reproduced

Table 20.2 Damage observations for S1 (wo/infill) at different damage levels (at the end of each drift range, dynamic tests were performed)

Drift range [%]	P_{max} range [kN]	Major features of the observed damages
0.075–0.2	28.3–56	No visible damage observed until the end 0.15 % drift. Within 0.2 % drift flexural hairline cracks observed at column (bottom ends) and beam ends
0.25–0.5	67–91	More tensile cracks (flexural) observed at beam and column ends (bottom ends), some hairline cracks observed at beam-column joint panel, some beam cracks extended to the slab
0.75–1.0	105–114	New flexural cracks observed along both columns, some hairline cracks emerged at beam-column joint panel, already existing flexural cracks on the columns are in the range of 2 mm, new flexural cracks observed at the slab midspan
1.4	119	Existing cracks propagated and extended on both columns and beam, new flexural cracks observed on the beam and slab, at some points on the columns concrete cover started falling off
1.75–2.2	122–109	Concrete crushing observed at column ends, some early cracks observed on the beam ends progressed to about 1 cm, plastic hinge mechanisms were visible at beam ends, concrete spalling observed at beam and column ends, uplift between foundation and right column observed, at the end of this drift range, stirrups at beam ends were visible. Maximum lateral strength of 122 kN was reached within 1.75 % drift
2.75–3.5	114–100	Significant amount of concrete crushing observed at column bottom ends, column cover concrete spalled off, buckling of beam longitudinal reinforcements on both ends and eventually rupturing occurred, long. reinforcements were clearly visible and the cracks were in the order of 1.5 cm

Table 20.3 Damage observations for S2 (w/infill) at different damage levels (at the end of each drift range, dynamic tests were performed)

Drift range [%]	P_{max} range [kN]	Major features of the observed damages
0.075–0.2	94–168	No visible damage observed until the end 0.15 % drift. Within 0.2 % drift masonry panel separated from the frame borders
0.25–0.5	184–250	First horizontal tension cracks (hairline) observed at column bottom ends
0.75–1.0	255–237	First diagonal cracks observed within the wall, existing flexural cracks propagated at both column ends, new flexural cracks on the columns emerged, horizontal cracks formed on the central zone of the wall, wall plaster started to spall-off. Maximum lateral strength of 255 kN was reached within 1.0 % drift level
1.4	213	Existing column cracks close to the mid-height propagated, significant wall plaster spalling observed, new flexural cracks appeared on the columns, diagonal cracks on the wall extended significantly and new horizontal panel cracks observed. Flexural cracks on beam ends observed. Uplift between column ends and foundation was visible
1.75–2.2	188–162	Excessive infill wall damage observed (some wall units fell off), plastic hinge mechanism developed on the beam ends, some beam bottom tension reinforcements were visible, and buckling of long. reinforcement at beam ends were slightly visible
2.75–3.5	140–120	Cover concrete on beam ends spalled off and the long. reinforcements were clearly visible, first buckling occurred, later some bottom tension reinforcements ruptured, crushing of concrete at column bottom ends were observed, long. reinforcement on column ends were visible

feedback signal is far from being broadband. Figure 20.3a shows the transfer function estimate of the shaker when the shaker was placed on different surfaces, namely: (i) on a firm ground, (ii) on the specimen's slab when the specimen (S1) was at undamaged state, and (iii) on the specimen (S1) when the specimen was damaged at 3.5 % drift level (similar results were obtained for S2 but not shown here for the sake of conciseness). It should be noted here that for the last two cases, there is specimen-shaker interaction. All three transfer functions are very similar to each other showing that there is almost negligible specimen-shaker interaction taking place at the excitation levels used for this study. On the other hand, the transfer function estimations clearly show that the shaker will not be able to reproduce the input signal (broadband nature) both in terms of amplitude and frequency content. Figure 20.3b clearly shows this problem; here "input" is the desired signal and "output" is the measured acceleration of the shaker platform by an accelerometer mounted there (Sta 8). In order to remedy this problem in a fast and efficient way, offline tuning technique is used. In this technique, the input signal (desired signal) is multiplied by the inverse of the estimated transfer function in the frequency domain, and then transferred back to time domain using the discrete time inverse Fourier transform. This modified input signal is used as the new input to the shaker. The result of this technique in signal reproduction fidelity is shown in Fig. 20.3c – zoomed to certain portions. It is very clear that this time the desired input and output signals (signal reproduced on the platform of the shaker) match both in time and frequency domains. Moreover, the output signal has the desired broadband nature. Transfer functions used for offline tuning can be various as noted above; but also at each damage state of the specimens a new transfer function can be estimated and used for tuning purposes. In order to investigate the effects of damage states on transfer functions estimations a systematic study was carried out. Tests shown in Table 20.1 indicated as "WN Tests w/offline-2" were performed to investigate this effect. In these tests, new transfer functions corresponding to the indicated drift levels were estimated and used for offline tuning; in other words transfer functions were adapted to the new conditions of the specimens. In "WN Tests w/offline-1" tests only a single transfer

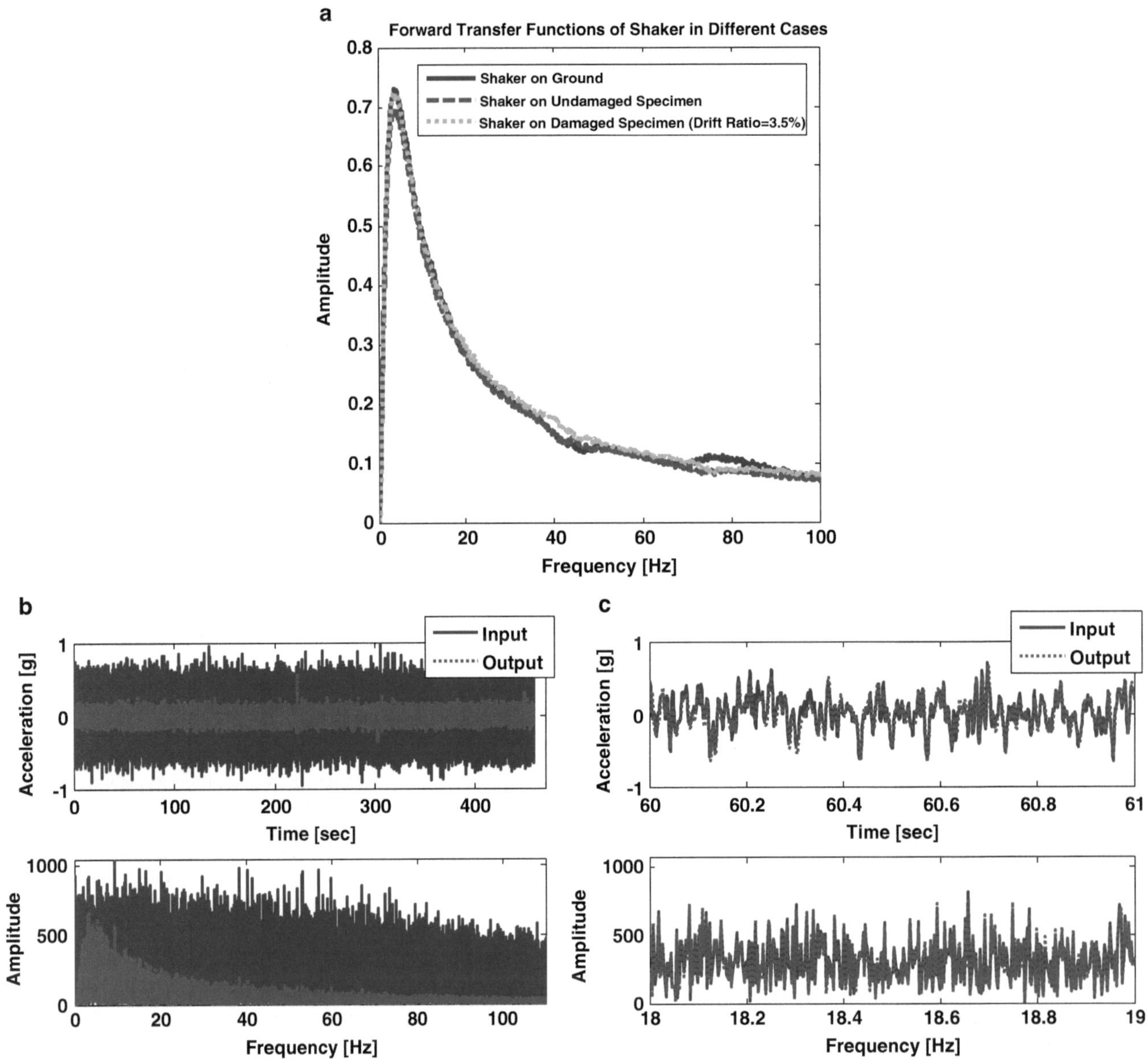

Fig. 20.3 (**a**) Forward transfer function estimations, (**b**) signal reproduction fidelity before offline tuning, (**c**) signal reproduction fidelity after offline tuning (zoomed)

function was used, which is the transfer function corresponding to the undamaged specimen condition. Signal reproduction fidelity is expected to affect modal parameter estimations. The modal parameter estimation results will be presented later in the paper.

20.5 System Identification Methods Used for Modal Parameter Estimation

Three different output-only system identification methods were used to estimate the modal parameters of specimens at different damage levels. These methods are (i) Data-driven Stochastic Subspace Identification (SSI-Data), (ii) Natural Excitation Technique with Eigensystem Realization Algorithm (NExT-ERA), and (iii) Enhanced Frequency Domain Decomposition (EFDD). Also, for impact hammer tests, ERA method was used, but these results are not included herein. In the paragraphs below, brief descriptions of the methods are presented.

20.5.1 SSI-Data

The SSI-DATA method obtains the mathematical model in linear state-space form based on the output-only measurements directly [5]. One of the main advantage of the method compared to two-stage time-domain system identification methods such as covariance-driven SSI and NExT-ERA, is that it does not require any pre-processing of the data to calculate auto/cross correlation functions or spectra of output measurements. One other advantage of the method is that QR factorization, singular value decomposition (SVD) and least squares are involved as robust numerical techniques in the identification process. Modal parameters such as natural frequencies, damping ratios, and mode shapes of the dynamic system can be identified. SSI-Data method is not programmed within the presented study, instead commercial ARTEMIS® operational modal analysis software is used where the method, among other system identification methods, is available. SSI-Data results obtained will serve as reference values to be compared with the results obtained using other two methods.

20.5.2 NExT-ERA

The underlying principle of NExT method is that cross-correlation function calculated between two response measurements (output) from a structure excited by broadband excitation has the same analytical form as the free vibration or impulse response of the structure. Therefore modal parameters can be estimated using cross-correlation functions [6]. One of the main advantages of NExT method is the ability to identify closely-spaced modes and corresponding modal parameters [6, 7]. Once the auto- and/or cross-power spectral density functions are estimated, correlation functions in time-domain are obtained by discrete inverse Fourier transform. These functions can either be used directly or be used at the next stage with ERA method to estimate modal parameters. There are numerous techniques used for identifying modal parameters from free vibration/impulse response data [2]. Here ERA method is used to extract modal parameters and model reduction of linear systems. The method is first proposed by Juang and Pappa (1985) to identify modal parameters of multi-degree-of-freedom system using free vibration response [8]. In ERA method, Hankel matrix is formed using free vibration response of the system. By applying singular value decomposition (SVD) to the Hankel matrix, the rank (i.e., model order) can be determined in the case of noise free measurement. Since real life measurements are always noisy, the model order can be determined by sorting the singular values in descending order, and performing partitioning of the SVD decomposition accordingly. Smaller singular values correspond to computational (nonphysical) modes whereas larger singular values correspond to the real physical modes. Once the model order is determined, based on the realization algorithm, the estimates of discrete-time state space matrices can be constructed. From the reduced order state space realization and sampling interval natural frequencies, damping ratios, and mode shapes can be obtained. In this study, NExT-ERA method is programmed in Matlab® programming environment.

20.5.3 EFDD

As an output-only method the Frequency Domain Decomposition (EFDD) method is an extension of classical pick picking (PP) technique and is based on the classical frequency domain approach also known as basic frequency domain technique. The classical technique has shortcomings such as identifying closely-spaced modes and furthermore frequency estimates are limited by the frequency resolution of power spectral density estimates and damping estimations are highly uncertain [9]. In EFDD method, *output* power spectral density (PSD) matrix is estimated at discrete frequencies then the spectral matrix is decomposed using SVD. The corresponding mode shapes can be extracted from the singular vectors. In EFDD, natural frequencies and damping ratios are estimated using PSD functions of single-degree-of-freedom systems by transforming them back to time domain using inverse discrete Fourier transform. Auto-correlation function for a SDOF system in time domain can be used for estimating frequency and damping values by using zero-crossing times and logarithmic decrement. EFDD method is programmed in Matlab® programming environment.

20.6 Modal Parameter Estimation Results

Modal parameter estimations for specimens S1 and S2 under different excitation conditions are given in Tables 20.4 and 20.5, respectively. WN w/offline-1 and WN w/offline-2 results are described above, the same description holds also for the following tables. Impact hammer test results were processed with ERA; but will be presented in another study. Results given in the tables are obtained with SSI-Data method. Results obtained using the other two methods could not be presented due to space limitation; but a general comparison will be presented later using bar plots. It should be noted that for S1, ambient vibration tests for the undamaged case is missing, therefore the results for these tests are indicated as "NA" in Table 20.4. In Table 20.5 for S2 (w/infill), for certain damage cases (undamaged, 1.0 % and 1.4 %) from ambient vibration tests no modal parameters could be extracted (indicated with the symbol "-").

From the results it can be observed that different excitation conditions resulted in similar values. One important result is that the smallest natural frequency estimates are obtained from the WN wo/ offline tests. This is due to the fact that, although slightly larger, RMS values for these tests are larger than the other three tests. From this perspective, also note that the highest frequency estimates are obtained from the AV tests. MAC values given in the tables are calculated between the modes corresponding to the *undamaged state* (using the test WN w/Offline-1 test – reference case and SSI-Data method) and the modes for the damaged cases (using all other tests and SSI-Data estimations). It can be observed that as the damage level increases the MAC values between the undamaged and damaged cases decreases. Notice that at 2.20 % drift, there is a dramatic decrease in the MAC value. Although not shown here, NExT-ERA method was able to estimate modal parameters under AV data better than SSI-Data and EFDD methods, meaning that modal parameters could be extracted successfully at all damage states even under very low level ambient vibrations.

Table 20.4 Modal parameter estimation results for S1 (wo/infill) under different excitation conditions using SSI-Data system identification method

Excitation type	Modal params	No damage	0.2 %	0.5 %	1.0 %	1.40 %	2.20 %	3.50 %
Ambient vibration	ω [Hz]	NA	14.17	13.33	12.28	11.81	10.55	9.03
	ξ [%]	NA	1.709	0.809	1.032	0.635	0.587	0.310
	MAC	NA	0.987	0.979	0.964	0.960	0.944	0.920
WN wo/offline	ω [Hz]	14.14	13.61	12.76	11.7	11.26	9.99	8.41
	ξ [%]	1.017	1.518	1.606	1.505	1.725	1.305	2.082
	MAC	0.997	0.986	0.931	0.951	0.944	0.943	0.910
WN w/offline-1	ω [Hz]	14.28	13.79	12.99	12.01	11.53	10.28	8.68
	ξ [%]	1.515	1.194	1.697	1.626	1.036	1.083	1.387
	MAC	1	0.987	0.917	0.958	0.959	0.945	0.917
WN w/offline-2	ω [Hz]	14.28	13.79	13.01	12.08	11.60	10.40	8.78
	ξ [%]	1.515	1.246	2.137	1.586	0.884	1.101	1.174
	MAC	1	0.987	0.953	0.964	0.960	0.946	0.918

Table 20.5 Modal parameter estimation results for S2 (w/infill) under different excitation conditions using SSI-Data system identification method

Excitation type	Modal params	No damage	0.2 %	0.5 %	1.0 %	1.40 %	2.20 %	3.50 %
Ambient vibration (AV)	ω [Hz]	–	17.97	18.13	–	–	16.51	11.82
	ξ [%]	–	3.947	1.998	–	–	1.992	1.442
	MAC	–	0.903	0.974	–	–	0.445	0.977
WN wo/offline	ω [Hz]	18.02	17.49	17.89	17.67	16.7	16.01	10.87
	ξ [%]	1.59	3.212	2.315	3.538	4.24	3.004	2.448
	MAC	0.999	0.983	0.980	0.938	0.966	0.607	0.795
WN w/offline-1	ω [Hz]	18.08	17.74	17.60	17.48	16.94	16.11	11.63
	ξ [%]	1.544	1.964	2.867	2.69	2.378	1.284	2.059
	MAC	1	0.986	0.913	0.898	0.951	0.817	0.820
WN w/offline-2	ω [Hz]	18.08	17.68	17.66	17.48	17.08	16.17	11.80
	ξ [%]	1.544	1.436	2.804	2.889	2.182	1.411	1.89
	MAC	1	0.988	0.898	0.966	0.903	0.835	0.811

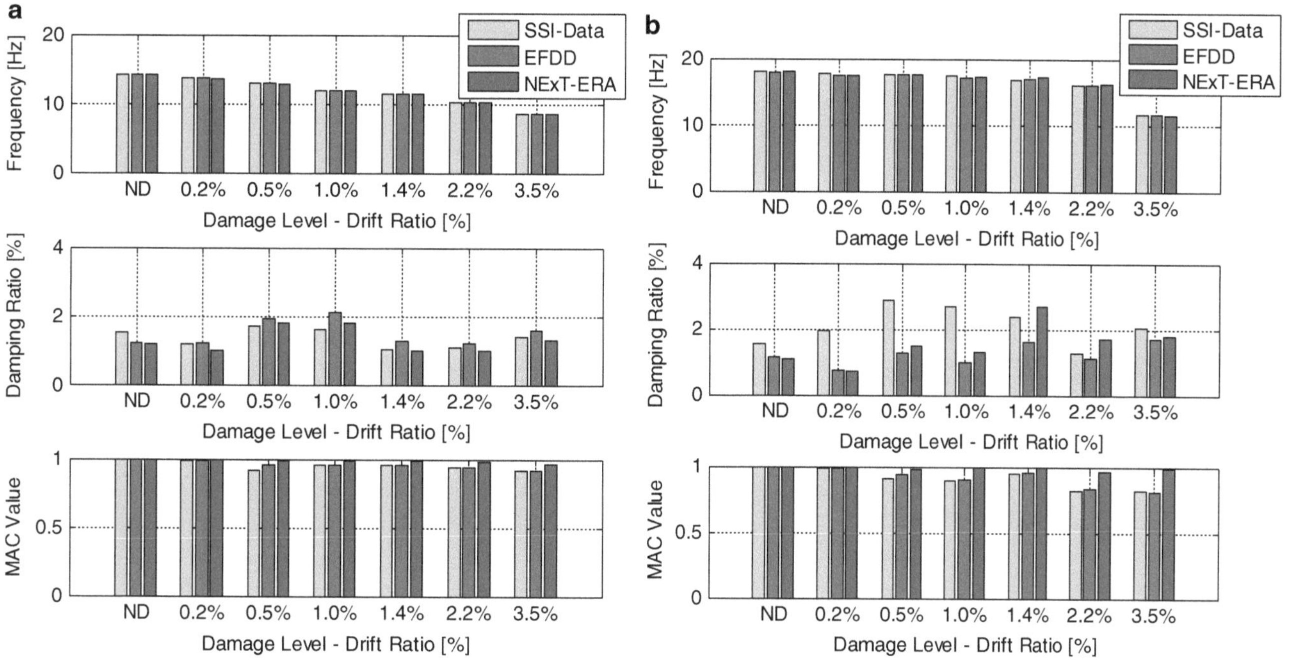

Fig. 20.4 Modal parameters evolution of (**a**) S1 and (**b**) S2 specimens estimated with SSI-Data for different damage levels under WN excitation with offline tuning using transfer function estimated for undamaged state

Figure 20.4a and b show the modal parameter estimations for S1 and S2 using different system identification methods but using only the WN w/offline-1 tests, respectively. For the natural frequency and MAC estimations, all methods give very similar results. It is clear that frequency values decrease as the damage level increases as expected. MAC values are calculated between the undamaged mode (estimated using WN w/offline-1 tests and SSI-Data method) and damaged modes (estimated using WN w/offline-1 tests and other methods). Largest variations in the results among the methods, especially for S2, are observed in damping estimations. Damping estimations by SSI-Data for S2 specimen are consistently larger than the estimations with other methods, namely NExT-ERA and EFDD.

Figure 20.5a shows a closer look at the changes observed in natural frequency and damping estimations for S1 and S2 as a function of damage levels. Undamaged longitudinal in-plane natural frequencies for S1 and S2 are estimated to be 14.28 and 18.08 Hz, respectively. Inclusion of infill wall increases the natural frequency about 27 %. From the frequency plots, it is very clear that there is a steady decrease in natural frequencies as the damage level increases both for S1 and S2. S1 (wo/infill) loses its lateral stiffness faster than S2 specimen (w/infill), this is somewhat an expected condition since, although gets damaged, the infill wall contributes substantially to the lateral stiffness of the frame. At the ultimate damage states at 3.5 % drift level, estimated natural frequencies for S1 and S2 are 8.68 and 11.63 Hz, respectively. Therefore the changes are 39 % and 36 %, respectively. The total change in natural frequency for S2 is slightly lower than S1. This result is somewhat counterintuitive since the frame-infill interaction was expected to be more prominent for S2 even at this damage level. If the curve for S2 is investigated more closely, it can be seen than the curve has a very small negative slope up to 2.2 % drift (a higher wall contribution up to this level), and a sharp drop right after that at 3.5 % (wall contribution suddenly drops). The plot for the damping ratio estimations shows that estimated ratios for S2 are larger than S1. This observation indicates that infill wall contributes to the damping level in RC frames. It is interesting to note that a similar trend is observed for S1 and S2 specimens, except that the values for S2 are shifted upwards. Again for 2.2 % drift level, there is a sudden drop in the estimated damping value. This observation may be a supporting evidence for the fact that at 2.2 % drift infill wall contribution falls down substantially.

In Fig. 20.5b damping value estimations for S1 and S2 are given under low and high level WN excitations. Here the low level excitations are WN w/offline-1 tests, and the high level excitations are WN wo/offline tests. It should be noted that although for the high level excitations the amplitude was higher (RMS level 0.44 g) it was not substantially higher than the low level excitations (RMS level 0.20 g) due to shaker's stroke limitations. It is interesting to note that for S1 (wo/infill), damping estimations are less sensitive to the excitation amplitude levels whereas for S2 (w/infill wall), damping estimations are more sensitive to the excitation amplitude. Again for the high level excitations there is not a clear trend in damping estimations as damage level increases.

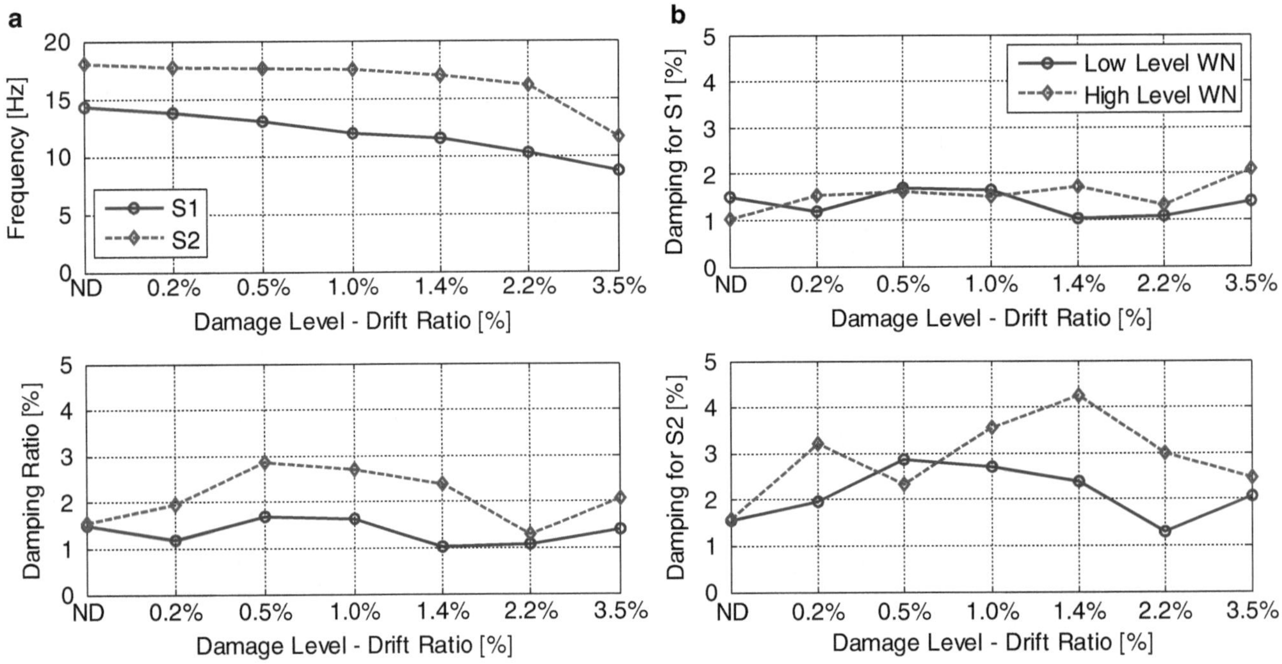

Fig. 20.5 (**a**) Change in frequency and damping estimations with increasing damage levels for S1 and S2 specimens, (**b**) effect of excitation level on damping estimations for S1 and S2 specimens

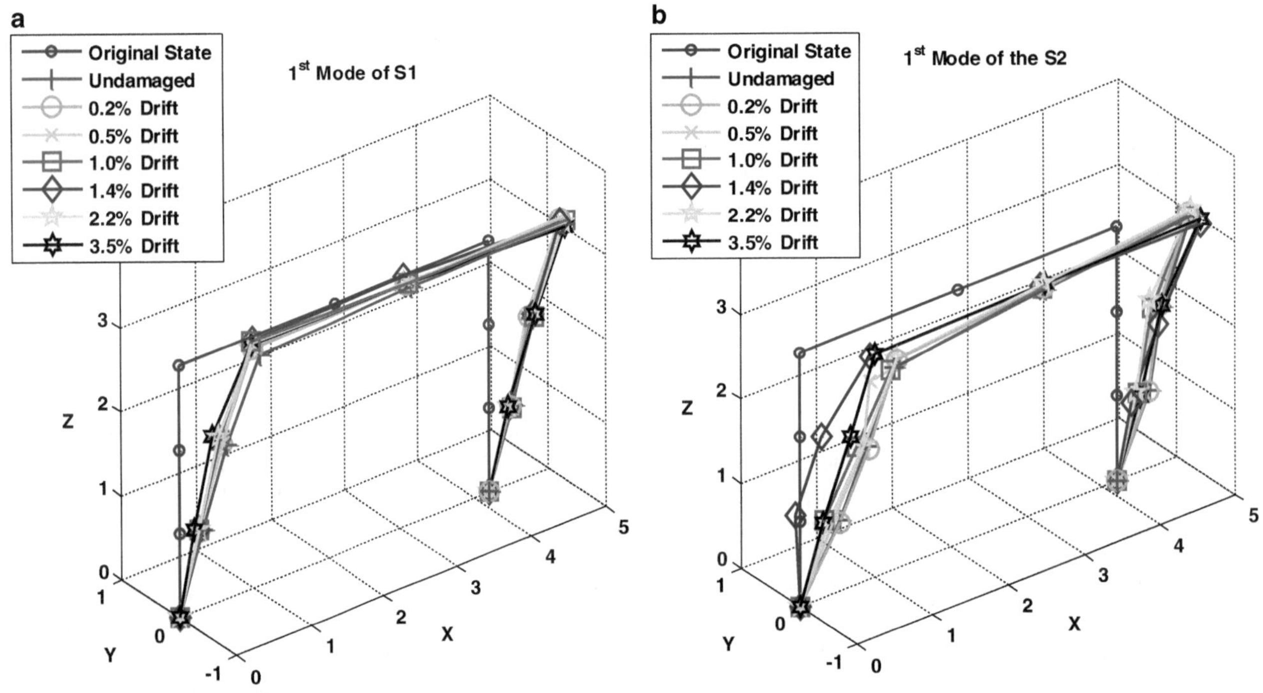

Fig. 20.6 Mode shape estimations at different damage levels using NExT-ERA method under WN w/offline-1 tests (offline tuning exists), (**a**) S1 and (**b**) S2

Figure 20.6a and b shows the longitudinal in-plane mode shape estimations for S1 and S2 using by NExT-ERA method using WN w/offline-1 data, respectively. It can be seen from the figures that mode shapes change as damage level changes. It is important to note that in S2 (w/infill) mode shapes seem a bit more sensitive to the damage level (notice the dramatic change at the end of drift level 1.4 %). It should also be noted that these modes are called in-plane lateral mode due to the

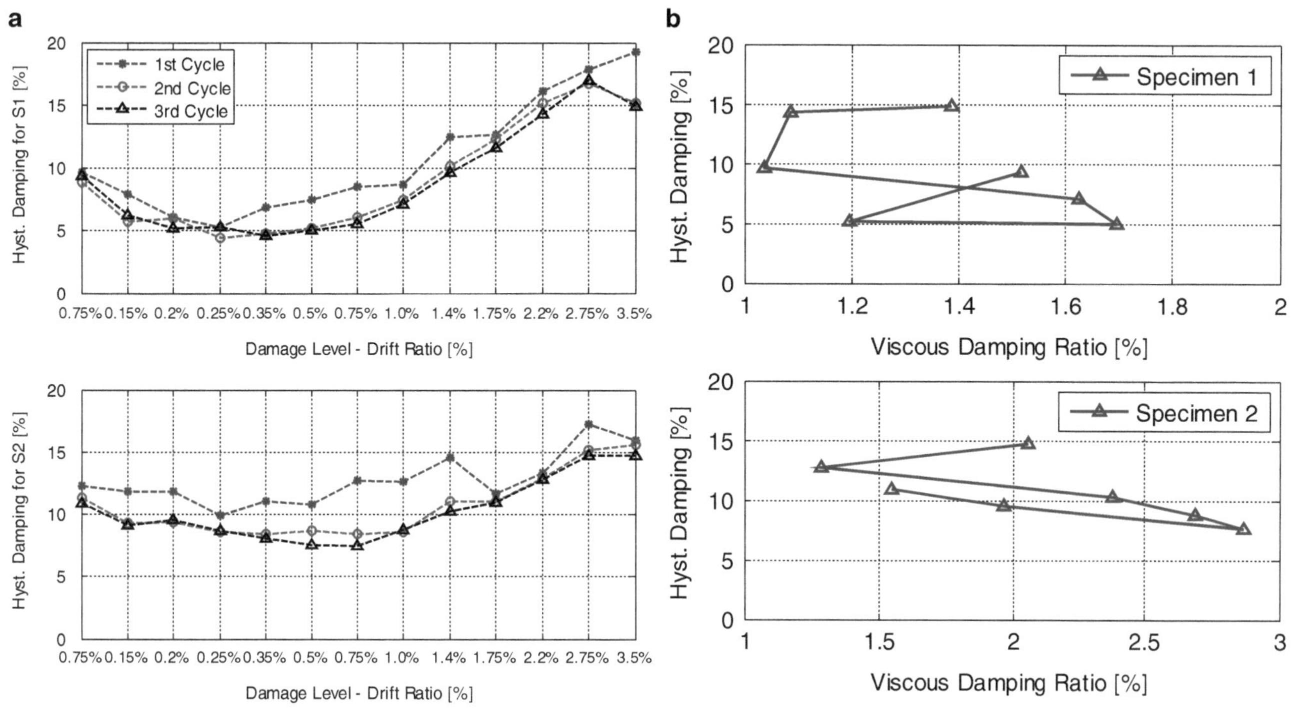

Fig. 20.7 (**a**) Hysteretic damping values vs. different damage level, (**b**) hysteretic damping vs. estimated modal damping values for S1 and S2

fact that main motion is along x-direction, but they have also components along y- and z-directions. In other words, there is no purely in-plane lateral mode but all the estimated modes (also the ones not shown here for space limitations) are coupled modes.

20.6.1 Hysteretic Damping

It is known that there is not a direct/strong relation between equivalent viscous damping and hysteretic damping (a function of the area under the hysteretic force-displacement curves) [10]. However, it would be insightful to show the change of hysteretic damping vs. estimated viscous damping coefficients as a function of damage levels. Figure 20.7a shows a plot of the calculated hysteretic damping ratios for S1 and S2 for each cycle for all drift levels, and Fig. 20.7b shows plots of hysteretic damping at the third cycle of some drift levels vs. estimated damping ratios for S1 and S2 using WN w/offline-1 tests by the SSI-Data method at that corresponding drift level. It is clear that there is not a clear trend between hysteretic damping and viscous damping estimations at this excitation levels; but at much higher excitation levels, which is not possible to induce with the available shaker in hand, a relation between these two damping values may exist.

20.7 Conclusions

In this study two half-scale reinforced concrete frame system with and without infill conditions were tested quasi-statically under increasing lateral drifts. At certain drift levels, dynamic tests were performed and modal parameters at different damage levels are estimated using three different output-only system identification methods, namely SSI-Data, NExT-ERA, and EFDD. Dynamic excitation was applied to the specimens from the slab level by an electro-dynamic shaker; also effects of offline-tuning technique is investigated which is a practical control technique to improve the signal reproduction fidelity of the shaker therefore providing a truly broad-band excitation. Main outcomes of the presented study are as follows: (i) although small, different excitations lead to different modal parameter estimations, specifically AV tests result in higher natural frequency estimates than WN wo/offline, WN w/offline-1, WN w/offline-2 tests, (ii) SSI-Data, NExT-ERA and

EFDD methods give very similar modal parameter estimation results under different excitation conditions, NExT-ERA gives the best results in AV tests -under very low level excitation as compared to the other two methods, (iii) infill wall increased the lateral stiffness of the frame substantially leading to 27 % increase for the frequency of the longitudinal in-plane lateral mode, (iv) at the end of ultimate drift level of 3.5 % decrease observed in the natural frequencies for S1 and S2 are 39 % and 36 %, respectively, (v) decreasing trend in the natural frequency as a function of damage for S2 is slower than S1, (vi) it is observed that the contribution of the wall to the lateral stiffness of the frame goes down substantially at 2.2 % drift, (vii) infill wall contributes also to the damping levels of RC frames, (viii) excitation level affects the damping estimations, meaning that higher excitation levels lead to higher damping values especially when infill wall exists, (ix) at the excitation levels considered, there is not a clear relation between hysteretic and estimated viscous damping values; therefore more has to be done to associate damage with damping values estimated with the methods studied here.

Acknowledgment Authors greatly acknowledge the financial support provided by The Scientific and Technological Council of Turkey (TUBITAK) under the grant no. 112 M203. Authors also would like to thank undergraduate students Muhammed Demirkiran, Filiz Vargun, Gokhan Okudan, Mazlum Yagiz, and Duygu Senol for their help in test preparations. Any opinions, findings, conclusions or recommendations expressed in this publication are those of the writers, and do not necessarily reflect the views of the sponsoring agency.

References

1. Doebling SW, Farrar CR, Prime MB (1998) A summary review of vibration-based damage identification methods. Shock Vib Dig 30(2):99–105
2. Sohn H, Farrar CR, Hemez FM, Shunk DD, Stinemates DW, Nadler BR (2003) A review of structural health monitoring literature: 1996–2001. Los Alamos National Laboratory report, LA-13976-MS, Los Alamos
3. Turkish Earthquake Code (2007) Specifications for structures to be built in disaster areas. Ministry of Public Works and Settlement, Ankara
4. APS 113 Electro-Seis Data Sheet (2014) APS 113 Electro-Seis long stroke shaker with linear ball bearings, Product Manual
5. Van Overschee P, De Moor BL (1996) Subspace identification for linear systems: theory, implementation, applications, vol 3. Kluwer academic publishers, Dordrecht
6. James GH III, Carne TG, Lauffer JP (1993) The natural excitation technique (NExT) for modal parameter extraction from operating wind turbines (No. SAND–92–1666). Sandia National Labs, Albuquerque
7. Farrar CR, James GH III (1997) System identification from ambient vibration measurements on a bridge. J Sound Vib 205(1):1–18
8. Juang JN, Pappa RS (1985) Eigensystem realization algorithm for modal parameter identification and model reduction. J Guid Control Dyn 8(5):620–627
9. Brincker R, Zhang L, Andersen P (2001) Modal identification of output-only systems using frequency domain decomposition. Smart Mater Struct 10:441–445
10. Chopra AK (2007) Dynamics of structures, 3rd edn. Prentice Hall, Upper Saddle River

Chapter 21
An Innovative Tool for Simulation and Control of a Flying-Cutting Machine

S. Cinquemani and H. Giberti

Abstract In the manufacturing field it happens that products are processed by high-speed production lines. They pass through different machines in order to be processed, according to a pre-established order, resting on moving supports. Usually they do not stop during this process: rather, the machines are able to track and handle them while they are moving.

The advantages of this kind of process stem from the continuous motion of the product, that results in an increased productivity since there is no downtime during the production. As a consequence, the tracking of goods is widely adopted to process, to manipulate and to package the products. As reported in literature, strong effort is being directed at the development of an effective control system allowing for the synchronization of the operations and the detection of the position of the incoming products. All of these systems are generally called "cutting systems on the fly".

For the design and the setup of these systems, it is critical to have a tool to simulate and analyse the behaviour of the machine.

This article proposed a systemic approach to model the main elements that characterize such systems aimed to develop the most effective control strategy to improve the performance of the machine. The simulation tool allows to predict the behaviour of the machine and to test different control strategies.

Keywords Automation • Flying-cutting machine • Simulation tool • Machine design • Automatic machines

21.1 Introduction

Very often, in the manufacturing sector, the products are processed by high-speed production lines [1]. They pass through different machines in order to be processed, according to a pre-established order, resting on moving supports (e.g. conveyor belts). Usually they do not stop during this process: rather, the machines are able to track and handle them while they are moving. The advantages of this kind of process stem from the continuous motion of the product, that results in an increased productivity since there is no downtime during the production. As a consequence, the tracking of goods is widely adopted to process, to manipulate and to package the products. As reported in literature [2–9], strong effort is being directed at the development of an effective control system allowing for the synchronization of the operations and the correct detection of the position of the incoming product.

In food production, examples of the application of the aforementioned method can be found in the manipulation of food to be properly well-worked (pick-and-place operation), in the cutting of the raw material in single portions and in the packaging of the finished products. One kind of particularly widespread machine is the wrapping machine [1, 10, 11]. It wraps and seals the products without stopping the line, allowing to save time during the processing phases and thus obtaining a larger productivity. As a consequence of the motion of the product during the cutting phase, the relative velocity between the cutting tool and the product has to be zero in order to obtain an accurate and sharp cut. This condition introduces some complications in the design of the cutting phase and can be achieved by a careful planning of the law of motion of the cutting tools.

Flying-cutting machines are of widespread use due to the above mentioned advantages. Various kind of machines employ this technology in very different fields. In the rolling and extrusion field, the flying cutting is applied to process bars or pipes manufactured [3, 5, 12]. In the paper industry a die cutter is used to obtain smaller products like labels or stickers from a moving paper sheet. A "linear version" of the flying cutting application is the circular saw mounted on linear drive used in the soap process to obtain single bars. Finally, flying cutting is adopted in the packaging process to weld and cut the single

S. Cinquemani (✉) • H. Giberti
Department of Mechanical Engineering, Politecnico di Milano, Via La Masa 1, Milano 20156, Italy
e-mail: simone.cinquemani@polimi.it

M. Mains (ed.), *Topics in Modal Analysis, Volume 10*, Conference Proceedings of the Society for Experimental Mechanics Series, DOI 10.1007/978-3-319-15251-6_21

product at the same time. Even if the layout of the machines above mentioned is very different, the cutting process can be modelled in the same way as demonstrated in [13]. In particular, it allows to describe the "flying machine" using an unified set of design parameters whose influence on the flexibility of the motor-reducer unit is the topic of the present work. This general view allows to overcome the limits of the few previous works that, referring to a particular cutting technology, focus their attention on the specific control system without any effort to generalize the problem [3–5].

The paper deals with the development of a tool to help the designer in the synthesis of automatic machines for flying-cutting operations. The paper is structured as follows: Sect. 21.2 gives an outline of the most common automatic flying-cutting machines, highlighting their operating principles and their main sub-systems. Section 21.3 shows the model developed for each of these main systems. Section 21.4 introduces the use of the modelling tool on a real machine, highlighting some numerical results and considerations. Finally conclusions are drawn in Sect. 21.5.

21.2 Flying-Cutting Machines

Flying cutting machine are based on the idea to perform a cut on the product flow without stopping it. To do this, the relative speed between the cutting tool and the product has to be zero.

These machine can be mainly subdivided into two groups. Those belonging to the first one are designed to move the cutting tool with the same speed of the product and can be profitably used to perform fast operations (Fig. 21.1a). The others are designed to slow and hold the product where the cut takes place, but without interrupting the continuous flow of material (Fig. 21.1b).

Both models shown in Fig. 21.1 have similar sub-systems: a "master" device moves the product flow with a known speed, while an unwinder machine unrolls it. Buffers are usually placed after these two sub-systems with the double purpose of allowing the accumulation of the material and maintaining its tension as constant.

In machines belonging to the first group (Fig. 21.1a), the speed of the product flow is always constant. It is well known that the cutting operation must be conducted by imposing a zero relative speed between the tool and the product, along the direction of the flow. For this reason, during the cutting operation, the tangential speed of the cutting tool must be equal to the one of the product. During all other phases, the angular speed must be adjusted according to the dynamics of the system and to the length of the product to be cut.

On the contrary, in the machines belonging to the second group (Fig. 21.1b), the product is pulled at a constant speed but, in correspondence of the cutting zone, its velocity is converted into intermittent. This operation is made possible by two sliders which, synchronous moving in the opposite direction of the product, allow the latter to have zero speed in correspondence of the cutting zone. When this operation has been carried out, the sliders return back to their initial position, accelerating the product.

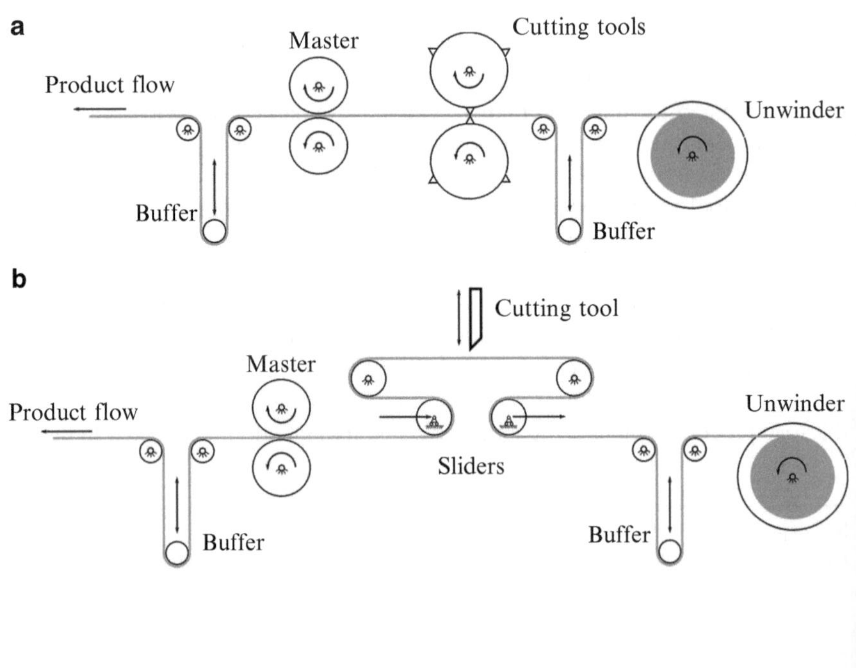

Fig. 21.1 Flying cutting machines

21.3 The Simulation Tool

According to the brief description of flying-cutting machines, some main subsystems can be highlighted. These are: the master, the unwinder, the buffer, the cutting tool and the sliders. For each of these elements a mathematical model is developed and the corresponding Simulink model is built. As these machines repeat the operation for each working cycle, the length of the simulation can be set equal to the cycle time t_c. To comply with designers habits, the independent variable is the so-called master angle. This variable is equal to 0 when the cycle starts and equal to 2π radians at the end. The new variable θ is then:

$$\theta = {}^{t}\!/_{t_c} \cdot 2\pi$$

21.3.1 *The Master*

The master subsystem is responsible for moving the product to the desired speed. Typically this speed is constant, but in some cases, it may be changed during the work cycle according to needs. Long machines may have multiple master systems. In these cases, only one is actually considered as master, while the others are assigned to the slave function. In all the cases, all the subsystems devoted to move the product are synchronous to avoid tensions within the product.

The master sub-system is made of a gear-motor to move two rolls (Fig. 21.2). The Simulink model of this device is depicted in Fig. 21.3. While the input of the model is the master angle, the output are the theoretical and the real displacements of the product flow and the corresponding speed. Theoretical and real quantities can be different if disturbances and delays are considered. Depending on the desired motion law, the user can select constant or variable speed.

21.3.2 *The Unwinder*

The unwinder subsystem is made of a gear-motor connected to a reel and it is devoted to provide the product flow to the flying cutting machine (Fig. 21.4). The speed of the unwinder is generally set as constant and it is adjusted according to the position of the buffer downstream.

The speed of the product flow is not constant and can be controlled through a PID regulator. The input of the buffer is the displacement of the buffer, while the output is the product flow displacement and speed (Fig. 21.5).

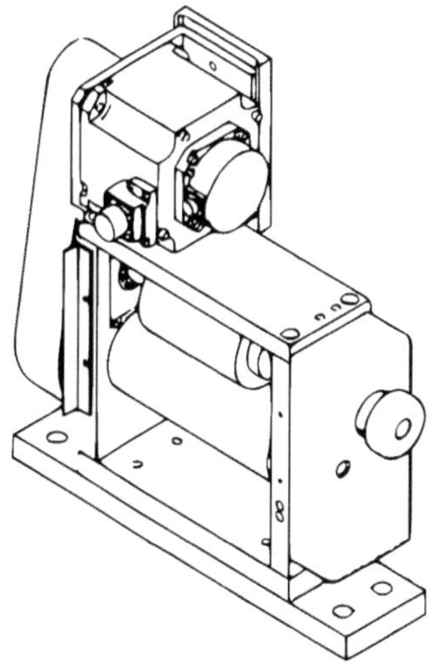

Fig. 21.2 Schematic of the master sub-system

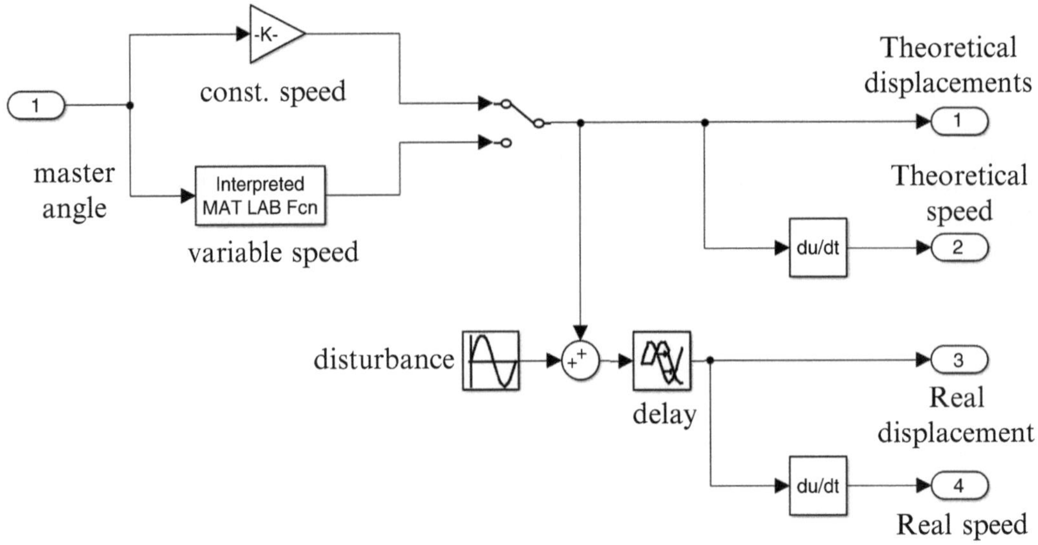

Fig. 21.3 Simulink model of the master sub-system

Fig. 21.4 Schematic of the unwinder sub-system

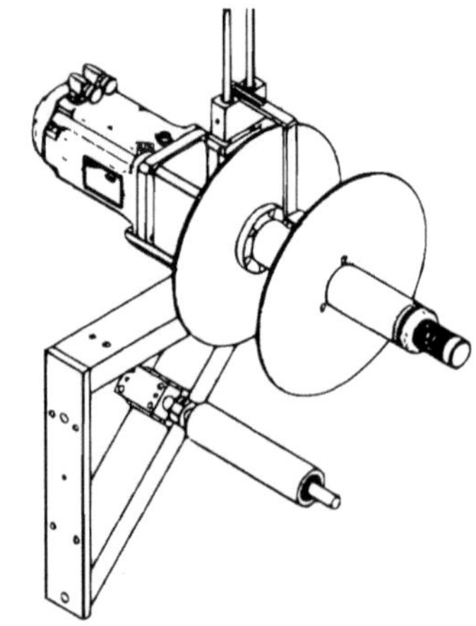

Fig. 21.5 Simulink model of the unwinder sub-system

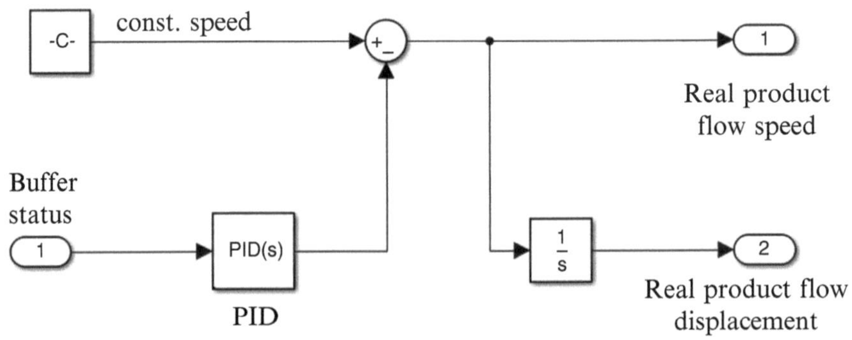

21.3.3 *The Buffer*

The buffer sub-system can be realized in different ways, but its main functions are to accumulate the product flow and to maintain its tension as constant (Fig. 21.6). The status of the buffer is measured usually through displacement sensors (e.g. Encoder, potentiometer, etc.). This information is used to regulate the speed of the flow (Fig. 21.7).

The input of the Simulink model are the displacements of the product flow upstream (incoming the buffer) and downstream (out coming the buffer), while the output if the position of the buffer and its status.

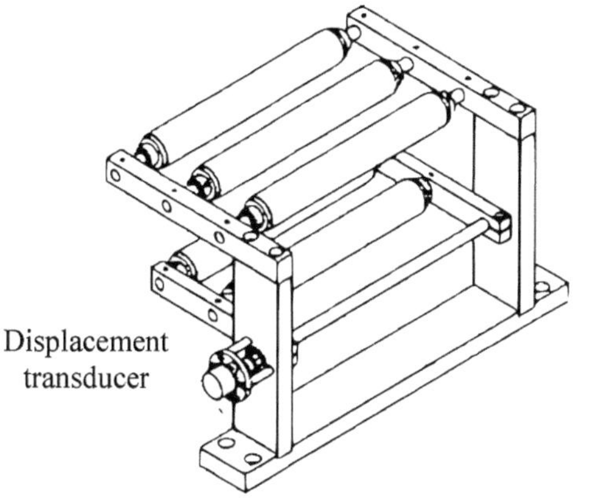

Fig. 21.6 Schematic of the buffer sub-system

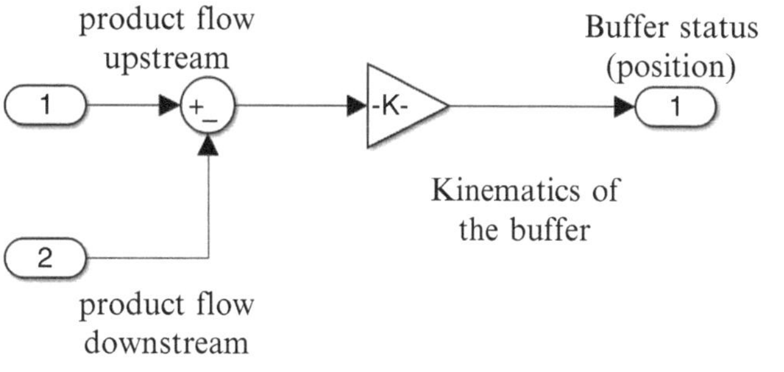

Fig. 21.7 Simulink model of the buffer sub-system

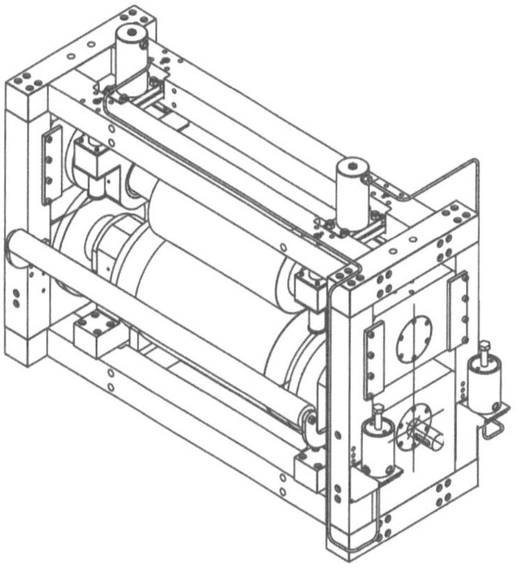

Fig. 21.8 Schematic of the cutting tool sub-system

21.3.4 Cutting Tools

According to different layouts introduced in Fig. 21.1, two different cutting tool sub-systems have to be considered. The first (Fig. 21.8) needs the cutting tool to have the same speed of the product flow. For this reason the Simulink model (Fig. 21.9a) takes the displacement of the product as input. The user can define different motion law of the cutting tool, as a function of the number of blades, the length of the final product, etc. The output is the speed of the motor.

On the contrary, cutting tool sub-systems operating while the product flow is still can be modelled simply considering the kinematic of the mechanism used to move the cutting blade (usually a crank mechanism). In this case, its position is not directly a function of the product position, but the designer has to check that the blade crosses the material only when it is still.

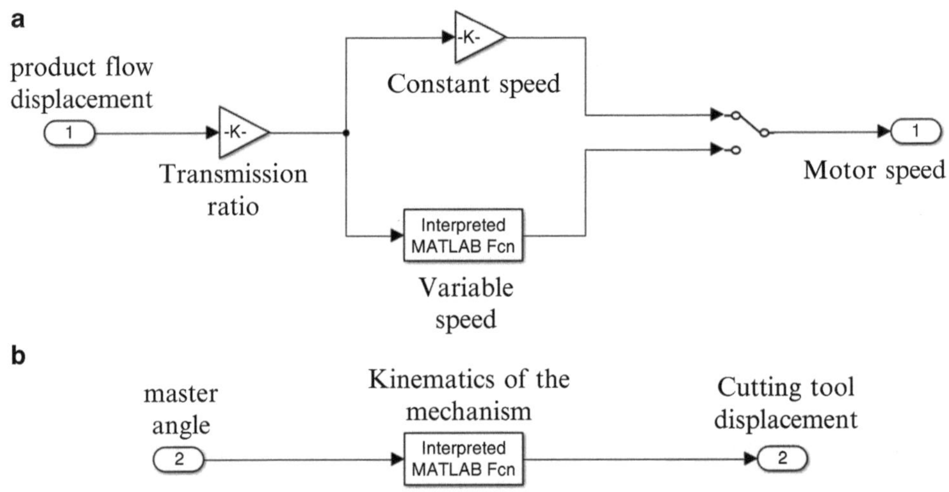

Fig. 21.9 Simulink model of the cutting tool sub-system

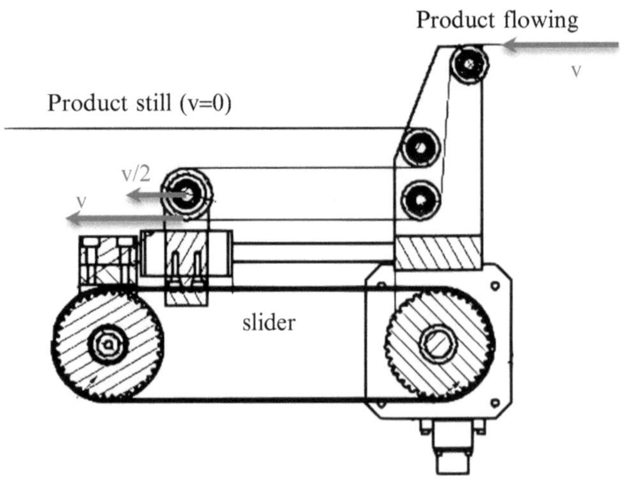

Fig. 21.10 Schematic of the slider sub-system

21.3.5 Sliders

Sliders subsystems are used to transform the motion of the product from continuous to intermittent. Supposing the product flow has a speed equal to v. According to Fig. 21.10, when the slider moves with half the speed of the incoming product ($v/2$), the outcoming one has no speed. However the motion of the slider has some limitations due to the available stroke and due to the dynamics of the system. For this reasons, the product can be maintained still only for a small time interval. Figure 21.11 shows the corresponding Simulink code to describe the kinematics of the slider. Input of this subsystems are the master angle θ and the displacement of the product flow. The motion of the slider can be changed according to different motion laws both for the forward and the backward motions. To generalize the problem, this speed is normalized between [0½] and it is multiplied by the speed of the product flow. The output of the model are the displacement and the speed of the slider.

The five main subsystems modelled in this work are the most common ones. Obviously they can not describe the totality of the cases, but the reader will understand that it is possible to model other subsystems [14–17] by using a logic similar to the one of the authors.

21.4 Test Case

The simulation tool can be profitably used to predict and analyse the behaviour of a flying-cutting machine. Figure 21.12 shows the layout of an automatic machine (see Fig. 21.1b).

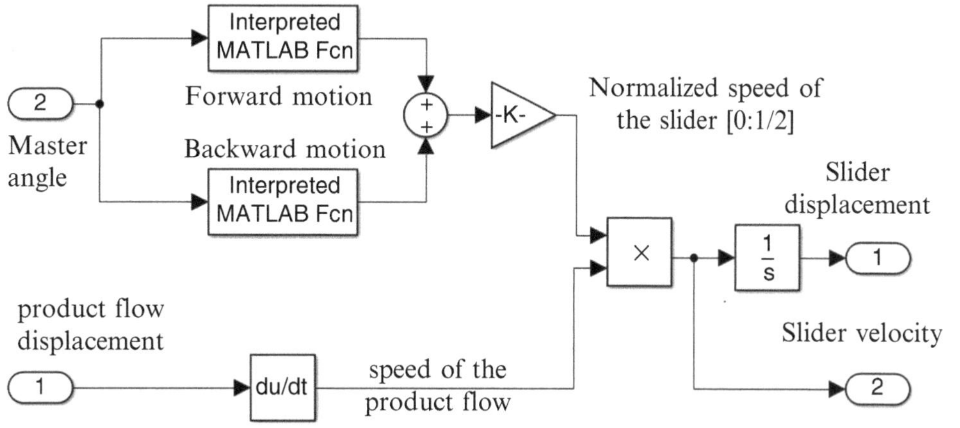

Fig. 21.11 Simulink model of the slider sub-system

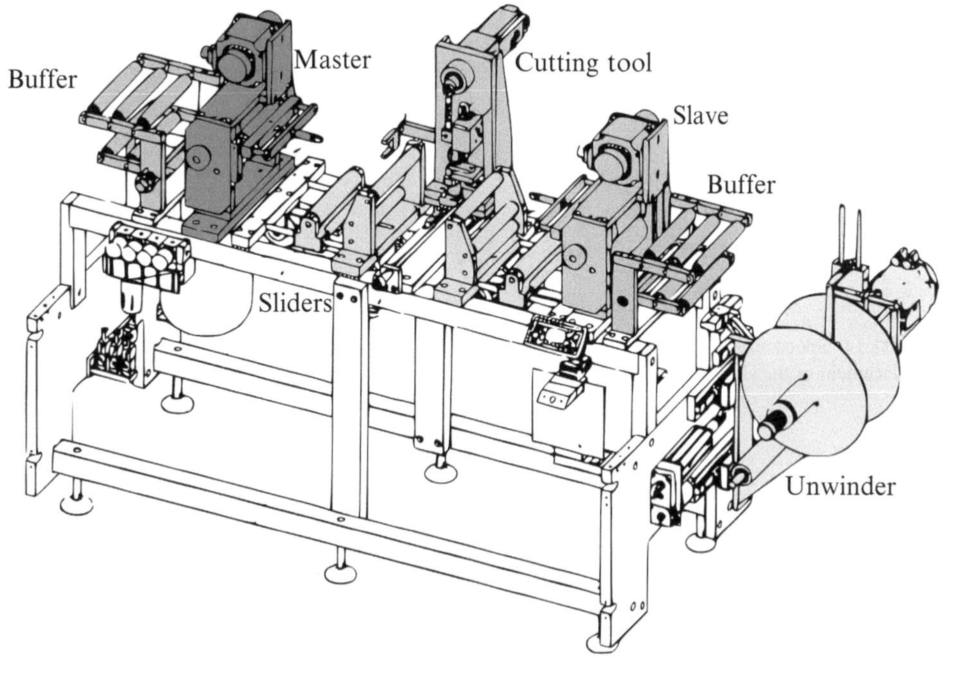

Fig. 21.12 Flying-cutting machine with main subsystems highlighted

The product moves from right to left and it is unwound from the unwinder. A proximity sensor monitors the reel is not finished. The two groups master and slave move the product by imposing a constant speed to avoid unwanted tension in the product flow. The two sliders transform the continuous motion into intermittent; in this way the cutting tool has enough time to cut the product while the flow is still. Finally, buffers are opportunely placed to recover the material.

According to the described layout, a model of the machine is carried out using the simulation tool developed. Some interesting graphs can be useful to properly setup the machine. Figure 21.13 shows the speed of the sliders and the position of the cutting tool during a cycle. It can be observed that the cutting tool performs the cut exactly when the slider has the required speed and the product is still. According to this graph, the designer can decide to change the motion law of the slider to extend/reduce the range where the product is still. At the same time (Fig. 21.14) knowing the displacement of the slider during the cycle, it is possible to properly design the stroke of the subsystem.

In addition to analysis on the right functioning of the machine, some failures conditions can be modelled. Figure 21.15 shows the length of the product required by the *Master* sub-system during a cycle (160 mm). If the unwinder was not able to supply the requested product, the buffer can compensate thus guarantying the right functioning of the machine.

These are just some of the analysis that can be performed. Obviously the user has the ability to customize his own model, depending on the variables he wants to monitor or the conditions of use that he would analyse.

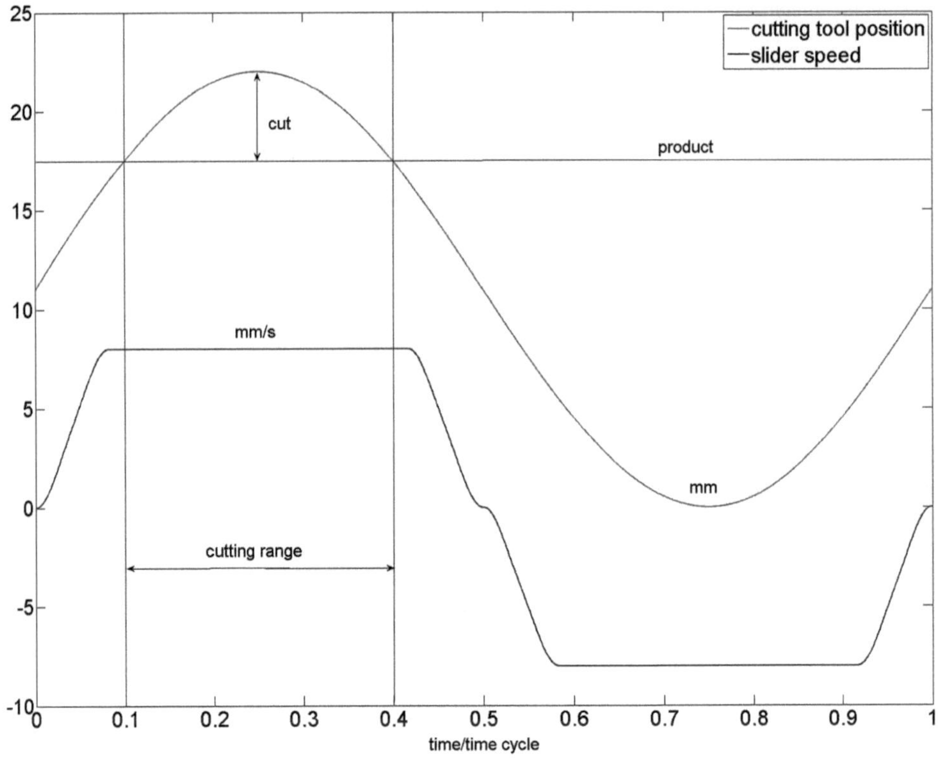

Fig. 21.13 Position of the cutting tool vs. speed of the product during a cycle

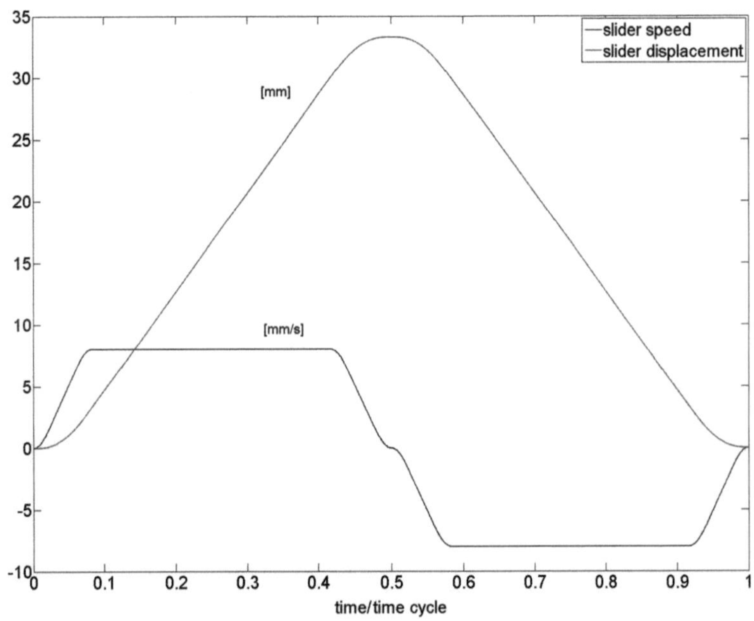

Fig. 21.14 Speed and displacement of the slider during a cycle

21.5 Concluding Remarks

The paper proposed a systemic approach to model the main elements that characterize flying cutting machines and introduced a tool to model these systems. Machine designers have the opportunity to use this tool to describe the kinematics of the machine and to setup all the motion of all the subsystems to guarantee the best performance.

Future works will allow to introduce the dynamics of the machine, thus allowing the right sizing of the gearboxes and the optimization of the motion laws.

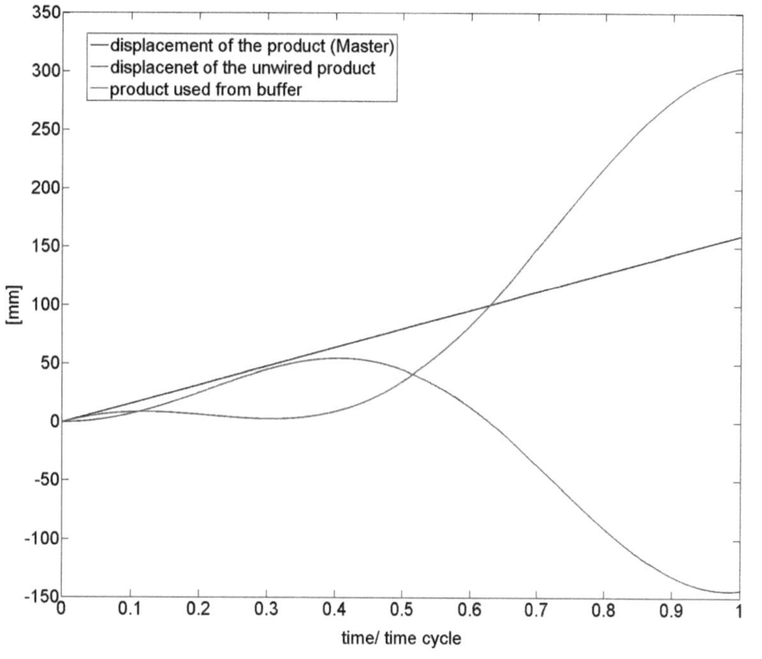

Fig. 21.15 Speed and displacement of the slider during a cycle

References

1. Liberopoulos G, Tsarouhas P (2005) Reliability analysis of an automated pizza production line. J Food Eng 69(1):79–96
2. Hansen D, Holtz J, Kennel R (2003) Cutter distance sensors for an adaptive position/torque control in cross cutters. IEEE Ind Appl Mag 9(5):33–39
3. Peric N, Petrovic I (1990) Flying shear control system. IEEE Trans Ind Appl 26(6):1049–1056
4. Shepherd R (1964) A computer-controlled flying shear. Stud Q J 34(135):143–148. doi:10.1049/sqj.1964.0006
5. Varvatsoulakis MN (2009) Design and implementation issues of a control system for rotary saw cutting. Control Eng Pract 17(1):198–202
6. Sun R-L (2000) A follow-up control system for flying cut. Proc Inst Mech Eng B J Eng Manuf 214(4):327–331
7. Visvambharan BB (1988) On-line digital control system for a flying saw cutting machine in tube mills. In: IECON'88. Proceedings 14 annual conference of Industrial Electronics Society, vol 2, pp 385–390. doi:10.1109/IECON.1988.665170
8. Machado C, Mendes J, Fonseca J (2003) Intelligent cutting-off of pipes and bars. In: IEEE international symposium on industrial electronics, ISIE'03, vol 1, pp 460–465. doi: 10.1109/ISIE.2003.1267293
9. Khare MR (2008) Automatic system for measuring and controlling the length of a moving product in industries. IEEE Trans Instrum Meas 57(4):781–790
10. Lin J-C, Tai J-C, Huang H-H (1995) Design and manufacture of an automatic wrapping machine. In: Proceedings of the international IEEE/IAS conference on industrial automation and control: emerging technologies, pp 698–701
11. Cromarty AD (1994) Automating a horizontal flow-wrapping machine. IEE Colloq Config Servo Contr Syst 176:2/1–5. http://ieeexplore.ieee.org/xpl/articleDetails.jsp?reload=true&arnumber=381724
12. Kolesnikov YN, Gritsenko SA, Eletskikh VI, Polivanov VA (2004) Improving high-capacity drum-type flying shears for cutting rolled bars to measured lengths. Metallurgist 48(11–12):565–567
13. Strada R, Zappa B, Giberti H (2012) An unified design procedure for flying machining operations. In: ASME 2012 11th biennial conference on engineering systems design and analysis, ESDA2012-82392, Nantes, France, vol 2, pp 333–342. doi: 10.1115/ESDA2012-82392
14. Giberti H, Cinquemani S, Legnani G (2011) A practical approach to the selection of the motor-reducer unit in electric drive system. Mech Based Des Struct Mach 39(3):303–319
15. Giberti H, Cinquemani S (2010) On brushless motors continuous duty power rate. In: ASME 2010 10th biennial conference on engineering systems design and analysis, ESDA2010, vol 3, pp 863–872
16. Giberti H, Cinquemani S (2012) The specific accelerating factor to compare brushless motors. In: ASME 2012 11th biennial conference on engineering systems design and analysis, ESDA 2012, vol 2, pp 409–417
17. Giberti H, Clerici A, Cinquemani S (2014) Specific accelerating factor: one more tool in motor sizing projects. Mechatronics 24(7):898–905

Chapter 22
Modal Analysis and Testing of Honeycomb Sandwich Composites

Adarsh Kumar and Ramesh S. Sharma

Abstract In the present work, the modal parameters such as fundamental natural frequency, mode shapes and damping are estimated for honeycomb sandwich composites by varying cell size and its effect on parameters have been studied. Experimental Modal tests were conducted on hexagonal cell honeycombs of sizes 6 and 12 mm maintaining the facing thickness constant at around 1 mm with three different boundary conditions viz C-F-F-F, C-F-C-F and C-C-C-C. The traditional "strike method" has been used to measure the vibration properties. The modal characteristics of the specimens have been obtained by studying its impulse response. Each specimen has been subjected to impulses through a hard tipped hammer which is provided with a force transducer and the response has been measured through the accelerometer.

Keywords Honeycomb • Modal testing • Aluminum • Cell size • Frequency

22.1 Introduction

Sandwich panels have been successfully used for many years in the aviation and aerospace industries, as well as in marine, and mechanical and civil engineering applications. This is due to attendant high stiffness and high strength to weight ratios of sandwich system. Yu and Cleghorn [1] comparatively studied the free flexural vibration of honeycomb panels using three different plate theories. Numerical results indicate that the classical and improved plate theories are not adequate for the flexural vibration analysis of honeycomb panel. Saitoa et al. [2] studied the parameter identification of aluminum honeycomb sandwich panels with the assumption that they can be treated as orthotropic continua. Boudjemai et al. [3] analyzed the frequency dependence of damping for honeycomb sandwich plate. The case of C-F-F-F plate had been studied and the effect of cell size, material nature and damping coefficient are highlighted. The results show that geometry and the type of material have an effect on the value of the honeycomb plate modal frequencies.

Luo et al. [4] conducted the dynamic analysis of the space craft structure made up of the honeycomb sandwich plates in ANSYS, the equivalent parameters of the honeycomb sandwich plates have been identified using three equivalent methods, which are called the sandwich theory, the equivalent-plate theory and the honeycomb-plate theory. The computational results show that the honeycomb-plate theory is the most suitable method to simplify the spacecraft structure. Liu and Zhao [5] studied Vibration frequencies and mode shape of honeycomb sandwich panels with various structural parameters using computational and experimental methods. Two computational models were used to predict the mode shapes and frequencies of honeycomb sandwich panels. In the first model, honeycomb core was assumed to be quasi-orthotropic; in the second model, plate elements were used for honeycomb cell walls to reflect the geometric nature of hexagonal cells.

The quantitative effect of the anisotropic core on the vibration properties of the sandwich panels was studied, and the mode shape and frequencies were presented. Objective of this paper is to find out influence of cell size on modal parameters of honeycomb sandwich composites.

A. Kumar (✉) • R.S. Sharma
Department of Mechanical Engineering, R.V. College of Engineering, Bangalore, India
e-mail: kumar.adarsh1234@yahoo.in; rssharma25@yahoo.com

M. Mains (ed.), *Topics in Modal Analysis, Volume 10*, Conference Proceedings of the Society for Experimental Mechanics Series, DOI 10.1007/978-3-319-15251-6_22

22.2 Experimental Procedure

22.2.1 Specimen Details

Two different cell sizes viz 6 and 12 mm of aluminium honey comb sandwich panels were prepared to study their influence on the dynamic characteristics. The face sheet and core both are aluminium here. The types of specimens investigated in this study are in the form of plates. The specimens are cut with effective dimensions 160 × 160 mm to obtain cantilever, two ends fixed and all ends fixed conditions. Similarly, specimens were prepared for C-F-F-F, C-F-C-F, C-C-C-C conditions.

22.2.2 Modal Test Method

The modal characteristics of the specimen have been obtained by studying its impulse response. The specimen was fixed at one end to simulate the clamped-free-free-free (C-F-F-F) condition shown in Fig. 22.1, (C-F-C-F) condition shown in Fig. 22.2, (C-C-C-C) condition shown in Fig. 22.3. The specimen has been subjected to impulses through a hard tipped hammer which is provided with a force transducer (PCB make) with a sensitivity of 2.25 mV/N and the response has been measured through the accelerometer (PCB make) with an accelerometer of sensitivity 10 mV/g. The impulse and the response are processed on a computer aided FFT analyzer test system in order to extract the modal parameters with the help of built in software. The sandwich specimen has been subjected to impulses at 25 station locations. The response has been measured by placing the accelerometer at station 1. Due to inherent damping in the specimen, the test was restricted to fundamental vibration mode with the impact hammer. The test was conducted for all the types of specimen with three different boundary conditions and the results recorded (Fig. 22.4).

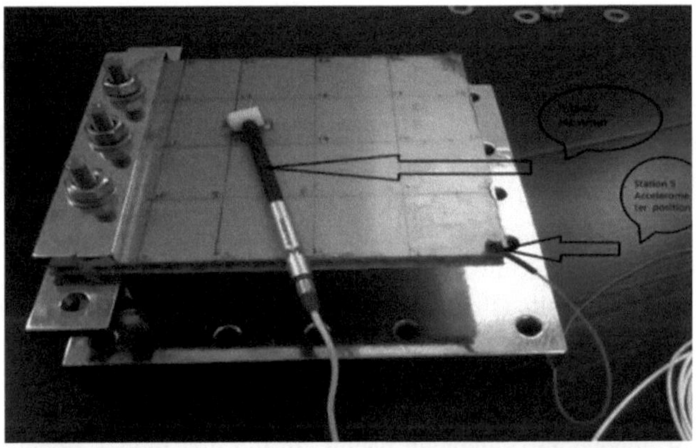

Fig. 22.1 C-F-F-F condition

Fig. 22.2 C-F-C-F condition

Fig. 22.3 C-C-C-C condition

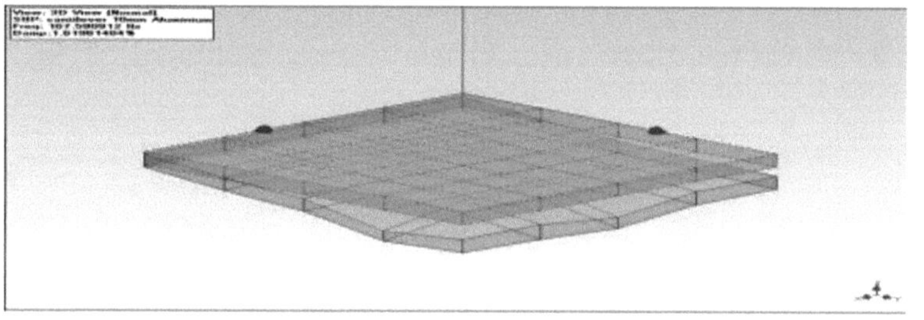

Fig. 22.4 Mode 1—symmetric bending

Table 22.1 Geometric details of sandwich panel (C-F-F-F)

Designation	Length × width (mm)	Facing thickness (mm)	Core cell size (mm)	Wall thickness (mm)	Core thickness (mm)
C−6	160 × 160	0.93	6	0.1	8.23
C−12	160 × 160	1	12	0.1	17.93

Table 22.2 Geometric details of Sandwich panel (C-F-C-F)

Designation	Length × width	Facing thickness (mm)	Core cell size (mm)	Wall thickness (mm)	Core thickness (mm)
C−6	160 × 160	0.98	6	0.1	7.98
C−12	160 × 160	1.23	12	0.1	18.11

Table 22.3 Geometric details of sandwich panel (C-C-C-C)

Designation	Length × width	Facing thickness (mm)	Core cell size (mm)	Wall thickness (mm)	Core thickness (mm)
C−6	160 × 160	1.1	6	0.1	8.01
C−12	160 × 160	0.98	12	0.1	17.98

22.3 Results and Discussions

Tables 22.1, 22.2, and 22.3 indicate the geometric details of sandwich panel for end conditions C-F-F-F, C-F-C-F and C-C-C-C conditions.

22.3.1 Experimental Results

22.3.1.1 Mode Shape

Figures 22.5, 22.6, and 22.7 indicates the different mode Shapes for honeycomb sandwich panel with cell size 6 and 12 mm with C-F-F-F condition

Similarly 18 mode shapes, 3 each for C-F-C-F, C-F-F-F and C-C-C-C conditions for cell sizes 6 and 12 mm have been extracted and the frequency values are compared for different cell sizes.

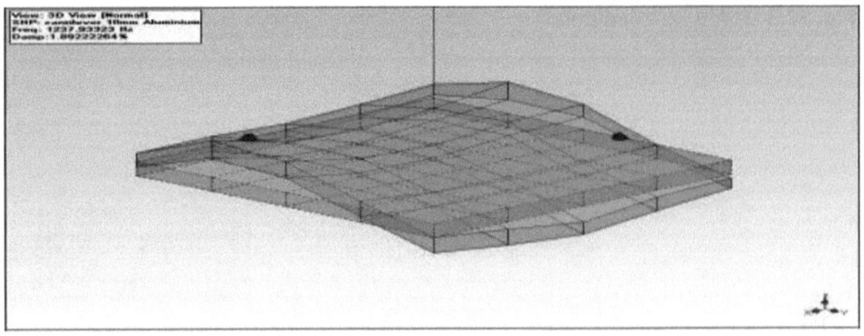

Fig. 22.5 Mode 2—anti symmetric bending

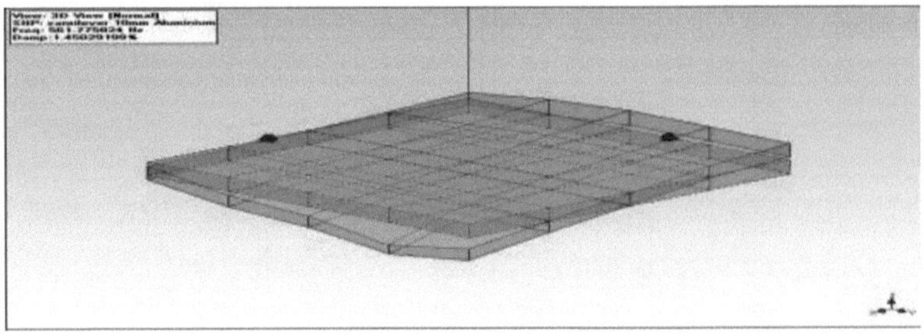

Fig. 22.6 Mode 3—torsion

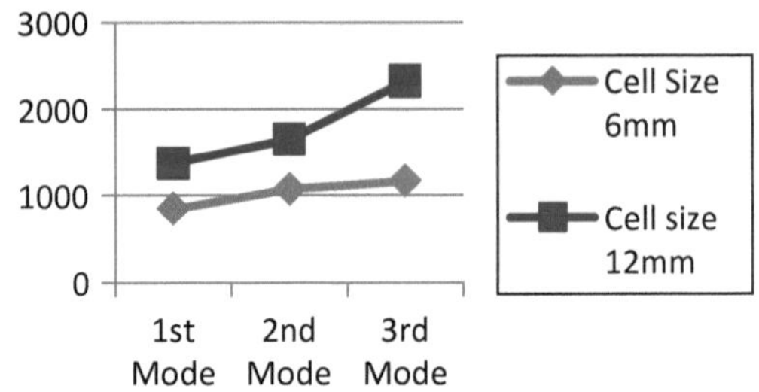

Fig. 22.7 C-F-C-F condition

Table 22.4 (C-F-C-F)

Sl No.	Cell size (mm)	f_1 (Hz)	% damping	f_2 (Hz)	% damping	f_3 (Hz)	% damping
1	6	841	4.63	1070	2.13	1,165	2.21
2	12	1, 371	1.44	1,648	2.60	2,316	3.34

Table 22.5 (C-F-F-F) condition

Sl No.	Cell size (mm)	f_1	% damping	f_2	% damping	f_3	% damping
1	6	167	1.6196	561	1.45029	1,237	1.892233
2	12	237	2.2068	261	7.6222	969	1.3710

Table 22.6 (C-C-C-C) condition

Sl No.	Cell size (mm)	f_1	% damping	f_2	% damping	f_3	% damping
1	6	1,354	2.6978	1,631.9	0.84	1,703.9	1.0958
2	12	1,706	0.8750	2,183	1.2301	2,324	1.0321

22.3.2 *Frequency and Damping*

22.3.3 *Graphs*

22.3.3.1 Frequency Variation with Varying Cell Size

22.3.4 *Discussions*

Modal analysis is performed to study the influence of cell size on modal parameters of honeycomb sandwich panels. Thus results of resonant frequencies, modal damping has been shown in Tables 22.4, 22.5, and 22.6 for three different boundary conditions. It can be noticed from the tables, that as cell size increases, there is steep increase in the frequency for different boundary conditions tested. Figures 22.7, 22.8, and 22.9 depicts the natural frequency versus mode number for three different boundary condition, it can be observed that the natural frequency decreases in case of C-F-F-F condition, emphasizing on

Fig. 22.8 C-F-F-F condition

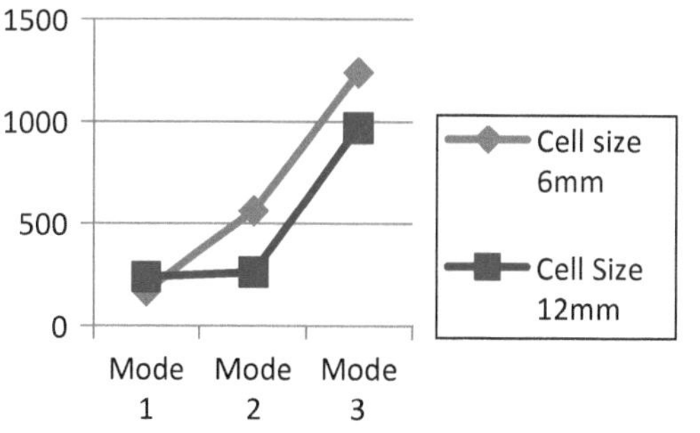

Fig. 22.9 C-C-C-C condition

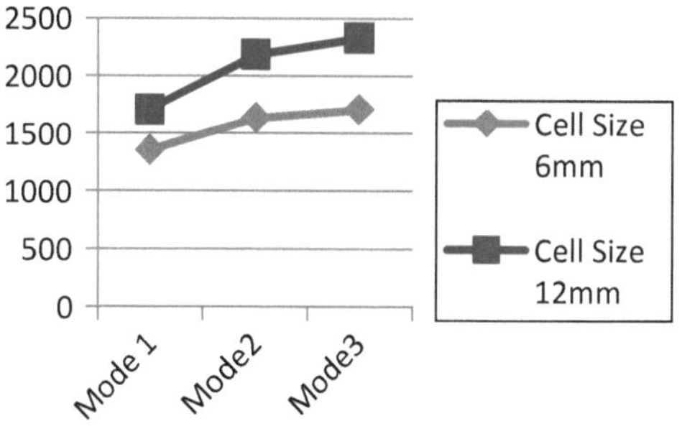

the fact that the structural damping may be dominant in this area. However, thus frequency increases upon increasing the cell size for all the modes captured for the other two conditions. Three different mode shapes recorded for C-F-F-F case are shown in Figs. 22.4, 22.5, and 22.6 respectively.

Conclusions

In this paper, an attempt has been made to evaluate the dynamic behavior of honeycomb sandwich panels with two different cell size. The conclusions are outlined as follows:

- The effect of cell size has a significant role to play in the dynamic behavior of honeycomb sandwich composites.
- It is observed that the fundamental natural frequency increases upon increasing the cell size for C-F-C-F and C-C-C-C conditions and decreases for C-F-F-F condition.
- Modal analysis technique adopt in this work has yielded good results which provides an insight to designers' for designing honeycomb sandwich panels subjected to dynamic loading conditions.

Acknowledgments Authors gratefully acknowledge the Management, Principal, Head of the Department of Mechanical Engineering, and Dean, PG Studies for their constant encouragement and support in carrying out this work.

References

1. Yu SD, Cleghorn WL (2005) Free flexural vibration analysis of symmetric honeycomb panels. J Sound Vib 284(1–2):189–204
2. Saitoa T, Parberyb RD, Kawnoa S (1997) Parameter identification for aluminium honeycomb sandwich panels based on orthotropic Timoshenko beam theory. J Sound Vib 208(2):271–287
3. Boudjemai A, Bouanane MH, Mankour RA, Salem H, Chouchaoui B (2012) MDA of hexagonal honeycomb plates used for space application. World Acad Sci Eng Technol 66:221–229
4. Luo H, Liu G, Ma S, Liu W (2011) Dynamic analysis of the spacecraft structure on orbit made up of honeycomb sandwich plates. In: 2011 IEEE international conference on computer science and automation engineering, vol 1, pp 83–87
5. Liu QL, Zhao Y (2002) Role of anisotropic core in vibration properties of honeycomb sandwich panels. J Thermoplast Compos Mater 15(1): 944–952

Chapter 23
Towards an Automatic Modal Parameter Estimation Framework: Mode Clustering

Majid Khorsand Vakilzadeh, Vahid Yaghoubi, Anders T. Johansson, and Thomas J.S. Abrahamsson

Abstract The estimation of modal parameters from a set of measured data is a highly judgmental task, with user expertise playing a significant role for distinguishing between physical and spurious modes. However, it can be very tedious especially in situations when the data is difficult to analyze. This study presents a new algorithm for mode clustering as a preliminary step in a multi-step algorithm for performing physical mode selection with little or no user interaction. The algorithm commences by identification of a high-order model from estimated frequency response functions to collect all the important characteristics of the structure in a so-called library of modes. This often results in the presence of spurious modes which can be detected on the basis of the hypothesis that spurious modes are estimated with a higher level of uncertainty comparing to physical modes. Therefore, we construct a series of data using a simple random sampling technique in order to obtain a set of linear systems using subspace identification. Then, their similar modes are grouped together using a new correlation criterion, which is called Modal Observability Correlation (MOC). An illustrative example shows the efficiency of the proposed clustering technique and also demonstrates its capability to dealing with inconsistent data.

Keywords FRF based N4SID • Modal observability correlation • Clustering • Modal parameters • QR- and singular value decomposition • Inconsistent data

23.1 Introduction

During the last decades a lot of effort has been put to develop efficient algorithms for identification of the modal parameters, both in time and frequency domain. A central problem in most of these algorithms is to determine the correct model order which, theoretically, equals the number of eigenvalues that constitute a state-space model [1].

In the framework of system identification, several model validation algorithms are available to automatically estimate the model order. The criteria, such as Akaike Information Criterion (AIC) [2] for Maximum likelihood Estimator and Singular Value Criterion (SVC) [3] for subspace algorithms, compare models of different orders in terms of the information gain or loss and search for the model with the best prediction capability. Although these criteria can perform well in order to validate models, they are inadequate to select the model which most fully captures the pertinent physical characteristics of the test structure in the operating frequency range [1].

On the contrary, in modal analysis community, the primary interest is to identify the model consists of physically relevant modes in the frequency range of interest, rather than a model with high prediction capability. To this target, a model with an order much larger than needed is usually identified to ensure that all modes present in the considered frequency band are captured [4, 5]. Unfortunately, this results in the appearance of computational, or spurious, modes in the model which have no physical relevance.

Probably, the so-called stabilization diagram is the most common tool to remove spurious modes from a model [6, 7]. This diagram is constructed using estimated eigenfrequencies of models with increasing order. The idea behind this approach is that for a physical mode, estimated eigenfrequencies appear near the same frequency while for a spurious mode are scattered [8]. However, in this technique selection of physical modes is a highly judgmental task and demands lots of user interaction. Moreover, highly noisy data results in a stabilization diagram which is hard to interpret and this, in turn, leads to distinct outcomes contingent upon the user.

The available literature on automated modal parameter estimation falls into two main categories [1]. The first category consists of techniques which are based on a single record of test data. Examples include genetic algorithm [9], neural network

M.K. Vakilzadeh • V. Yaghoubi (✉) • A.T. Johansson • T.J.S. Abrahamsson
Department of Applied Mechanics, Chalmers University of Technology, Göteborg SE-412 96, Sweden
e-mail: khorsand@chalmers.se

M. Mains (ed.), *Topics in Modal Analysis, Volume 10*, Conference Proceedings of the Society for Experimental Mechanics Series, DOI 10.1007/978-3-319-15251-6_23

[10], and clustering. The majority of these studies group similar modes together using clustering techniques. Hierarchical clustering or connectivity based clustering begins by assigning one cluster to each mode in the stabilization diagram. Then, it proceeds by merging closest clusters together until the distance between all clusters exceeds a user-defined threshold. Finally, physical modes are clusters in which the number of their modes is larger than a threshold [11]. In centroid based clustering, for instance k-means and fuzzy c-means [4, 8], clusters are represented using a mean vector, or value, which is not needed to be a member of data set. The main drawback of this approach is that the number of clusters is assumed to be a priori known value, which is often not the case.

The second category involves strategies that are based on several data records from one structure, using fuzzy c-means clustering [12], or a group of similar structures, using self-learning algorithms [7]. In the latter study, a simple distance based clustering algorithm groups similar modes together in a stabilization diagram. Then, the final decision on the nature of a cluster, either physical or spurious, is made by a self-learning Support Vector Machine (SVM) algorithm. Once the SVM algorithm is trained using a set of data obtained from a synthetic experiment, the algorithm would automatically estimate modal parameters for other test data.

This study presents a new algorithm for mode clustering as a preliminary step in a multi-step algorithm for autonomous modal parameter estimation. We begin the paper with a brief review of the relevant aspects of linear systems theory. Next, a new correlation metric proposed which circumvent the difficulty of existing correlation metrics to deal with spatial aliasing, coalescent modes, and coherent noise induced modes. Then, after a short statement of the existing problems in automation of modal parameter estimation algorithm, an overview of the proposed autonomous procedure is given. Afterwards, different building blocks of the proposed mode clustering algorithm are explained in details. This section also demonstrates how singular value decomposition can be effectively employed to automate the process of detection and mitigation of inconsistencies in the test data. The effectiveness of the proposed algorithm is illustrated using synthetic data from an academic example.

23.2 Theory

This section begins with a brief review of the relevant features of linear systems theory. Next, a new correlation criterion is proposed to ease the problem of consistency analysis between different estimates of a modal vector.

23.2.1 *Background*

A continuous, linear, time-invariant (LTI) system can be represented in the state-space form as,

$$\begin{cases} \dot{\boldsymbol{x}}(t) = \boldsymbol{A}\boldsymbol{x}(t) + \boldsymbol{B}\boldsymbol{u}(t) \\ \boldsymbol{y}(t) = \boldsymbol{C}\boldsymbol{x}(t) + \boldsymbol{D}\boldsymbol{u}(t) \end{cases} \tag{23.1}$$

where $\boldsymbol{x}(t) \in \mathbb{R}^{n_\mathrm{x}}$ is the state vector, $\boldsymbol{A} \in \mathbb{R}^{n_\mathrm{x} \times n_\mathrm{x}}$, $\boldsymbol{B} \in \mathbb{R}^{n_\mathrm{x} \times n_\mathrm{u}}$, $\boldsymbol{C} \in \mathbb{R}^{n_\mathrm{y} \times n_\mathrm{x}}$ and $\boldsymbol{D} \in \mathbb{R}^{n_\mathrm{y} \times n_\mathrm{u}}$ are internal evolution, input, output and feedthrough matrices, respectively, $\boldsymbol{u}(t) \in \mathbb{R}^{n_\mathrm{u}}$ is the system input and $\boldsymbol{y}(t) \in \mathbb{R}^{n_\mathrm{y}}$ is the system output. We denote this system using $\boldsymbol{\Sigma}$ throughout. Whenever the input vector is exponentially bounded, the Laplace transform can be used to write the transfer function of system $\boldsymbol{\Sigma}$ as $\boldsymbol{G}(s) = \boldsymbol{C}(s\boldsymbol{I} - \boldsymbol{A})^{-1}\boldsymbol{B} + \boldsymbol{D}$ for $s \in \mathbb{C}$.

In general, the state-space realization of a given LTI system is not unique. It means that a different choice of basis for the state vector of $\boldsymbol{\Sigma}$ leads to another state-space representation while both describe the same input–output behavior. More precisely, given the transformation $\boldsymbol{x}_\mathrm{T}(t) = \boldsymbol{T}\boldsymbol{x}(t)$, the quadruple $(\boldsymbol{A}, \boldsymbol{B}, \boldsymbol{C}, \boldsymbol{D})$ of $\boldsymbol{\Sigma}$ will be transferred to a new state-space representation as,

$$\boldsymbol{\Sigma}_\mathrm{T} = (\boldsymbol{A}_\mathrm{T}, \boldsymbol{B}_\mathrm{T}, \boldsymbol{C}_\mathrm{T}, \boldsymbol{D}) = \left(\boldsymbol{T}\boldsymbol{A}\boldsymbol{T}^{-1}, \boldsymbol{T}\boldsymbol{B}, \boldsymbol{C}\boldsymbol{T}^{-1}, \boldsymbol{D}\right) \tag{23.2}$$

where $\boldsymbol{\Sigma}_\mathrm{T}$ is called equivalent to $\boldsymbol{\Sigma}$. This results to several possible representations of a dynamical system with distinct properties. One common representation in system theory, and also structural dynamics, is the balanced representation. Roughly speaking, the idea of balanced representation is to represent a system using a state vector such that the degree of controllability and observability of each state are the same. This goal is achieved using the so-called balancing transformation, $\widetilde{\boldsymbol{T}}$. Assuming that the system, $\boldsymbol{\Sigma}$, is controllable and observable the balancing transformation makes its controllability Gramian,

$$\boldsymbol{W}_{\mathcal{C}} = \int_0^{\infty} e^{\boldsymbol{A}t}\boldsymbol{B}\boldsymbol{B}^{\mathrm{H}}e^{\boldsymbol{A}^{\mathrm{H}}t}dt \tag{23.3}$$

and observability Gramian,

$$\boldsymbol{W}_{\mathcal{O}} = \int_0^{\infty} e^{\boldsymbol{A}^{\mathrm{H}}t}\boldsymbol{C}^{\mathrm{H}}\boldsymbol{C}e^{\boldsymbol{A}t}dt \tag{23.4}$$

to be equal and diagonal as follows

$$\widetilde{\boldsymbol{W}}_{\mathcal{C}} = \widetilde{\boldsymbol{W}}_{\mathcal{O}} = \operatorname{diag}(\sigma_1, \sigma_2, \ldots, \sigma_n) \tag{23.5}$$

Here, $\widetilde{\boldsymbol{W}}_{\mathcal{C}} = \widetilde{\boldsymbol{T}}\boldsymbol{W}_{\mathcal{C}}\widetilde{\boldsymbol{T}}^{\mathrm{H}}$ and $\widetilde{\boldsymbol{W}}_{\mathcal{O}} = \widetilde{\boldsymbol{T}}^{-\mathrm{H}}\boldsymbol{W}_{\mathcal{O}}\widetilde{\boldsymbol{T}}^{\mathrm{H}}$ are the corresponding gramians for the balanced system $\widetilde{\boldsymbol{\Sigma}} = \left(\widetilde{\boldsymbol{A}}, \widetilde{\boldsymbol{B}}, \widetilde{\boldsymbol{C}}, \boldsymbol{D}\right)$, and $(.)^{\mathrm{H}}$ denotes the Hermitian of a matrix. In Eq. (23.5), $\sigma_i,\ i = 1, 2, \ldots, n$ are the so-called Hankel singular values, where $\sigma_1 \geq \sigma_2 \geq \cdots \geq \sigma_n$. The larger a Hankel singular value, the larger is the contribution of the associated state to the input–output behavior of the system [13, 14].

Another common representation in structural dynamics is modal representation. Assume that the dynamical system $\boldsymbol{\Sigma}$ is diagonalizable and let $\boldsymbol{A}$ be a Hermitian matrix which has the eigenvalue set of $\lambda = \{\lambda_1, \lambda_2, \ldots, \lambda_{n_x}\}$ and the corresponding eigenvector matrix $\boldsymbol{\Phi} = [\phi_1, \phi_2, \ldots, \phi_{n_x}]$. Using the transformation $\overline{\boldsymbol{x}}(t) = \boldsymbol{\Phi}^{-1}\boldsymbol{x}(t)$, with $\overline{\boldsymbol{T}} = \boldsymbol{\Phi}^{-1}$, the dynamical system $\boldsymbol{\Sigma}$ can be projected onto modal coordinates in form of the quadruple

$$\overline{\boldsymbol{\Sigma}} = \left(\overline{\boldsymbol{A}},\ \overline{\boldsymbol{B}},\ \overline{\boldsymbol{C}},\ \boldsymbol{D}\right) \tag{23.6}$$

where $\overline{\boldsymbol{x}}(t) \in \mathbb{R}^{n_x}$ is the modal coordinate vector, $\overline{\boldsymbol{A}} = \boldsymbol{\Lambda} = \operatorname{diag}(\lambda_1, \lambda_2, \ldots, \lambda_{n_x}) = \boldsymbol{\Phi}^{-1}\boldsymbol{A}\boldsymbol{\Phi}$, $\overline{\boldsymbol{B}} = \boldsymbol{\Phi}^{-1}\boldsymbol{B} = \left[\overline{\boldsymbol{b}}_1, \overline{\boldsymbol{b}}_2, \ldots, \overline{\boldsymbol{b}}_{n_x}\right]$ is the modally projected input matrix, and $\overline{\boldsymbol{C}} = \boldsymbol{C}\boldsymbol{\Phi} = \left[\overline{\boldsymbol{C}}_1, \overline{\boldsymbol{C}}_2, \ldots, \overline{\boldsymbol{C}}_{n_x}\right]$ is the modally projected output matrix. In particular, it should be noted that the ith column of $\overline{\boldsymbol{C}}$, $\boldsymbol{C}\phi_i$, is the projection of the ith eigenvector of $\boldsymbol{A}$ to the space spanned by the output $\boldsymbol{y}$. In this representation, let

$$\boldsymbol{G}_{-i}(s) = \frac{\overline{\boldsymbol{c}}_i\overline{\boldsymbol{b}}_i}{s - \lambda_i} \tag{23.7}$$

be the transfer function of the error system resulting from deflation of the ith modal coordinate from $\overline{\boldsymbol{\Sigma}}$.Then, the contribution of the ith modal coordinate to the input–output relation of the system in $\mathcal{H}_2$ and $\mathcal{H}_\infty$ senses can be defined as [15].

$$\begin{aligned} \left\|\boldsymbol{G}_{-i}(\mathrm{j}\omega)\right\|_{\mathcal{H}_\infty} &= \max_{\omega\in\Re}\left|\boldsymbol{G}_{-i}(\mathrm{j}\omega)\right| \\ \left\|\boldsymbol{G}_{-i}(\mathrm{j}\omega)\right\|_{\mathcal{H}_2}^2 &= \tfrac{1}{\pi}\operatorname{trace}\left\{\overline{\boldsymbol{b}}_i^{\mathrm{H}}\overline{\boldsymbol{c}}_i^{\mathrm{H}}\overline{\boldsymbol{c}}_i\overline{\boldsymbol{b}}_i\right\}\left[\tfrac{1}{|\Re(\lambda_i)|}\arctan\left(\tfrac{\omega}{|\Re(\lambda_i)|}\right)\right]_{-|\Im(\lambda_i)|}^{\infty} \end{aligned} \tag{23.8}$$

It can be shown that the $\mathcal{H}_\infty$ norm defined in Eq. (23.8) is equivalent to the $\mathscr{L}_2$ induced norm of the associated input–output map [16]. In general it can be assumed that the more important a mode is, the larger will be its contribution to the input–output relation. Thus, the measures in Eq. (23.8) can be decisive measures for discriminating between physical and spurious modes.

23.2.2 Modal Observability Correlation

Historically, Modal Assurance Criterion (MAC) was developed to alleviate the need for a measure of consistency (linearity) between different estimates of a modal vector. For systems with well separated resonance frequencies and with many response measurement stations for representation of modal vector, MAC typically discriminates modal vectors associated to different resonance frequencies with no problem. However, too few sensors in the experimental determination of modal vectors can result in high MAC value between two modal vectors with different resonance frequencies, which is called spatial aliasing. Furthermore, a MAC value near unity can be due to the consistency of a modal vector with a noise vector. The aforementioned problems motivated us to develop Modal Observability Correlation metric which was first proposed in [17].

The spatial aliasing effect arises when two modes of the structure are observed similarly through the system outputs. This raises special interests to consider the relationship between the system's observability and modal properties. One test to assess the observability of the system, $\boldsymbol{\Sigma}$, involves evaluation of the rank of the observability matrix, $\mathcal{O}$, i.e.,

$$\mathcal{O} = \begin{bmatrix} C \\ CA \\ CA^2 \\ \vdots \\ CA^{n-1} \end{bmatrix} \tag{23.9}$$

If $\mathcal{O}$ has full rank, then the system is fully observable. The observability matrix can also be formulated using the modal representation of the system, Eq. (23.6), as

$$\overline{\mathcal{O}} = \begin{bmatrix} C\Phi \\ C\Phi\Lambda \\ C\Phi\boldsymbol{\Lambda}^2 \\ \vdots \\ C\Phi\boldsymbol{\Lambda}^{n-1} \end{bmatrix} = \begin{bmatrix} C\Phi_1 & \cdots & C\Phi_n \\ \vdots & \ddots & \vdots \\ C\Phi_1\lambda_1^{n-1} & \cdots & C\Phi_n\lambda_n^{n-1} \end{bmatrix} = \left[\overline{\mathcal{O}_1} \cdots \overline{\mathcal{O}_n}\right] \tag{23.10}$$

This equation shows that each column of modal observability matrix, $\overline{\mathcal{O}}$, only contains information related to only one vibrational mode. This evidence suggests a departure from traditional MAC formulation and replacement of the modal vectors by the columns of the modal observability matrix, i.e.,

$$b\mathrm{MOC}\left(\overline{\mathcal{O}_i}, \overline{\mathcal{O}_j}\right) = \frac{\left|\overline{\mathcal{O}_i}^{\mathrm{H}}\overline{\mathcal{O}_j}\right|^2}{\left(\overline{\mathcal{O}_i}^{\mathrm{H}}\overline{\mathcal{O}_i}\right)\left(\overline{\mathcal{O}_j}^{\mathrm{H}}\overline{\mathcal{O}_j}\right)} \tag{23.11}$$

We call this new metric basic Modal observability Correlation (*b*MOC). It should be noted here that the *i*th and *j*th columns of $\overline{\mathcal{O}}$ contain not only information about eigenvectors, but also contain information about the associated eigenvalues. This allows for discriminating between two consistent modal vectors with different eigenvalues which makes *b*MOC more preferable than MAC in dealing with the problem of spatial aliasing

In general, *b*MOC is numerically computable for discrete time systems. For continuous time systems, the eigenvalues, $\boldsymbol{\lambda}$, are transformed to the discrete form using

$$\lambda_{\mathrm{disc}} = \exp\left(\boldsymbol{\lambda}.\tau\right) \tag{23.12}$$

where time constant, τ, can be selected arbitrarily. In this study, it is selected as

$$\tau = \pi/\Delta f \tag{23.13}$$

to distribute the discrete-time eigenvalues associated to the resonance frequencies in the considered frequency range, Δf, within the complex plane unit disc.

Since *b*MOC is normalized by the magnitude of the columns of the modal observability matrix, it is only a measure of consistency. This means that it may reflect the consistency between two modal vectors or a true modal vector with a coherent noise vector. Moreover, the modal vectors are arbitrarily scaled while it is known that the random noise vectors often indicate small contributions to the input–output relation of the structure [6]. From these two considerations, we take a step forward and use a coordination transformation such that the modal scale factor now reflects the degree of controllability and observability for a certain mode. Toward this end, let the subsystem associated to each eigenvalue λ_j be written as follows

$$\begin{cases} \dot{\overline{x}}_j(t) = \lambda_j \overline{x}_j(t) + \overline{\boldsymbol{b}}_j \boldsymbol{u}(t) \\ \boldsymbol{y}_j(t) = \overline{\boldsymbol{c}}_j \overline{x}_j(t) + \boldsymbol{D}\boldsymbol{u}(t) \end{cases} \tag{23.14}$$

Using the balancing transformation $\tilde{T} \in \mathbb{R}^1$, the quadruple $\left(\lambda_j, \tilde{\overline{\boldsymbol{b}}}_j, \tilde{\overline{\boldsymbol{c}}}_j, \boldsymbol{D}\right) = \left(\lambda_j, \tilde{T}^{-1}\overline{\boldsymbol{b}}_j, \overline{\boldsymbol{c}}_j \tilde{T}, \boldsymbol{D}\right)$ is then the internally balanced representation of the system associated to the *j*th mode which now is equally controllable and observable, i.e.,

$$\left\| \widetilde{T}^{-1} \overline{\boldsymbol{b}}_j \right\| = \left\| \overline{\boldsymbol{c}}_j \widetilde{T} \right\| \tag{23.15}$$

Here, $||\cdot||$ denotes the magnitude of a vector. Then, the *j*th column of the modal observability matrix can be written as

$$\tilde{\overline{\mathcal{O}}}_j = \widetilde{T}\overline{\mathcal{O}}_j \tag{23.16}$$

Note that, the uniqueness property of the balancing transformation render a meaningful and fixed scaling factor for modal vectors which indicates the extent in which a certain mode contributes to the system's input–output behavior. Acknowledging this fact, we define a new correlation metric as

$$MOC\left(\tilde{\overline{\mathcal{O}}}_i, \tilde{\overline{\mathcal{O}}}_j\right) = \frac{\left| \tilde{\overline{\mathcal{O}}}_i^{\mathrm{H}} \tilde{\overline{\mathcal{O}}}_j \right|^2}{\max\left(\tilde{\overline{\mathcal{O}}}_i^{\mathrm{H}} \tilde{\overline{\mathcal{O}}}_i, \tilde{\overline{\mathcal{O}}}_j^{\mathrm{H}} \tilde{\overline{\mathcal{O}}}_j \right)^2} \tag{23.17}$$

This metric can further be used to compare and contrast consistent modal vectors with close eigenvalues but rather they have strong dissimilarities in the input–output contribution.

23.3 Problem Statement

The ultimate goal of the present study and it's consecutive study is to develop a fully automated modal parameter estimation algorithm such that the following criteria are satisfied: (a) the involving parametric system identification algorithm allows for fast and robust identification of MIMO systems with many outputs, (b) provides the uncertainty bounds on the estimated modal parameters, (c) treats data inconsistencies between resonance frequencies of consecutive test data, (d) needs no user-specified parameters or thresholds, and (e) provides a framework where the identified physical modes that constitute a model with acceptable prediction capacities.

To meet the first criterion, the frequency domain subspace identification is selected in this study as the parametric system identification algorithm. Given N_{f} noise corrupted samples of the frequency response function $\{\boldsymbol{G}_k^*\}$, $k = 1, \ldots, N_{\mathrm{f}}$, with $\boldsymbol{G}_k^* \in \mathbb{C}^{n_{\mathrm{y}} \times n_{\mathrm{u}}}$, the subspace identification algorithm estimates discrete time LTI state-space models of the form

$$\begin{cases} \boldsymbol{x}_{k+1} = \boldsymbol{A}_{\mathrm{d}} \boldsymbol{x}_k + \boldsymbol{B}_{\mathrm{d}} \boldsymbol{u}_k \\ \boldsymbol{y}_k = \boldsymbol{C}_{\mathrm{d}} \boldsymbol{x}_k + \boldsymbol{D}_{\mathrm{d}} \boldsymbol{u}_k \end{cases} \tag{23.18}$$

or equivalently the continuous LTI system of $\boldsymbol{\Sigma}$ using projections of the subspaces spanned by the rows and columns of the block Hankel matrix consisted of the coefficients of the IDFT from the samples of the frequency response function [18]. Thus, this algorithm eliminates the need for expensive optimization of a non-convex cost function as in frequency domain maximum likelihood estimator.

However, the subspace identification does unfortunately not meet the second criterion which is a valuable and decisive feature [1, 4] for discriminating among physical and spurious modes. Here, we circumvent this problem by means of identifying a set of linear models, $\boldsymbol{\Sigma}_i$, $i = 1, \ldots, N_{\mathrm{s}}$, of the same order, n_{r},using N_{s} sets of random sampling from the noisy measurements of the frequency response function. Afterwards, a clustering algorithm, based on a new correlation criterion, collects similar modes together where the statistical information of each cluster will now yield the uncertainty bounds on the estimated model parameters.

In many practical cases, the data obtained from a series of consecutive tests which may result in inconsistency between resonance frequencies. When a model is fitted to this type of data multiple close modes can be identified instead a single mode near a single resonance. To the authors' knowledge, no automated modal parameter estimation algorithm in the literature tried to alleviate this problem. Nevertheless, a pragmatic remedy is presented in this study.

It should be mentioned that the fourth and fifth criteria would be addressed in a subsequent paper where a self-learning classification algorithm will be employed for distinguishing between physical and spurious clusters and it, hence, avoids the need for user specified parameters or thresholds. The consecutive paper also introduces a new technique to improve the prediction capability of the model constituted from the identified individual physical modes. To the authors' knowledge, identification methods developed in modal analysis do not often address this particular model identification criterion.

23.4 Autonomous Modal Parameter Estimation

As shown in Fig. 23.1, the approach proposed for clustering algorithm consists of three main steps, which will be discussed in details in next sections.

(1) Identification of a high order model using frequency domain subspace identification algorithm and, subsequently, establishing a library of modes and estimation of model order for repeated model realizations.
(2) Repeated random sampling of data and identification of new model realizations with the model order found in the previous step.
(3) Clustering similar modes of the repeated realizations using the new correlation criterion.

23.4.1 Library of Vibrational Modes

Assume noise corrupted samples of the frequency response function, $\{G_k^*\}$, $k = 1, \ldots, N_\mathrm{f}$, with $G_k^* \in \mathbb{C}^{n_\mathrm{y} \times n_\mathrm{u}}$, are available and a linear model, $\boldsymbol{\Sigma}_\mathrm{e}$ of order n_e has been estimated using subspace algorithm. Assume further that from this model $\boldsymbol{\Sigma}_\mathrm{e}$ a set of modes are obtained and characterized by a frequency, damping value and corresponding modal vector, $\varphi_i, i = 1, \ldots, n_\mathrm{e}$.

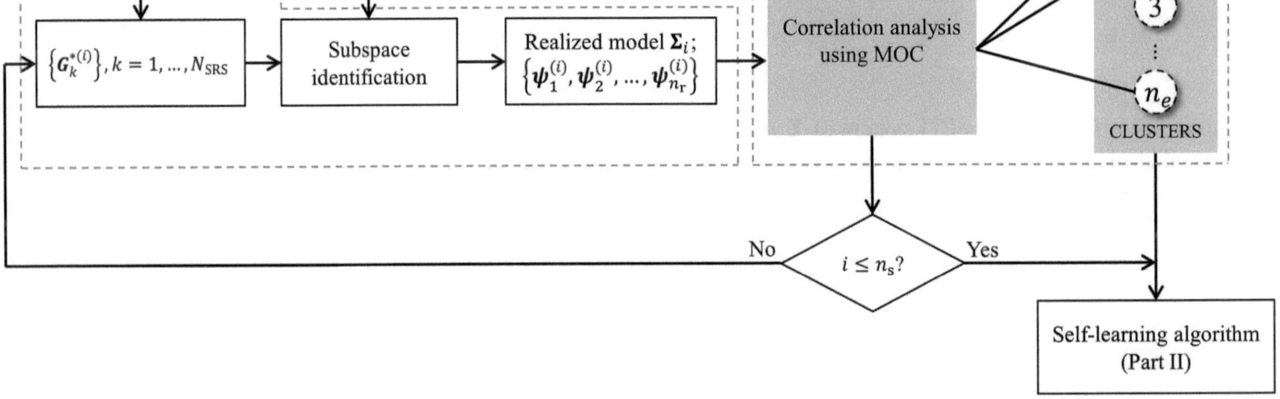

Fig. 23.1 Flowchart of the proposed mode clustering algorithm

In subspace identification, the model in Eq. (23.1) is obtained using projections of rows and columns of so-called block Hankel matrices containing the inputs and outputs of the system. These projections can typically be calculated using basic tools as QR- and singular value decompositions and making the method inherently robust [18].

This model serves two significant roles in the proposed clustering algorithm of Fig. 23.1. First, an exhaustive model is identified with an order much higher than necessary to capture all pertinent physical characteristics of the test structure in the considered frequency band. Therefore, the physical eigenvector space of the structure is a subspace of the space spanned by the eigenvectors of the exhaustive model. In other words, the exhaustive model produces a Library, $\mathscr{L}$, consisting a very large number of eigenvectors, library members. Then, each member of this library is representative for one cluster. Later, we will show that one can adaptively scan the library, $\mathscr{L}$, to cluster modes obtained from the set of realized models with the library member which shows the highest correlation.

Secondly, evaluation of uncertainty bounds for estimation of modal parameters using repeated random sampling of data can be exceedingly expensive if a high order model, with the same order as the exhaustive model, is fitted to each random data set. This motivates the need for reduction of the dimension of the model in some way such that it is still higher than the number of physical modes. Here, we take the information criterion, Singular Value Criterion (SVC), proposed in [3] to estimate the model order, n_r, in the context of subspace identification. This criterion is defined as;

$$SVC(n) = \sigma_{n+1}^2 + \frac{C\left(N_f\right) d(n)}{N_f}, \quad n = 1, 2, \ldots, n_e \tag{23.19}$$

where $d(n) = n(m+s) + ns + sm$ is the number of parameters of a state space model of order n, $C(N_f)/N_f$ is the weight for the penalty term, N_f is number of data points and σ_{n+1} is the estimated Hankel singular values of balanced Gramian. This formulation compares the significance of adding another state to the model to a penalty term. The choice $C(N_f) = \log N_f$ is the lowest bound on the penalty term to have a consistent estimator of model order while it also results in the overestimation of the model order. The model order which minimizes this criterion function will be subsequently used as the order of models fitted to random sampling datasets.

23.4.1.1 Dealing with Inconsistent Data

At this point, the library of modes is in place and it remains to adjust these modes, as the representative for the clusters, such that enables the autonomous modal parameter estimation procedure to deal with inconsistent data. Here, data inconsistency refers to inconsistency between resonance frequencies which is mostly due to the inertia loading of the structure by accelerometers or force transducers. The most noticeable situation that this type of data inconsistency is encountered is when a series of consecutive shaker excitations needs to be executed and, specifically, the resonance frequencies would differ among different patches of data. Presence of such phenomenon in the test data introduces closely spaced eigenvalues to the identified system, where there should only exist a single eigenvalue, with similar eigenvectors [19]. Here, we introduce an approach to eliminate such similar modes from library and only keep the dominant, average, mode shape. This merges the ensemble of identified modes corresponding to a resonance frequency, with a slight change from patch to patch, into one cluster.

To treat such a problem, assume $\{\varphi_j\}, j = 1, \ldots, n_e$ represents the sorted mode shapes in the library based on the magnitude of their eigenvalues and $\{\overline{\mathcal{O}}_j\}, j = 1, \ldots, n_e$ denotes the corresponding modal observability columns which are calculated using Eq. (23.10). Assume further that $\boldsymbol{b}\mathbf{MOC} \in \mathbb{R}^{n_e \times n_e}$ demonstrates the unitary auto-MAC matrix for all the modal observability columns $\{\overline{\mathcal{O}}_j\}$. Then the Singular Value Decomposition (SVD) of $\boldsymbol{b}\mathbf{MOC}$ can be defined as

$$\boldsymbol{b}\mathbf{MOC} = \boldsymbol{U}\,\boldsymbol{\Xi}\,\boldsymbol{V}' \tag{23.20}$$

where the column of $\boldsymbol{U} = \left[\boldsymbol{u}_1 \cdots \boldsymbol{u}_{n_e}\right] \in \mathbb{R}^{n_e \times n_e}$ are left singular vectors, columns of $\boldsymbol{V} = \left[\boldsymbol{v}_1 \cdots \boldsymbol{v}_{n_e}\right] \in \mathbb{R}^{n_e \times n_e}$ are right singular vectors, $\Xi = \mathrm{diag}\left(\varepsilon_1, \ldots, \varepsilon_{n_e}\right)$, $\varepsilon_1 \geq \cdots \geq \varepsilon_{n_e} \geq 0$ denotes singular value matrix of $\boldsymbol{b}\mathbf{MOC}$, and $(.)'$ denotes the transpose of real valued matrix.

Using the definition in Eq. (23.20), the following elegant properties hold:

(a) Let N_d be the number of singular values for which $\varepsilon_1^2 \geq \cdots \geq \varepsilon_{N_d}^2 > 1$. Then, there exists N_d groups of consistent modal observability columns in the library.
(b) Let $g_i = \left[\varepsilon_i^2\right] + 1$, then g_i demonstrates the number of consistent columns in each group, [.] denotes the round down of a value.

(c) The indices for the modal observability columns in the ith group, associated to ε_i, is the same as the indices of the g_i largest elements in the left singular vector.

Effectively, one can use these properties to find the groups of similar modal observability columns associated to the modes in the library. It should be noted that the present consistency of modes in each group can be due to either data inconsistency, or coherent noise vectors, where this theory treats both of them.

Now, let $\left\{\overline{\mathcal{O}}_j^{\mathrm{c}}\right\}$, $j = 1, \ldots, N_{\mathrm{d}}$, represents the groups of consistent modal observability columns. A SVD is performed on each group such that

$$\overline{\mathcal{O}}_j^{\mathrm{c}} = \boldsymbol{U}_j^{\mathrm{c}} \boldsymbol{\Xi}_j^{\mathrm{c}} \boldsymbol{V}_j^{\mathrm{c}\,'} \tag{23.21}$$

If we assume that $\boldsymbol{\Xi}_j^{\mathrm{c}}$ has N_j nonzero singular values, it can be shown that the first N_j columns of $\boldsymbol{U}_j^{\mathrm{c}}$ are the dominant basis for the column space of $\overline{\mathcal{O}}_j^{\mathrm{c}}$ and can hence substitute $\overline{\mathcal{O}}_j^{\mathrm{c}}$. The main advantage of this method is that the cluster that will be represented by the mode with the most dominant observability column will collect all the identified modes associated to one resonance frequency with inconsistency from patch to patch.

23.4.2 Repeated Random Sampling of Data

It can be expected that the level of uncertainty on the estimated modal properties is small for the physical modes and much larger for the spurious ones. Indeed, Verboven et al. [4] reported that the standard deviation of the estimated spurious eigenvalues is usually 10–100 times larger than that of the physical eigenvalues. As a consequence, uncertainty bound on the estimated modal parameters is one of the valuable and decisive information for discriminating between physical and spurious modes. Unfortunately, subspace identification does not allow for estimation of the level of uncertainty on the estimated modal parameters. However, there exist statistical techniques operating on the measured data to infer the level of uncertainty on any parametric estimator. One of such techniques, and probably the most basic method of sampling, is the Simple Random Sampling (SRS) without replacement. The idea behind SRS without replacement is that in finite population sampling, sampling the same data more than once provides no additional information and data samples of size N_{SRS} is generated such that every possible subset of size N_{SRS} from the empirical data has the same probability of being selected as sample data. In design of an SRS, one must decide about two parameters; the sample size, N_{SRS}, and number of generated sample data sets, N_{S}. The selection of the sample size depends on the tolerable amount of sampling error in the modal parameter estimates and the desired computational efficiency, see [20] for further information.

In this study, a set of SRS samples of size N_{SRS}, $\left\{\boldsymbol{G}_k^{*(q)}\right\}$, $k = 1, \ldots, N_{\mathrm{SRS}}$, $q = 1, \ldots, N_{\mathrm{s}}$ is yielded from noise corrupted frequency response function $\{\boldsymbol{G}_k^*\}$, $k = 1, \ldots, N_{\mathrm{f}}$. Subsequently, a set of linear models, $\boldsymbol{\Sigma}_q, q = 1, \ldots, N_{\mathrm{s}}$, of order n_r is estimated using subspace identification on each SRS sample. From now on we call the estimated models using each sample data realized models, $\boldsymbol{\Sigma}_q$. From each realized model, $\boldsymbol{\Sigma}_q$, a set of modes with the following properties can be computed.

$$\begin{aligned} \omega_j^{(q)} &= \Im\left(\lambda_j^{(q)}\right), \\ \xi_j^{(q)} &= -\Re\left(\lambda_j^{(q)}\right) / \left|\lambda_j^{(q)}\right|, \\ \boldsymbol{\psi}_j^{(q)} &= \overline{\boldsymbol{C}}^{(q)}(:, j) = \overline{\boldsymbol{C}}_j^{(q)} \end{aligned} \tag{23.22}$$

Here, $\omega_j^{(q)}$, $\xi_j^{(q)}$, $\boldsymbol{\psi}_j^{(q)}$ are the jth damped natural frequency, damping coefficient and mode shape of $\boldsymbol{\Sigma}_q$, respectively, and $\Im(.)$, $\Re(.)$ and $|.|$ denote imaginary, real and absolute value of a complex value and j ranges from 1 to n_r for each identified model.

23.4.3 Clustering Based on Correlation Analysis

In general, the target of clustering is to put a set of objects into different groups such that the objects in a group are more similar to each other, in some sense, than to the other objects. However, the measure of similarity cannot be uniquely defined which is one reason for development of many different clustering algorithms in statistical data analysis and machine learning fields. Some of these algorithms such as hierarchical clustering [9, 21], k-means algorithm and fuzzy C-means clustering [4] are used in field of modal analysis to automate the interpretation of stabilization diagram. As the main drawbacks, these algorithms rely on user-defined parameters and thresholds and they are mostly computationally intensive.

This study proposes a straightforward procedure for clustering of vibrational modes. In this procedure, each cluster is represented by a mode from the constructed library, $\mathscr{L}$. Hence, the number of clusters is fixed by the number of modes in $\mathscr{L}$ which is overestimated to eliminate the dependency of the proposed clustering algorithm on the initial estimation of the number of clusters. The clustering procedure starts with an adjustment step where the observability columns associated to the representative modes are modified such that the clustering algorithm can also deal with inconsistent data, as explained in Sect. 23.4.1.1. Then, the procedure proceeds with an assignment step for each realized model. In this step, each mode of a realized model is assigned to the cluster whose the representative mode gives the highest correlation to this mode. The correlation criterion used in this step is the Modal Observability Correlation (MOC) [17], mentioned in the theory part of this study. Therefore, given a set of modes, $\boldsymbol{\psi}_p^{(q)},\ p = 1, \ldots, n_\mathrm{r}$, corresponding to the realized system, $\boldsymbol{\Sigma}_q$, the proposed clustering algorithm aims to partition these modes into sets of clusters, $\mathcal{S}_i,\ i = 1, \ldots, n_\mathrm{e}$, such that

$$\mathcal{S}_i = \left\{\boldsymbol{\psi}_p^{(q)} \middle| MOC\left(\boldsymbol{\psi}_p^{(q)}, \varphi_i\right) > MOC\left(\boldsymbol{\psi}_p^{(q)}, \varphi_j\right),\ \forall j : 1 \leq j \leq n_\mathrm{e}\right\}, \qquad p = 1, \ldots, n_\mathrm{r} \tag{23.23}$$

This step would then be repeated for all realized models, $\boldsymbol{\Sigma}_q, q = 1, \ldots, N_\mathrm{s}$. The most apparent advantages of this algorithm are; first, this procedure can be implemented very efficiently due to the simplicity of the decision process in the assignment step. Secondly, the algorithm is not iterative as is the case for instance for typical k-means algorithm. Finally, the proposed algorithm does not rely on any user-defined parameters or thresholds.

23.4.3.1 Dealing with Coalescent Eigenvalues

A central problem in modal analysis is that of coalescent or closely spaced eigenvalues. Let $\boldsymbol{\Phi}_j \in \mathbb{C}^{n \times n_\mathrm{c}}$ represents the matrix which contains the linear independent eigenvectors associated a coalescent eigenvalue, λ_j, of dimension n_c. Theoretically, n_c cannot exceed the number of actuators, n_u. The space spanned by the columns of $\boldsymbol{\Phi}_j$, $\boldsymbol{\mathcal{V}}$, is an $\boldsymbol{A}$-invariant[1] subspace of $\mathbb{C}^n$ where any vector that lies in this subspace can be considered as a possible eigenvector corresponding to λ_j. This might lead to situations that repeated realizations of the corresponding eigenvector matrix give different variants of linearly independent bases for $\boldsymbol{\mathcal{V}}$. Here, the interest is thus to search for a fixed orthogonal basis for $\boldsymbol{\mathcal{V}}$ such that

$$\boldsymbol{\mathcal{V}} = \boldsymbol{\mathcal{N}}_1 \oplus \boldsymbol{\mathcal{N}}_2 \oplus \cdots \oplus \boldsymbol{\mathcal{N}}_{n_\mathrm{c}} \tag{23.24}$$

In which, $\oplus$ represents the direct sum of basis vectors, $\boldsymbol{\mathcal{N}}_i$, that are fixed and complement of each other in $\boldsymbol{\mathcal{V}}$. Additionally, it is known that the projection of a vector onto a subspace is unique. Acknowledging this fact, we propose to use the information about the spatial distribution of the system inputs, actuators, which is enclosed in the column vectors of input matrix, $\boldsymbol{B}$, in Eq. (23.1). The projection of columns of $\boldsymbol{B}$ onto $\boldsymbol{\mathcal{V}}$, $\boldsymbol{P}_{\boldsymbol{\mathcal{V}}}(\boldsymbol{B})$, is defined as

$$\boldsymbol{P}_{\boldsymbol{\mathcal{V}}}(\boldsymbol{B}) = \Phi_j \left(\Phi_j^\mathrm{H} \Phi_j\right)^{-1} \Phi_j^\mathrm{H} \boldsymbol{B} \tag{23.25}$$

where $\boldsymbol{P}_{\boldsymbol{\mathcal{V}}}(\boldsymbol{B}) \in \mathbb{C}^{n \times n_\mathrm{u}}$. The QR decomposition of $\boldsymbol{P}_{\boldsymbol{\mathcal{V}}}(\boldsymbol{B})$ gives

$$\boldsymbol{P}_{\boldsymbol{\mathcal{V}}}(\boldsymbol{B}) = \boldsymbol{Q}\boldsymbol{R} \tag{23.26}$$

Here, $\boldsymbol{Q} \in \mathbb{C}^{n \times n}$ is a unitary matrix and $\boldsymbol{R} \in \mathbb{C}^{n \times n_\mathrm{u}}$ is an upper triangular matrix. The first n_c columns of $\boldsymbol{Q}$ forms the desired fixed orthogonal basis for $\boldsymbol{\mathcal{V}}$, i.e.,

$$\boldsymbol{\mathcal{V}} = \boldsymbol{\mathcal{N}}_1 \oplus \boldsymbol{\mathcal{N}}_2 \oplus \cdots \oplus \boldsymbol{\mathcal{N}}_{n_\mathrm{c}} = \boldsymbol{Q}_1 \oplus \boldsymbol{Q}_2 \oplus \cdots \oplus \boldsymbol{Q}_{n_\mathrm{c}} \tag{23.27}$$

[1]If $\boldsymbol{F} \in \mathbb{C}^{p \times q}$ is a matrix whose columns $f_1, f_2, \ldots, f_q$ span a q-dimensional subspace, $\mathcal{F}$. Then, $\mathcal{F}$ is $\boldsymbol{A}$-invariant if and only if there exists a matrix $\boldsymbol{Z} \in \mathbb{C}^{q \times q}$ that satisfies $\boldsymbol{A}\boldsymbol{F} = \boldsymbol{F}\boldsymbol{Z}$.

where each $\boldsymbol{Q}_i$ is also $\boldsymbol{A}$-invariant and can subsequently be used as the eigenvectors associated to the considered coalescent eigenvalues, λ_j [22]. This algorithm would implemented for multiple modes present in the library and also for those present in the realized models prior to computation of modal observability columns using Eq. (23.10) or Eq. (23.16).

23.5 Numerical Evidence

To provide evidence for the feasibility of the proposed clustering algorithm, the accelerance FRF data from the finite element model of a square aluminum plate is simulated, described further in [15]. As shown in Fig. 23.2, the plate is simply supported at the four corner nodes and is subjected to two force inputs. The acceleration of the vertical motion of the plate captured at eight nodes. The rest of this section presents the performance of the proposed clustering algorithm for three datasets obtained from different simulation scenarios. All datasets include FRFs in the frequency range of 0–200 Hz. Thus, target of the proposed algorithm is to estimate modal parameters associated to 24 modes present in this frequency range. Figure 23.3 demonstrates the six double mode shapes of the plate in the considered frequency range.

23.5.1 Experiment with No Frequency Shift and No Measurement Noise

In the first analysis, the problem of model order estimation is illustrated. One dataset is generated with no added noise and no inconsistency in data. An exhaustive model of order 250 is identified, using subspace identification method, to the dataset to yield a model library with 125 eigenmodes. Contribution of each mode to the input–output map of the identified system in the $\mathcal{H}_\infty$ sense is computed and is also normalized using the maximum contribution. To remove the spurious modes, the common practice in modal analysis context is to order modes in descending order of modal contributions and search for a clear drop. As it can be seen in Fig. 23.4a, there is no sudden drop in the normalized contribution even though the original simulation data is used for identification. A second way to estimate the true model order for subspace algorithm is to use the information criterion of Eq. (23.19). The value of this criterion for different model order is plotted in Fig. 23.4b where the estimated model order is 74 which is higher than the true model order of 48. Two important conclusions can be made here; first, a more sophisticated algorithm is needed to make an unbiased estimate of the true model order. Secondly, this result also confirms the claim that SVC provides an overestimation of the model order. Thus, SVC estimation of model order can be used later as the model order of realized models identified from random datasets.

23.5.2 Experiment with Noisy Measurements

In the second analysis, the FRFs are contaminated by a small amount of noise, 1 % RMS noise-to-signal ratio. This dataset is also identified by a model, the exhaustive model, of order 250 using subspace method. Here, the library, $\mathscr{L}$, includes 125 eigenmodes. As it can be seen in Fig. 23.5, the model order estimated by SVC criterion is 134 which is much higher than the true model order. However, this model order would be used as the model order for realized models.

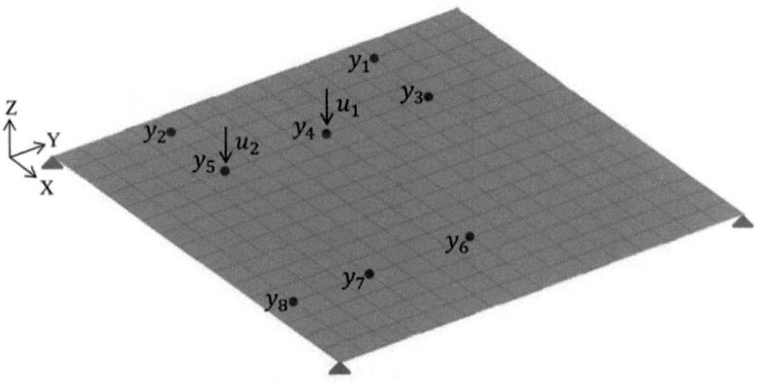

Fig. 23.2 Schematic view of the FE model of the Aluminum plate

Fig. 23.3 Mode shapes corresponding to double modes of plate

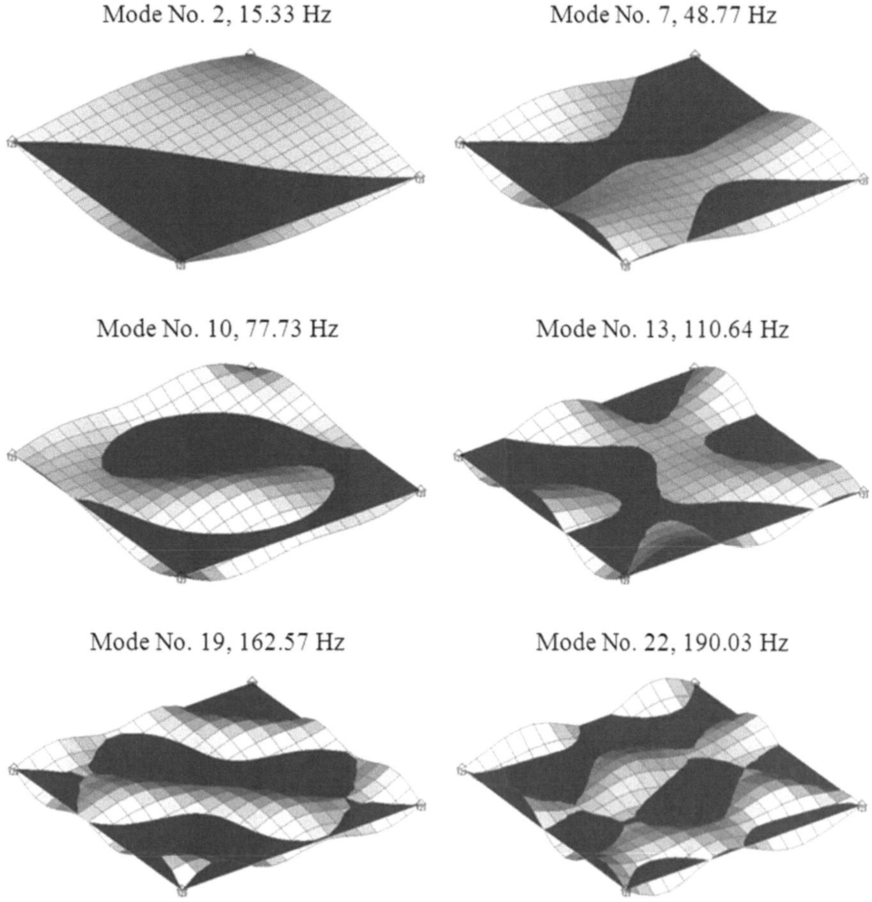

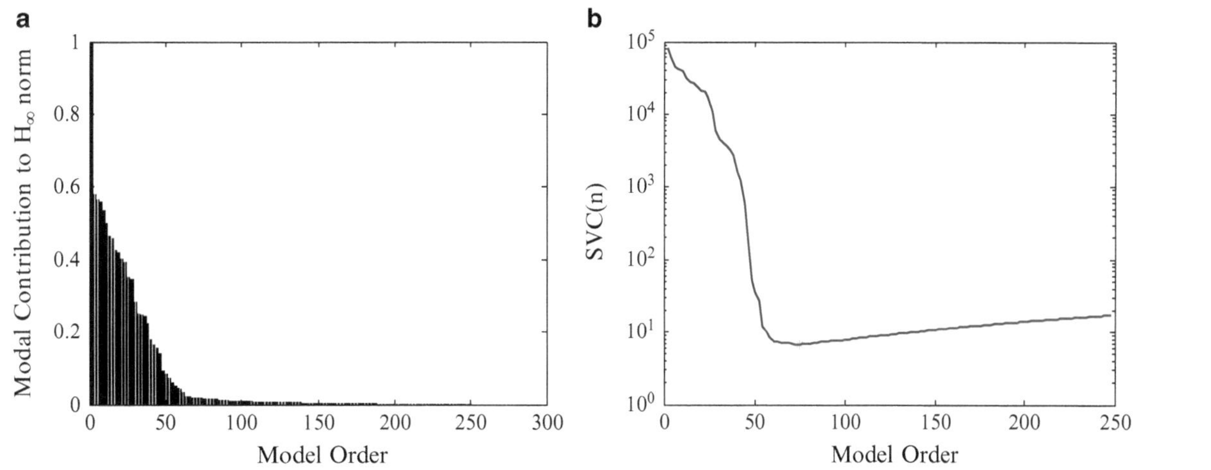

Fig. 23.4 (**a**) Contribution of modes in input–output relation in $\mathcal{H}_\infty$ sense, and (**b**) Singular Value Criterion (SVC) against the model order for data without noise, the *red star* shows the minimum of SVC

In the next step of the mode clustering algorithm, an ensemble of 100 datasets is generated using SRS technique. The size of each SRS sample is equal to 60 % of the original dataset. Figure 23.6a demonstrates the noisy FRF from system inputs to the output number 8 and Fig. 23.6b shows one instance of the resampled datasets. All data sets are identified by state-space models of order 134.

The next step in the clustering algorithm is the adjusting step where we modify either the modal vectors representing the clusters or their associated modal observability columns. The modal vectors associated to the double eigenvectors showed in Fig. 23.2 are fixed using the information in the input matrix, i.e. see Eq. (23.25). However, since there is no inconsistency in data in this example, the modal observability columns are computed for the adjusted modal vectors with no further adjustment.

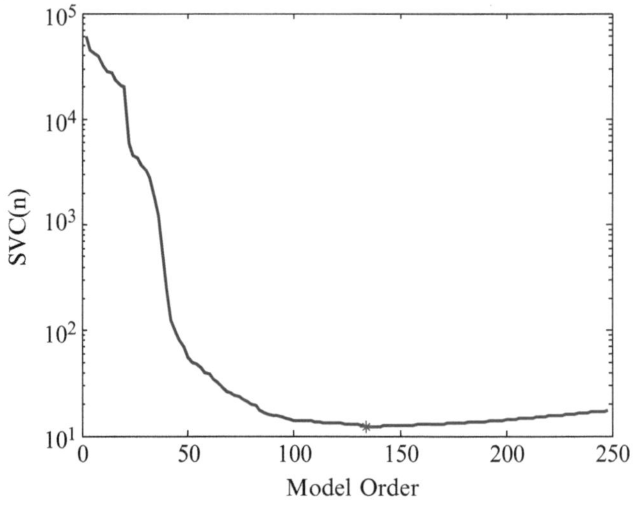

Fig. 23.5 Singular Value Criterion (SVC) versus the model order for noise corrupted dataset, the *red star* shows the minimum of SVC

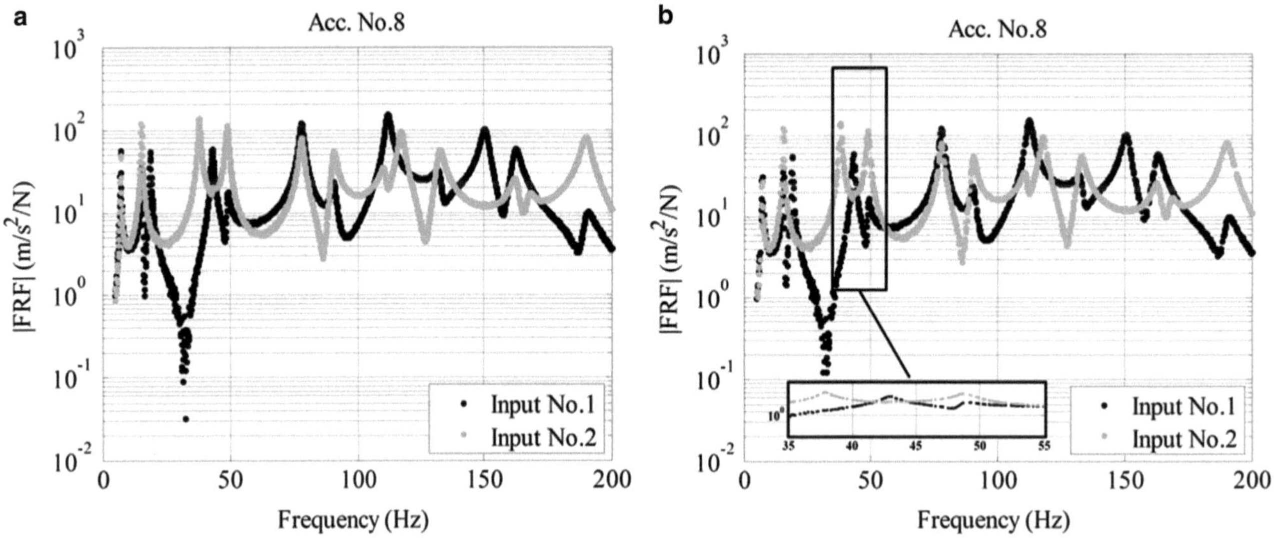

Fig. 23.6 Simulated FRFs with 1 % noise-to-signal ratio between two inputs to accelerometers (**a**) No. 8, and (**b**) one instance of the corresponding resampled data generated by SRS

Now, the adjusted mode library is in place to represent the clusters. Then similar modal parameters are grouped to each other using the proposed clustering algorithm. In this algorithm, each mode of realized models belongs to the cluster which gives the highest correlation. The correlation between modes is measured using the newly proposed correlation metric, *bMOC*.

Figure 23.7 demonstrates the statistics obtained for 125 clusters. Since the data is noisy, with no surprise, no cluster is empty, see Fig. 23.7a. The statistics, mean value and Shanon entropy[2] [23], for contribution in $\mathcal{H}_\infty$ sense and the correlation criterion of the modes inside each cluster are shown in Fig. 23.7b–d. These figures demonstrate how clusters consisted of spurious modes are gathered in a close vicinity of each other in a region with low expected value for MOC and modal contribution, see Fig. 23.7d. Investigation of entropy is motivated based on the fact that noise vectors do not appear in the realized models as often as physical modes and this leads to high level of uncertainty which in turns results in large values for Shanon entropy. On the other hand, the noise vectors, or spurious modes, do not contribute to the system input–output behavior. This means that the expected value of the modal contribution must be small. Thus, we can expect that these statistics can be s potential feature space for the supervised-learning algorithm to divide modes to spurious and physical classes.

[2]Shanon entropy is a measure of the uncertainty of a random variable. Let X be continuous random variable with probability density function $P(X)$. The entropy of this variable is defined as, $H(X) = -E\left[\ln\left(P(X)\right)\right]$. Here, $E[.]$ denotes the expected value.

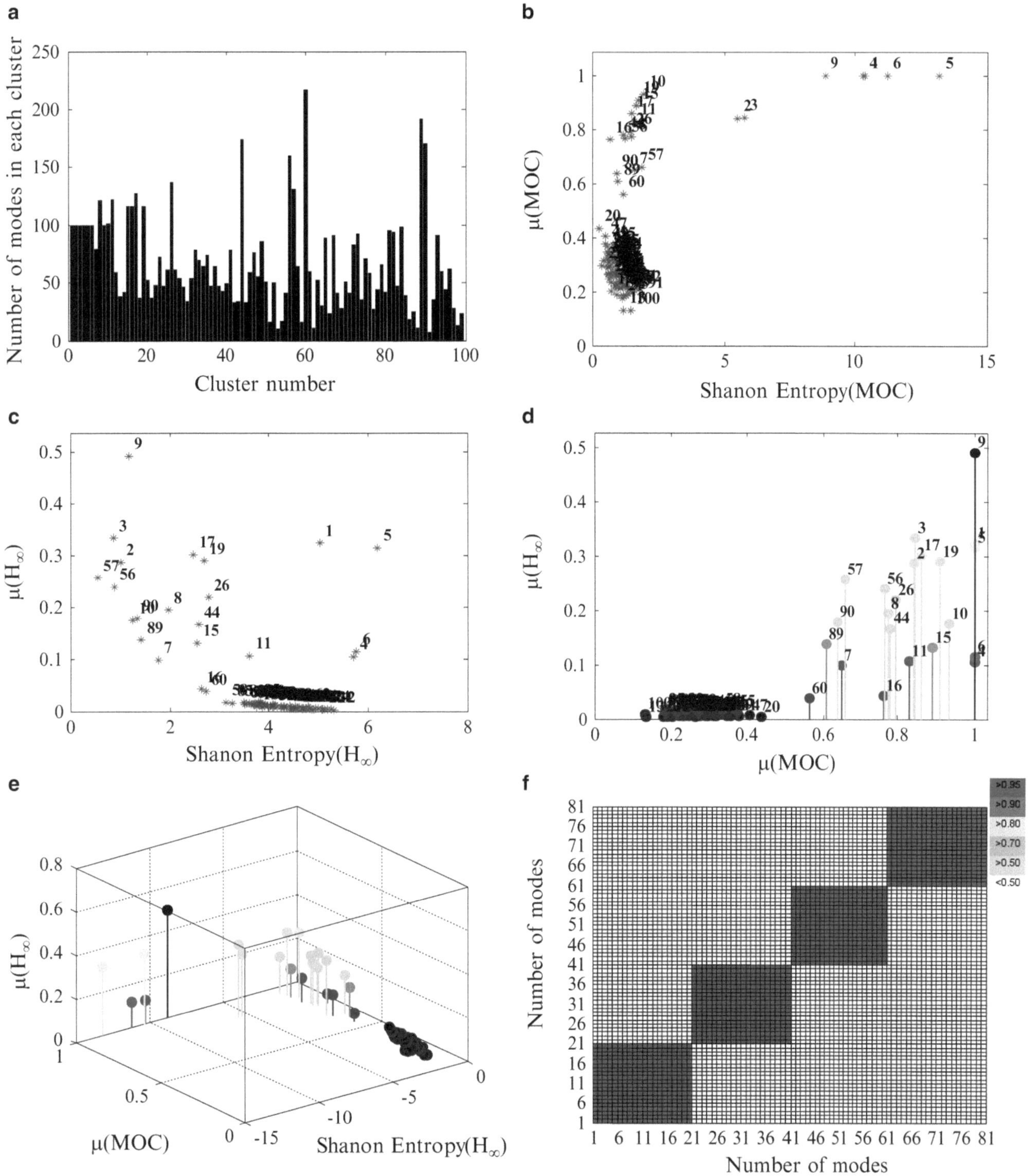

Fig. 23.7 Statistics of clusters for noisy data; (**a**) Number modes in each cluster, (**b**) The expected value of correlation criterion versus its Shanon entropy, (**c**) The expected value of modal contribution in $\mathcal{H}_\infty$ sense versus its Shanon entropy, (**d**) The expected value of modal contribution in $\mathcal{H}_\infty$ sense versus the expected value of modal observability correlation, (**e**) Relation between the expected value of modal contributions and MOC and the Shanon entropy of modal contributions for clusters, and (**f**) The Auto-MAC of the modal observability columns of clusters 1–4 considering only the first 20 modes in each clusters

However, presence of multiple modes remains a challenge for most of clustering approaches which demands an extra step after clustering to divide clusters consisted of multiple modes into multiple clusters. However, the QR factorization technique used in the proposed algorithm alleviates this challenge and the clusters corresponding to multiple eigenmodes are identified during the process. This fact is illustrated in Fig. 23.7f where the correlation analysis for the first 4 clusters shows no that the clusters are not correlated or consistent, the second and third clusters are associated to the first double mode showed in Fig. 23.3.

23.5.3 *Experiment with Frequency Shift and No Measurement Noise*

The last analysis concerns with inconsistent data. The plate is subjected to two consecutive shaker excitations and FRFs obtained from each excitation is saved in one patch of data. Then identification of a model using all patches of data would result in identification of coalescent modes where there should exist only one single mode. This phenomenon is demonstrated in Fig. 23.8 where shows the FRFs from two patches of data, i.e. from two different inputs, on top of each other. This figure shows that there is a slight frequency shift between resonance frequencies. To alleviate this problem, we proposed to modify the observability columns associated to these modes in the library such that there is just one cluster that can collect all the realized modes for a resonance frequency with inconsistency. Then this mode would be realized once in the final model. To this target, Sect. 23.4.1.1 suggested an SVD based algorithm to automatically identify and adjust the consistent modes realized for inconsistent resonance frequencies. Figure 23.8b demonstrates a double mode at 162 Hz which has inconsistency between patches. Modes 70–75 in the auto-MAC matrix between modal observability columns are the consistent modes in the library associated to this mode, see Fig. 23.9 for a closer look to the correlation of these modes. After applying SVD based

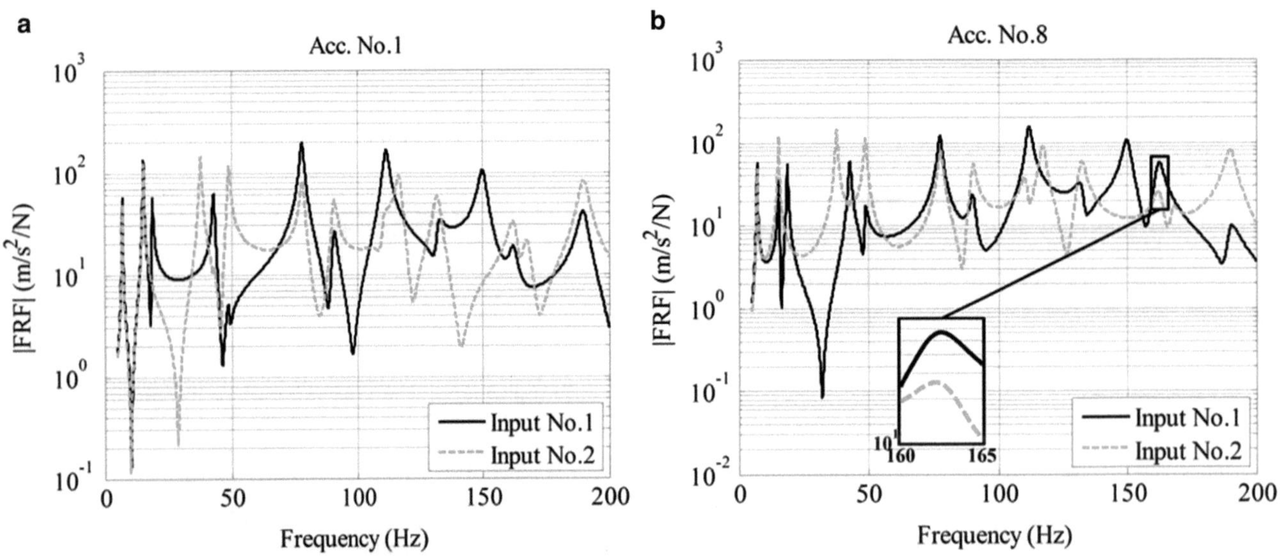

Fig. 23.8 FRFs measured with inconsistency between resonance frequencies for two inputs and accelerometers (**a**) No. 1, and (**b**) No. 8

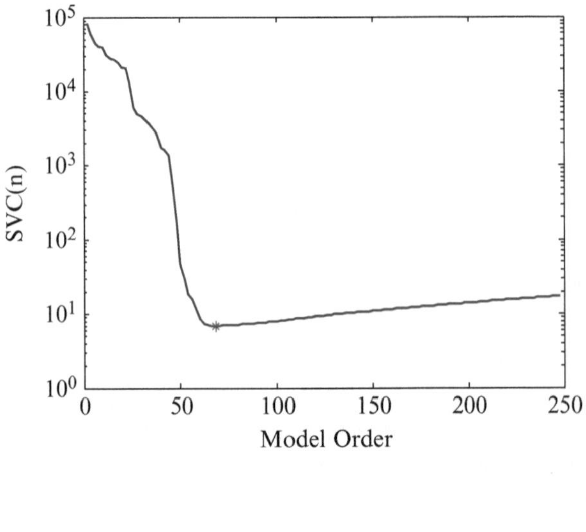

Fig. 23.9 Auto-MAC for modal observability columns associated to modes in library before application of SVD

algorithm to these modes, Fig. 23.10 demonstrates that how the modes are now uncorrelated together. This modification of the modes in the library results in collection of the corresponding mode in one cluster with the dominant singular vector.

After adjustment of the modes in the library, 100 data sets are randomly sampled from the data. These data sets are realized using modes of order 72, see Fig. 23.11. Figure 23.12 demonstrates the statistics for the clusters. As it can be

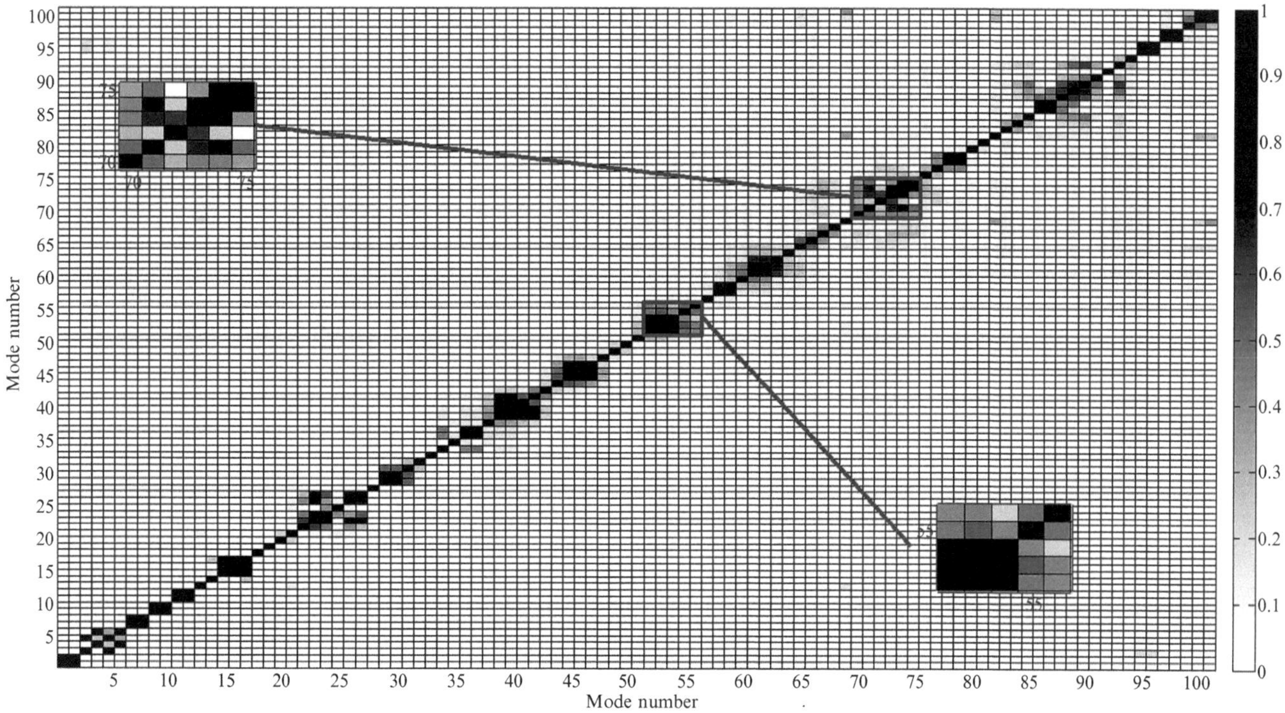

Fig. 23.10 Auto-MAC for modal observability columns associated to modes in library after application of SVD

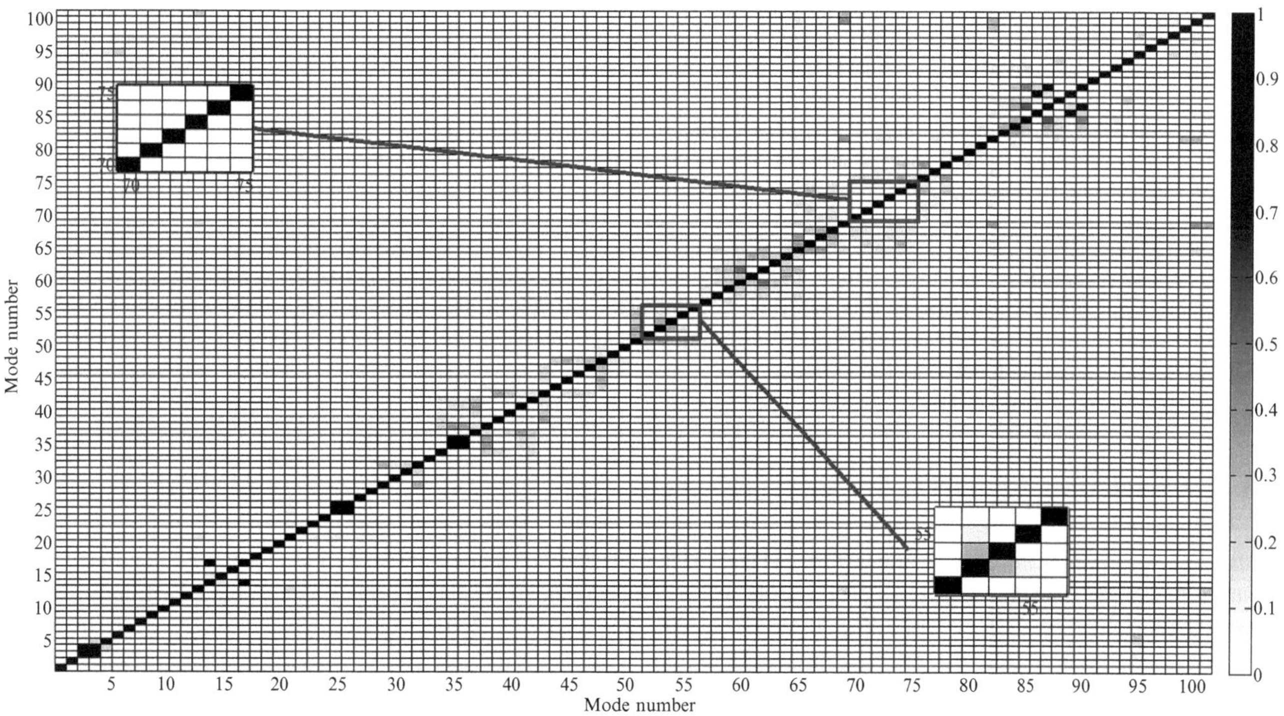

Fig. 23.11 Singular Value Criterion (SVC) against the model order for data with inconsistencies in the resonance frequencies, the *red star* shows the minimum of SVC

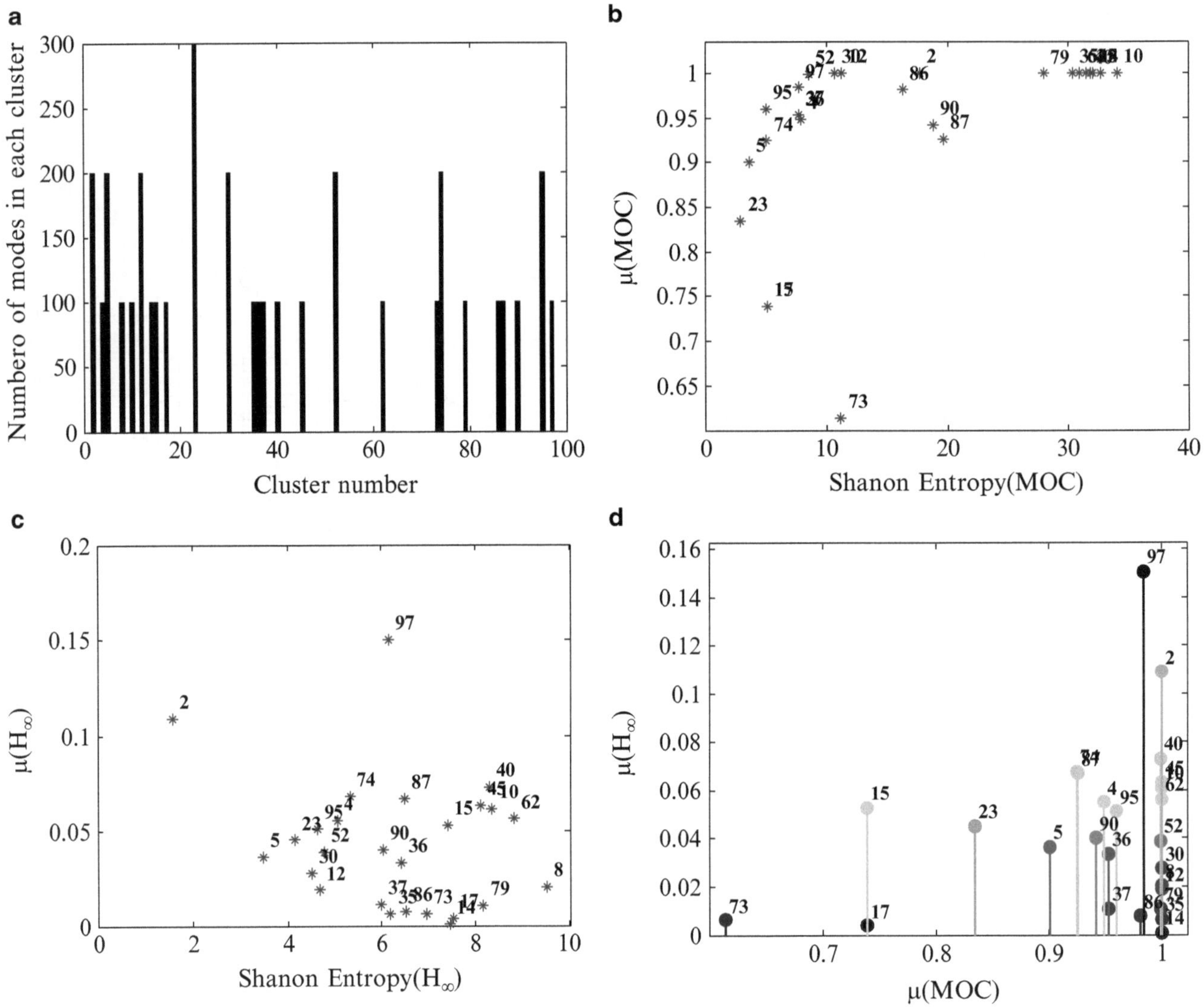

Fig. 23.12 Statistics of clusters for data with inconsistencies in the resonance frequencies; (**a**) Number modes in each cluster, (**b**) The expected value of correlation criterion versus its Shanon entropy, (**c**) The expected value of modal contribution in $\mathcal{H}_\infty$ sense versus its Shanon entropy, (**d**) The expected value of modal contribution in $\mathcal{H}_\infty$ sense versus the expected value of modal observability correlation

seen in Fig. 23.12d, many clusters remain empty after assignment step of the proposed clustering algorithm. The reason is two folded; first, there is no noise in this analysis. Secondly, the modes in the cluster are adjusted using the SVD based technique such that only the cluster represented with dominant modal vector collects modes realized for one single mode with inconsistency between patches (Fig. 23.13).

23.6 Concluding Remarks

This paper presents a novel and straightforward approach for mode clustering illustrated by means of simulated FRFs from the example of Aluminum plate. The results are very promising since the approach is capable of grouping similar modes together even in difficult cases of multiple eigenvalues and data with inconsistencies between resonance frequencies. The proposed approach provides the basis for the next step, i.e. classification of clusters into physical and spurious using supervised learning algorithms, which would be discussed in a subsequent paper. Using the proposed clustering approach based on modal observability correlation criterion, the need for user-defined thresholds and parameters is alleviated.

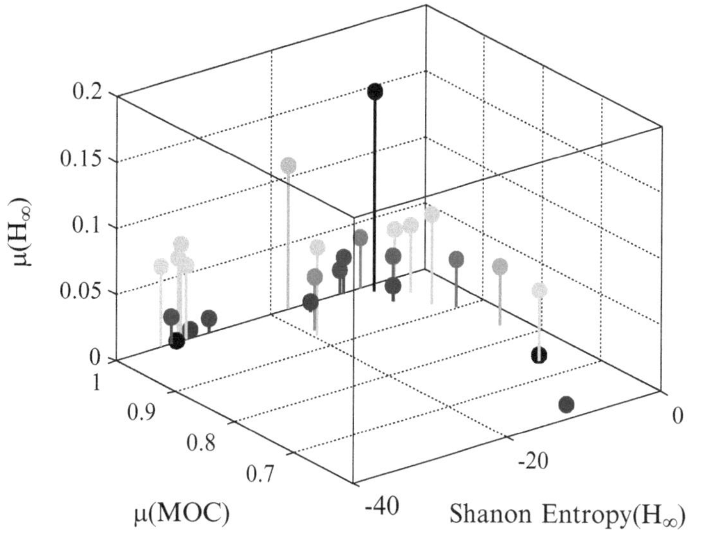

Fig. 23.13 Relation between the expected value of modal contributions and MOC and the Shanon entropy of modal contributions for clusters for data with inconsistency between resonance frequencies

Acknowledgment The present study is made possible through the financial support of the Swedish Wind Power Technology Center (SWPTC).

References

1. Reynders E, Houbrechts J, De Roeck G (2012) Fully automated (operational) modal analysis. Mech Syst Signal Process 29:228–250
2. Akaike H (1974) A new look at the statistical model identification. IEEE Trans Autom Control 19(6):716–723
3. Bauer D (2001) Order estimation for subspace methods. Automatica 37(10):1561–1573
4. Verboven P, Parloo E, Guillaume P, Van Overmeire M (2002) Autonomous structural health monitoring – part I: modal parameter estimation and tracking. Mech Syst Signal Process 16(4):637–657
5. Verboven P, Parloo E, Guillaume P, Van Overmeire M (2001) Autonomous modal parameter estimation based on a statistical frequency domain maximum likelihood approach. IMAC XIX, Orlando (FL), USA
6. Goethals I, De Moor B (2002) Model reduction and energy analysis as a tool to detect spurious modes. In: Conference proceedings of the 2002 international conference on noise and vibration engineering, ISMA, Leuven, Belgium, pp 1307–1314
7. Goethals I, Vanluyten B, De Moor B (2004) Reliable spurious mode rejection using self learning algorithms. In: Conference proceedings of the 2004 international conference on noise and vibration engineering, ISMA, Leuven, Belgium, pp 991–1003
8. Vanlanduit S, Verboven P, Guillaume P, Schoukens J (2003) An automatic frequency domain modal parameter estimation algorithm. J Sound Vib 265(3):647–661
9. Chhipwadia KS, Zimmerman DC, James Iii GH (1999) Evolving autonomous modal parameter estimation. In: Conference proceedings of IMAC XXVII, Orlando (FL), USA
10. Lim TW, Cabell RH, Silcox RJ (1996) On-line identification of modal parameters using artificial neural networks. J Vib Acoust Trans ASME 118(4):649–656
11. Magalhães F, Cunha Á, Caetano E (2009) Online automatic identification of the modal parameters of a long span arch bridge. Mech Syst Signal Process 23(2):316–329
12. Carden EP, Brownjohn JMW (2008) Fuzzy clustering of stability diagrams for vibration-based structural health monitoring. Comput-Aided Civ Infrastruct Eng 23(5):360–372
13. Glover K (1984) All optimal Hankel-norm approximations of linear multivariable systems and their L, ∞ -error bounds†. Int J Control 39(6):1115–1193
14. Moore B (1981) Principal component analysis in linear systems: controllability, observability, and model reduction. IEEE Trans Autom Control 26(1):17–32
15. Vakilzadeh MK, Rahrovani S, Abrahamsson T (2013) Modal reduction based on accurate input-output relation preservation. IMAC XXXI, Garden Grove
16. Antoulas AC (2005) Approximation of large-scale dynamical systems. Society for Industrial and Applied Mathematics, Philadelphia
17. Yaghoubi V, Abrahamsson T (2013) The modal observability correlation as a modal correlation metric. In: Conference proceedings of IMAC XXXI, Los Angeles, CA, USA, vol 45, pp 487–494
18. Mckelvey T, Akçay H, Ljung L (1996) Subspace-based multivariable system identification from frequency response data. IEEE Trans Autom Control 41(7):960–979
19. Van Der Auweraer H, Leurs W, Mas P, Hermans L (2000) Modal parameter estimation from inconsistent data sets. IMAC XVIII, San Antonio, Texas, USA
20. Lohr S (2009) Sampling: design and analysis. Cengage Learning
21. Chauhan S, Tcherniak D (2009) Clustering approaches to automatic modal parameter estimation. IMAC XXVII, Orlando
22. Laub AJ (2005) Matrix analysis for scientists and engineers. Siam, Philadelphia
23. Cover TM, Thomas JA (2012) Elements of information theory. John Wiley & Sons, Hoboken

MIX
Papier aus verantwortungsvollen Quellen
Paper from responsible sources
FSC® C105338

If you have any concerns about our products,
you can contact us on
ProductSafety@springernature.com

In case Publisher is established outside the EU,
the EU authorized representative is:
Springer Nature Customer Service Center GmbH
Europaplatz 3, 69115 Heidelberg, Germany

Printed by Libri Plureos GmbH
in Hamburg, Germany